Essentials of Igneous and Metamorphic Petrology

Second Edition

All geoscience students need to understand the origins, environments, and basic processes that produce igneous and metamorphic rocks. This concise textbook, written specifically for one-semester undergraduate courses, provides students with the key information they need to understand these processes. Topics are organized around the types of rocks to expect in a given tectonic environment, rather than around rock classifications: this is much more interesting and engaging for students, as it applies petrology to real geologic environments. This textbook includes nearly 300 illustrations and photos, and is supplemented by additional color photomicrographs made freely available online. Application boxes throughout the text encourage students to consider how petrology connects to wider aspects of geology, including economic geology, geologic hazards, and geophysics. End-of-chapter exercises allow students to apply the concepts that they have learned and to practice interpreting petrologic data.

B. Ronald Frost is an emeritus professor of geology at the University of Wyoming, where he performs wide-ranging research on igneous and metamorphic petrology as well as ore deposits. He has authored more than 120 scientific papers on topics ranging from serpentinization and the metamorphism of serpentinites, ocean-floor metamorphism, granulites, thermobarometry, the geochemistry of granites, and melting of sulfide ore deposits. He has conducted extensive field research in the Precambrian basement of Wyoming, as well as in Siberia, Greenland, northern Canada, and the Broken Hill area of Australia. He received the Alexander von Humboldt Research Award from the German government. He has been an associate editor for the *Journal of Metamorphic Geology, Geochimica et Cosmochimica Acta*, and the *Journal of Petrology*. He is a member of the American Geophysical Union, the Society of Economic Geologists, and the Geochemical Society, and a fellow of the Mineralogical Society of America. He taught mineralogy, petrology, optical mineralogy, and ore deposits for more than 35 years.

Carol D. Frost is a professor in the Department of Geology and Geophysics at the University of Wyoming. She investigates the origin and evolution of the continental crust, the provenance of clastic sedimentary rocks, and granite petrogenesis, and she applies isotope geology and geochemistry to environmental issues, including water coproduced with hydrocarbons and geological sequestration of carbon dioxide. She is the author of more than 130 scientific papers. She is President and fellow of the Mineralogical Society of America and fellow of the Geochemical Society and Geological Society of America. She has served as the science editor for the Geological Society of America's journal, *Geosphere*. She was awarded the CASE Wyoming Professor of the Year award in 2001. In 2008, she received her university's highest faculty award, the George Duke Humphrey medal, recognizing teaching effectiveness, distinction in scholarly work, and distinguished service to the university and to the state. She has served in the administration of the University of Wyoming as director of the School of Energy Resources, associate vice president for research and economic development, associate provost, and vice president for special projects. From 2014 to 2018 she was director of the Division of Earth Sciences, National Science Foundation.

The two authors are unrelated.

Essentials of Igneous and Metamorphic Petrology

Second Edition

B. Ronald Frost

UNIVERSITY OF WYOMING

Carol D. Frost

UNIVERSITY OF WYOMING

CAMBRIDGE
UNIVERSITY PRESS

University Printing House, Cambridge CB2 8BS, United Kingdom

One Liberty Plaza, 20th Floor, New York, NY 10006, USA

477 Williamstown Road, Port Melbourne, VIC 3207, Australia

314–321, 3rd Floor, Plot 3, Splendor Forum, Jasola District Centre, New Delhi – 110025, India

79 Anson Road, #06–04/06, Singapore 079906

Cambridge University Press is part of the University of Cambridge.

It furthers the University's mission by disseminating knowledge in the pursuit of education, learning, and research at the highest international levels of excellence.

www.cambridge.org
Information on this title: www.cambridge.org/9781108482516
DOI: 10.1017/9781108685047

First published 2014
Second edition 2019

A catalogue record for this publication is available from the British Library.

ISBN 978-1-108-48251-6 Hardback
ISBN 978-1-108-71058-9 Paperback

Contents

Preface

Petrology, from the Greek words *petra*, meaning rock, and *logos*, system of understanding, is the study of rocks and the conditions in which they form. It includes igneous, metamorphic, and sedimentary petrology. Igneous and metamorphic petrology are commonly taught together because both disciplines depend on the use of chemistry and phase diagrams. In contrast, sedimentary petrology is often combined with stratigraphy because both of these sciences depend on understanding the physical processes that accompany the deposition of sediments. Igneous and metamorphic petrology share common foundations: for example, both use phase diagrams to understand the conditions that control the crystallization of various minerals. However, there are important differences between the disciplines. In igneous petrology, the bulk composition of the rock is important because it gives clues to the tectonic environment in which it formed. Metamorphic petrology is concerned mostly with the use of mineral assemblages and textures to determine the conditions under which the rock crystallized. Because igneous rocks may later be transformed into metamorphic rocks, this book begins with igneous petrology and takes up metamorphic petrology second.

In contrast to many petrology textbooks, which are written for the upper-level undergraduate and graduate student audience, this book is accessible to introductory-level geology students who may have taken few earth science courses beyond physical geology and mineralogy. It aims to convey the essential petrologic information that is needed by all geoscientists no matter what their eventual specialization, be it geophysics, geochemistry, economic geology, geohydrology, or indeed any aspect of the Earth system.

This book focuses on the fundamental principles that govern the mineralogy of igneous and metamorphic rocks. For igneous petrology, this involves an understanding of how the mineralogy of igneous rocks reflects the equilibria that govern the crystallization of minerals from magma and how the geochemistry of a rock reflects its magmatic differentiation. The book uses several major-element discrimination diagrams, including the iron-enrichment index, the modified alkali–lime index, and the aluminium-saturation index, to compare and contrast magmatic suites that form in different tectonic environments. These simple geochemical parameters effectively highlight the different magmatic processes that create magmatic suites formed at oceanic and continental divergent plate boundaries, in arcs formed at oceanic and continental convergent margins, and in oceanic and continental intraplate tectonic settings.

In metamorphic petrology, the mineral assemblages in metamorphic rocks depend fundamentally upon the protolith of the rock as well as on the mineral reactions that take place at successively higher temperatures and pressures. Starting with mafic and ultramafic protoliths, which are the simplest, the text describes how pressure, temperature, and fluid composition affect the mineral assemblages in progressively more complex systems, including pelitic and calcareous protoliths. This book emphasizes chemographic projections as a way to determine the metamorphic mineral assemblages that occur together at specific metamorphic conditions. In addition, the text discusses the environments where various types of metamorphism are found and the tectonic significance of different types of metamorphic belts.

Throughout the textbook the authors have provided examples of how petrology relates to other areas of geology, including economic geology, geologic hazards, and geophysics. These short vignettes help students make connections between the study of igneous and metamorphic rocks and other fields of geology and illustrate the value of a fundamental understanding of petrology.

The first half of the book is a study of igneous petrology – magmas and the rocks that solidify from magma. Chapters 1–6 describe the fundamentals of igneous petrology. Chapter 1 presents the classification of igneous rocks, and the crystallization of magmas and resulting igneous textures and structures. In Chapter 2, we introduce igneous phase diagrams and how these are used to understand fundamental processes that produce the mineral assemblages, textures, and rock associations that occur in nature. Chapter 3 is an introduction to silicate melts and magmas, including their physical properties and mechanisms that produce differentiated igneous rock suites. Chapter 4 describes how the major- and trace-element chemistry of igneous rocks reflect the processes by which igneous magmas form and differentiate, and Chapter 5 introduces the application of stable and radiogenic isotopes to igneous petrology. This portion of the text concludes with Chapter 6, in which we explore melt generation from the mantle and the tectonic environments where magmas are generated. Chapters 7–10 examine the igneous rock suites that form in different tectonic environments: the oceanic floor (Chapter 7), convergent margins (Chapter 8), and intracontinental rifting (Chapters 9 and 10). Chapter 11 examines the granitic rock suites that compose the continental crust, and whose mineralogy, geochemistry, and isotopic compositions preserve information about their magma sources and tectonic environment of formation.

The second half of the text concerns metamorphic petrology. Chapter 12 explores how metamorphic petrologists use mineral assemblages to determine the parent, or protolith, of the rock and how metamorphic textures are used to determine whether solid-phase recrystallization took place in a static environment or during deformation. Chapter 13 introduces metamorphic phase diagrams and petrogenetic grids. Chapters 14–17 describe how pressure, temperature, and fluid composition affect the metamorphism of major protoliths, including mafic rocks (Chapter 14), peridotites (Chapter 15), pelitic rocks (Chapter 16), and calc-silicate rocks (Chapter 17). Chapter 18 covers thermobarometry, the quantitative estimation of metamorphic conditions. Finally, Chapter 19 describes the various tectonic environments where igneous and metamorphic rocks are found and the characteristic igneous rock suites and types of metamorphism found in each.

Acknowledgments

This textbook is the result of several decades of experience teaching igneous and metamorphic petrology at the University of Wyoming. The authors began writing this material when what had been two separate, semester-long courses in igneous and metamorphic petrology were combined into one and the existing textbooks were more exhaustive than the new course format could accommodate. They would like to acknowledge the hundreds of students who used successive versions of the igneous and metamorphic petrology course packet and provided edits and suggestions. They are especially grateful to those former students who went on to become geoscience faculty members and who have encouraged the authors to convert the course packet into a commercially published textbook.

The authors also wish to thank the external reviewers for their helpful suggestions and the editors and staff at Cambridge University Press for their expertise and patience in seeing the book through to publication. Also, they warmly acknowledge their colleagues at the University of Wyoming and elsewhere for providing a stimulating and rewarding environment in which to pursue petrologic teaching and research. Last, Carol Frost acknowledges with gratitude her family's forbearance while this textbook underwent repeated revision over many evenings, weekends, and holidays.

What is New in the Second Edition

To produce this second edition, we have modified the first edition in the following ways:

In response to suggestions from reviewers, we have added a new chapter on application of stable and radiogenic isotopes in petrology. This chapter introduces the concept of isotopic fractionation and the use of stable isotopes in geothermometry and as tracers of magmatic processes. It also describes the process of radioactive decay and the application of radioactive isotopes and their daughter products in geochronology and isotopic petrogenesis.

We have added sections to explain more fully the petrologic significance of phase diagrams, both igneous and metamorphic. Our examples draw explicit connections between the reactions shown on phase diagrams and the mineralogy and textural relations preserved in igneous and metamorphic rocks. Our goal is to demonstrate the usefulness of phase relations of simple systems in predicting and understanding mineral equilibria in natural igneous and metamorphic systems that are chemically more complex.

The second edition places increased emphasis on the connections between igneous and metamorphic rock suites and the tectonic environment in which they form. In addition to new sections throughout the text, we have rewritten the final chapter and present an integrated review of major tectonic environments in which we describe the igneous rock suites and the metamorphic facies typical of each.

The chapter on convergent margin magmatism has been revised to include an updated and more complete description of the origin of arc magmas and mechanisms of magma ascent and emplacement within the crust.

Throughout the text, we have rewritten sections to improve clarity, added new problems, and updated the suggestions for further reading. A new glossary defines bolded terms.

Chapter 1

Introduction to Igneous Petrology

1.1 Introduction

Igneous petrology is the study of magma and the rocks that solidify from magma. Thus, igneous petrologists are concerned with the entire spectrum of processes that describe how magmas are produced, how they ascend through the mantle and crust, their mineralogical and geochemical evolution, and their eruption or emplacement to form igneous rocks. Igneous petrology requires a working knowledge of mineralogy. Readers who wish to review the characteristics of the major rock-forming igneous minerals will find a concise summary in the appendix, which emphasizes the identification of rock-forming minerals in hand sample and in thin section. In addition, the appendix includes descriptions of minerals found in minor abundance but commonly occurring in igneous rocks, including accessory minerals that contain trace amounts of uranium and are important geochronometers.

Before geologists can understand the origin of igneous rocks, they must classify and describe them. This chapter introduces the classification of igneous rocks using the mineralogical classification system recommended by the International Union of Geological Sciences (IUGS) Subcommission on the Systematics of Igneous Rocks, which has the advantage that it is relatively simple and can be applied in the field. For rocks that are too fine-grained to name using this classification, a geochemical classification can be employed instead. The simplest of these, the total alkali versus silica classification, is introduced in this text.

Finally, this chapter introduces basic terminology that describes the textural and structural features of igneous rocks. Descriptions of igneous textures document crystal shape, size, and the arrangement of the various minerals, glass, and cavities in the rock. Igneous structures are larger-scale features that are the result of rock-forming processes. The textures and structures preserved in igneous rocks provide information about their evolution, emplacement, and crystallization, all of which are fundamental goals of igneous petrology.

1.2 The Scope of Igneous Petrology

All rocks are ultimately derived from magmas, which solidify to form igneous rocks. Consider, for example, the history of a shale. Such a rock is now composed of clay minerals. These clay minerals may have formed by weathering of a sedimentary rock that contained rock fragments and mineral grains. These components in turn may have been produced by erosion of a granitic gneiss. Before it was metamorphosed, this gneiss may have been a granodiorite, which is an igneous rock formed by crystallizing magma. As this example illustrates, the study of igneous petrology forms a foundation from which to study metamorphic and sedimentary rocks.

Igneous petrology is the study of the classification, occurrence, composition, origin, and evolution of rocks formed from magmas. The discipline can be divided into two components: **igneous petrography**[1], which is the description and classification of igneous rocks; and **igneous petrogenesis**, which is the study of the origin and evolution of igneous rocks. There are many different ways to approach the study of igneous petrology. *Field geology* is very important to the study of igneous petrology because important information is contained in the field relationships between rock units, the structure of an igneous rock, and its texture and physical appearance. For example, volcanologists depend heavily on their field observations during an eruption, and on the distribution of ash, lava, and other volcanic ejecta formed as the result of the eruption, to model the processes that occurred within a volcano before and during an eruption. *Laboratory identification* of the minerals in a thin section of an igneous rock, along with the chemical composition and age of a rock, are important means of classifying and relating it to other rocks with which it is spatially associated.

Another important way to study igneous rocks is through geochemistry. *Major-element geochemistry* can determine whether a suite of rocks is related through a process such as magmatic differentiation or mixing. *Trace-element geochemistry* is used to identify the role various minerals may have played as either crystallizing phases or residual phases in a suite of rocks. *Isotope geochemistry*, which can involve both radiogenic and stable isotopes, can determine if a suite of rocks formed from a single magma, or if a more complex, multi-source process was involved.

Because magmas that crystallize beneath Earth's surface are not observable and lavas erupted on the surface are hot and often dangerously explosive, geologists find it difficult to study the formation of igneous rocks directly. Therefore, *experimental petrology* is an important aspect of igneous petrology in which the pressures and temperatures required for igneous rocks to form and evolve are reproduced in the laboratory. For many rocks, a field and petrographic description does not provide conclusive proof of the process by which they formed. For these rocks, data gathered from experimental petrology are essential.

1.3 Classification of Igneous Rocks

One of the most tedious aspects of igneous petrography is the mastery of terminology. Innumerable, and often inscrutable, names have been applied to igneous rocks over the past few centuries as petrology grew in importance and sophistication. Much igneous terminology is arcane because in the early days of the science, petrologists did not have access to experimental data, phase diagrams, isotopic systems, or thermodynamic data and thus their work was mainly descriptive as opposed to quantitative. One way they described rocks was to name them. Among the more picturesque names is **charnockite**, which was named after the rock that formed the tombstone of Job Charnock, the founder of Calcutta (now Kolkata), India. Charnockite is a name given to an orthopyroxene-bearing granite, but there is no way to determine that from the origin of the name unless one was to desecrate Job Charnock's tombstone by sampling it for thin section and chemical analysis.

Unfortunately, like charnockite, most of the rock names that arose early in the development of igneous petrology do not provide much insight into the origin or evolution of the rock they describe. Many of the rock names based on type locality were given in the nineteenth or early twentieth century. Over time, geologists recognized the necessity of a more systematic rock classification scheme. In 1972, the IUGS Subcommission on the Systematics of Igneous Rocks published a rock classification system that has been widely adopted and use of many of the old rock names has been abandoned (Streckeisen, 1976; LeBas and Streckeisen, 1991; Le Maitre et al., 2005).

There are two basic approaches to the naming of rocks. A rock can be classified either according to the minerals it contains or by its chemical composition. The first approach has the benefit that geologists can name rocks in the field by identifying their mineralogy; however, it is not very helpful for classifying fine-grained rocks. Alternately, a chemical classification requires analytical data, and therefore is not useful in the field, but it does provide a means of naming fine-grained or glassy rocks. The compositions of most igneous rocks can be expressed in nine oxides: SiO_2, TiO_2, Al_2O_3, Fe_2O_3, FeO, MgO, CaO, Na_2O, and K_2O. These combine to form the major rock-forming igneous minerals, which include pyroxene, olivine, garnet, amphibole, mica, quartz, plagioclase, alkali feldspar, feldspathoid, magnetite, and ilmenite. Most rocks contain only a few of these minerals. The IUGS classification uses both mineralogical and chemical data but emphasizes classification on the basis of mineralogy.

1.3.1 Preliminary Classification

Igneous rocks are divided into the general categories of **plutonic**, **hypabyssal**, and **volcanic**, depending on their grain size. Plutonic rocks characteristically have coarse or medium grain sizes (>1 mm) and are inferred to have crystallized deep in the crust. Hypabyssal and volcanic rocks are fine-grained to glassy. Volcanic rocks crystallize at the surface and hypabyssal rocks crystallize at shallow depths, typically less than a kilometer. Because the grain size of an igneous rock is determined in part by the cooling rate of the magma, and this is a function both of magma temperature and the ambient temperature of the rocks into which the magma was emplaced, grain size generally increases with depth but there is no specific depth associated with the transition from plutonic to hypabyssal rocks.

In addition to classification according to grain size, the general composition of a rock can be described using the terms **felsic**, **mafic**, and **ultramafic**. Rocks that are rich in quartz, feldspars, or feldspathoids are light-colored and are called *felsic*. The term *felsic* combines parts of the words *fel*dspars (and *fel*dspathoids) and *si*lica. Darker-colored rocks, rich in ferromagnesian minerals, are called *mafic*. The term *mafic* reflects the enrichment of these rocks in *ma*gnesium and iron (*Fe*). Ultramafic rocks are essentially free of any felsic minerals.

1.3.2 IUGS Classification of Plutonic Rocks

Because plutonic rocks are relatively coarse-grained, so that their constituent minerals can be easily identified, either in hand specimen or in thin section, they are the most straightforward group of igneous rocks to classify.

The IUGS classification is based on the abundance of the common minerals in volume percent (modal mineralogy, or **mode**), which are divided into five groups:

Q	quartz
A	alkali feldspar, including albite with up to 5 mole percent anorthite ($<An_5$)
P	plagioclase, with composition An_5 to An_{100}
F	feldspathoids (foid): nepheline, sodalite, analcite, leucite, cancrinite
M	mafic minerals: olivine, pyroxenes, amphiboles, micas, and opaque minerals, and accessory minerals such as zircon, apatite, sphene, allanite, garnet, and carbonate

Rocks containing less than 90 percent mafic minerals (M<90) are classified according to the amounts of Q, A, P, and F minerals they contain, whereas rocks containing more than 90 percent mafic minerals are classified according to the proportions of major mafic minerals. Felsic and mafic rocks typically have far less than 90 percent mafic minerals and ultramafic rocks far more.

Because rocks never contain both quartz and feldspathoids, felsic and mafic rocks can be classified in terms of three components, either QAP or FAP. Triangular plots of the three components are shown in Figure 1.1 along with the names assigned to rocks containing particular proportions of Q, A, P, and F minerals. However, some rocks are not uniquely defined by QAP or FAP alone. For example, both diorite and gabbro fall in the same portion of the QAP triangle. They are distinguished primarily on the basis of plagioclase composition: plagioclase in diorite is more sodic than An_{50}, whereas that in gabbro is more calcic. Because the IUGS classification does not consider the composition of the plagioclase, it cannot distinguish these two rock types. A third rock name is assigned to the gabbro/diorite portion of the QAP triangle: anorthosite. Anorthosite is a special name applied to rocks that contain more than 90 percent plagioclase. Because the IUGS classification is based only on the proportion of Q, A, P, and F minerals, it does not distinguish between rocks with only 10 percent ferromagnesian minerals and rocks with up to 90 percent ferromagnesian minerals. Therefore, anorthosite occupies the same part of the triangle as do the diorites and gabbros that have considerably higher mafic mineral contents. This classification scheme can be further specified by adding the names of the major mafic minerals present, with the most abundant placed closest to the rock name.

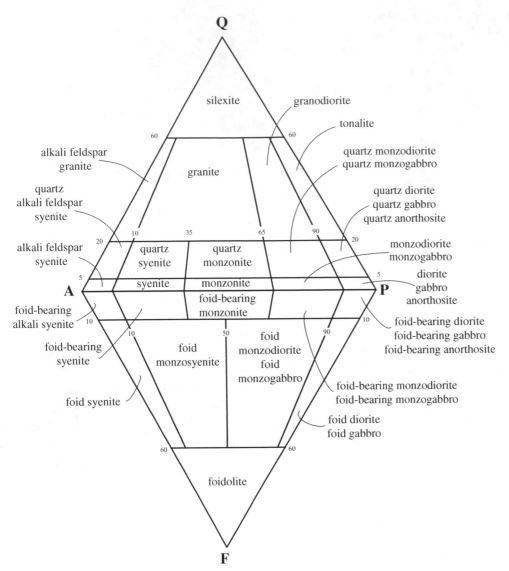

Figure 1.1 IUGS classification of plutonic rocks, based upon modal mineralogy. A = alkali feldspar, F = feldspathoid, P = plagioclase, Q = quartz. After Le Maitre et al. (2005).

For example, a biotite–hornblende tonalite contains more hornblende than biotite.

Mafic rocks can be further subdivided according to the proportion of plagioclase, orthopyroxene, clinopyroxene, olivine, and hornblende they contain (Figure 1.2). Strictly speaking, the term *gabbro* applies to a rock consisting of augite and calcic plagioclase, although the term is also broadly applied to any rock consisting of calcic plagioclase and other ferromagnesian minerals. For example, troctolite, a rock with olivine + calcic plagioclase, and norite, a rock with orthopyroxene + calcic plagioclase, are included in the gabbro family. Though not shown in Figure 1.2, rocks consisting of calcic plagioclase and hornblende are, quite logically, called hornblende–gabbros. Most gabbroic rocks contain between 35 and 65 percent mafic minerals. If they contain less than this,

the rock name may be prefixed by *leuco-*, meaning light. If they contain more than 65 percent mafic minerals, they may be prefixed by *mela-*, meaning dark.

Ultramafic rocks contain little or no plagioclase and thus require their own classification scheme, based on ferromagnesian mineral content. Ultramafic rocks containing more than 40 percent olivine are called *peridotites*, whereas ultramafic rocks containing more than 60 percent pyroxene are called *pyroxenites* (Figure 1.3). Peridotites and pyroxenites are further divided depending on the relative proportions of orthopyroxene, clinopyroxene, and olivine. The presence of other mineral phases can be used to further specify the name of the ultramafic rock; for instance, lherzolite that contains garnet is called garnet lherzolite.

Charnockites (orthopyroxene-bearing granitic rocks), lamprophyres (mafic and ultramafic rocks with mafic

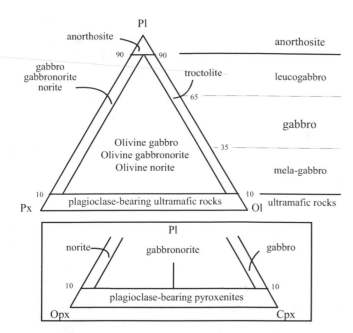

Figure 1.2 IUGS classification of gabbroic rocks. Ol = olivine, Pl = plagioclase, Px = pyroxene. Inset shows classification with regards to the type of pyroxene. Opx = orthopyroxene, Cpx = clinopyroxene. After Le Maitre et al. (2005).

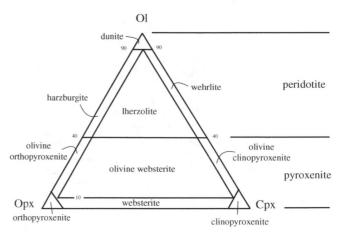

Figure 1.3 IUGS classification of ultramafic rocks. Ol = olivine, Opx = orthopyroxene, Cpx = clinopyroxene. After Le Maitre et al. (2005).

1.3.3 IUGS Classification of Volcanic and Hypabyssal Rocks

Whenever possible, the IUGS recommends that volcanic rocks be classified on the basis of modal mineralogy. The names for volcanic and hypabyssal rocks determined in this way are given in Figure 1.4. There are a few plutonic rock types for which there are no volcanic equivalents, such as anorthosite and ultramafic rocks. These plutonic rocks usually represent accumulations of crystals, and no liquid of that composition ever existed. The only ultramafic lava solidifies to form a rare rock called **komatiite**, which occurs almost exclusively in ancient Archean terrains. It is the volcanic equivalent of peridotite.

If the volcanic rocks are so fine-grained that minerals cannot be identified, then they must be classified on the basis of chemical composition. The IUGS has recommended that volcanic rocks be classified based upon their total alkali and silica contents (TAS) (LeBas et al., 1986) (Figure 1.5). The TAS diagram has as its x-axis the weight percent of SiO_2 of the rock, and as its y-axis the weight

phenocrysts), carbonatites (igneous carbonate-rich rocks), and pyroclastic rocks have their own classification schemes (Le Maitre et al., 2005).

percent $Na_2O + K_2O$ of the rock. The diagram is then divided into 15 fields. Classification using this chemical approach gives rock names that are typically consistent with the names based on the QAPF diagram.

1.4 Igneous Textures

Petrologists use textures and structures to interpret how igneous rocks crystallized. The terms texture and structure are nearly interchangeable, although *texture* of a rock refers to the small-scale appearance of the rock: the size, shape, and arrangement of its constituent phases, including minerals, glass, and cavities. The *structure* of a rock refers to larger-scale features, recognizable in the field, such as banding, variations in mineral abundances, or jointing. Textures may provide information about cooling and crystallization rates and the phase relations between minerals and magma at the time of crystallization. Structures indicate the processes active during the formation of rocks and the mechanisms of differentiation.

1.4.1 The Crystallization of Igneous Melts

The formation of igneous rocks by crystallization of melts can be understood from basic thermodynamic principles. Thermodynamics is a field of chemistry that is designed to determine what assemblage of solids, liquid, and gas would be stable from a set of chemical compounds at a given temperature and pressure. In igneous petrology, thermodynamics predicts the minerals that will crystallize from a melt. A key observation of thermodynamics is that a chemical reaction such as crystallization can occur

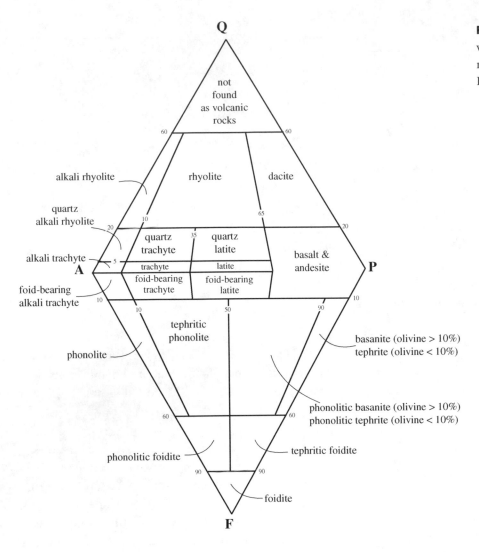

Figure 1.4 IUGS classification of volcanic rock based on modal mineralogy. Abbreviations as in Figure 1.1. After Le Maitre et al. (2005).

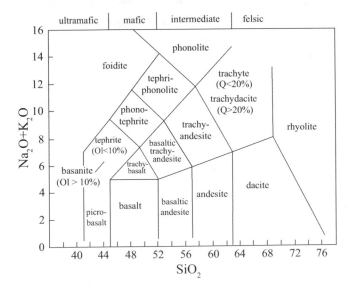

Figure 1.5 IUGS classification of volcanic rocks based on chemical composition, in weight percent oxide. Q = quartz, Ol = olivine. After Le Maitre et al. (2005).

spontaneously only if it evolves heat. Heat is tied up in minerals (and other substances) in two ways, **enthalpy** and **entropy**.

Enthalpy is the heat that is tied up in the chemical bonds of a substance. Because the bond strengths in each substance are different, a reaction from substance A to substance B will release (or consume) heat. *Entropy* is a measure of "randomness." It is an indication of how tightly bonded atoms are in the structure. The entropy of an element in a gas is much higher than the entropy of the same element in a melt, which in turn is higher than the entropy of that element in a solid. Similarly, the entropy of an element that is bonded by weak van der Waals' bonds is higher than that of the same element bonded by covalent bonds. The amount of energy released by a chemical reaction that can be used for chemical work is called **free energy**. It is symbolized by G, after J. Willard Gibbs, the person who formulated it, in

1873. The equation for the change in free energy between the products and reactants in a chemical reaction (symbolized by the Greek letter Δ) is given as:

$$\Delta G = \Delta H - T\Delta S \tag{1.1}$$

This equation says that the amount of heat released by a reaction equals the amount of heat released by the change in bonding strength between the products and reactants minus the amount of heat consumed as a result of the change in configuration of the atoms. For any reaction there are three possibilities for the value of ΔG. It can be positive, in which heat must be added to make the reaction proceed. This means that the reactants in the reaction are stable. ΔG can be zero, in which the reactants and products are in **equilibrium**. The other possibility is that ΔG is negative, which means that the reaction may occur spontaneously. A spontaneous reaction may not take place even if $\Delta G < 0$ because there may be an energy barrier to overcome. In this case, the system is said to be **metastable**.

1.4.2 Crystal Size

For crystallization to begin during cooling of a silicate melt, the ΔG between the melt and the minerals in the solid state must be less than 0. However, crystallization of an igneous melt involves more than favorable thermodynamic conditions. It also involves **nucleation** of new crystals and growth of the crystals that have nucleated. Both nucleation and growth are dependent on **diffusion**, the rate at which ions of various elements can move through the melt. In general, the diffusion rate is faster for ions that are small or have low valence and slower for those that are larger and have high valence (Zhang, 2010). For example, Ca^{2+}, Mg^{2+}, or Fe^{2+}, which having the same charge and similar size, diffuse roughly at the same rate. The larger, singly charged Na^+ ion will diffuse more quickly and Al^{3+} will diffuse more slowly. The diffusion rate of ions in a melt is also a function of the **viscosity** (resistance to flow) of the melt. Diffusion is faster in less viscous melts like basalt, and slower in viscous melts such as rhyolite (Zhang et al., 2010).

Consider a simple example such as crystallization of ice from water. When the temperature drops below 0 °C, ice crystals begin to form. Ice is more stable than water below 0 °C because ΔG for the reaction water ⇌ ice is less than zero. Water doesn't freeze when the temperature is exactly 0 °C even though at this temperature water and ice are in equilibrium. At temperatures slightly below

0 °C, H_2O molecules in the water may bond together to make a tiny crystallite of ice. The first few H_2O molecules to bond together are metastable. Although energy is released when the molecules are bonded together, the ions that lie on the margins of the ice crystallite are not bonded to the ions of other ice molecules. Instead, they are bent back into the crystallite, producing energy that is called **surface energy**. The amount of energy released when a few molecules of ice combine together is the difference between the amount of free energy released because of the bonding of the molecules and the amount of surface energy from the unfilled bonds on the margins of the ice cluster. This difference will be large when the crystallite is small because the surface energy is large relative to the volume of the crystallite. The difference decreases with increases in crystallite size because the free energy released during the formation of a crystallite is a function of the volume of the crystal, whereas the surface energy is a function of the area. As the radius (r) of the crystallite increases, the volume increases by a factor of r^3 whereas the surface area increases by a factor of r^2. Thus, as the radius increases, the free energy of the crystallite increases at a much faster rate than the surface energy, which means that there is a certain radius at which the crystallite becomes stable. When this happens, the crystallite becomes a **nucleus**.

Because a crystallite must become large enough to overcome the surface energy barrier before it can become a nucleus, it is difficult to nucleate a crystal from a melt when the temperature is close to the temperature of crystallization. However, nucleation becomes energetically favored if the melt is cooled below the crystallization temperature, a condition called undercooling. With low amounts of undercooling the ΔG between the crystals and melt is small. ΔG increases with increased undercooling, which means that the nucleation rate of crystals in a melt will increase as a melt is undercooled (Figure 1.6A). With a certain amount of undercooling, nucleation rate reaches a maximum. Nucleation rate drops with further undercooling because diffusion becomes increasingly difficult as the melt cools and becomes more viscous. At some temperature where the melt is nearly solidified, diffusion becomes very slow, and the molecules necessary for growth of a new crystal cannot move to a nucleation site.

Once a crystal has nucleated, its growth from a melt involves two steps: the diffusion of ions through the melt, and the adherence of these ions onto the growing face of the crystal. The diffusion rate through the melt will be

A

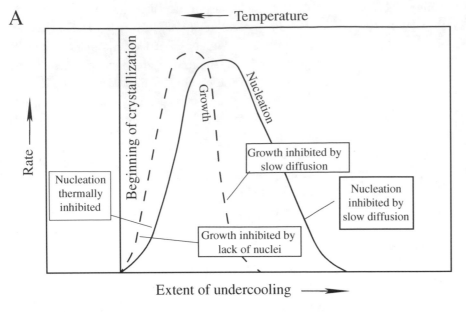

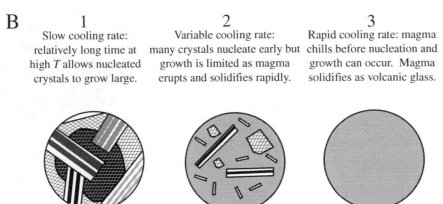

Figure 1.6 (A) Schematic representation of the interplay between growth and nucleation during cooling of melt. (B) Inset 1 shows textures formed in a plutonic rock in which the melt cooled slowly through temperatures at which the first crystals form. In these conditions growth is faster than nucleation, resulting in a rock with relatively few, large crystals. Inset 2 shows textures from a volcanic rock where the magma erupted after initial crystal nucleation and growth, and cooling took place rapidly, limiting further crystal growth. Inset 3 shows textures in a volcanic rock that cooled rapidly and solidified before nucleation and crystal growth could occur.

fastest at temperatures near the beginning of crystallization of the melt and will decrease with cooling as the melt becomes increasingly viscous. With low undercooling the adherence of ions, commonly called *nutrients*, to the face of the crystals will be slow, because only a few nuclei are present and hence only a limited crystal surface area is available on which the nutrients may adhere. The growth rate will increase with falling temperature as the crystals grow and an increasing area is available for the nutrients to populate. The maximum nucleation rate occurs at lower temperatures than the maximum growth rate because during nucleation the nutrients must migrate only short distances. In contrast, as the crystal grows and nearby nutrients are depleted, for the crystal to grow further nutrients must migrate longer distances. The total growth rate, therefore, will be zero at the temperature of melting, will rise with some degrees of undercooling, and will fall again as diffusion becomes increasingly difficult

in the progressively more viscous, solidifying melt (Figure 1.6A).

The textures observed in igneous rocks reflect the interplay between nucleation, crystal growth, and the rate of cooling. If the magma cools slowly and crystallizes in an environment where the rate of growth was faster than nucleation, then it would solidify as a rock with large crystals, as for example 1 in Figure 1.6B. Such a rock would be called **phaneritic.** If the magma cools so that during crystallization it resides at conditions where nucleation was faster than growth, then it would consist of many small crystals. If the magma then erupted and solidified, the matrix would be composed of glass (example 2 in Figure 1.6B). Such a texture would be called **aphanitic.** If the melt chilled quickly before nuclei could form it would solidify to a **glassy** texture where there are no crystals at all (example 3 in Figure 1.6B).

Another variable that affects grain size is the presence of volatile components or elements, such as H_2O or F, that decrease the viscosity of the melt and, hence, enhance the diffusion of essential elements to the face of a growing crystal. Melts with an abundance of these elements may crystallize extremely coarse-grained crystals, forming an intrusive rock called **pegmatite**. Pegmatites may have grain sizes up to a meter or more.

1.4.3 Crystal Shape

Petrologists use the shape of crystals and how the various minerals are arranged in an igneous rock to decipher the crystallization history of a rock. A mineral growing in a melt will tend to have grain boundaries that are **euhedral**, that is, they are bounded by well-formed crystal faces. The thin section of nepheline basalt shown in Figure 1.7A is composed of euhedral crystals of augite and olivine, contained in a fine-grained matrix. The textures shown in the thin section suggest that the augite and olivine began to crystallize from the melt and had grown to sizes of 1–5 mm before the lava erupted. The fine-grained matrix indicates that the melt in which the crystals were entrained chilled quickly and solidified as volcanic glass. A close examination of Figure 1.7A shows that the matrix

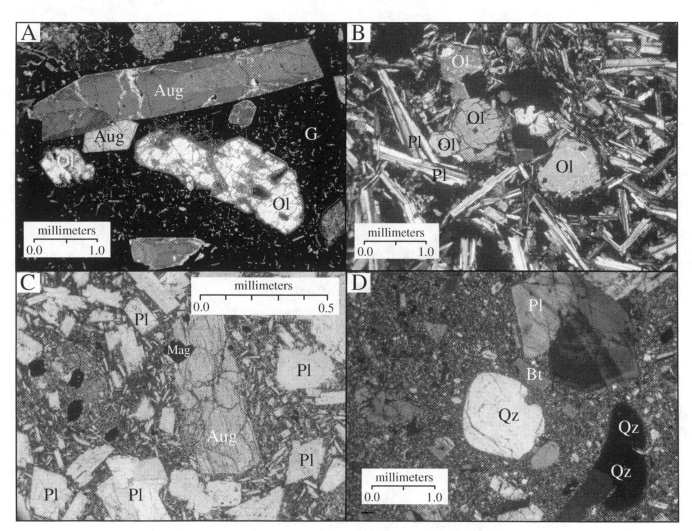

Figure 1.7 Photomicrographs showing textures in volcanic rocks. (A) Glassy nepheline basalt containing phenocrysts of olivine (Ol), augite (Aug), and glass (G), erupted near the Kaiserstuhl, southern Germany. Crossed polarized light (XPL). (B) Olivine tholeiite containing phenocrysts of olivine (Ol) and plagioclase (Pl) in a matrix of fine-grained olivine, augite, plagioclase, and glass, from the Snake River Plain, Idaho, USA. XPL. (C) Andesite with phenocrysts of augite (Aug) plagioclase (Pl) and magnetite (Mag) in matrix of fine-grained plagioclase, augite, and glass, from Soufriere volcano, St. Vincent. Plane-polarized light (PPL). (D) Dacite consisting of quartz (Qz), plagioclase (Pl), and biotite (Bt) in a matrix of quartz, plagioclase, and glass. XPL.

is not all glass; a few extremely small grains of augite and plagioclase are also present. These probably nucleated shortly after the basalt erupted but before it completely solidified.

A crystal that is relatively large compared to the minerals composing the matrix of an igneous rock is called a **phenocryst**. In Figure 1.7A, the contrast in size between the phenocrysts and the matrix is obvious. However, few igneous rocks have a matrix so dominated by glass. More typically, the matrix undergoes some degree of crystallization. For example, the basalt shown in Figure 1.7B contains phenocrysts of equant olivine and elongate plagioclase in a matrix of finer-grained olivine, augite, plagioclase, and glass. Relations are similar in the andesite shown in Figure 1.7C, except the plagioclase in the andesite is stubbier than the plagioclase in the basalt. Phenocrysts of quartz may occur in highly siliceous melts, such as dacite (Figure 1.7D) and rhyolite, and the presence of quartz phenocrysts is one way to identify these rocks in the field.

Many of the textures characteristic of volcanic rocks also help petrologists interpret plutonic rocks. The early crystallizing minerals form a matrix of interlocking euhedral grains, in a texture called **cumulate** texture. The minerals that formed later are constrained to grow in the interstices of these cumulus grains. These are called **postcumulate** grains and they are **anhedral**, which means they are not bounded by crystal faces. Examples of cumulate texture are shown in Figure 1.8A, a gabbro consisting of cumulus plagioclase and magnetite and postcumulus augite, and in Figure 1.8B, a pyroxenite with cumulus orthopyroxene and postcumulus plagioclase. Some granitic rocks contain tabular plagioclase or potassium–feldspar; for example, the granodiorite shown in Figure 1.8C contains distinctly tabular plagioclase. The plagioclase has the same stubby aspect ratio as plagioclase of similar composition in the volcanic rock shown in Figure 1.7C. The concentric zoning in this plagioclase records changes in composition as the plagioclase grain grew in the granodioritic melt.

In some plutonic rocks, the magma solidifies after relatively coarse-grained minerals have formed, making a rock called a **porphyry**. This rock has a texture that is characterized by euhedral grains dispersed in a finer-grained matrix (Figure 1.8D). A porphyritic texture tells a geologist that the rock underwent a complex cooling history. First, it cooled slowly, during which time the phenocrysts grew, followed by sudden cooling that caused the rapid solidification of the rest of the melt.

1.5 Igneous Structures

Igneous rocks exhibit a wide variety of forms. Mafic volcanic rocks occur mostly as flows; felsic volcanic rocks may also form flows, but also commonly form **pyroclastic** rocks, or rocks fragmented while still hot. Hypabyssal rocks may form as **lava domes**, **dikes**, or **sills**, and plutonic rocks occur as **plutons** and **batholiths**, as well as dikes and sills.

1.5.1 Structures in Volcanic Flows

Lava flows may range in thickness from less than a meter to more than 10 meters. Mafic lava flows are often divided into two types: blocky lava is known as **aa** (Figure 1.9A), and massive lava with a ropey surface is called **pahoehoe** (Figure 1.9B). Pahoehoe texture forms on relatively hot lavas but as the lava cools, the surface breaks apart, making aa. These names are etymologically Hawaiian; abundant lava flows in Hawaii allowed native Hawaiians ample opportunity to develop a terminology, comparing the textures of the flows. In cross-section, many flows, particularly those that ponded before completely crystallizing, show **columnar jointing** (Figure 1.9C). Columnar jointing forms by contraction that cracks the rock as heat dissipates from upper and lower margins of the flow. The vertically oriented columns, which are typically hexagonal in cross-section, commonly form a relatively wide set of columns at the base of the flow and a colonnade of narrower columns at the top.

Where basalts erupt under or flow into water, they form **pillows** (Figure 1.9D). The magma that contacts water is chilled and quenches, forming a distinctive lobate, or "pillow," shape. As lava continues to flow, it breaks the solidified crust of the initial pillow to form another lobe. A pillow basalt is constructed of hundreds of these nested lobes. In cross-section, the pillows have a rounded top and a tail that points downward. Pillow basalts are diagnostic of subaqueous volcanism and because they are well preserved in the geologic record, they allow geologists to identify underwater eruptions that may be up to billions of years old.

Commonly, gas bubbles exsolved from the magma gather at the top of a flow. Solidification of the melt will produce a rock pocked by holes from these exsolved gas bubbles. The holes are called **vesicles**, and they are key evidence of lava flows because gas bubbles are unlikely in plutonic rocks. Vesicles are also important markers of the

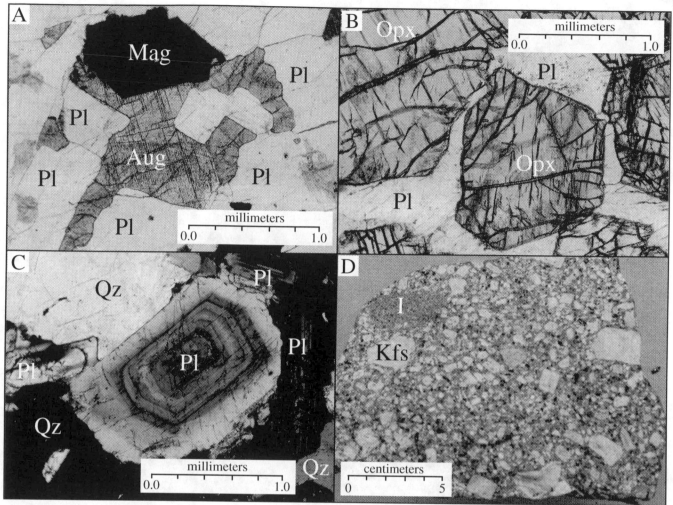

Figure 1.8 Photomicrographs showing textures in plutonic igneous rocks. (A) Plane-polarized light (PPL) photomicrograph showing euhedral (i.e. cumulus) plagioclase (Pl) and magnetite (Mag), surrounded by anhedral (i.e. postcumulus) augite (Aug). Gabbro from the Skaergaard intrusion, Greenland. (B) Photomicrograph in PPL showing euhedral orthopyroxene (Opx) and anhedral plagioclase (Pl) in a feldspathic pyroxenite from the Stillwater intrusion, Montana, USA. (C) Photomicrograph in crossed polarized light (XPL), showing compositionally zoned, euhedral plagioclase (Pl) surrounded by anhedral quartz (Qz). Biotite hornblende granodiorite from Blue Mountains, Oregon, USA. (D) Photograph of a granite porphyry dike containing phenocrysts of K-feldspar (Kfs) in a fine-grained matrix of K-feldspar, plagioclase, quartz, biotite, and hornblende. The rock also contains an inclusion (I), which is interpreted as an autolith, composed of chilled material from the margin of the dike. Willow Creek pass, Colorado, USA.

top of a flow, something that may be difficult to recognize in complexly deformed volcanic rocks.

1.5.2 Structures in Pyroclastic Deposits

Pyroclastic deposits are classified according to two factors: the size of the fragments within the deposit and the relative abundance of glass, crystals, and rock fragments (Figure 1.10). Fragments larger than 32 mm in diameter are called either **bombs** or **blocks**. Bombs are

clots of magma that were partly or entirely plastic when erupted. Shapes of bombs are controlled by the initial fluidity of the magma, length and velocity of flight through the air, and deformation on impact. Blocks are erupted fragments of solid rock. Solid or liquid materials between 4 mm and 32 mm in size at the time of eruption are called **lapilli**. Finely spun glass threads are called *Pele's hair*; *accretionary lapilli* are spheroidal, concentrically layered pellets, formed by accretion of ash and dust

Figure 1.9 Structures of volcanic rocks. (A) Blocky or aa lava flow in Snake River Plain, Idaho, USA. United States Geologic Survey photo library I.C. 738. (B) Ropey or pahoehoe lava from 1972–1974 eruption of Kilauea volcano, Hawaii, USA. United States Geological Survey photo library HVO 221ct. (C) Columnar jointing in basalt, San Miguel Regla, Hidalgo, Mexico. United States Geologic Survey photo library Fries, C.4. (D) Pillow in basalt from Curaçao, Netherlands Antilles. Note the rind on the pillow.

by condensed moisture in eruption clouds. **Ash** (Figure 1.11A) is incoherent ejecta less than 4 mm in diameter and may be vitric, crystal, or lithic ash depending on the proportion of glass, crystals, or rock fragments. **Pumice** and **scoria** are ejecta of melt that have a porosity of 30–80 percent. Scoria is from magma that is andesitic or basaltic in composition, whereas pumice comes from magma of intermediate to siliceous composition. Because the vesicles in pumice are isolated, pumice may have a density less than that of water and can float. **Tuff** is consolidated volcanic ash. The crystal-vitric tuff shown in Figure 1.11B contains both glassy material – ash and pumice – and crystals of quartz. The vitric tuff in Figure 1.11C contains pumice fragments, flattened by the weight of the overlying pyroclastic material.

Pyroclastic deposits are also classified by their areal extent and their structure and give geologists information on the eruption process. One type of pyroclastic deposit is a **pyroclastic fall deposit** that forms from pyroclastic material that falls directly out of the sky. Because of their mode of formation, pyroclastic fall deposits mantle topography with a uniform thickness of ash over a local area. Over large areas, pyroclastic fall deposits show systematic decreases in thickness and grain size away from the source. An isopach map of pyroclastic fall deposit

A

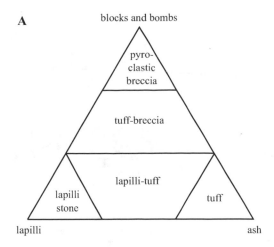

B

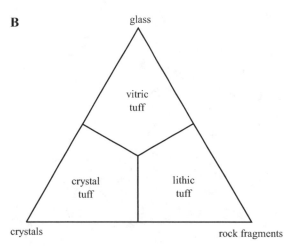

Figure 1.10 Classification of pyroclastic rocks. After (A) Fisher (1966) and (B) Pettijohn (1975).

explosive exsolution of volatiles that blows apart the magma. If a vent is situated where water has ready access, the mechanism of explosion changes fundamentally. Magma is torn apart by exsolving gases and mixes with water. Rapid vaporization triggers a thermal explosion and further fragmentation. These **phreatomagmatic explosions** are more violent and produce fine-grained deposits, composed of glassy ash or **hyaloclastite**.

Another type of pyroclastic deposit is a **pyroclastic flow deposit**. These deposits form from avalanches of pyroclastic fragments that move down topographic lows and fill depressions. Their movement is broadly analogous to other natural debris flows, such as rock flows and mud flows. The deposits are characterized by poorly sorted material with a continuum of sizes, from large blocks to fine ash, because there is little room and time for sorting in a fast-moving avalanche of closely packed particles. In contrast, air-fall deposits are usually well sorted because during transport through the high atmosphere the particles are sorted according to size and density. Because pyroclastic flow deposits are gravity controlled, they infill topographic lows instead of mantling topography. As with air-fall deposits, pyroclastic flow deposits vary by several orders of magnitude in their volume and dispersal.

Pyroclastic flow deposits may also form when a growing lava dome collapses (Figure 1.13). Growing lava domes are unstable and commonly break up to form landslides. If the melt is close to water saturation at the time the landslide forms, sudden decompression of the underlying magma could lead to explosion, which triggers an avalanche of hot blocks, ash, and gas. These deposits are typically monolithologic. Transported individual blocks can reach tens of meters in diameter. The pyroclastic flow deposits of Mont Pelée that formed on the island of Martinique in 1902 originated by collapse of a lava dome.

If the temperature of emplacement is sufficiently hot, pyroclastic deposits sometimes undergo processes of **welding** after deposition (Figure 1.11C). Welding occurs when particles are fused together by solid-state diffusion at particle contacts. For rhyolitic glass, the minimum temperature for welding is 625 °C at 1 atm and 590 °C at 10 atm. If the glass is sufficiently ductile (i.e. hot), the pumice and ash particles deform as they weld under the weight of the overlying deposit. The end result is a rock in which all porosity is removed and pumices are deformed in streaks or **fiamme**.

thickness can show the location of the vent, the wind direction, and the height of the eruption column. We can define two end members of a spectrum of pyroclastic fall deposits. In a **strombolian eruption**, the eruption column is low (1–3 km) and the fragments accumulate around the vent, forming the cone. This type of eruption is named after Stromboli, a volcano north of Sicily that has had frequent, rather quiet eruptions since historical times. In a **plinian eruption**, the eruption column is high (20–50 km) and pumice and ash are spread as a thin sheet, covering areas up to 10^6 km^2 (Figure 1.12). This type of eruption is named after Pliny the Younger, who, in 79 CE, wrote elaborate letters describing the eruption of Vesuvius that destroyed Pompeii (and killed his uncle, Pliny the Elder). Plinian and strombolian deposits are generally well-sorted and are produced by an

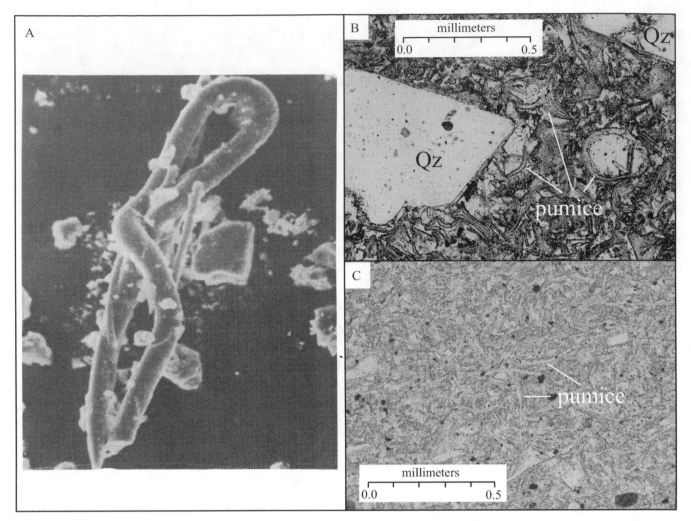

Figure 1.11 Photomicrographs of pyroclastic rocks. (A) Ash from Mount Saint Helens collected in Laramie, Wyoming, after the May 18, 1980 eruption. Length of the glass strand is 200 μm. (B) Crystal–vitric tuff showing crystals and crystal fragments of quartz (Qz) in a matrix of pumice. Bandolier, New Mexico. PPL, field of view (FOV) = 1.25 mm. (C) Vitric tuff, showing pumice fragments that are more compressed than those in Figure 1.11B. Lava Creek Tuff, Yellowstone, Idaho, USA. PPL, FOV = 1.25 mm.

1.5.3 Structures in Hypabyssal Rocks

Hypabyssal rocks are rocks that crystallized at shallow depths. Magmas emplaced near the surface cool relatively quickly, and hypabyssal rocks are, therefore, typically fine-grained but lack evidence that they ever erupted on the surface. Examples of hypabyssal rocks include lava domes, **volcanic necks**, dikes, and sills.

Lava domes include both hypabyssal and eruptive classes of igneous structures. They form from highly viscous lava, which produces bulging, dome-shaped bodies that may be several hundred meters high (Figure 1.14). The surface of the dome may be made of fragmented lava (much like aa) that erupted on the surface of the dome but didn't manage to flow far. Beneath the surface, the dome consists of magma that shallowly solidified and was pushed into the domal shape by magma intruding from below.

Some volcanoes erupt easily eroded, fragmented rocks. As such, the volcano itself may not survive long as a topographic feature. As the volcanic edifice erodes away, the vent of the volcano, which is made of rock that is more resistant to erosion, may remain. This irregularly shaped spire of hypabyssal rock is called a *volcanic neck* (Figure 1.15). Volcanic necks are common features in some volcanic terrains.

Dikes are tabular bodies of igneous rock that form when magma solidifies within a subterranean fracture.

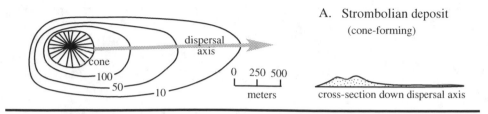

A. Strombolian deposit
(cone-forming)

Figure 1.12 Sketch showing the relative scale of aerial distribution of pyroclastic air-fall deposits from strombolian and plinian eruptions.

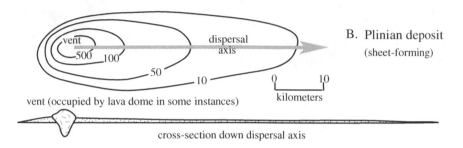

B. Plinian deposit
(sheet-forming)

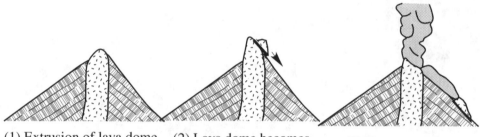

Figure 1.13 Diagram of the formation of a pyroclastic flow deposit by the collapse of a lava dome.

(1) Extrusion of lava dome

(2) Lava dome becomes unstable and top slides off

(3) Decompression causes explosions in underlying magma

Figure 1.14 Photo of two lava domes within the caldera of Mount Saint Helens. The dome in the foreground occurred as part of the eruptive activity of March–April 1982. The "whaleback" dome behind grew during renewed seismic and volcanic activity in September–October 2004. United States Geological Survey photograph taken by Steve Schilling on February 22, 2005.

Dikes can range from centimeters to kilometers in thickness, although the thickness of hypabyssal dikes tends to be on the order of meters. Dikes can form on a local scale during the eruption of single volcanoes. Some volcanic necks have dikes radiating out from them that may extend for more than 10 kilometers (Figure 1.15). These dikes indicate that, in addition to the eroded fragmental rocks, the fossil volcano erupted magma that was supplied by fissures now occupied by dikes.

When magma intrudes sedimentary rocks, it is commonly injected parallel to sedimentary bedding, rather than being forced through fractures across bedding planes. Such intrusions are called *sills*. The term sill also applies to dikes that have intruded parallel to metamorphic layering in metamorphic rocks.

Chilled margins are a common, distinctive feature of hypabyssal sills and dikes (Figure 1.16). When magmas are emplaced at fairly shallow depths, the ambient

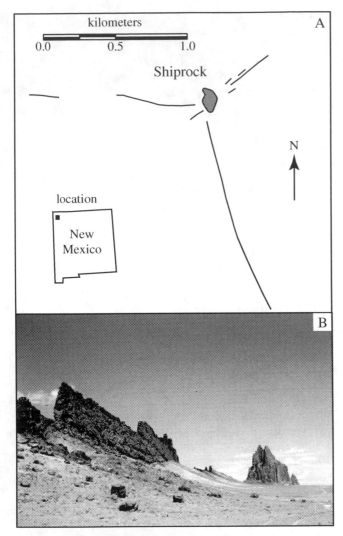

Figure 1.16 Photo of chilled Tertiary dike, intruding and baking adjacent Cretaceous sedimentary rocks, southern Colorado, USA. Photo by Eric Erslev.

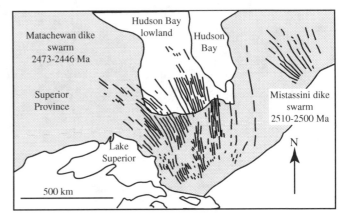

Map 1.1 Geologic map of the Matachewan and Mistassini dike swarms in the southern portion of the Canadian Shield. Dikes in the Hudson Bay lowland are mapped aeromagnetically. Modified after Buchan et al. (2007).

Figure 1.15 (A) Geologic sketch map of Shiprock in New Mexico, USA, showing dikes (linear features) radiating out of a volcanic neck (irregular gray shape). The volcanic neck, called Shiprock, is about 600 m high. (B) View of Shiprock and dikes from the southeast. Photo from the United States Geological Survey photo library, McKee, 1007ct.

temperature is not very high and the magma on the margin of the dike or sill may chill very rapidly and be fine-grained. The fine-grained margins of the dike or sill insulate the magma in the interior of the dike or sill, allowing it to cool more slowly, becoming coarser-grained.

When the crust fractures in an extensional tectonic environment, intrusion of magma into the resulting faults produces a **dike swarm**. A dike swarm consists of many dikes with similar orientation and chemistry that extend over tens to hundreds of kilometers. Dike swarms are best exposed in Precambrian terrains (Map 1.1) where erosion has stripped away the sedimentary cover. The compositions, dates, and orientations of Precambrian dike swarms may be used to reconstruct Precambrian continental configurations.

1.5.4 Structures in Plutonic Rocks

Plutonic rocks occur as irregularly shaped bodies known as plutons. A pluton that is larger than 100 km² (40 mi²)

Figure 1.17 (A) Biotite-rich xenolith in granodiorite dike, cutting the Laramie anorthosite complex, Wyoming, USA. (B) Fine-grained granodiorite autolith in granite, floor of main terminal building, Denver International Airport.

area of about 600 km × 200 km in eastern California, consists of hundreds of separate plutons that were emplaced over a time period that ranges over most of the Mesozoic Era, though the bulk of the batholith was emplaced throughout the Cretaceous Period. The term *batholith* is usually applied to granitic rocks. Large plutons that are composed of mafic rocks are more commonly referred to as **intrusions**.

Plutons emplaced in shallow environments may preserve chilled margins, although those intruded deeper in the crust may not. Igneous intrusions commonly contain blocks of exotic rock that range from centimeters to kilometers in size. If it is unclear whether the inclusion is related to the host rock, the term **enclave** can be used. In some occurrences, the inclusions are fragments plucked off the country rock during the intrusion of the magma or that foundered into the magma from the roof of the intrusion. Such fragments of country rock are called **xenoliths** (Figure 1.17A); the term *xeno-* means foreign. In some plutons, inclusions consist of pieces that were clearly part of the same magma sequence as the host rock. These types of inclusions are called **autoliths**. Autoliths are fragments of the older, solidified parts of an intrusion that were ripped off and incorporated into a younger magma (Figure 1.17B; see also Figure 1.8D).

Dikes are also present in plutonic rocks, although because they solidified at greater depth (and hence relatively high temperatures), they seldom show chilled margins. Because dikes in plutonic environments tend to be emplaced into a relatively warm environment, they may also be as coarse-grained as the rocks they intrude, unlike hypabyssal dikes.

in outcrop is called a batholith, although large batholiths are composed of many individual plutons. For example, the Sierra Nevada batholith, which is exposed over an

Summary

- Igneous rocks form by solidification of magma, either on Earth's surface (volcanic rocks), near the surface (hypabyssal rocks), or at depth (plutonic rocks).

- Igneous rocks are classified either on the basis of the proportions of quartz, feldspars, and mafic minerals or by their geochemical composition.

- A magma will crystallize spontaneously when the reaction will evolve heat. The crystallization of a magma is governed by the nucleation of crystals and diffusion of nutrients to the crystal faces of the growing crystal.

- The texture and structures preserved in igneous rocks allow geologists to interpret the environment in which the rocks formed.

Questions and Problems

Problem 1.1. Determine the rock names for coarse-grained rock samples with the following proportions of alkali feldspar, plagioclase, and quartz.

	A	B	C
Alkali feldspar	0.55	0.37	0.10
Plagioclase	0.22	0.36	0.49
Quartz	0.23	0.27	0.41

Problem 1.2. Determine the rock names for coarse-grained rock samples with the following mineral proportions.

	A	B	C
Alkali feldspar	15	3	0
Plagioclase	46	64	92
Quartz	21	2	3
Biotite	3	5	0
Hornblende	13	15	5
Other	2	11	0

Problem 1.3. A coarse-grained rock sample consists of 15 percent plagioclase, 35 percent augite, and 50 percent enstatite. What is the name of this rock according to the IUGS classification?

Problem 1.4. Determine the rock names for volcanic rock samples with the following compositions: ($FeO^* = $ total iron expressed as FeO)

	A	B	C
SiO_2	54.7	70.97	60.06
TiO_2	1.71	0.33	0.83
Al_2O_3	16.34	13.72	16.51
FeO^*	11.58	3.11	7.68
MnO	0.24	0.06	0.17
MgO	2.36	0.32	0.38
CaO	6.75	1.5	3.14
Na_2O	4.14	3.67	4.57
K_2O	1.47	5.22	5.58
P_2O_5	0.71	0.06	0.25

Problem 1.5. Compare and contrast the nucleation and crystal growth histories of the rocks shown in Figure 1.7A and 1.7B. Refer to Figure 1.6 in your explanation.

Problem 1.6. Figure P1.1 is a photomicrograph of a trachyandesite with augite (Aug), magnetite (Mag), olivine (Ol), and plagioclase (Pl). Determine the order in which these minerals crystallized and explain your reasoning.

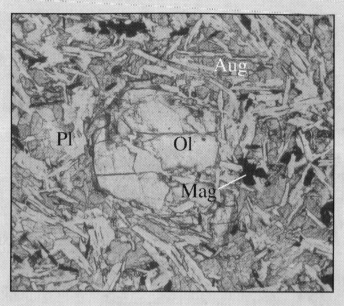

Further Reading

Jerran, D. A. and Davidson, J. P., 2007, Frontiers in textural and microgeochemical analysis. *Elements*, 3, 235–8.

LeBas, M. J., Le Maitre, R. W., Streckeisen, A., and Zanettin, B. A., 1986, Chemical classification of volcanic rocks based on the total alkali–silica diagram. *Journal of Petrology*, 27, 745–50.

Le Maitre, R. W., Streckeisen, A., Zanettin, B., LeBas, M. J., Bonin, B., and Bateman, P., 2005, *Igneous Rocks: A Classification and Glossary of Terms*, 2nd edn. Cambridge University Press, Cambridge.

MacKenzie, W. S., Donaldson, C. H., and Guilford, C., 1982, *Atlas of Igneous Rocks and Their Textures*. John Wiley & Sons, New York, NY.

Thorpe, R. S. and Brown, G. C., 1993, *The Field Description of Igneous Rocks*. John Wiley & Sons, New York, NY.

Wilson, C. N., 2017, Volcanoes: Characteristics, tipping points, and those pesky unknown unknowns. *Elements*, 13, 41–7.

Note

1 All terms in boldface are defined in the glossary.

Chapter 2

An Introduction to Igneous Phase Diagrams

2.1 Introduction

Silicate melts are chemically complex. Not only does it take more than nine elements to characterize most igneous rocks, melts also contain a number of volatile components, including H_2O, CO_2, HF, and HCl. Despite the complexity of natural igneous melts, phase diagrams of mineral relations in chemically simple systems provide a way to understand the processes by which igneous rocks crystallize. These phase diagrams may not duplicate the melting or crystallization process exactly, but they can help identify the factors that control the melting or crystallization of minerals in igneous melts. Surprisingly, many of the relations illustrated in simple phase diagrams can be extended to igneous rocks, despite their chemical complexity. This chapter begins with a review of the phase rule and lever rule, both of which are prerequisite to an understanding of phase diagrams. It then covers how to interpret the crystallization and melting relations in the various types of binary phase diagrams. Finally, this chapter provides a brief introduction to ternary and pseudoternary phase diagrams. Throughout these discussions the emphasis is on how relations in simple phase diagrams can be used to interpret crystallization and phase relations in more complex igneous rocks.

2.2 The Phase Rule

In the previous chapter the concept of free energy, G, was used to define the conditions in which crystals and melt are in equilibrium ($\Delta G = 0$), and when crystal growth is favored ($\Delta G < 0$). In this section the applications of thermodynamics will be mostly graphical, through the use of phase diagrams. A phase diagram is a projection from free-energy space onto a pressure (P)–temperature (T), temperature–composition, or composition–composition plane, showing the locus of points for given reactions where $\Delta G = 0$. The number of phases (or minerals) that may occur together in a phase diagram is given by the phase rule. Some thermodynamic definitions provide a necessary background for understanding the phase rule and phase diagrams. Among them are:

System – That part of the universe that is under consideration. Systems typically are described by their chemical constituents. For example, crystallization of olivine from a melt can be modeled by the system Mg_2SiO_4–Fe_2SiO_4. Because minerals may have complex chemical formulae and referring to geological systems by their chemical constituents can be cumbersome, geologists commonly refer to systems informally with the abbreviated names of the mineral end members (see Table A.1). For example, the system Mg_2SiO_4–Fe_2SiO_4 may be referred to as the system Fo–Fa, and the system $CaMgSi_2O_6$–$CaAl_2Si_2O_8$ may be called the system Di–An.

Phase – A homogeneous, mechanically separable part of a system. Phases are separated from one another by interfaces. Geologists usually think of the phases in a rock as the minerals that are present; however, during the evolution of an igneous rock, a silicate melt phase may be present as well as, perhaps, an H_2O-rich or a CO_2-rich fluid phase.

Component – A chemical constituent of a system. The phase rule quantifies the minimum number of components, or chemical constituents, needed to define all the phases in the system. The components of a system are often listed in terms of chemical formulae, as in the model system Mg_2SiO_4–Fe_2SiO_4 (or Fo–Fa). However, in applying the phase rule, it is important to identify the *minimum* number of components that define a system. For example, a system containing the phases andalusite, sillimanite, and kyanite can be described as having a single component: Al_2SiO_5. It would be a

mistake to call this a two-component system (i.e. Al_2O_3 and SiO_2). In this book, components are given either as chemical compositions (e.g. Mg_2SiO_4) or as abbreviations for mineralogic end members (for example, Fo for Mg_2SiO_4).

Variance – Variance refers to the number of variables that have to be constrained before the equilibrium conditions of a system can be known. It is also known as *degrees of freedom*.

As noted in the previous chapter, the measure of stability of a phase or series of phases in a system is determined by the change of free energy that occurs when a phase is formed. The change of free energy in a system is a function of temperature and pressure as well as the number of compositional variables (or number of components, C) in the system. If there are C components in the system, then there are $C + 2$ unknowns (i.e. the number of components plus temperature and pressure). A system with $C + 2$ unknowns requires $C + 2$ reactions to solve it uniquely. Because each phase in a system can be used to fix one of the compositional unknowns, a system with P phases will have P equations. Thus, if there are $C + 2$ phases (i.e. $P = C + 2$), then the system is uniquely determined. If there are fewer phases, then the system is under-determined and has some degrees of freedom (F). The equation relating the variance (or degrees of freedom) in a system to the number of phases and components is:

$$P + F = C + 2 \tag{2.1}$$

This relationship is known as the **phase rule**. (A mnemonic for the phase rule is Police Force = Cops + 2.) The number 2 represents pressure and temperature. If pressure is fixed (as it is in many igneous phase diagrams) the phase rule becomes:

$$P + F = C + 1 \tag{2.2}$$

A good way to understand the phase rule is to consider a phase diagram for a one-component system. The phase diagram for the system Al_2SiO_5 is a perfect example (Figure 2.1). If all three aluminosilicates are in equilibrium in a rock, then the crystallization conditions of the rock are known; it must have formed at the aluminosilicate triple point (i.e. point a in Figure 2.1 at 550 °C and 4.5 kbar). Such an assemblage has no degrees of freedom and is known as **invariant**. If only two aluminosilicates are present in equilibrium, for example, kyanite and

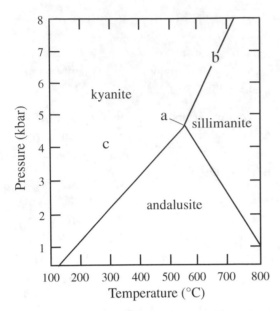

Figure 2.1 Phase diagram for the one-component system Al_2SiO_5 (from Pattison, 1992). Points a, b, and c are examples of invariant, univariant, and divariant assemblages, respectively.

Figure 2.2 Diagram showing olivine solid solutions. Point a refers to fayalite (pure Fe_2SiO_4), b refers to forsterite Mg_2SiO_4, and c refers to olivine with 75 percent of the forsterite component and 25 percent of the fayalite component or $Mg_{1.5}Fe_{0.5}SiO_4$.

plots at a, whereas pure forsterite plots at b. Intermediate olivine plots somewhere between a and b. The variable that describes the composition of olivine is called **mole fraction** and is abbreviated as $X_{Fe_2SiO_4}$, X_{Fa}, X_{Fe}^{Ol}, or sometimes $X_{Fe_2SiO_4}^{Ol}$. Mole fraction is the molar ratio of components in a solution. For olivine, the ratio is:

$$X_{Fe}^{Ol} = \frac{Fe_2SiO_4}{Fe_2SiO_4 + Mg_2SiO_4} \tag{2.3}$$

Since all olivine compositions share the SiO_4 framework, Equation 2.3 simplifies to:

$$X_{Fe}^{Ol} = \frac{Fe}{(Fe + Mg)} \tag{2.4}$$

Olivine is usually a binary solution of iron and magnesium that can be expressed as:

$$X_{Fe}^{Ol} = \left(1 - X_{Mg}^{Ol}\right) \tag{2.5}$$

Since point a in Figure 2.2 has no magnesium, it has $X_{Fe}^{Ol} = 1.0$ and $X_{Mg}^{Ol} = 0.0$. Likewise, since point b has no iron, it has $X_{Fe}^{Ol} = 0.0$ and $X_{Mg}^{Ol} = 1.0$. As X_{Fe} increases, the olivine plots closer to a and further from b. The lever rule indicates the exact location that an olivine with a given X_{Fe} will plot on Figure 2.2, expressed as a distance of X_{Fe^*}, the total distance from a toward b. For example, point c is located 75 percent of the way from a toward b and 25 percent of the way from b toward a. It has the composition $X_{Fe}^{Ol} = 0.25$ or $X_{Mg}^{Ol} = 0.75$.

sillimanite (point b, Figure 2.1), then the rock must have crystallized along the reaction curve kyanite = sillimanite. This assemblage has one degree of freedom (i.e. it is **univariant**), meaning that if one variable, either pressure or temperature, is known, then the other is defined by the univariant curve kyanite = sillimanite. Finally, an assemblage with only one aluminosilicate (such as point c in Figure 2.1) reveals only that the rock crystallized in one of the fields in the phase diagram. This assemblage has two degrees of freedom (it is **divariant**), and it is necessary to define both temperature and pressure to better constrain its crystallization conditions.

It is important to remember that the phase rule describes the state of the system at equilibrium. Therefore, phases that formed during later alteration of the rock cannot be included.

2.3 The Lever Rule

In addition to indicating pressure and temperature, phase diagrams communicate information about composition of the phases. The **lever rule** is used to locate compositions on a phase diagram. For example, olivine is a solid solution that consists of compositions that vary from Fe_2SiO_4 to Mg_2SiO_4 (Figure 2.2). All olivine compositions will plot somewhere on the line a–b, depending on the ratio of iron to magnesium in the olivine. Pure fayalite

A phase composed of three components can be plotted on a ternary diagram, where each apex of the triangle is the composition of a component. Consider phase E in Figure 2.3, which has the formula x_2yz. By normalizing the ions so they total to one (i.e. there are four ions, so dividing the stoichiometric coefficient of each by four), phase E can be represented by the following molar ratios $x:y:z = 0.5:0.25:0.25$. Point E in the x–y–z triangle (Figure 2.3) must lie on a line connecting all points that

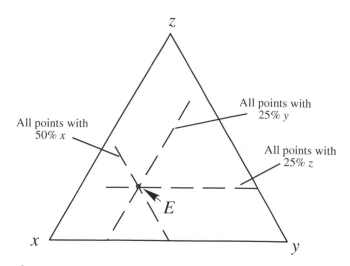

Figure 2.3 A ternary diagram for the system x–y–z showing where phase E with the composition of x_2yz plots.

contain 50 percent of x. The same operation can be performed to find lines marking the locus of points containing 25 percent y and 25 percent z, and the intersection of the three lines represents the composition of phase E.

2.4 Two-Component Systems Involving Melt

The phase rule and the lever rule provide the basis for interpreting phase diagrams. The simplest phase diagrams are one-component systems. Melting relations in these systems are simple: because the melt has a fixed composition, it can be treated the same as any other phase in the system. For example, adding a field for Al_2SiO_5 melt in Figure 2.1 requires only curves at high temperature for the melting of sillimanite and kyanite. On the other hand, it is not so simple to show melt relations in a two-component system because the melt is a solution involving both components. As a result, phase diagrams showing melting relations in two-component systems are usually drawn at fixed pressure, most commonly 1 bar, with the variables of composition and temperature. As a result, the phase rule for igneous phase diagrams is $P + F = C + 1$, where the 1 stands for temperature. To understand how melt interacts with various types of solids, five types of binary phase diagrams with melts are recognized. These are:

(1) binary systems with a eutectic;

(2) binary systems with a peritectic;

(3) binary systems with a thermal barrier;

(4) binary systems with solids that have a complete solid solution; and

(5) binary systems with solids that have a partial solid solution.

On each of these phase diagrams it is possible to identify the paths the melt would follow during the following processes:

(1) **Equilibrium crystallization** is the process whereby the crystallizing minerals remain in contact with the melt throughout the crystallizing process.

(2) **Fractional crystallization** is a process in which crystallizing minerals are immediately extracted from the melt and do not react with it further. In nature, fractional crystallization can occur by one of two mechanisms: (i) crystals may be removed from communication with the melt when they sink to the bottom of a magma chamber, or (ii) crystals may be left behind as the melt moves away, in a process called **filter pressing**.

(3) **Equilibrium melting** is a process whereby the melt and the residuum remain in communication throughout the melting process.

(4) **Fractional melting** models a system where melt is extracted from a system as soon as it forms and does not react further with the residuum.

2.4.1 Binary Systems with a Eutectic

An example of a two-component system with a melt is the system H_2O–NaCl, which is composed of the phases ice, salt, and melt (i.e. water). Ice is a solid that contains very little salt, and salt is a solid that contains very little H_2O. Liquid water, on the other hand, can contain a large amount of salt. At surface conditions, salt is not completely miscible with water, but this system is so familiar that it is a good one with which to introduce binary phase diagrams. A simple rule for the melting of most substances is that the temperature at which the solid will melt is highest when the melt has the same composition as the solid. In other words, addition of any component to a melt will reduce the melting temperature of solids in equilibrium with that melt.

The system H_2O–NaCl offers a well-known example of this rule. What happens when salt is spread on an icy

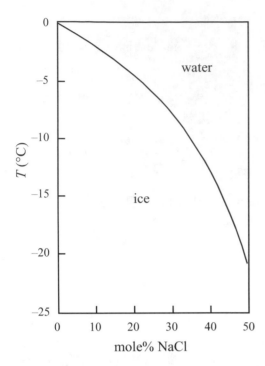

Figure 2.4 Phase diagram showing a portion of the system H_2O–NaCl.

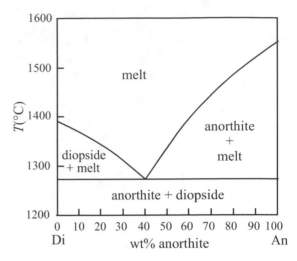

Figure 2.5 Phase diagram for the system An–Di at 1 bar (Bowen, 1915, used by permission of the *American Journal of Science*).

sidewalk? Of course, the ice melts. Figure 2.4 shows the effect of salt on the melting temperature of ice. This effect is not a magical property of NaCl – anything that can dissolve in water will depress the freezing temperature. Why is this?

When ice is in equilibrium with its melt, the molecules of H_2O are leaving the surface of the ice crystal and entering the water (i.e. the ice is melting) at the same rate as the molecules of H_2O are leaving the water and adhering to the ice (i.e. the water is freezing). What happens with the addition of a component, such as NaCl, that can dissolve in the water but not in ice? The rate at which H_2O molecules in the water adhere to the ice depends on the rate at which these molecules collide with the surface of the ice. The addition of NaCl to the water affects the rate at which H_2O molecules collide with the surface because some of the H_2O molecules will be bonded to Na^+ or Cl^- and will not be available to bond to the surface of the ice. As a result, the rate at which water freezes to form ice is lower than the rate at which ice melts to form water. If the temperature stays the same, the ice will melt. To equalize the rate of melting with the rate of freezing would require lowering the temperature (Figure 2.4).

The $CaAl_2Si_2O_8$–$CaMgSi_2O_6$ (An–Di) Phase Diagram. The system An–Di (Figure 2.5) is another good

example of how an additional component depresses the melting point of any phase. Pure anorthite melts at 1553 °C. Addition of even a small amount of $CaMgSi_2O_6$ to the melt will cause anorthite to melt at lower temperatures. Similarly, diopside melts at 1392 °C, and the addition of $CaAl_2Si_2O_8$ to a diopside-rich system will cause diopside to melt at lower temperatures. The curve showing the freezing point depression of anorthite (or of diopside) is the **liquidus**. The two liquidus curves meet at 1274 °C. This point is a **eutectic** and represents the lowest temperature at which melt may be present in this system. It is important to note that the melting temperature for this system (and other silicate phase diagrams shown in this chapter) is much higher than the temperature of most silicate melts because the additional components in a typical igneous melt further depress the melting temperature.

Relationships in Figure 2.5 are best understood by applying the phase rule. Since the system is isobaric, the phase rule should be: $P + F = C + 1$. The system has two components, so when three phases are present there are no degrees of freedom (i.e. an invariant point), when two phases are present there is one degree of freedom (i.e. a univariant line), and when only one phase is present there are two degrees of freedom (i.e. a divariant field). There is only one place on the diagram where three phases are present, and that is the eutectic where diopside, anorthite, and melt coexist. At the eutectic, there are no degrees of freedom, the temperature is 1274 °C, and the melt has a fixed composition (40 percent anorthite). There are three places on the diagram where two phases occur. These are the fields labeled diopside + anorthite, diopside + melt, and anorthite + melt in Figure 2.5. The diopside + anorthite field lies at temperatures below the eutectic. Diopside and anorthite have fixed composition (located at either end of the diagram). The univariant nature of the field is represented by the fact that at any temperature below 1274 °C diopside and anorthite of fixed composition will coexist. In the field anorthite + melt, the two phases are anorthite (of fixed composition) and a melt of variable composition. The univariant nature of this field is represented by the fact that, at a given temperature, the composition of the melt is fixed at the point where the isotherm for that temperature intersects the liquidus. Alternatively, if a melt composition is specified to be in equilibrium with anorthite, then the temperature at which anorthite + melt occurs is fixed. The one-phase field in Figure 2.5 is labeled melt. The composition of the melt is not constrained in this field. Even if the temperature is fixed, the composition of the melt is not. Likewise, if the melt composition is fixed, the temperature is not constrained.

Equilibrium Crystallization. During equilibrium crystallization of a melt in the system An–Di, the first phase to crystallize depends on the starting composition. For example, consider a melt with the composition 40 percent diopside and 60 percent anorthite (composition X in Figure 2.6). During cooling, the composition of this melt is unchanged until it hits the anorthite-melt liquidus at 1400 °C and anorthite begins to crystallize. Extraction of a small amount of anorthite makes the melt richer in

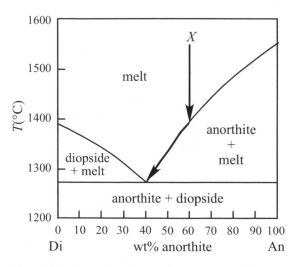

Figure 2.6 Phase diagram for the system An–Di showing the crystallization path for a melt with composition X.

$CaMgSi_2O_6$, and crystallization ceases unless the temperature cools further. As the temperature continues to fall, more anorthite crystallizes out of the melt, making the melt progressively richer in $CaMgSi_2O_6$. Eventually, at the eutectic (1274 °C), the melt becomes saturated in diopside and diopside crystallizes out along with anorthite. At this point, a small drop in temperature will cause the remaining melt to crystallize in a mixture of 60 percent diopside and 40 percent anorthite. If the initial melt has more diopside than the eutectic composition, then diopside is the initial mineral to crystallize. The composition of this melt becomes richer in $CaAl_2Si_2O_8$ and migrates toward the eutectic composition, where the final melt will crystallize.

Equilibrium Melting. Equilibrium melting follows the same path as equilibrium crystallization but in reverse. A rock with diopside + anorthite begins melting at 1274 °C, regardless of the proportion of diopside and anorthite in the rock. Rocks with different amounts of diopside and anorthite generate different amounts of melt at the eutectic, but all rocks in this system begin melting at the same temperature. At the eutectic, a small increase in temperature produces extensive melting. If the rock contains more than 40 percent $CaAl_2Si_2O_8$, all the diopside melts during this eutectic event and, with increasing temperature, the melt composition moves up the anorthite + melt liquidus as more anorthite melts. If the rock contains less than 40 percent $CaAl_2Si_2O_8$, anorthite is depleted and the melt moves along the diopside + melt liquidus as diopside melts.

Fractional Crystallization and Fractional Melting. In fractional crystallization, a crystal is removed from communication with the melt as soon as it forms. In systems such as An–Di, the fractional crystallization path is the same as the equilibrium crystallization path because there is no reaction between the crystals and the melt. In fractional melting, the melt is extracted from the solids as soon as it forms. In the system An–Di the melt forms at the eutectic temperature (1274 °C) with a composition of 60 percent $CaMgSi_2O_6$ and 40 percent $CaAl_2Si_2O_8$. This melt composition is constant as long as diopside and anorthite are present. Once either diopside or anorthite is depleted, melting ceases until the temperature reaches the melting temperature of the residual phase, be it anorthite (at 1553 °C) or diopside (at 1392 °C).

Petrologic Importance. Igneous rock textures commonly record the sequence of crystallization, preserving early-formed phases as euhedral crystals and late-forming minerals as anhedral crystals. Simple phase diagrams help explain the observed sequence in which mineral phases crystallize, how this sequence depends on the composition of the initial melt, and why multiple mineral phases crystallize simultaneously at the eutectic temperature until the melt is exhausted. The diagram also explains why, upon heating, a large amount of melt can be formed over a small temperature range at eutectic-like compositions.

2.4.2 Binary Systems with a Peritectic

Most minerals melt to form a liquid of the same composition as the solid, a process called **congruent melting**. For congruent melting, the reaction can be written:

$$\underset{\text{diopside}}{CaMgSi_2O_6} \rightleftharpoons \underset{\text{melt}}{CaMgSi_2O_6} \qquad (2.6)$$

However, not all minerals melt congruently. When enstatite, for example, is heated to its melting point, the melt is more silica-rich than enstatite, and the remaining solid converts from enstatite to olivine. This process is called **incongruent melting**. The reaction for the melting of enstatite is called a **peritectic** reaction and can be written:

$$\underset{\text{enstatite}}{Mg_2Si_2O_6} \rightleftharpoons \underset{\text{forsterite}}{Mg_2SiO_4} + \underset{\text{melt}}{SiO_2} \qquad (2.7)$$

The phase diagram for the system Mg_2SiO_4–SiO_2 (Fo–silica) is shown in Figure 2.7. In this diagram and several others in this chapter, the term "silica" is used because at high temperatures other polymorphs may be stable instead of quartz. As with the An–Di diagram, the last melt disappears at a eutectic where melt reacts to enstatite and silica exists. In this system, however, an additional invariant point occurs where forsterite, enstatite, and melt coexist. This point is the peritectic and it represents the equilibrium expressed by Equation 2.7. The difference between a eutectic and a peritectic is that at a eutectic, a melt reacts to form two solid phases, whereas at the peritectic, a solid phase reacts with the melt to form another solid phase. As the temperature falls through a peritectic, the solid phases in equilibrium with the melt change, but all the melt is not necessarily consumed.

Equilibrium Crystallization. Figure 2.8 shows an enlargement of the portion of the phase diagram for the system Fo–silica that contains the peritectic and eutectic. Equilibrium crystallization of melts that are more silica-rich than point P will follow very similar paths to those of melts in the system Di–An. On cooling, the melt hits a

liquidus, either the silica + melt liquidus or the enstatite + melt liquidus, at which point either silica or enstatite crystallize out, eventually driving the melt to the enstatite + silica eutectic (point E). However, if the bulk composition lay on the Fo side of point P, then the crystallization process will be very different. For example, a melt with the composition X in Figure 2.8 begins crystallizing forsterite and, as crystallization proceeds, the composition of

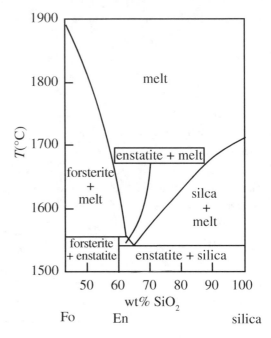

Figure 2.7 Phase diagram for the system Fo–silica (Bowen and Andersen, 1914, used by permission of the *American Journal of Science*).

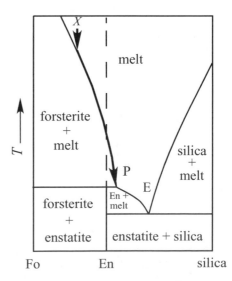

Figure 2.8 Phase diagram for the system Fo–silica, showing the equilibrium crystallization path for a melt with composition X.

the melt is driven to progressively more silica-rich compositions until the melt becomes more siliceous than enstatite. Even so, forsterite continues to crystallize until the melt reaches the composition of the peritectic. At this temperature (1556 °C), forsterite reacts with the melt to form enstatite by Equation 2.7. This reaction proceeds until one phase is completely consumed. If the bulk composition of the melt lies between Fo and En (left of the dashed line on Figure 2.8), the melt is used up resulting in the final assemblage of forsterite + enstatite. Alternately, if the bulk composition of the melt lies between that of enstatite and P, the reaction at the peritectic consumes forsterite and a small amount of melt remains. Continued cooling causes more enstatite to crystallize, driving the melt to the eutectic (1543 °C), at which point silica and enstatite crystallize together until the melt is consumed.

Equilibrium Melting. Equilibrium melting follows the inverse path of equilibrium crystallization. Melting of the assemblage quartz + enstatite begins at the eutectic (1543 °C). If the solid assemblage has more quartz than the eutectic composition, enstatite is depleted by melting at the eutectic and the melt moves up the silica + melt curve until all the quartz is depleted from the rock. If the solid assemblage has more enstatite than the eutectic composition, quartz is depleted by melting at the eutectic and the melt moves up the enstatite + melt curve. If the rock contained less enstatite than the composition of the peritectic, then this mineral is depleted from the solid assemblage before reaching the peritectic. If the rock contained more enstatite than the peritectic composition (point P in Figure 2.8), then at 1556 °C enstatite is entirely removed from the assemblage by reaction with the melt to form forsterite. Melt then traverses the forsterite + melt curve until attaining the same bulk composition of the original assemblage. If the original assemblage contained forsterite + enstatite, then melting begins at 1556 °C. After enstatite is depleted from the rock by peritectic melting, the melt composition moves along the forsterite + melt curve. Melting ceases when the melt composition is the same as the original bulk composition of the system.

Fractional Crystallization. In the system diopside–anorthite, it is unimportant whether the system undergoes equilibrium crystallization or fractional crystallization; the end product of crystallization is the same. In a system with a peritectic, however, the path followed during fractional crystallization is different from the trajectory followed during equilibrium crystallization.

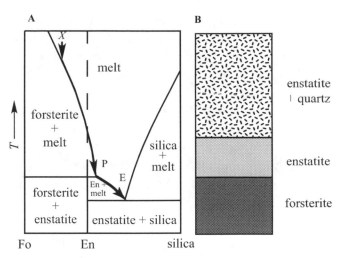

Figure 2.9 (A) Phase diagram for the system Fo–silica showing the path followed by fractional crystallization. (B) The sequences of mineral assemblages encountered in a hypothetical magma chamber of composition X that underwent fractional crystallization.

Equilibrium crystallization of a melt with a composition X will cease at the peritectic (point P in Figure 2.8), where all the melt is consumed by Equation 2.7. During fractional crystallization, the forsterite is entirely extracted from the system as soon as it forms. As a result, when the melt reaches P, forsterite is not available to react with the melt. Since forsterite is absent, the melt moves down the liquidus, past P, and enstatite begins to crystallize in place of forsterite. Enstatite crystallization drives the melt to E, the enstatite–silica eutectic (Figure 2.9A). Thus, in fractional crystallization, all melts reach the eutectic regardless of initial composition.

Whereas magmas that crystallize by the process of equilibrium crystallization form a rock with the same bulk composition as the original magma, magmas undergoing fractional crystallization form a series of rocks of different compositions. Consider what rock sequence would result if a melt of composition X was emplaced into a magma chamber and crystallized entirely by fractional crystallization (Figure 2.9B). If the first grains to crystallize, forsterite, sink to the bottom of the magma chamber, they form a layer of pure dunite. As the subsequent melt crosses the peritectic, enstatite crystallizes and may accumulate above the forsterite-rich layer. At the eutectic, the remainder of the melt crystallizes to an assemblage of 60 percent enstatite and 40 percent quartz. The sum of the abundances of forsterite, enstatite, and quartz in this theoretical magma chamber will equal the initial bulk composition of the magma. Applying the lever rule defines the exact proportions of the minerals formed in Figure 2.9B. (In this calculation, keep in mind that the proportions calculated will be in weight percent (wt%) because those are the units in which the phase diagram is plotted.)

Fractional Melting. In fractional melting, for a rock with the original assemblage enstatite + silica, melting begins at 1543 °C (the eutectic temperature; Figure 2.7). If enstatite is depleted during melting, quartz melts next when the temperature reaches 1712 °C. If quartz is depleted during melting, the next melting occurs at the peritectic (1556 °C) where enstatite converts to forsterite. Further melting occurs only when the temperature increases to the melting point of forsterite (1890 °C).

Petrologic Importance. The peritectic reaction for the system Fo–silica is critical in igneous petrology because it provides a mechanism by which olivine-saturated melt can evolve to form silica-saturated rocks. Magnesian olivine and quartz are incompatible, yet it is not uncommon for olivine-bearing igneous rocks to evolve toward silica-saturated, residual melts. Because, during fractional crystallization, a melt (even one with a complex composition) can pass through the orthopyroxene peritectic, it may produce a silica-enriched melt while leaving an olivine-rich residue behind.

2.4.3 Binary Systems with a Thermal Barrier

If a binary system has an interior phase that melts congruently, the resulting phase diagram has a **thermal barrier**. A phase diagram with a thermal barrier can be conceptualized as two binary phase diagrams, each of which has a eutectic, that are joined at the interior phase. A good example is the system $NaAlSiO_4$–SiO_2 (Nph–silica), which has the interior phase albite (Figure 2.10). Albite melts congruently, meaning this system has two eutectics, one where the melt crystallizes to nepheline + albite and another where the melt crystallizes to silica + albite. As noted previously, the highest temperature at which any phase melts is when it melts to its own composition. Thus, if a melt in the Nph–silica system has the composition of pure albite, it will crystallize directly to albite. If, however, a small amount of silica or nepheline is added to the system, the melt will crystallize at a lower temperature than the pure albite system.

Equilibrium Crystallization. Figure 2.11 details the system Nph–silica to show the effect of a thermal barrier on equilibrium crystallization. Consider two melts, X and

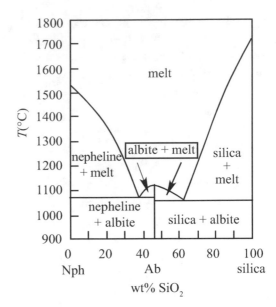

Figure 2.10 Phase diagram for the system Nph–silica (Schairer and Bowen, 1956, used by permission of the *American Journal of Science*).

Y, that have very similar compositions; X is slightly less siliceous than albite whereas Y is slightly more siliceous. Upon crystallizing albite, melt X moves toward nepheline-rich compositions and eventually reaches the nepheline + albite eutectic. In contrast, crystallization of albite from melt Y drives the melt composition to more silica-rich compositions and eventually to the albite + silica eutectic. Because crystallization of the melt always involves extraction of albite, there is no way a melt can move from the silica-saturated field to the nepheline-saturated field, making the albite composition a thermal barrier. Since a system with a thermal barrier is simply a system with two eutectics, the paths followed by fractional crystallization, equilibrium melting, and fractional melting will be similar to those observed in a system with a single eutectic.

Petrologic Importance. The thermal barrier represented by the albite composition plays an important role in the evolution of igneous rocks. As noted previously, in the system Fo–silica, fractional crystallization may drive melts across the olivine–orthopyroxene peritectic, allowing melts to evolve from an olivine-saturated composition to a quartz-saturated one. However, because albite imposes a thermal barrier, there is no way a nepheline-saturated melt can evolve to quartz-saturated compositions by fractional crystallization. The thermal barrier means that two basaltic melts with only slightly

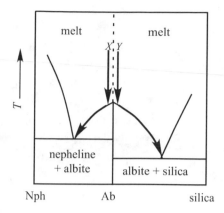

Figure 2.11 Phase diagram for the system Ne–silica, showing how crystallization of two melts with similar compositions can produce very different residual melts.

different compositions (one slightly silica-saturated and one slightly nepheline-saturated) will evolve two distinctly different residual melts, one rhyolitic (equivalent to the albite–silica eutectic in Figure 2.11) and the other phonolitic (equivalent to the nepheline–albite eutectic in Figure 2.11).

2.4.4 Binary Systems with Solid Solution

A binary system involving two end members that have complete solid solution can have only two phases, a melt and a solid, both of which are solutions between compositional extremes. The system never has three phases, so it cannot have a eutectic because a binary eutectic requires two solid phases and the melt phase. A phase diagram for this system displays two univariant lines: one is the liquidus, which gives the composition of the liquid at any given temperature, the other is the **solidus**, which gives the composition of the solid. Good examples of binary systems with solid solutions are the melting relations of the plagioclase (An–Ab) and olivine (Fo–Fa) series (Figure 2.12). When the plagioclase solid solution melts, the melt is always more sodium-rich than the coexisting plagioclase mineral (Figure 2.12A). Similarly, an olivine in equilibrium with a melt is always more magnesium-rich than the associated melt (Figure 2.12B). The crystallization and melting relations can be illustrated in the olivine system, although the relations are the same for the plagioclase system.

Equilibrium Crystallization. In the system Fo–Fa, a cooling melt of the composition X (Fo_{75}) will begin to crystallize at around 1790 °C (see Figure 2.13A). The olivine that crystallizes from this melt has the composition o_1 (approximately Fo_{90}). As cooling proceeds, the extraction of magnesian olivine enriches the melt in iron. In equilibrium crystallization, olivine remains to react with the melt, so the olivine present becomes more iron-rich as well. Both the olivine and the melt become more iron-rich (see heavy arrows in Figure 2.13A). Although both olivine and melt progressively become more iron-rich, the bulk composition of the system does not change because the abundances of olivine and melt change in tandem. Olivine becomes more abundant and melt becomes less abundant until, at 1628 °C, olivine (o_2) matches the composition of the bulk system (Fo_{75}). The last melt to crystallize at this temperature (m_2) has the composition Fo_{44}.

Fractional Crystallization. If the forsterite–fayalite system undergoes fractional crystallization, each batch of crystallizing olivine is removed from reaction with the melt. As with equilibrium crystallization, the first olivine to crystallize from a melt of Fo_{75} will be Fo_{90}. Subsequent removal of this Fo_{90} olivine will make the melt more iron-rich. However, because the olivine does not react with the melt, fractional crystallization of increasingly iron-rich olivine causes the melt composition to move all the way to Fo_0. After crystallization ceases, rather than having a rock with olivine of a single composition, the rock contains olivine with a composition that ranges across the spectrum, from Fo_{90} to Fo_0 (Figure 2.13B).

Equilibrium and Fractional Melting. Melting is the inverse of the crystallization processes. During equilibrium melting, olivine of Fo_{75} generates a melt with Fo_{44} and proceeds with olivine becoming simultaneously less abundant and more magnesian. Melting ceases when the last bit of olivine has melted, at which point the olivine will be around Fo_{90}. Fractional melting is complementary to fractional crystallization. As with equilibrium melting, the first melt to form from an olivine of Fo_{75} will be Fo_{44}. This melt, however, will be extracted from the system so any olivine that melts as temperature increases will be more magnesium-rich than Fo_{75}. The residual olivine becomes more magnesian as melting progresses until the last olivine to melt has a composition of Fo_{100}.

Petrologic Importance. Fractional crystallization of olivine (or any ferromagnesian mineral) extracts magnesium preferentially from a melt, leaving the melt relatively enriched in iron. Because of this phenomenon, most differentiated rocks evolved in this system are enriched in iron relative to magnesium. Petrologic

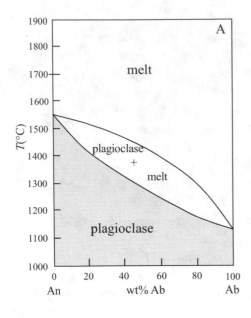

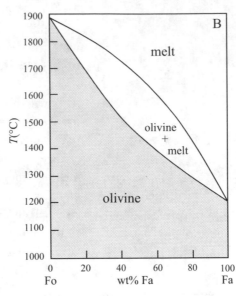

Figure 2.12 (A) Phase diagram for the system An–Ab (Bowen, 1913, used by permission of the *American Journal of Science*). (B) Phase diagram for the system Fo–Fa (Bowen and Schairer, 1935, used by permission of the *American Journal of Science*).

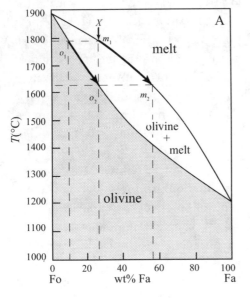

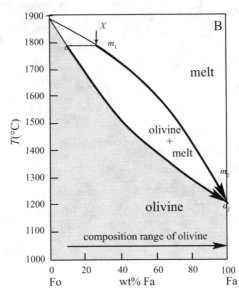

Figure 2.13 Phase diagram for the system Fo–Fa, showing the path followed by a melt with composition X during (A) equilibrium crystallization and (B) fractional crystallization.

evidence of fractional crystallization of olivine, however, is rarely seen in thin section because olivine easily equilibrates with the melt by simple exchange of iron and magnesium. Therefore zoned olivine is rare in igneous rocks. Fractional and equilibrium crystallization of plagioclase follows processes similar to fractional and equilibrium crystallization of olivine. The first plagioclase to crystallize is more calcic than the coexisting melt (Figure 2.12A), and both the plagioclase and the melt will become more sodic as crystallization proceeds. As a result of this process, evolved rocks are enriched in sodium relative to calcium. Plagioclase commonly does not

equilibrate easily with the melt, because doing so requires the exchange between calcium and aluminum in the plagioclase and sodium and silicon in the melt. Since both the aluminum and silicon are tightly bound in the tetrahedral site, equilibration is very slow. This slow diffusion explains petrologic evidence of zoned plagioclase, such as the one shown in Figure 1.8C, is abundant in igneous rocks.

2.4.5 Binary Systems with Partial Solid Solution

Systems with partial solid-solution phases have phase diagrams similar to those that show complete solid

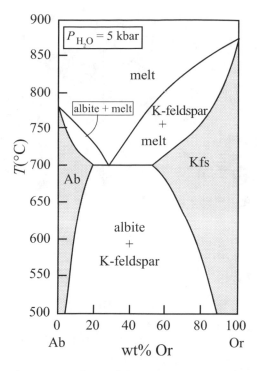

Figure 2.14 Phase diagram for the system Ab–Or at high water pressures (P_{H2O} = 5 kbar) (Morse, 1970).

solution; both have a liquidus and a solidus. A good example is shown in Figure 2.14, which illustrates the system $NaAlSi_3O_8$–$KAlSi_3O_8$ (Ab–Or) at high water pressure. This system contains a eutectic corresponding to the point where the last melt crystallizes to a mixture of K-feldspar and albite, as well as two liquidus lines (one for the melt in equilibrium with albite and one for the melt in equilibrium with K-feldspar), two solidus lines, and two **solvus** lines that reflect the extent of the solid solution of sodium in orthoclase and potassium into albite. The K-feldspar that crystallizes at the eutectic contains about 45 percent albite and albite that crystallizes at the eutectic contains 20 percent orthoclase. If the original melt contained less than 45 percent albite, the melt crystallizes to K-feldspar solid solution in much the same way that a melt in the system An–Ab crystallizes to a plagioclase solid solution. Likewise, if the melt contained less than 20 percent K-feldspar, an albite solid solution will crystallize without the melt ever reaching the eutectic.

The last melt present during crystallization of a melt that originally had a composition with between 20 percent and 55 percent Or will reach the eutectic. Unlike the system shown in Figure 2.6, where the solid phases have fixed compositions, both albite and K-feldspar will

change composition during crystallization. If the melt composition lies to the orthoclase-rich side of the eutectic, the first phase to crystallize will be K-feldspar. As crystallization proceeds, both K-feldspar and the melt become enriched in sodium until arriving at the eutectic. At the eutectic, albite with the composition $Ab_{80}Or_{20}$ begins to crystallize in equilibrium with orthoclase composed of $Ab_{45}Or_{55}$. If this system is rapidly cooled, as in volcanic rocks, the feldspars will retain their high-temperature compositions. However, if the system is slowly cooled, the feldspars will re-equilibrate along the solvus, with albite becoming more sodic and K-feldspar becoming more potassic.

Petrologic Importance. The solvus in the phase diagram for the system Ab–Or explains the presence of **perthite** in alkali feldspar. If the feldspars cool slowly, the Na^+ ions will diffuse to form sodium-rich regions and the K^+ ions will diffuse to form potassium-rich regions within the feldspar. This unmixing, or exsolution, results in intergrown lamellae of one phase within the other, producing perthitic texture. The exsolution lamellae are observable in thin section and sometimes by the naked eye. When visible in hand sample, the intergrowths are known as macroscopic perthite (Figure 2.15). Macroscopic perthite is common in microcline from pegmatites.

2.5 Phase Diagrams of Ternary Systems

Ternary systems involving a melt include four variables: temperature and three chemical components. Phase relations in this system are displayed on a three-dimensional diagram (Figure 2.16A) in which the base is an equilateral triangle that shows the compositional variation in the system. The temperature axis is perpendicular to this base. Each of the three sides of the diagram consists of a binary diagram. In Figure 2.16A these are the binary systems X–Y, Y–Z, and X–Z (note that the system X–Z lies at the back of Figure 2.16A). Each of the binary diagrams in Figure 2.16A is a simple binary with a eutectic. The alternative way to show phase relations in a ternary diagram is to use a **polythermal projection**, wherein temperature variation is drawn as contours (Figure 2.16B).

Just as an additional component causes depression of a freezing point in a one-component system, addition of a third component to a binary system causes depression of the temperature of a binary eutectic (Figure 2.16B). When

Figure 2.15 Macroscopic perthite, composed of white exsolution lamellae of albite in pink microcline (gray in black and white image).

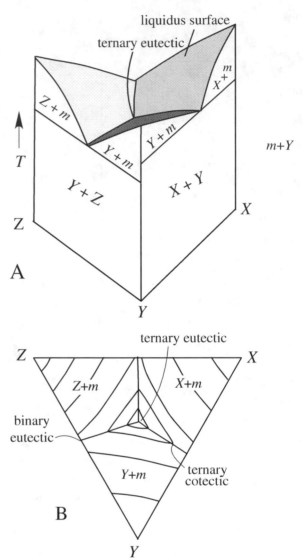

Figure 2.16 Ternary phase diagram for the system X–Y–Z (m = melt). (A) Perspective view of ternary temperature-composition diagram. (B) Liquidus projection with thermal contours.

a new component is added, the binary eutectic is no longer invariant; instead, it becomes univariant. This univariant curve is called a **cotectic.** Figure 2.16 shows three ternary cotectics that intersect at a ternary eutectic. As is the eutectic in a binary diagram, the ternary eutectic is an invariant point that reflects the lowest temperature at which a melt phase can exist.

2.5.1 The Ternary System CaAl$_2$Si$_2$O$_8$–CaMgSi$_2$O$_6$–Mg$_2$SiO$_4$

The ternary system $CaAl_2Si_2O_8$–$CaMgSi_2O_6$–Mg_2SiO_4 (An–Di–Fo) is an analog for a basaltic melt. Figure 2.17 shows that the system has four, rather than three, fields of primary crystallization. In addition to fields of forsterite + melt, diopside + melt, anorthite + melt, the system requires an additional field for spinel + melt. The composition of spinel ($MgAl_2O_4$) lies outside of the An–Di–Fo compositional triangle. A system that has one or more phases that lie outside of the plane of a ternary diagram is known as a **pseudoternary diagram**. Even though, strictly speaking, the system An–Di–Fo is pseudoternary, for most compositions it behaves like a true ternary system.

Equilibrium Crystallization. Consider how crystallization proceeds for a melt with the composition X on Figure 2.17. The first crystals, forsterite, form at

temperatures slightly below 1600 °C. As the temperature falls, the crystallization of olivine drives the melt composition directly away from the forsterite apex. At temperatures slightly above 1400 °C, spinel joins forsterite as a crystallizing phase. As temperature falls to 1317 °C (at point D), spinel reacts with the melt to make forsterite + anorthite. Point D is a ternary peritectic that is analogous to the binary peritectic in the system Fo–silica. A melt that was originally rich in spinel component is exhausted at the peritectic, whereas melts with a composition

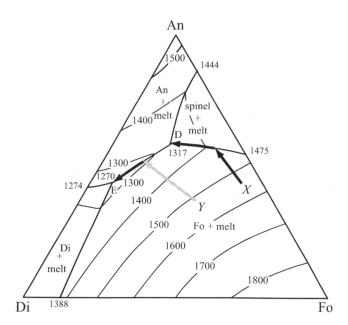

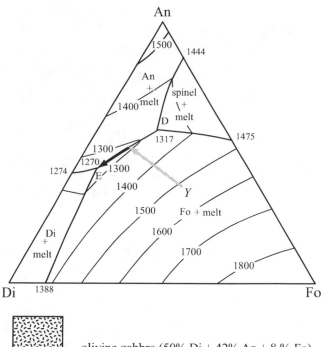

Figure 2.17 Ternary phase diagram for the system anorthite–diopside–forsterite, showing the crystallization path followed by melts of the composition X and Y. D = ternary peritectic and E = ternary eutectic. After Osborn and Tait (1952), used by permission of the *American Journal of Science*.

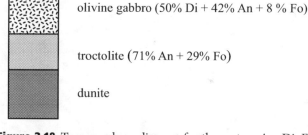

olivine gabbro (50% Di + 42% An + 8 % Fo)

troctolite (71% An + 29% Fo)

dunite

Figure 2.18 Ternary phase diagram for the system An–Di–Fo, showing the sequence of rocks produced from a melt with a composition Y that has undergone fractional crystallization and crystal settling. Mineral proportions are determined by the lever rule.

like X consume spinel by the peritectic reaction. At temperatures below 1317 °C, forsterite and anorthite are the crystallizing phases and their crystallization drives the melt along the cotectic toward the eutectic (point E). At the eutectic (1270 °C), diopside joins the crystallizing phases and forsterite + diopside + anorthite crystallize together until the melt is entirely consumed.

If the melt started with the composition Y on Figure 2.17, olivine crystallization forces the melt composition directly away from the Fo apex and toward the anorthite–forsterite cotectic (gray line in Figure 2.17). Melts of this composition miss the peritectic at point D and intersect the anorthite–forsterite cotectic at temperatures slightly below 1300 °C. Anorthite then joins the crystallizing assemblage, driving the melt composition to the eutectic.

Fractional Crystallization. As in simple binary systems, such as that illustrated in Figure 2.9, fractional crystallization produces a range of rocks from a magma of a fixed composition. Consider what happens to a melt of composition Y in Figure 2.18 during fractional crystallization. The forsterite formed during initial crystallization stages sinks to the bottom of the magma chamber, accumulating a layer of dunite. Once the melt reaches the cotectic, fractional crystallization deposits a layer of anorthite + forsterite,

producing a rock called a troctolite. Once the eutectic is reached, the residual melt crystallizes to an assemblage of diopside, anorthite, and forsterite, which together compose olivine gabbro. Although the layered sequence of rocks formed by this process (shown schematically in Figure 2.18) is an extreme simplification, it provides a good illustration of the kinds of crystallization processes that may form layered mafic intrusions discussed in Chapter 9.

Petrologic Significance. Ternary phase diagrams, though representing relatively simple systems, help to explain crystallization sequences and igneous textures found in nature. Figure 2.19A, a photomicrograph of anorthosite from the Laramie anorthosite complex, shows large plagioclase tabular crystals and interstitial olivine. This crystallization sequence is predicted in Figure 2.18 for a melt lying in the anorthite field at the top of the ternary

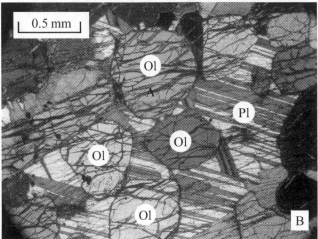

Figure 2.19 Photomicrographs showing crystallization sequences in gabbroic rocks. (A) Photomicrograph in crossed polarized light (XPL) showing tabular plagioclase (Pl) and interstitial olivine (Ol) in anorthosite from the Laramie anorthosite complex, Wyoming, USA. (B) Photomicrograph in crossed polarized light (XPL) showing euhedral olivine and interstitial plagioclase in olivine gabbro from the Rum intrusion of Scotland. Photo by Michael Cheadle.

diagram. Figure 2.19B, a photomicrograph of an olivine gabbro from the Rum intrusion of Scotland, shows euhedral olivine, and interstitial plagioclase, indicating a crystallization sequence analogous to that described in Figure 2.18 for melt of composition Y. These examples demonstrate how the simple systems portrayed in phase diagrams help petrologists interpret the crystallization of igneous rocks.

2.6 Implications for Petrology

Phase diagrams from simplified systems provide considerable insight into how igneous melts crystallize and evolve.

Most important, these display how igneous melts crystallize over a wide range of temperatures. Primitive basaltic lavas erupt at temperatures around 1200 °C, whereas granite melts may remain liquid at temperatures below 700 °C, and melts of alkaline rocks may survive to even lower temperatures. Even in simple systems, crystallization occurs over a range of temperatures before the eutectic is reached; for example, melts with a composition X or Y in Figure 2.17 will crystallize over a range of 200 °C to 300 °C. Phase diagrams show how eutectic melting can form a substantial amount of melt over a very small temperature range. Because natural silicate melts contain a large number of elements, true eutectic melting is probably a rare occurrence in nature but nevertheless, partial melting of silicate rocks may still generate a large amount of melt over a narrow temperature range.

The range of temperatures over which crystals and melts coexist provides ample opportunity for fractional crystallization. As noted earlier, two types of reactions allow melts to change composition during fractional crystallization. One is by means of peritectic reactions, the most important of which is the reaction between olivine and melt to make orthopyroxene (Figure 2.7). If olivine is preferentially extracted from a melt, it will cause silica to be enriched in the residual melt (Figure 2.9). The other is through continuous reactions such as those shown in Figure 2.12. Fractional crystallization of plagioclase will preferentially deplete calcium from the melt and leave sodium behind. Similarly, fractional crystallization of ferromagnesian silicates will enrich the melt in iron. For this reason, fractional crystallization of most melts will tend to produce residual melts enriched in SiO_2 and that have higher FeO/MgO and Na_2O/CaO than the original melt.

Finally, phase diagrams show that albite forms a thermal divide that cannot be crossed by fractional crystallization (Figure 2.10). This provides important insights into how nepheline-bearing rocks may form. Melts relatively poor in Na_2O will crystallize toward silica-saturated compositions (Figure 2.11). However, fractional crystallization of alkalic melts, which are rich in Na_2O, will follow a different trend. Rather than becoming more silica-rich, alkalic melts may evolve to progressively lower silica contents, with the result that nepheline will eventually crystallize (Figure 2.11). Similarly to melts that are relatively poor in Na_2O, fractional crystallization enriches alkaline melts in FeO/MgO and Na_2O/CaO.

Summary

- The melting temperature of most phases is decreased when additional components are added to the melt.

- In eutectic melting, large volumes of melt may be generated with small changes in temperature. Although true eutectic melts probably do not occur in nature, many common melts, such as basalt and granite melts, may be "eutectic-like."

- Olivine-bearing melts may differentiate to silica-bearing melts, but nepheline-bearing melts cannot differentiate to quartz-bearing melts. Thus, there are two distinct series of evolved igneous rocks, those that are silica-saturated and those that are nepheline-saturated.

- Differentiation via fractional crystallization enriches residual melts in iron over magnesium and sodium (and potassium) over calcium.

Questions and Problems

Problem 2.1. The figure below shows the phase diagram for the system diopside–anorthite.

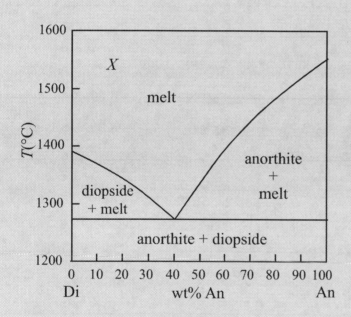

a. Draw the path of crystallization for a melt with composition X.

b. At what temperature does crystallization begin?

c. At what temperature does crystallization cease?

d. What is the composition of the melt at the eutectic?

Problem 2.2. Answer the following questions, assuming fractional crystallization in the system illustrated on the figure above.

a. What proportion of the original melt will crystallize out as diopside before anorthite appears?

b. Sketch the layers and label the minerals found in each, assuming that as the melt underwent fractional crystallization the minerals that crystallized settled out and were removed from contact with the remaining melt.

Problem 2.3. The figure below shows the phase diagram for a portion of the system Fo–silica. For both *equilibrium* and *fractional* crystallization:

a. Show the path followed by the liquid composition; and

b. What are the final minerals to crystallize? In what proportion are they?

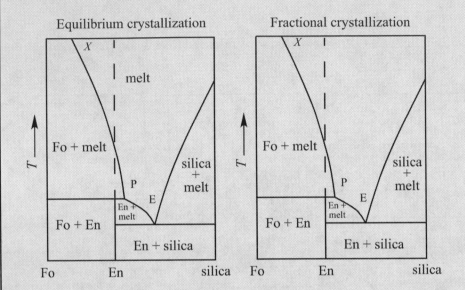

Problem 2.4. The figure below shows phase diagrams for the system forsterite–fayalite. Show the path of the melt composition for both equilibrium crystallization and fractional crystallization.

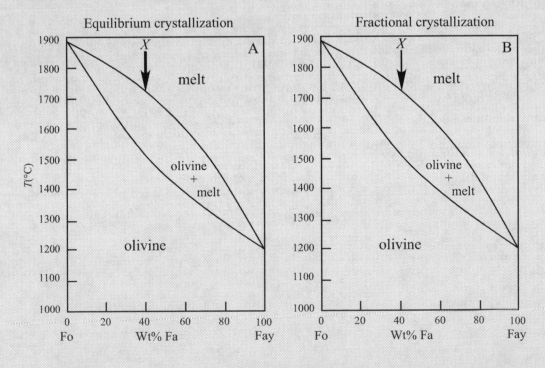

Problem 2.5. The figure below shows a ternary diagram for the hypothetical system *XYZ*.

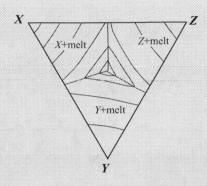

a. Label:

1. The ternary eutectic

2. A binary eutectic

b. On the phase diagram shown in Problem 2.5 above, show the composition of the initial melts *a*, *b*, and *c* that will crystallize on each of the following paths:

1. Melt $a \rightarrow Z$ + melt $\rightarrow X + Y$+ melt $\rightarrow X + Y + Z$

2. Melt $b \rightarrow Y$ + melt $\rightarrow Y$+ Z + melt $\rightarrow X + Y + Z$

3. Melt $c \rightarrow X$ + melt $\rightarrow X$+ Y + melt $\rightarrow X + Y + Z$

c. One of the crystallization paths listed on question (b) above is impossible. Which is it?

Problem 2.6. The figure below shows the system An–Di–Fo and two melt compositions, X and Y.

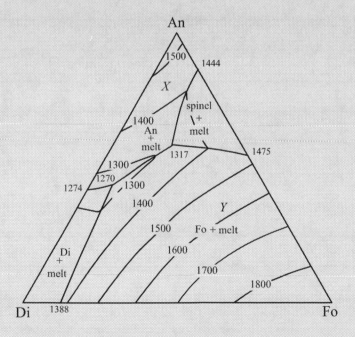

a. Show the crystallization path followed by each melt during fractional crystallization.

b. Assume that fractional crystallization occurred and the crystals that formed sank to the bottom of a magma chamber. Show the sequence of rocks and relative proportion of minerals in each layer of rock formed by this process from each melt composition.

Further Reading

Bowen, N. L., 1928, *The Evolution of the Igneous Rocks*. Dover, Mineola, NY.

Cox, K. G., Bell, J. D., and Pankhurst, R. J., 1979, *The Interpretation of Igneous Rocks*. Allen and Unwin, London, pp. 42–144.

Morse, S. A., 1980, *Basalts and Phase Diagrams*. Springer-Verlag, Berlin.

Chapter 3

Introduction to Silicate Melts and Magmas

3.1 Introduction

Igneous rocks form by crystallization of silicate melts.[1] As melt moves through the mantle and crust, it almost invariably carries crystals. These crystals may be minerals that crystallized from the melt as it cooled or they may be **xenocrysts**, foreign crystals incorporated into the melt from the rocks through which the melts ascended. This mixture of melt and crystals is called **magma**.

Silicate minerals are highly ordered at the atomic scale, consisting of an oxygen framework in which the cations sit in two major types of sites: tetrahedral sites and octahedral sites. Tetrahedral sites contain silicon (and to a lesser extent aluminum) and, depending on the mineral, they may occur as isolated tetrahedra or as linked tetrahedra (polymers) that form chains, sheets, or frameworks. The octahedral sites lie elsewhere in silicate structure and bond various tetrahedrally bonded polymers together. Cations typically found in octahedral coordination include Fe^{2+}, Fe^{3+}, Mg^{2+}, Ca^{2+}, and Na^+.

Silicate melts also have these structures at the molecular scale. When a silicate mineral melts, the long-range order disappears, but, unless the melt is taken to temperatures far above the liquidus, the short-range order is maintained. The octahedral and tetrahedral sites remain in the melt, but the polymers formed from the interlocked silica tetrahedra are discontinuous (Figure 3.1). The smaller ions, primarily Si^{4+}, Al^{3+}, and P^{5+}, are called **network-forming ions** because they occupy the tetrahedral sites that help polymerize the melt. The larger ions (Fe^{2+}, Ca^{2+}, Mg^{2+}, Na^+, etc.) are referred to as **network-modifying ions** because their presence will tend to depolymerize the melt. Because granites are rich in network-forming tetrahedra such as SiO_2 and Al_2O_3, granitic melts consist of many linked tetrahedra and hence have a high degree of polymerization, making them viscous. In contrast, rocks rich in network-modifying ions, such as basalt, are not strongly polymerized. Such melts will be more fluid than the granitic melts.

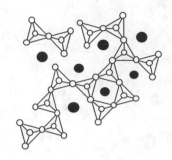

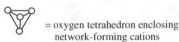

= oxygen tetrahedron enclosing
network-forming cations

= octahedrally coordinated
network-modifying cations

Figure 3.1 Schematic diagram showing the structure of a silicate melt. Small ions (Si^{4+} and Al^{3+}) occupy the silica tetrahedra and are called the network-forming ions; larger ions occur in sites outside the tetrahedral polymers and are called network-modifying ions.

= oxygen tetrahedron enclosing
network-forming cations

Figure 3.2 Diagram showing how the solution of H_2O into a silicate melt breaks silica polymers, replacing the bridging oxygen with OH.

3.2 Role of Volatiles

The presence of volatile components (most commonly H_2O and CO_2) plays an important role in silicate melts. These volatiles not only affect the melting temperature of rock and the viscosity of the melt, they affect the processes that accompany melt ascent and solidification and provide an important mechanism for eruption of lavas.

3.2.1 Role of H_2O

Water plays a critical role in igneous petrology. Not only does water pressure determine the presence or absence of hydrous ferromagnesian silicates, the presence of H_2O has a major effect on the properties of silicate melts. Water reacts with the bridging oxygens to break the silicate network (Figure 3.2) by a reaction such as:

$$H_2O + Si\text{-}O\text{-}Si \rightleftharpoons 2(Si\text{-}OH)$$

Because of this behavior, H_2O is more soluble in highly polymerized granitic melts than in less polymerized basaltic melts. Furthermore, because solution of H_2O into a melt depolymerizes the melt, addition of H_2O to a silicate melt will decrease its viscosity.

The addition of water to a silicate melt also decreases the temperature of crystallization. The effect of water on the melting of albite is shown in Figure 3.3. Albite is a

fairly good model for the melting of granite because both albite and granite melts are dominated by network-forming ions. In a dry system, granite and albite will melt at very high temperatures (900 °C or above). Increasing water pressure will cause melting to occur at progressively lower temperatures. This has an important effect on the behavior of melt in the crust. For example, a dry melt, generated deep in the crust, such as at point *A* in Figure 3.3B, will become superheated as it rises in the crust because decompression leads it away from the dry solidus. Thus, it may be erupted at temperatures as high as 75 °C above the solidus. In contrast, although a water-saturated melt (point *B* in Figure 3.3B) may form at low temperatures, it cannot move to much shallower levels without crystallizing. This means that any granitic melt emplaced into the crust or erupted on the surface must have been undersaturated in water at the depth where it was originally formed (point *C*, Figure 3.3B).

When a silicate melt crystallizes in the crust, the H_2O originally bound into the melt structure is released as fluid. To understand the importance of this process, one must understand how the volume of water vapor changes as a function of pressure (Figure 3.4). At pressures above about 3000 bar, the molar volume of H_2O in the melt is nearly the same as that in the fluid. Water exsolved from the melt at these pressures should disperse through the grain boundaries of the country rocks without much dilational effect. Melts emplaced at shallower levels (i.e. at lower pressures) will exsolve H_2O that has a much higher molar volume. For example, melt crystallized at 1000 bar will exsolve fluid with a molar volume more than three times larger than

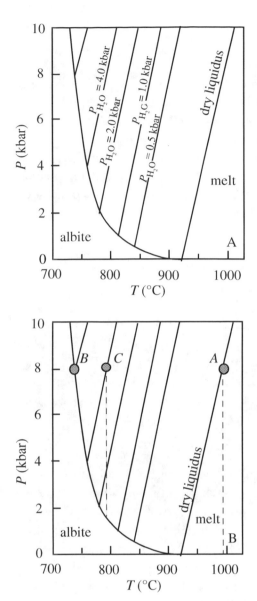

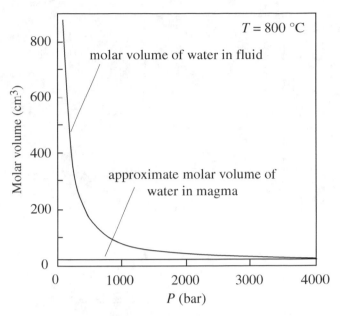

Figure 3.4 Comparison of molar volume of water in melt and in fluid at 800 °C. Data from Burnham et al. (1969).

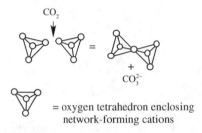

= oxygen tetrahedron enclosing network-forming cations

Figure 3.5 Diagram showing how the solution of CO_2 into a silicate melt enhances polymerization.

Figure 3.3 (A) Effect of H_2O on the melting temperature of albite. (B) How the phase relations control the melting and movement of melts in the crust. The behavior of melts at points A, B, and C are discussed in the text. After Burnham (1979).

that of the H_2O dissolved in the melt. When such a melt crystallizes, the H_2O released will hydrofracture the rock and produce a halo of veins around the intrusion. A melt crystallizing at 100 bar will exsolve an aqueous fluid that occupies a volume nearly 50 times that of the H_2O dissolved in the melt. This volume change is so huge that melts crystallizing at shallow levels may exsolve water explosively, producing caldera-type eruptions.

3.2.2 Role of CO_2

Silicate melts contain dissolved H_2O and CO_2, as well as other volatile components such as halogens. CO_2 may be the dominant volatile constituent in melts at depth, whereas H_2O may be incorporated into melts mainly as they pass through the crust. CO_2-rich fluid inclusions are found in many volcanic and plutonic rocks. Therefore, it is important to consider the effect of adding CO_2 on the viscosity of a melt. CO_2 dissolves into the melt by the following reaction:

$$2(Si - O)^- + CO_2 = Si - O - Si + CO_3^- \qquad (3.1)$$

As shown schematically in Figure 3.5, solution of CO_2 into a melt increases the polymerization of the melt.

Carbon dioxide has a greater solubility in melts with low polymerization, such as basalts, than in ones with few network modifiers, such as albite-quartz melts, because the latter have few non-bridging oxygens with which the

42   Introduction to Silicate Melts and Magmas

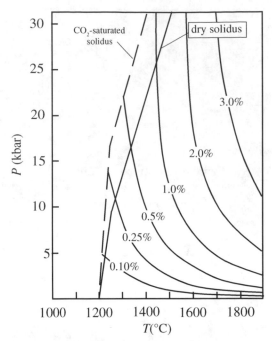

Figure 3.6 Solubility of CO_2 (in weight percent) into tholeiitic basalt. After Spera and Bergman (1980).

CO_2 may bond. Like H_2O, solubility of CO_2 into a melt lowers its melting temperature (Figure 3.6). However, the effect of CO_2 on the melting temperature of a silicate rock is far less extreme than that of water because CO_2 has a much lower solubility. CO_2 is most soluble in basaltic melts at high pressure. This means that as a basaltic melt rises in the crust it will exsolve CO_2. By the time the melt erupts it will consist of a mixture of silicate melt and CO_2-rich vapor. This vapor can become one of the major driving forces for eruption of basaltic melts, and indeed the most abundant volatiles emitted during volcanic eruptions are H_2O and CO_2.

3.3 Physical Properties of Magma

3.3.1 Temperature

Magma temperatures are difficult to measure directly. The temperature of flowing lavas or lava lakes can be measured with a pyrometer or with a thermocouple, and these direct measurements are supplemented with experimental determinations of silicate melting temperatures in the laboratory. The temperature of intrusive magmas is even harder to determine; for these, petrologists rely on experimental results and temperatures calculated from the compositions of coexisting minerals.

Extrusion temperatures for lavas are mostly in the range of 800 °C to 1200 °C. The temperature of basaltic magma is higher than that of rhyolitic magma. Basalts are rarely much hotter than the temperature at which their first minerals crystallize; that is, they are rarely superheated. In contrast, some rhyolites may erupt at temperatures up to 200 °C above their liquidus temperature. The temperature of intrusive magmas is probably lower than that of extrusive lavas. Water plays a role in this relationship because the solubility of water in silicate melts increases with increasing pressure. The increased water content of plutonic rocks has the additional effect of depressing liquidus temperatures to the extent that some granitic magmas may have been intruded at temperatures around 700 °C.

3.3.2 Heat Capacity and Heat of Fusion

Heat capacity of a substance is its capability to absorb heat energy. Specific heat capacity, C_p, is defined as the amount of heat energy that must be added to raise the temperature of one kilogram of the substance by one degree Celsius. Mathematically heat capacity is expressed as:

$$C_p = dH/dT \qquad (3.2)$$

where dH is the amount of heat (enthalpy) that must be added and dT is the change of temperature. Water has a very high specific heat capacity of 4184 J kg^{-1} °C^{-1}, which means it can absorb a large amount of heat without getting much hotter. Metals, such as copper with C_p = 385 J kg^{-1} °C^{-1}, quickly become hot when heat is applied to them. Silicate rocks and melts have intermediate heat capacities. The heat capacity of silicate rocks is approximately 1000 J kg^{-1} °C^{-1} and anhydrous silicate melts have a C_p of 1300–1400 J kg^{-1} °C^{-1}. Unsurprisingly, addition of water to the melt increases heat capacity: the C_p of dry granite melt is 1375 J kg^{-1} °C^{-1} whereas granite melt with 2 percent H_2O has a C_p of 1600 J kg^{-1} °C^{-1}.

Another heat term involved in producing igneous melts is the **heat of fusion**, dH_f. This variable defines the amount of heat that must be added to a rock, already at the melting temperature, to produce one gram of melt. The heat of fusion depends on the composition of the rock, varying from 220 J g^{-1} for granite to 400 J g^{-1} for gabbro and 580 J g^{-1} for peridotite. This means that for a given heat input, felsic crustal rocks will undergo more melting than more mafic rocks. It explains why partial melting and assimilation of felsic wall rocks occurs as magmas ascend through the crust.

One unusual feature of silicate melts is the great difference in these two quantities, C_p and dH_f. This means it takes about the same amount of heat to melt a given mass of rock as it takes to raise its temperature by 200 °C or 300 °C. This huge heat of fusion makes ascending magmas an efficient means for moving heat through the crust. Crystallization of magmas in the deep crust releases heat, necessary for partial melting and high-grade metamorphism, whereas at shallow crustal levels, crystallization may provide the heat that drives the hydrothermal circulation that is associated with many ore deposits.

3.3.3 Viscosity

The **viscosity** of a melt is a measure of its resistance to flow. It is a function of a number of properties, most importantly the composition of the melt, including the types and amounts of dissolved gases such as H_2O and CO_2, and its temperature. The viscosity of *magmas* is more complicated than that of silicate *melts* because most magmas contain crystals suspended in the melt. These crystals vary in size and shape, which also affect viscosity. The density of suspended crystals is particularly important: viscosity may be very different when crystals are abundant (greater than >40 percent by volume) than in more dilute suspensions of crystals in melt (Petford, 2009). Although quantification is difficult, viscosity is an important control on fundamental igneous processes, including the rate of magma transport in dikes and sills.

Disregarding the complexities introduced by crystal-bearing magmas, it is possible to gain a sense of the variations in viscosities by considering melts of different compositions and temperatures. The compositional control on viscosity of a given melt may be predicted to an extent by determining the number of network-modifying ions. The extent of polymerization is given by the term *SM*, which is the molar oxide sum of all the network modifiers in the melt (Giordano et al., 2006):

$$SM = \sum (Na_2O + K_2O + CaO + MgO + MnO + FeO^{tot}/2) \qquad (3.3)$$

Smaller values of *SM* indicate more polymerized melts, which will be more viscous than melts with larger *SM* values, which indicate melts that are less polymerized and therefore more fluid. The range of *SM* extends from zero for pure silica melts to 53 for a pure forsterite melt.

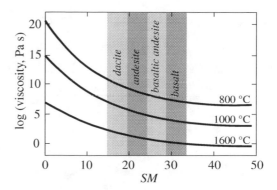

Figure 3.7 Relationship between viscosity, temperature, and composition of melts at 1 bar. *SM* is a measure of polymerization, where smaller values indicate more polymerized melts. After Giordano et al. (2006).

Viscosity is also temperature-dependent: the higher the temperature, the less viscous (more fluid) the melt. Temperature is related to viscosity (η) by the equation:

$$\eta = Ae^{E/RT} \qquad (3.4)$$

where A is a constant, E is the activation energy for viscous flow, R is the gas constant, and T is temperature. Activation energy is higher for more polymerized melts (Scarfe, 1973). Experimental measurements of the viscosity of melts with different values of *SM* show that rhyolitic melts have a viscosity around 1000 times higher than that of a tholeiite at the same temperature (Figure 3.7). Figure 3.7 also shows that the viscosity of a melt increases as temperature falls. Between 1400 °C and 1200 °C the viscosity of andesitic melt doubles (Giordano et al., 2006).

The differences in viscosity of silicate magmas affect the shapes and processes associated with volcanic rocks. For example, basaltic magmas, which have high *SM* values and which are extruded at high temperatures, tend to be fluid (similar to ketchup; see Table 3.1) and may flow for long distances, forming shield volcanoes or flood basalts. In contrast, rhyolite magmas, which have low *SM* values and are erupted at low temperature, tend to have high viscosity, and their behavior is more akin to Silly Putty® (Table 3.1). Rhyolitic magmas may form steep-sided domes or, if the magma is too viscous for gas to escape, they may erupt explosively.

3.3.4 Density

Density is an important property that affects the behavior of magmas in various ways: it is one of the factors controlling whether magmas rise through the crust,

Table 3.1 *Viscosities of magmas and common substances*

Material	Viscosity (Pa s)	Weight percent SiO$_2$	Temperature (°C)
ASE motor oil	2×10		20
Ketchup	$\sim 5 \times 10$		20
Basalt	$10\text{--}10^2$	45–52	1200
Peanut butter	$\sim 2.5 \times 10^2$		20
Crisco®	2×10^3		20
Andesite	$\sim 3.5 \times 10^3$	58–62	1200
Silly Putty®	$\sim 10^4$		20
Rhyolite	$\sim 10^5$	73–77	1200
Rhyolite	$\sim 10^8$	73–77	800

Source: From Philpotts and Ague (2009) and sources therein.

whether crystals settle out, and whether ions diffuse readily. The density of magmas varies from around 2.2 g cm^{-3} to 3.1 g cm^{-3}. Density rises with increasing pressure; it falls with increasing temperature. In general, mafic magmas are denser than felsic ones, mainly because mafic magmas are typically richer in heavy oxides such as CaO and FeO, whereas felsic magmas are richer in lighter oxides such as SiO$_2$, Al$_2$O$_3$, and Na$_2$O.

3.4 Ascent of Magmas

Most magmas are less dense than the surrounding country rocks, and therefore tend to ascend. The ascent velocity of a body of magma depends on its density and viscosity. At depth, magmas may ascend as buoyant diapirs, plastically deforming as they push aside the country rocks. The velocity of ascent is approximated by Stoke's law:

$$V = \frac{2g\Delta\rho r^2}{9\eta_w} \tag{3.5}$$

where V is the velocity, g = the force of gravity, r is diapir radius, $\Delta\rho$ is the density contrast between wall rock and magma, and the viscosity subscript refers to the wall rock (w). From this equation it is clear that velocity is greatest for magma bodies of large size and when magmas ascend through wall rocks with low viscosity.

Magmas may also rise through fractures, forming sheet-like intrusions known as dikes. In this situation, the buoyancy force causing the magma to rise is sufficient to fracture the rock. The fractures may be the site of

Figure 3.8 Stoped block of gabbro that sank through the magma chamber, where it deformed layers of anorthosite accumulating on the magma chamber floor. Laramie anorthosite complex, USA.

multiple intrusions of magma, which may be of similar or contrasting composition. A set of sub-parallel dikes, usually composed of basalt, is referred to as a dike swarm (Map 1.1). Radial dikes may emanate from a central magma body, such as the conduit feeding a volcano (Figure 1.15). Magmas cool as they fill fractures to form dikes because dikes commonly have a very large ratio of surface area to volume. If enough magma travels along a fracture, the wall rock may be thermally eroded, and a more cylindrical conduit may form. Cylindrical geometry is much more favorable for transporting magma with a minimum of heat loss, and it is the common shape of kimberlite diatremes.

At shallow levels, magma ascent may be primarily by **stoping**. In this process, wall rocks are fractured and founder into the magma (Figure 3.8). The magma loses heat both to the wall rock and by assimilation of the stoped blocks. For this reason, it is unlikely that magmas ascend great distances by this process, and ascent by stoping may be limited to the uppermost few kilometers of the magma's rise.

Because not all magma bodies will be of the appropriate temperature, size, density, or viscosity to rise all the way to the surface of the earth, it is worth remembering that those magmas extruded on the surface may not be representative of magmas formed at depth.

3.5 Magmatic Differentiation

Magmatic differentiation refers to all the mechanisms by which a parent magma may give rise to igneous rocks of various compositions. These processes include the crystallization of mineral phases and separation of the residual liquid (fractional crystallization), separation of two melts (liquid immiscibility), mixing of separate magmas, and assimilation of country rock either as a solid or as a partial melt into the parent magma. It is also possible to produce a diversity of rock types by partially melting solid rock and removing the melt, leaving behind a refractory restite. Because magmas originate by processes of melting, these mechanisms will be examined first.

3.5.1 Partial Melting

Partial melting, also referred to as *partial fusion* or **anatexis**, is the process by which melt is produced in a proportion less than the whole. Partial melting can be considered by examining the two end-member models introduced in the discussion of phase diagrams in Chapter 2:

Equilibrium melting: The partial melt, which forms continually, reacts and equilibrates with the remaining solid until the moment the melt is removed. Up to the point of segregation, the bulk composition of the system remains constant.

Fractional melting: The partial melt is removed in infinitely small increments so that it cannot interact with the residual solid. The bulk composition of the solid continually changes because melt, which is of a different composition than the initial solid, is lost from the system.

Which of these end-member processes most closely approximates partial melting in nature will depend on the ability of the melt to separate from the residual crystals. This in turn depends on many factors that control the microscopic geometry of the partial melt within the solid rock. Models of melt production and segregation in the mantle suggest melts can escape from the solid matrix after relatively small degrees of melting. Assuming that compaction drives expulsion of partial melt, McKenzie (1984) suggested melt will be expelled once the melt represents 3 percent or more of the total volume. McKenzie and O'Nions (1991) present evidence, based on the inversion of rare-earth element concentrations, that melt segregation in the mantle can occur when melt fractions are less than 1 percent. More recently, Rabinowicz and Toplis (2009) considered the combined effect of shear segregation related to the ductile flow of the mantle and compaction, and determined that, depending on the viscosity of the solid mantle, basaltic melts will be expelled at melt fractions between 3.5 and 10 percent.

3.5.2 Crystallization Processes

Crystallization of solid phases from a melt also presents opportunities to form rocks of a different composition than the original melt. Like partial melting processes, crystallization can be considered by examining the two end-member models discussed in Chapter 2:

Equilibrium crystallization: Crystals remain in contact with the residual liquid after they form and continually react and equilibrate with the liquid. In this case, the bulk composition of the final solids is the same as the original melt composition, and no magmatic differentiation takes place.

Fractional crystallization: Crystals are removed from the residual liquid as soon as they are formed by processes such as gravitational settling or floating. In this process, the bulk composition of the remaining liquid changes as crystals form and are removed.

It is possible to segregate liquid from crystals by mechanisms other than gravitational separation. The liquid remaining before crystallization may be complete and can be squeezed out in a process called *filter pressing*. A buoyant liquid in a mush of loosely packed crystals may migrate to a zone of lower pressure, just as water is driven out of a pile of accumulating and compacting sediments. Another mechanism, **flow segregation**, may separate crystals from the remaining melt during flow through a dike or along the walls of a pluton. Nearest the contact with country rock, the velocity gradient is steepest and a zone of maximum shear is present. This shearing results in a force that tends to drive crystals out of the zone of maximum shear and toward the interior of the flow. Crystals are thus found in the areas of lowest velocity gradient; in a dike they may be concentrated in the center, and the margins of the dike are much finer-grained.

3.5.3 Liquid–Liquid Fractionation

Liquid–liquid fractionation does not appear to be a major process by which silicate magmas become differentiated. However, certain melt compositions may separate into two or more *immiscible* (i.e. unmixable) liquids. Immiscibility is thought most common in mafic melts, which may separate into a sulfide liquid and a mafic silicate liquid. Immiscibility is also likely to occur in alkaline melts, rich in CO_2, and which may separate into a high-alkali silicate liquid and a carbonate-rich liquid. Iron-rich basaltic melts may separate into a felsic silica-rich liquid and a mafic iron-rich liquid. If the two liquids formed have very different densities, then they may separate very effectively, as oil does from water. If the magma is very viscous and crystal-rich, the two liquids may not separate as well, and small droplets may be evident in the interstices between early crystallizing minerals. Evidence of this process is preserved as solid inclusions of (Na, K)Cl in K-feldspar in the monzosyenites of the Laramie anorthosite complex, Wyoming, USA, which suggests these rocks formed from a magma that contained immiscible droplets of a chloride-rich melt (Frost and Touret, 1989).

3.5.4 Assimilation

Magmas may also change composition by assimilating material from their country rocks. **Assimilation**, sometimes referred to as *contamination*, may occur in two ways:

Bulk assimilation occurs when blocks of wall rock are stoped into the magma and completely melt.

Assimilation of partial melts occurs when the wall rocks are heated to their solidus and begin to melt. It is important to remember that the composition of the first melts are usually different than the bulk composition of the rock, and hence the compositional effect on the magma is not identical to that from bulk assimilation.

Both assimilation of bulk rock and partial melt require thermal energy for melting. The melt necessarily becomes

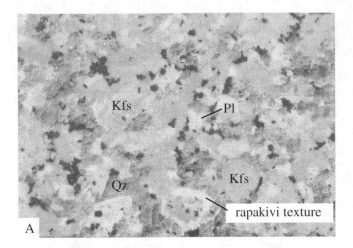

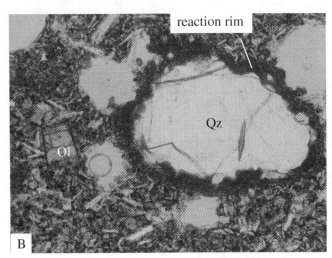

Figure 3.9 (A) Photo of alkali feldspar mantled with plagioclase, a texture known as rapakivi texture. Mantled feldspars indicate a change in crystallization conditions. Kfs = alkali feldspar, Pl = plagioclase, Qz = quartz. Sherman batholith, Wyoming, USA. (B) Resorbed crystal of quartz (Qz) in an olivine-bearing (Ol) groundmass, Capulin Volcano, New Mexico. The anhedral crystal shape and reaction rim around the quartz suggest that this crystal was entrained into a magma in which quartz was not stable.

cooler, so much so that it may begin to crystallize. As was noted above in Section 3.3.2, the energy input required to melt rock is large because the heat of fusion for silicate rocks is high. This heat energy is released upon crystallization and may provide the energy for assimilation to proceed. Therefore, assimilation is usually accompanied by crystallization (assimilation with fractional crystallization, often abbreviated as AFC).

Various textural features may be evidence for assimilation. Xenocrysts, crystals inherited from the country rock that would not normally be expected to crystallize from the melt, may be present. *Reversely zoned crystals*, such as plagioclase with calcic rather than sodic rims, suggest that the melt has changed its composition and that the composition of the phases crystallizing from the melt has changed in a manner not normally associated with differentiation by simple fractional crystallization (Figure 3.9A). *Resorbed crystals*, those that show textural evidence of remelting may (but do not always) result from changes in melt composition, which may cause a previously crystallizing phase to be out of equilibrium with the contaminated melt (Figure 3.9B).

3.5.5 Magma Mixing

If two magmas are introduced into the same magma chamber, they may mix and form a magma of intermediate composition. The magmas may be independently derived, or they may form when magma in a reservoir differentiates, and then the reservoir is recharged with another pulse of initial magma. If the two magmas do not completely homogenize, then evidence of commingled melts may remain. Intrusion of denser mafic magma into a felsic magma chamber sometimes forms a sheet of

Figure 3.10 The Hortavær complex in north central Norway was constructed by multiple injections of diorite (dark lenses) intruded into a syenitic magma chamber (light-colored rock).

mafic magma near the floor of the magma body. Thermal contrasts between the magmas cause the mafic magma to chill, forming lobate margins and other structures recording the presence of both mafic and felsic magmas within a chamber (Figure 3.10).

Summary

- Silicate melts are composed of network-forming ions, such as Si^{4+}, Al^{3+}, and P^{5+}, which occupy tetrahedral sites. These form linked tetrahedra that polymerize the melt. Larger ions, including Fe^{2+}, Fe^{3+}, Ca^{2+}, Mg^{2+}, and Na^+, form network-modifying ions that tend to depolymerize the melt. Granitic melts are richer in network-forming ions and are more strongly polymerized (and more viscous) than basaltic melts.

- The presence of H_2O in silicate melts tends to break the silicate networks and depolymerize the melt. CO_2 has the opposite effect and increases the polymerization of the melt.

- In general, mafic magmas are denser than felsic ones.

- Magma viscosity is a function of composition, the presence of H_2O or CO_2, temperature, and the kinds, quantities, and geometries of crystals present.

- Magmas are typically less dense than the surrounding rock, and hence have a tendency to rise. Magmatic ascent is an important mechanism to transport deep heat to the shallow crust.

- A diversity of igneous rocks can be produced by partially melting solid rock, extracting the melt, and leaving behind a more refractory restite. Partial melting may be fractional, in which infinitely small increments are removed from the remaining solid. Equilibrium melting describes the process by which partial melt continually reacts and equilibrates with the solid until melting is complete.

- A suite of igneous rocks with a variety of compositions may also form by fractional crystallization.

- Other processes that differentiate magmas include immiscible liquid–liquid fractionation, assimilation, and magma mixing. Assimilation is commonly coupled with crystallization because the latent heat released during crystallization provides energy for partial melting or digestion of stoped blocks of country rock.

Questions and Problems

Problem 3.1. Referring to the albite–H_2O system as an analog for granite (Figure 3.3), answer the following:

a. How does increasing pressure affect the amount of H_2O that can be dissolved in a granitic melt?

b. How does addition of H_2O affect the melting point of the granitic rock?

c. What is the effect of dissolution of H_2O on the viscosity of the granitic melt?

d. How does the ascent of a rising H_2O-saturated magma compare to that of a rising H_2O-undersaturated magma?

Problem 3.2. Aqueous fluids can be released during the late stages of crystallization of a magma in a magma chamber.

a. Refer to Figure 3.3 to explain how crystallization at a constant pressure can lead a magma to become H_2O-saturated.

b. At what pressures might the aqueous fluid released hydrofracture the rock? Discuss your answer in terms of the molar volume of H_2O.

c. Why might porphyritic texture result (like that shown in Figure 1.8D) when H_2O is lost from a magma body?

Problem 3.3. Evaluate the assertion in Section 3.3.2 that crystallization of magmas releases heat necessary for partial melting with the following calculation.

a. How much energy, in $J\,g^{-1}$, is required to heat dry granitic rock from 200 °C to its melting temperature of 800 °C? Given that the heat of fusion for granite is 220 $J\,g^{-1}$, what is the total energy required to heat and melt the rock? How does the amount of energy for heating to the melting temperature compare to the amount required to melt the rock once at its melting temperature?

b. How different would the energy requirement be to heat a gabbroic rock from 200 °C to its melting temperature of 1100 °C and melt it? (For gabbro, the specific heat capacity is 1000 $J\,kg^{-1}\,°C^{-1}$ and heat of fusion is 400 $J\,g^{-1}$.)

c. How many grams of a felsic magma with heat capacity of 1400 $J\,kg^{-1}\,°C^{-1}$ at its liquidus temperature would be required to crystallize to melt 1 g of cold (200 °C) country dry granitic rock? To melt 1 g of gabbroic rock?

Problem 3.4. Use Stoke's law to estimate the ascent velocity (in cm s^{-1}) of a magma diapir:

$$V = \frac{2g\Delta\rho r^2}{9\eta_w}$$

a. Calculate ascent velocity for a 10-km-diameter diapir with a density of 2.6 g cm^{-3} through mid-crustal wall rocks with a density of 3.0 g cm^{-3} and a viscosity of 10^{21} Pa s. (Recall that $g = 980$ cm s^{-2} for Earth and 1 Pa s = 10 g cm^{-1} s^{-1}.)

b. What will affect the ascent rate more: decreasing the diapir diameter to 1 km, or increasing viscosity of the wall rocks to 10^{24} Pa s, a value representative of upper crustal rocks? Present calculations to support your answer. On the basis of these simple calculations, does diapiric ascent appear to be an effect mechanism for moving magmas through the upper crust?

Further Reading

Hess, P. C., 1980, Polymerization model for silicate melts. In *Physics of Magmatic Processes*, ed. R. B. Hargreaves. Princeton University Press, Princeton, pp. 3–48.

Petford, N., 2009, Which effective viscosity? *Mineralogical Magazine*, 73, 167–91.

Winter, J. D., 2010, *Principles of Igneous and Metamorphic Petrology*, 2nd edn. Prentice Hall, New York, NY, Chapter 11: Magma diversity.

Note

1 Almost all igneous rocks form from silicate melts, but there are rare exceptions. Carbonatites, for example, form from carbonate melts and some high-temperature (and even low-temperature) sulfide deposits form from sulfide melts or from polymetallic melts. Iron oxide or iron–titanium oxide melts have been postulated to form in some environments.

Chapter 4

Chemistry of Igneous Rocks

4.1 Introduction

An important way of classifying suites of igneous rocks is by the variation in their chemical compositions. Indeed, igneous geochemistry is a complex field wherein major, minor, and trace-element characteristics, as well as isotopic compositions, help determine the origin and evolution of igneous rocks. Most of this subject is beyond the scope of an introductory course. However, because geochemistry is integral to the classification of igneous rocks, it is essential to provide an introduction to this topic.

The chemical formulae of the common rock-forming minerals are composed of relatively few elements referred to as **major elements**. Most rocks contain more than 1.0 weight percent of each of the major-element oxides SiO_2, Al_2O_3, FeO and Fe_2O_3, MgO, CaO, Na_2O, and K_2O. Because the major-element composition of a rock reflects its mineralogy, major-element oxide concentrations can be used to calculate what is referred to as the **normative mineralogy** of a rock. This topic is explained in the following section. **Minor elements** typically make up 0.1 to 1.0 weight percent of a rock. These include TiO_2, MnO, and P_2O_5. **Trace elements** comprise less than 0.1 percent of a rock; they are expressed as elements as opposed to oxides and their concentrations are reported in parts per million by weight. **Isotope geochemistry** is discussed in Chapter 5.

4.2 Modal Mineralogy versus Normative Mineralogy

The relative abundance of minerals in a rock is known as its mode. Modal abundances are expressed in volume percent. In rocks with individual minerals that can be distinguished by color, or that can be stained to mark minerals with distinct colors, the mode can be determined using image analysis programs. Modes can also be determined by *point-counting* under the petrographic microscope. In this method, the mineral under the cross-hairs is identified. The thin section of the rock is advanced systematically a specific distance across the microscope stage, and successive minerals under the cross-hairs are identified in a grid of points. The total number of points occupied by each mineral is converted to a volume percent. Although it is relatively easy to obtain modal mineralogy for plutonic rocks using these methods, it is very difficult to do so with extrusive rocks, which are finer-grained and may partially consist of glass.

To help overcome the difficulties in obtaining modes for fine-grained rocks, four petrologists, C. Whitman Cross, Joseph P. Iddings, Louis V. Pirsson, and Henry S. Washington, introduced the **norm** (Cross et al., 1902). Their calculated mineralogical composition, based on the major-element oxide abundances of the rock, is called the CIPW norm after the initial letters of their surnames. The CIPW norm calculations assume the melt crystallizes to anhydrous phases (olivine, pyroxenes, and iron–titanium oxides); no hydrous phases are allowed. Biotite would be expressed as normative orthopyroxene + K-feldspar, hornblende by normative orthopyroxene + clinopyroxene + plagioclase. Therefore, unless the rocks are anhydrous, the norm will not include the same assemblage as the mode. Moreover, unlike the mode and the IUGS classification system, both of which are reported in volume percent, the CIPW norm is reported in weight percent. Nevertheless, normative calculations can be very helpful. Not only does the norm provide a means to compare plutonic rocks with volcanic rocks, it can also help demonstrate how the geochemistry of a rock reflects its mineralogy. For example, a corundum-normative granite contains more alumina than can be accommodated in feldspar alone, and a nepheline-normative rock does not contain enough silica to form quartz.

4.3 Variation Diagrams Based on Major Elements

Suites of igneous rocks, such as a series of lavas erupted from a single volcano, can be inferred to originate from a common parent magma. Thus, the variation in their compositions arises from magmatic differentiation. One of the main mechanisms by which a suite of rocks of different compositions may crystallize from a common parental magma is by crystal–liquid fractionation. As is clear from the phase diagrams examined in Chapter 2, crystallizing minerals incorporate certain elements into their structure and, in doing so, leave behind melt of different composition.

Two common terms describe the behavior of elements during the differentiation of a melt. A **compatible element** is one that is preferentially fractionated from a melt into the crystallizing phases. In other words, the element is compatible with the crystallizing minerals. Because these elements are preferentially extracted during crystallization, the abundances of compatible elements in the melt should decrease with increasing fractionation (Figure 4.1, path *A*). In contrast, **incompatible elements** are not compatible with the crystallizing phases. Because they are preferentially retained in the melt rather than incorporated into crystallizing phases, the abundances of these elements in the residual melt should increase with increasing differentiation (Figure 4.1, path *B*).

Because each element behaves differently during magmatic differentiation, one way to assess differentiation in a suite of igneous rocks is to plot the weight percent of the various oxides in the rocks against some monitor of differentiation. Because the first minerals that crystallize out of a mafic melt are silica-poor (olivine has <40 percent SiO_2, calcic plagioclase has <50 percent SiO_2), differentiation typically causes the residual melts to become enriched in SiO_2 (i.e. SiO_2 behaves as an incompatible element). Therefore, examining the variation of each element against the weight percent SiO_2 is a common strategy for

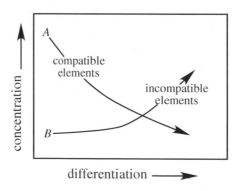

Figure 4.1 Diagram showing the behavior of compatible (path *A*) and incompatible (path *B*) elements during differentiation of a melt.

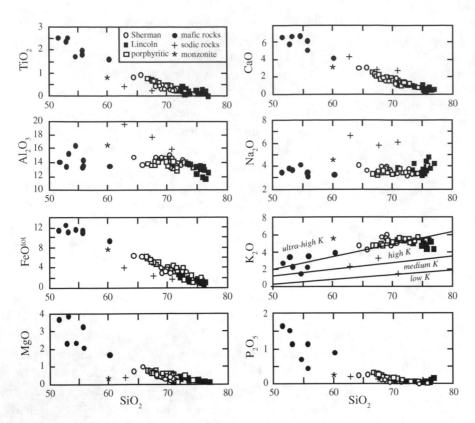

Figure 4.2 Harker diagrams showing the compositional variation in the Sherman batholith, Wyoming, USA. After Frost et al. (1999). FeO^{tot} is total iron expressed as FeO.

monitoring differentiation in a series of related igneous rocks (Figures 4.2 and 4.3). Such diagrams are called Harker diagrams, after their inventor, Alfred Harker (Harker, 1909). In Harker diagrams, the weight percent of elements incorporated into early crystallizing phases, such as ferromagnesian minerals and calcic plagioclase (CaO, FeO and Fe_2O_3, MgO, and TiO_2), tends to decrease with increasing SiO_2 content. Weight percent Al_2O_3 may also decrease with increasing SiO_2, but the change commonly is not as extreme as for CaO or MgO. In contrast, the weight percent of the alkalis (Na_2O and K_2O) typically increases with increasing SiO_2 as these accumulate in the magma until alkali feldspars crystallize.

Harker diagrams cannot always illustrate the processes occurring during differentiation of basaltic magmas because substantial differentiation may take place in basalt without discernibly affecting the SiO_2 content. In studying basalts, petrologists may choose to plot weight percent oxides as a function of weight percent MgO instead of SiO_2 (see Figures 7.9, 9.2). Fractional crystallization of olivine will enrich a melt in FeO while depleting it in MgO (see Figure 2.12B), thus weight percent MgO is a good monitor of differentiation in mafic magmas that undergo minimal changes in SiO_2. It is important to note that whereas increasing differentiation

is marked by *increases* in SiO_2, increasing differentiation is marked by *decreases* in MgO. Both chemical variations as a function of SiO_2 and chemical variations as a function of MgO give petrologists important information on suites of igneous rocks. If a suite of rocks forms a simple array on the diagram, there is good reason to suspect the rocks are all related, by differentiation of a common magma, from melting of a similar source, or through mixing of magmas. Conversely, if the samples scatter across the diagram instead of forming an array, the rocks are probably unrelated.

Variation diagrams can be used to distinguish suites of rocks formed by fractional crystallization from those formed by magma mixing. Figure 4.4 shows how fractional crystallization of olivine and anorthite would affect the Al_2O_3 and SiO_2 contents of a suite of rocks. Consider a melt of composition M_1. As olivine begins to crystallize, it enriches the residual magma in both SiO_2 and Al_2O_3 and drives the melt composition along the vector (V_1) that extends directly away from the olivine composition. Lava erupted after some olivine crystallization may have composition M_2. If plagioclase then begins to crystallize and olivine ceases to crystallize, the residual liquid will follow a trajectory along vector V_2. In all likelihood, olivine and plagioclase will crystallize together. If they

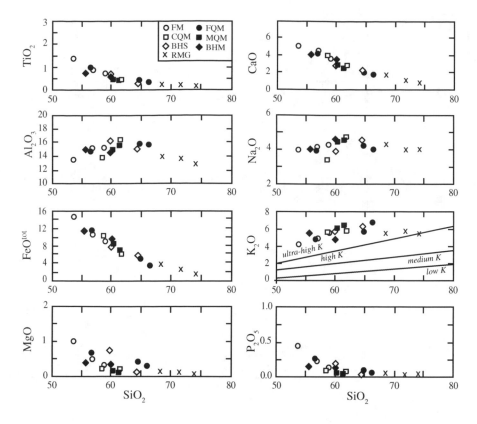

Figure 4.3 Harker diagrams for the Red Mountain pluton, Wyoming, USA. FM = fayalite monzonite, CQM = clinopyroxene quartz monzonite, BHS = biotite hornblende syenite, RMG = Red Mountain pluton, FQM = fine-grained quartz monzonite, MQM = medium-grained quartz monzonite, BHM = biotite hornblende monzonite. After Anderson et al. (2003).

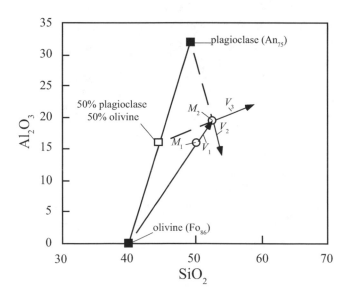

Figure 4.4 Diagram showing the effect of fractional crystallization on the trend followed by magmas on a Harker diagram. See text for discussion.

shows how a suite of rocks related by fractional crystallization will form curved arrays on a Harker diagram. The trajectories of these arrays will change in tandem with the composition and abundances of the crystallizing phases.

Harker diagrams can also determine whether a suite of rocks is related by mixing. For example, consider an andesite volcano that has a few dacite and rhyolite domes on its flanks. In a Harker diagram, the Al_2O_3, CaO, and K_2O contents of these three rocks form a linear array (Figure 4.5). This configuration suggests the dacite may have formed by mixing the andesite magma, perhaps ascending from depth, and a rhyolite magma, which may have been derived by partial melting of crust.

Harker diagrams cannot *prove* that any particular process has taken place. Rather, they simply indicate whether a certain process is *consistent* with the major-element data. To be certain of the process, the trends inferred from major elements must agree with trace-element, isotopic, and field evidence. Two good examples of suites formed by magma mixing and fractional crystallization are the Sherman batholith and the Red Mountain plutons (Figures 4.2 and 4.3). Both are 1.43-Ga plutons that occur in the Laramie Mountains of southeastern Wyoming, USA. Although the trends look similar, there are

do, then the residual magma will evolve along a path determined by the relative abundances of olivine and plagioclase. Figure 4.4 shows the situation in which olivine and plagioclase crystallize in equal abundances, driving the melt along vector V_3. This demonstration

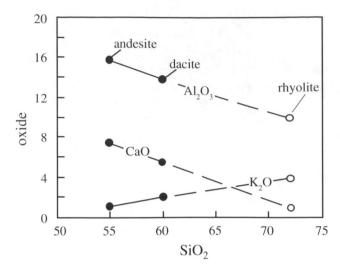

Figure 4.5 Diagram showing that the geochemical composition of dacite is consistent with its origin by mixing of rhyolite and andesite magmas.

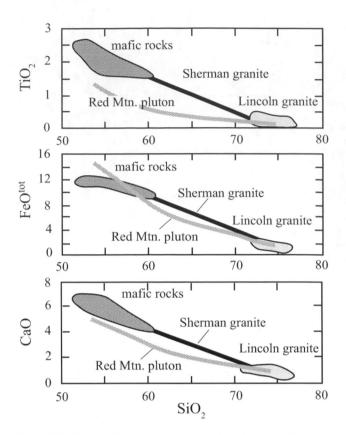

Figure 4.6 Harker diagrams comparing trends in the Sherman batholith and the Red Mountain pluton, Wyoming, USA. Data from Figures 4.2 and 4.3. See text for discussion.

important differences between the two. Observe how the samples from the Sherman batholith form linear trends, whereas those from the Red Mountain pluton form curves (Figure 4.6). This suggests that the rocks of intermediate composition in the Sherman batholith may have formed as the result of magma mixing, whereas the suite of rocks in the Red Mountain pluton may have formed by fractional crystallization.

These conclusions are supported by field evidence. The Sherman batholith consists of the eponymous Sherman granite, a porphyritic granite, a fine-grained granite called the Lincoln granite, and mafic dikes and enclaves (Figure 4.7). Mafic and Lincoln granite magmas commingled (Figure 4.7A) and mixed to form less mafic hybrid magma (Figure 4.7B). Mafic enclaves (Figure 4.7C), assimilated into the Sherman granite, produced variable mafic mineral contents (Figure 4.7D). Rims of plagioclase on crystals of K-feldspar indicate that the composition of the Sherman magma and the composition of the stable, crystallizing feldspar were changed during magma mixing (Figure 4.7E). The porphyritic granite was produced by mixing of the Sherman granite with the Lincoln granite, and also contains K-feldspar rimmed with plagioclase (Figure 4.7F). The Red Mountain pluton, in contrast, is a small, compositionally zoned pluton cut by fine-grained dikes that appear to represent magma expelled during crystallization (Anderson et al., 2003).

4.4 Major-Element Indices of Differentiation

In addition to Harker diagrams, other chemical variation diagrams can help identify the differentiation history of a magma. Many of the commonly used indices of differentiation are based on major elements. The alkali–lime index, iron-enrichment index, aluminum-saturation index, and alkalinity index have been used for many decades to categorize series of rocks and to identify the processes that produced their spectrum of compositions. One additional index, the feldspathoid silica-saturation index, is used together with the alkalinity index to identify types of alkaline rocks.

4.4.1 Modified Alkali–Lime Index

An important chemical control on the differentiation history of a magma is the relative abundances of CaO, Na_2O, and K_2O because these elements are important constituents of feldspars. If CaO is high relative to Na_2O and K_2O, then the first feldspar to crystallize will

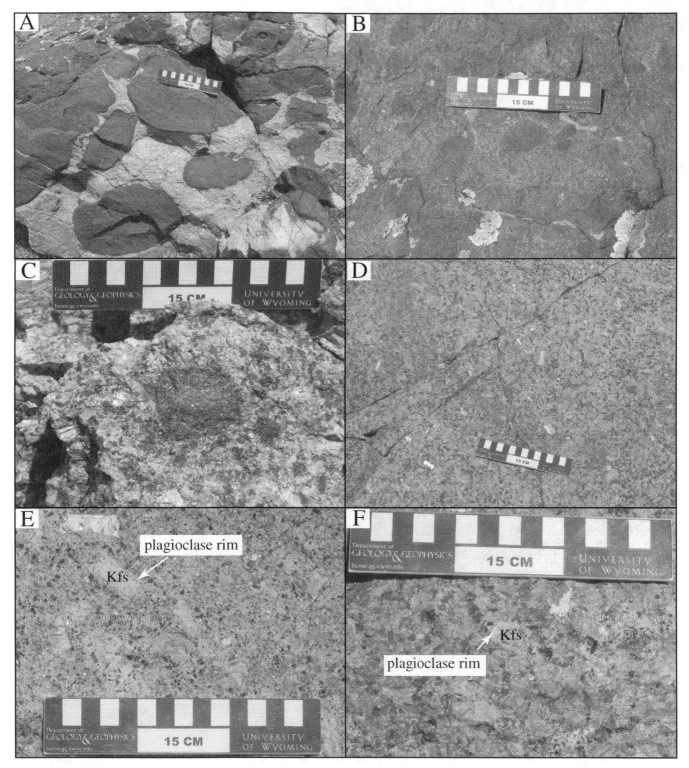

Figure 4.7 Photographs showing the relationships between the rock units composing the Sherman batholith, Wyoming, USA. (A) The fine-grained Lincoln granite and mafic magmas commingle, forming pillows of mafic magma and lobate and cuspate contacts between the two magmas. (B) Intermediate magma is formed by mixing of mafic and granite magmas; mafic enclaves are visible within the hybrid host rock. (C) Mafic enclaves in the Sherman granite. (D) Heterogeneous assimilation of mafic material produces Sherman granite with variable mafic mineral contents. (E) Rims of plagioclase on crystals of K-feldspar (rapakivi texture; arrow) indicate that magma mixing changed the composition of the Sherman magma, and also changed the composition of the stable, crystallizing feldspar. (F) The porphyritic granite, which is interpreted to form by mixing of the Sherman granite with the Lincoln granite, also contains K-feldspar rimmed with plagioclase (arrow).

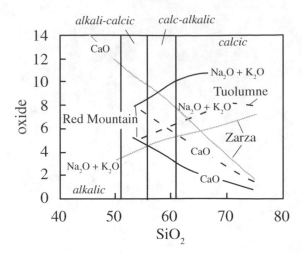

Figure 4.8 Harker diagram showing the variation of CaO and $Na_2O + K_2O$ in three batholiths. Dark solid lines = Red Mountain pluton, Wyoming, USA (Anderson et al., 2003). Dashed lines = Tuolumne pluton, Sierra Nevada batholith, California, USA (Bateman and Chappell, 1979). Gray solid lines = Zarza pluton, Baja California, Mexico (Tate et al., 1999).

be rich in anorthite ($CaAl_2Si_2O_8$), which is relatively rich in Al_2O_3 and poor in SiO_2. Crystallization of this feldspar, therefore, will enrich the residual melt in SiO_2 and deplete it in Al_2O_3. In contrast, if the alkalis are abundant relative to CaO then the first feldspar to crystallize will be rich in $NaAlSi_3O_8$ and $KAlSi_3O_8$. The alkali feldspar components are richer in SiO_2 and poorer in Al_2O_3 than is anorthite. Thus, if the first feldspars to crystallize from a melt are rich in alkali components, they will deplete the melt in SiO_2 and enrich it in Al_2O_3. As a result, alkaline rocks commonly differentiate to silica-depleted compositions, whereas calcic rocks tend to differentiate to silica-enriched compositions. Decades ago, Peacock (1931) recognized the importance of the relative abundances of CaO to the alkalis by introducing the **alkali–lime index**.

For most igneous rock series, CaO decreases with increasing SiO_2 on a Harker diagram, whereas Na_2O and K_2O increase. Consequently, with increasing SiO_2 on a Harker diagram the curves for CaO and $Na_2O + K_2O$ will intersect (Figure 4.8). If the rocks are from a relatively alkalic suite, the intersection will occur at relatively low SiO_2; whereas if the rocks are from a relatively calcic suite, the curves will intersect at relatively high SiO_2. As a monitor of the relative abundances of lime and alkalis in a suite of rocks, Peacock (1931) defined the alkali–lime index as the SiO_2 content at which the two curves intersect. He coined four terms to describe the relative alkalinity of a magma suite: **alkalic**,

Table 4.1 *Alkali–lime classification for igneous rocks*

Alkali–lime index (wt% SiO_2 where CaO = $Na_2O + K_2O$)	Name
< 51	Alkalic
51–56	Alkali-calcic
56–61	Calc-alkalic
>61	Calcic

alkali-calcic, **calc-alkalic**, and **calcic** (Table 4.1). The terms derived by Peacock (1931) have attained wide acceptance in the petrologic community but have been used so loosely that the original meaning has been all but lost. It is common to apply the term *calc-alkalic* to describe magmas associated with island arc magmatism, irrespective of their alkali–lime Index. This book uses the term in its strict geochemical sense.

There are several problems with the alkali–lime index as proposed by Peacock. One problem is the difficulty in comparing many suites in a single diagram because each suite is defined by two lines, so only a few suites will overwhelm the diagram with lines. Another problem is that this index is easily applicable only in rock suites with a range in SiO_2 content that covers the value where CaO and $Na_2O + K_2O$ intersect (for example, note that the CaO and $Na_2O + K_2O$ curves for the Red Mountain pluton do not intersect in Figure 4.8). Finally, it is difficult to apply this analysis to single samples or to rock suites with little variation in SiO_2 contents. To address this problem, Frost et al. (2001) introduced the variable $Na_2O + K_2O - CaO$, which they called the **modified alkali–lime index (MALI)**. At the SiO_2 content where the CaO and $Na_2O + K_2O$ curves cross, the modified alkali–lime index is 0.0. At higher SiO_2 contents it is positive, and at lower values it is negative. Alkalic, alkali-calcic, calc-alkalic, and calcic rocks occupy distinct fields on a plot of the MALI against SiO_2. Four examples are plotted in Figure 4.9: the Red Mountain pluton, which is alkalic; the Sherman batholith, which is alkali-calcic; the Tuolumne pluton, which is calc-alkalic; and the Zarza pluton, which is calcic.

The differences in alkali–lime index are reflected mineralogically in the compositions of the feldspars. In a calcic rock suite, the first plagioclase to crystallize is relatively calcic (for example, An_{80}), and K-feldspar joins the crystallization trend rather late in the differentiation sequence. A plutonic suite following a calcic

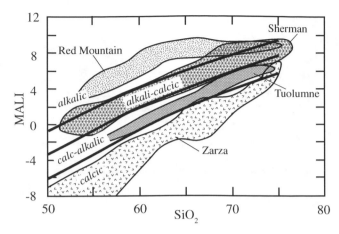

Figure 4.9 Plot of MALI (Na$_2$O + K$_2$O – CaO) vs. SiO$_2$ showing where calcic, calc-alkalic, alkali-calcic, and alkaliic rocks plot. The composition ranges for the Red Mountain, Sherman, Tuolumne, and Zarza plutons are shown for comparison. After Frost et al. (2001).

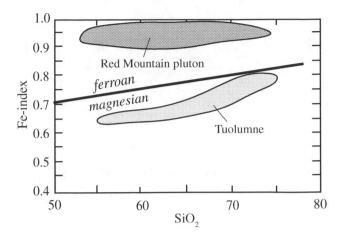

Figure 4.10 Fe-index (FeO^{tot}/[FeO^{tot} + MgO]) vs. SiO$_2$ diagram comparing the ferran Red Mountain pluton, Wyoming, USA, with the magnesian Tuolumne batholith, California, USA.

differentiation trend tends to span compositions from gabbro to diorite to quartz diorite to granodiorite (see Figure 1.1). A volcanic suite may follow the trend basalt–andesite–dacite–rhyolite (see Figure 1.4). In contrast, an alkalic suite first forms plagioclase that is relatively sodic (for example, An$_{50}$), and K-feldspar crystallizes relatively early in the differentiation trend. A plutonic, alkalic suite may follow the trend gabbro–monzonite–syenite–quartz syenite–granite (Figure 1.1), whereas an alkalic volcanic suite may include basalt–trachyte–quartz trachyte–rhyolite (Figure 1.4).

4.4.2 Iron-Enrichment Index

Decades ago petrologists recognized that differentiation in some suites leads to distinctive iron enrichment, whereas in other suites this iron enrichment is modest or lacking (Nockolds and Allen, 1956). The **iron-enrichment index (Fe-index)** [(FeO + 0.9Fe$_2$O$_3$)/(FeO + 0.9Fe$_2$O$_3$ + MgO)] measures the extent to which total iron became enriched during the differentiation of a magma. It is important to note that the "iron enrichment," measured by the Fe-index, is relative to magnesia. Although the absolute abundance of FeO, Fe$_2$O$_3$, and TiO$_2$ increase with differentiation during the early fractionation of most basalts, once Ti-magnetite and ilmenite begin to crystallize these oxides become compatible and decrease with differentiation. Thus, iron enrichment could take place in some suites even as the absolute abundance of iron decreases,

provided that iron abundance decreases at a slower rate than the depletion of MgO from the melt.

Differentiation associated with iron enrichment was originally referred to as the *tholeiitic* or *Skaergaard trend*, whereas that lacking iron enrichment was called the *calc-alkalic* or *Cascade trend* (Miyashiro, 1974). Because nothing in the definition of either a tholeiite or a calc-alkalic rock pertains to iron, Frost et al. (2001) proposed renaming these trends *ferroan* and *magnesian*, respectively, so that the names reflect the chemical variables on which the distinction is based. Figure 4.10 compares the compositional trend followed by the Red Mountain pluton, which is strongly iron-enriched, to the magnesian Tuolumne suite. Rocks following a ferroan trend undergo iron enrichment before becoming enriched in alkalis, whereas those following a magnesian trend show only minimal iron enrichment.

A number of differentiation processes may cause iron enrichment, including the fractional crystallization of olivine from a magma. In contrast, fractional crystallization of magnetite will deplete the melt in iron and enrich it in silica. For this reason, relatively reduced melts, which inhibit magnetite crystallization, will follow a ferroan crystallization trend, whereas oxidized rocks will follow a more magnesian trend. Additionally, crystallization of biotite and hornblende may play an important role in determining the degree of iron enrichment. Unlike olivine and the quadrilateral pyroxenes, which cannot accommodate much ferric iron (Fe^{3+}), both biotite and hornblende can contain considerable

Table 4.2 *Classification of rocks by aluminum saturation*

ASI (in moles)	Rock name
Al/(Ca + Na + K) > 1.0	Peraluminous
Al/(Ca + Na + K) < 1.0 and (Na + K) < Al	Metaluminous
(Na + K) >Al	Peralkaline

Fe^{3+} (as well as [ferrous] Fe^{2+}). Therefore, crystallization of hornblende and biotite represents an additional means for extracting iron from a melt, thus inhibiting iron enrichment.

4.4.3 Aluminum-Saturation Index

Yet another parameter for characterizing igneous rocks is the **aluminum-saturation index (ASI)** (Al/[Ca – 1.67P + Na + K]) (Table 4.2). The index was originally defined by Shand (1943); the phosphorous component was added by Zen (1988) to take into account the calcium residing in apatite. This change was needed so that rocks with an ASI of >1.0 would also be corundum-normative. The major hosts for aluminum in igneous rocks are the feldspars. This parameter indicates whether the alkalis needed to make feldspars balance the abundance of aluminum, or whether there is excess of either alkalis or aluminum. Most mafic rocks are **metaluminous** and have neither excess aluminum nor alkalis. In such rocks the alkalis are mostly accommodated in feldspars, and the remaining calcium (and minor sodium) is found in hornblende or augite.

Granitic rocks may be metaluminous, peraluminous, or peralkaline. If there is an excess of aluminum over alkalis (i.e. the molecular ratio of Al/(Ca + Na + K) > 1.0), then the rock is said to be **peraluminous.** Figure 4.11A compares the composition of two granites from the western United States. The Harney Peak granite in the Black Hills of South Dakota is strongly peraluminous, whereas the Tuolumne pluton in the Sierra Nevada of California is metaluminous and ASI increases with increasing SiO_2 content. Figure 4.11B shows where common minerals in granites plot on an ASI diagram. By construction, feldspars have ASI = 1.0 (i.e. the alkalis and the aluminum are balanced). Augite and hornblende have low ASI indices because they contain substantial amounts of calcium that is not balanced by aluminum. Orthopyroxene has widely varying ASI, but it is not

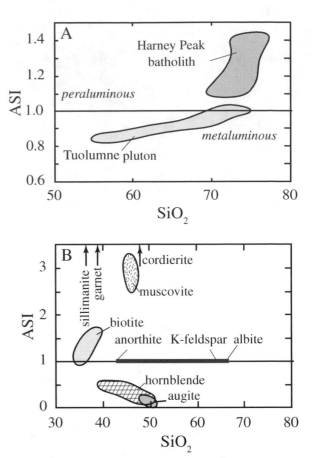

Figure 4.11 ASI vs. SiO_2 diagrams. (A) Comparison of the peraluminous Harney Peak granite, South Dakota, USA (Nabelek et al., 1992) and the metaluminous Tuolumne pluton, California, USA (Bateman and Chappell, 1979). (B) Diagram comparing where common minerals from granites plot. Heavy gray line illustrates the composition range for the feldspars.

shown on Figure 4.11B because it usually has only small amounts of either alkalis or aluminum, and hence doesn't affect the abundance of those elements in rocks. In its ideal formula, $KFe_3AlSi_3O_{10}(OH)_2$, biotite would have ASI = 1.0, but natural biotites usually contain excess aluminum substituting for iron and are hence weakly peraluminous. Muscovite is strongly peraluminous, as are sillimanite, garnet, and cordierite, which typically contain little or no alkalis.

Thus, the ferromagnesian minerals found in a granite (or a rhyolite) reflect the aluminum saturation of that rock. Metaluminous rocks contain minerals with low ASI, such as augite and hornblende. Augite and hornblende should be absent in peraluminous rocks. In weakly peraluminous rocks, the extra aluminum may be accommodated in biotite, but strongly peraluminous rocks should be marked by the presence of an aluminum-rich mineral in the rock. Common aluminous minerals found in granitic rocks are muscovite and garnet; cordierite and sillimanite are rarer.

4.4.4 Alkalinity Index

The **alkalinity index (AI)** (Al – [Na + K]) measures the relative abundances of aluminum and alkalis. **Alkaline** rocks are defined as those that have higher sodium and potassium contents than can be accommodated in feldspar alone (Shand, 1943). If there is an excess of alkalis over aluminum, the rocks are **peralkaline** and the AI will be less than zero. Sørensen (1974) recognized three subgroups of alkaline rocks. In the first, SiO_2 is adequate but Al_2O_3 is deficient, so the minerals that accommodate the excess alkalis include sodic pyroxenes and sodic amphiboles, resulting in the formation of peralkaline granites and their volcanic equivalents, pantellerite and comendite. In the second subgroup, Al_2O_3 is adequate but SiO_2 is deficient. These rocks contain feldspathoids along with micas, hornblende, and/or augite and form rocks such as metaluminous nepheline syenite. In the third subgroup, both Al_2O_3 and SiO_2 are deficient and both feldspathoids and sodic pyroxenes and/or amphiboles crystallize. These rocks include peralkaline nepheline syenites.

4.4.5 Feldspathoid Silica-Saturation Index

To distinguish these three types of alkaline rocks, Frost and Frost (2008) defined one additional index, the **feldspathoid silica-saturation index (FSSI):**

$$FSSI = (Q - [Lct + 2Nph])/100 \qquad (4.1)$$

where Qz, Lct, and Nph are the normative quartz, leucite, and nepheline contents, respectively. This index is positive for quartz-saturated rocks, for which it expresses the excess amount of SiO_2. For silica-undersaturated rocks, it represents the amount of SiO_2 that must be added to make it silica-saturated. The equation for FSSI involves multiplying the normative nepheline content by 2 because two formula units of quartz must be added to nepheline to make albite:

$$NaAlSiO_4 + 2SiO_2 = NaAlSi_3O_8 \qquad (4.2)$$

Only one formula unit of quartz must be added to leucite to make K-feldspar:

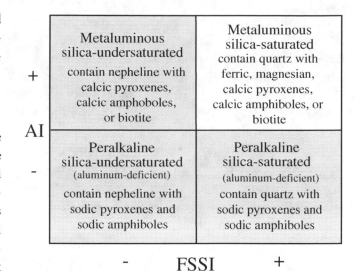

Figure 4.12 Classification of alkaline rocks, shown on a plot of AI versus FSSI. Alkaline rocks occupy the shaded portions of the diagram. They may be silica-undersaturated (top left corner), alumina-deficient (bottom right corner), or both silica- and alumina-deficient (bottom left corner). From Frost and Frost (2008).

$$KAlSi_2O_6 + SiO_2 = KAlSi_3O_8 \qquad (4.3)$$

Using the FSSI and AI indices, Frost and Frost (2008) developed a matrix showing how the subgroups of alkaline rocks are related to "normal" granitic rocks (Figure 4.12). The shaded quadrants of this diagram are occupied by the three subgroups of alkaline rocks, namely the metaluminous, silica-undersaturated rocks; the peralkaline, silica-undersaturated rocks; and the peralkaline, silica-saturated rocks. The fourth quadrant is occupied by metaluminous and peraluminous, silica-saturated rocks.

4.5 Identification of Differentiation Processes Using Trace Elements

Trace elements, which are present in rocks in concentrations of less than 0.1 percent by weight, include the *transition metals* Sc, Ti, V, Cr, Mn, Co, Ni, Cu, and Zn, the **rare-earth elements (REEs)** La, Ce, Nd, Sm, Eu, Gd, Tb, Dy, Ho, Er, Tm, Yb, and Lu, and other trace elements, including Cs, Rb, Ba, Sr, Y, Zr, Hf, Nb, Ta, Pb, Th, and U. Trace elements are useful fingerprints of the origin of igneous rocks and igneous processes because they exhibit a range in concentration far greater than for major elements. For example, in most igneous rocks, CaO varies between 0 and 10 weight percent. However, Sr, which behaves chemically much like Ca, may vary from tens of ppm to thousands of ppm in those same igneous rocks. Therefore, a process that affects the concentration of Ca may affect the concentration of Sr by a much greater magnitude, making it more likely that the process can be detected and identified by considering the Sr content.

Because of their wide variations in abundance, trace elements can be used to identify and quantify processes of crystallization and partial melting. For elements with low concentration (that is, for trace elements), the element is partitioned between minerals and melt according to the relationship:

$$D = C_i^{mineral} \Big/ C_i^{melt} \tag{4.4}$$

where D is called a **partition coefficient** and C_i is the concentration of any trace element, i, in either the melt or the crystal. D can be measured from experiments or by comparing the concentration of a trace element in phenocrysts in volcanic rocks with that of its glassy matrix. D values are strongly composition-dependent, so different values apply to mafic magmas than to felsic magmas.

Magmas will be in equilibrium with more than one mineral phase as they crystallize. To describe this situation, we define a **bulk-distribution coefficient**, D_B, which is calculated from the weight proportions (w) of each mineral in the assemblage:

$$D_B = \sum_{i=1}^{n} w_i D_i \tag{4.5}$$

Those elements that behave compatibly (i.e. those that preferentially concentrate in the minerals) have D_B greater than 1. In contrast, those elements that are incompatible and concentrate in the melt have D_B less than 1. In general,

compatible elements have ionic radii and charge similar to those of major elements like magnesium, iron, calcium, and sodium and therefore can substitute in crystallizing minerals for them. Incompatible elements may have either much larger or smaller radii or too high a charge to occupy major sites in the lattices of common rock-forming minerals.

4.5.1 Use of Trace Elements to Model Melting and Crystallization Processes

A number of simple models can resolve melting and crystallization processes from trace-element concentrations of rocks, and the reader is referred to more complete treatments of this topic in Shaw (2006). This section will consider a single example to illustrate the power of this approach: a model of trace-element partitioning during fractional crystallization. The theoretical variation of trace-element concentrations during fractional crystallization can be calculated from the Rayleigh equation:

$$C_i / C_i^o = F^{(D-1)} \tag{4.6}$$

where C_i is concentration of element i in the melt after a certain amount of fractionation has taken place, C_i^o is the concentration of element i in the initial melt, and F is melt fraction remaining. The Rayleigh equation allows petrologists to calculate how the concentrations of trace elements change during fractional crystallization (Figure 4.13). From Figure 4.13 it is clear that highly compatible elements (such as those with $D > 10$) are quickly removed from the melt, whereas highly incompatible elements (such as those with $D < 0.5$) can accumulate very high abundances in the last fraction of the melt to crystallize. This is the reason pegmatites, which form from the last remaining granitic melt, commonly contain high concentrations of incompatible elements such as lithium, boron, and beryllium.

BOX 4.1 | TRACE ELEMENTS AND TECHNOLOGY

Trace elements are essential to the manufacture of modern technology. For example, consider your smart phone. The display contains *gallium* and *germanium* in LEDs, *tin* is used in the circuit-board solder, and *indium* forms part of the conductive coating of the screen. *Copper* is essential to the electronic circuitry, *silver* is used in inks on the composite boards, *arsenic* is used in amplifiers, *tantalum* is part of capacitors, and *tungsten* provides the mass for mobile phone vibration. In addition, *rare-earth elements* (*REEs*) are used to produce magnets in speakers, microphones, and vibration motors. Smart phones also use lithium-ion batteries, which require both *lithium* for the anodes and *cobalt* for the cathodes. All these elements are mined from the Earth and assembled into a device in your pocket that enables modern communication (USGS, 2016).

Lithium-ion batteries are now ubiquitous, but they are a relatively recent invention, appearing on the market in 1991 in Sony camcorders. After lithium-ion batteries were introduced in hybrid and electric cars in 2008, demand accelerated and continues to increase. In a recent application, lithium-ion batteries are being used in electricity-grid storage systems to store power from intermittent renewable energy sources such as wind or solar. As a result, lithium prices have quadrupled and cobalt prices have doubled since 2015 (Economist, 2017).

Although trace elements are found in silicate minerals, the high melting points and low solubilities of silicates make it hard to separate elements within them. In general, trace elements can be found in higher concentrations in oxides or sulfides and it is more efficient to mine elements from these sources. Lithium, which is enriched in some evaporite deposits and continental brines, can be concentrated further by solar evaporation. This is a less complicated and energy-intensive process than extraction of lithium from silicate minerals such as spodumene. Other trace elements may be produced as a byproduct during the mining of a different element. Cobalt, for example, is a byproduct of copper and nickel production. Even though most trace elements are not particularly rare, favorable deposits for mining are located in only a few countries. For example, the majority of lithium is produced in Chile, Argentina, and Bolivia. More than half the cobalt production comes from Democratic Republic of Congo, and over 80 percent of rare earth elements are produced in China. Concern for secure supplies of critical elements and rising prices could lead to development of new mines. Alternatively, these factors could drive development of substitute materials and encourage recycling. Have you considered the trace elements they contain when you decide how to dispose of rechargeable batteries and your old smart phone?

Box Figure 4.1 The average smart phone may contain up to 62 different elements, including eight or more rare-earth elements (Rohring, 2015). Photo credit: Natee Meepian/EyeEm, via Getty Images.

The abundances of trace elements in minerals and rocks can be used to test various hypotheses for the origin of igneous rocks. For example, consider the idea that crystallization and removal of plagioclase from the lunar tholeiite produces residua that form the basaltic lunar rock type known as KREEP. This lunar material, first found during the Apollo 12 mission, is named for its unusually high contents of potassium (K), REEs, and phosphorus (P). Given that the REE cerium has $D = 0.1$ between plagioclase and basaltic liquid, and that KREEP has 200 ppm Ce and the initial tholeiitic parental magma had 18 ppm, how much plagioclase must be removed from a magma with the composition of the lunar tholeiite to produce a KREEP basalt? Substituting the concentrations into Equation 4.6:

$$200/18 = F^{(0.1-1)}$$

$$11 = F^{(-0.9)}$$

$$0.07 = F, \text{ the fraction of liquid remaining}$$

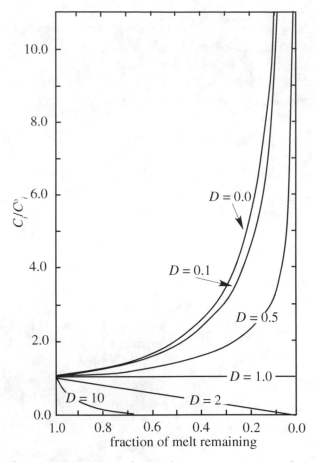

Figure 4.13 Variation of trace-element concentrations during fractional crystallization of a magma, according to the Rayleigh law. The diagram illustrates the rapid removal of highly compatible elements from the magma during crystallization, and the resulting enrichment of highly incompatible elements in the magma.

This result indicates that if 93 percent of the magma crystallized as plagioclase, then the 7 percent residual liquid would have the correct concentration of Ce. This calculation suggests that if KREEP is formed by removal of plagioclase from a tholeiitic magma, then KREEP is a highly differentiated basaltic rock type.

4.5.2 Graphical Representations of Trace-Element Compositions

Trace-element compositions may be portrayed graphically in a number of ways. First, they can be plotted on Harker diagrams, similar to how major elements are portrayed. With trace elements, percent SiO_2 is on the x-axis and the trace-element concentration, usually in ppm, is on the y-axis. An example of this graphical

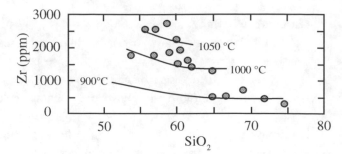

Figure 4.14 Variation in zirconium content of rocks from the Red Mountain pluton, Laramie anorthosite complex, Wyoming, USA. The contours show the temperature at which zircon saturates in the Red Mountain pluton melts (Watson and Harrison, 1983). The Red Mountain rocks contain fayalite and clinopyroxene and crystallized from hot magmas, but they probably did not begin to crystallize at temperatures much above 1000 °C. Zircon crystallized early, accumulating in the solid phase in those samples that have greater than 1000 ppm Zr and lie above the 1000 °C isotherm. From Anderson et al. (2003).

treatment is shown in Figure 4.14. Zirconium contents in rocks of the Red Mountain pluton, a monzonitic to granitic pluton associated with the Laramie anorthosite complex, decrease from 2000–3000 ppm in the most mafic parts of the pluton to less than 300 ppm in the most siliceous granites. This decrease is related to crystallization of zircon, which incorporates zirconium, thereby depleting the remaining magma in this element.

REEs are among the most commonly used trace elements in petrology. They are typically plotted as a group, arranged by increasing atomic number along the x-axis in a REE diagram. The y-axis is the element concentration in the sample divided by its concentration in primitive chondritic meteorites. Because even-numbered atomic elements are more abundant in the Solar System than odd-numbered atomic elements, normalizing to chondritic composition smooths the saw-tooth pattern that would be obtained if the sample concentrations alone were plotted. An example of the REE diagram is shown in Figure 4.15, in which REE compositions of lunar anorthosites and basalts are plotted. The normalized abundances make smooth curves except for excursions for the element europium. Of the REEs, europium is the only element that can be present in magmas in the 2+ oxidation state, which allows Eu^{2+} to substitute for Ca^{2+} in plagioclase. The large positive Eu anomaly

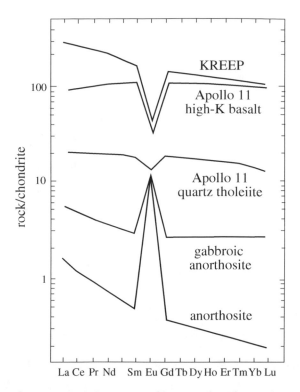

Figure 4.15 REE patterns of lunar rocks. The gap between neodymium and samarium is occupied by promethium, which has no stable isotopes. From Taylor (1975).

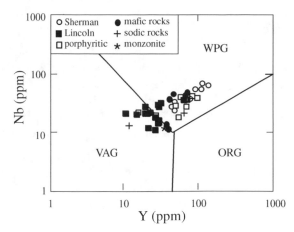

Figure 4.16 Niobium and yttrium contents of rocks from the Sherman batholith, Laramie Mountains, Wyoming, USA. Most samples plot in the within-plate granite (WPG) field, but the highly differentiated Lincoln granite samples (which probably have assimilated some continental crust) extend into the volcanic-arc granite (VAG) field. ORG denotes ocean ridge granites. From Frost et al. (1999).

in the REE pattern of anorthosite reflects incorporation of Eu^{2+} in feldspar. The negative Eu anomaly in the REE pattern of high-potassium and KREEP basalts are consistent with their interpretation as the residual magmas following plagioclase crystallization and formation of lunar anorthosite. The quartz tholeiite, with its intermediate REE abundances and nearly flat pattern, could approximate the composition of the parent magma from which anorthosite and residual KREEP basalt formed.

Different tectonic environments involve different conditions of melting or different source rocks, and they tend to generate magmas with different trace-element compositions. For example, mid-ocean ridge basalt tends to be depleted in the light REEs relative to the heavy REEs, and are typically depleted REEs in niobium. This has led to the use of trace-element abundances to indicate the origin for rocks that have subsequently been deformed or removed from their original setting (e.g. Pearce et al., 1984a; Pearce and Peate, 1995). It is important to remember that this

is an empirical approach, and that variables including the exact composition of the source rocks, the extent of differentiation, magma mixing and assimilation, and subsequent metamorphism may lead to incorrect interpretations. Consider one of the widely used *discrimination diagrams* for granitic rocks shown in Figure 4.16, which plots the niobium and yttrium contents of the granitic rocks from the Sherman batholith. The plot shows that the Sherman granite samples plot within the field for within-plate granite (WPG), whereas the Lincoln and a few of the porphyritic granites trend into the field for volcanic-arc granite (VAG). Because both these granites are part of a single batholith, they must have formed in a single tectonic setting. The fact that they plot in two different parts of the discrimination diagram could reflect the fact that the Lincoln granite probably originated from a different source than the Sherman. If Lincoln magmas were produced by partial melting of the country rock and these country rocks originally formed in a volcanic arc, then the Lincoln granite may plot in the VAG field even though the batholith formed far from an active volcanic arc.

Summary

- The *norm* is the calculated mineral abundances that would be present in a rock if it were anhydrous. It is used to compare fine-grained rocks with coarse-grained rocks and to classify fine-grained rocks.

- Suites of rocks derived from a common parent magma obtain their various compositions by *magmatic differentiation*, most commonly through crystallization and removal of minerals from a magma.

- *Compatible elements* are preferentially incorporated into phases crystallizing from a melt and decrease in abundance in the magma as differentiation proceeds.

- *Incompatible elements* are incompatible with phases crystallizing from a melt and increase in abundance in the magma during differentiation.

- *Harker diagrams* are a common way of graphically illustrating igneous rock composition. They can identify crystal fractionation or magma mixing.

- Indices of differentiation include:

  ○ *Modified alkali–lime index (MALI)*: categorizes rock suites as alkalic, alkali-calcic, calc-alkalic, and calcic.

  ○ *Iron-enrichment index (Fe-index)*: identifies ferroan versus magnesian suites.

  ○ *Aluminum-saturation index (ASI)*: distinguishes between peraluminous, metaluminous, or peralkaline rocks.

  ○ *Alkalinity index (AI)*: distinguishes whether rocks are peralkaline.

  ○ *Feldspathoid silica-saturation index* (FSSI): along with the AI, helps distinguish the various types of alkaline rocks.

  ○ It is important to distinguish between alkalic, peralkaline, and alkaline rocks because these terms, though similar, describe different conditions. Alkalic rocks have (Na_2O + K_2O) that is high relative to CaO as defined by the alkali–lime index. These rocks tend to have K-feldspar and albitic plagioclase. Alkaline rocks are rocks that contain more potassium and sodium than can be accommodated in feldspar. Peralkaline rocks have an excess of potassium and sodium compared to aluminum and therefore contain sodic pyroxenes or amphiboles.

  ○ Because variations of trace-element concentrations in igneous rocks are of much greater magnitude than the variation in major-element concentrations, trace elements provide more sensitive indicators of igneous processes, and can be used to quantitatively model those processes.

Questions and Problems

Problem 4.1. What is the difference between normative and modal mineralogy?

Problem 4.2. Using the normative calculation software available for download from the Volcano Hazards Program of the United States Geological Survey (https://volcanoes.usgs.gov/observatories/yvo/jlowenstern/other/NormCalc_JBL.xls), calculate the norm for the four samples (oxides in weight percent) on the table below. Which rocks are silica-saturated? If the rocks were anhydrous, what mafic minerals would be present?

	Shasta	Tuolumne	Brome	Mount Megantic
	82–91a	4	BR15	MG21
SiO_2	63.14	62.78	61.13	62.15
TiO_2	0.57	0.70	0.87	0.38
Al_2O_3	16.88	15.74	18.16	17.71
Fe_2O_3	1.54	2.07	1.25	0.37
FeO	2.43	3.22	2.35	3.68
MnO	0.07	0.09	0.17	0.09
MgO	3.59	2.50	0.69	0.30
CaO	5.89	4.80	1.58	1.35
Na_2O	4.11	3.25	7.01	5.26
K_2O	1.18	3.22	5.37	6.75
P_2O_5	0.17	0.17	0.00	0.00
LOI	0.3	0.35	0.00	0.00
Sum	99.57	98.89	98.58	98.04

Data from Bateman and Chappell (1979), Eby (1985), and Grove et al. (2005). LOI = loss on ignition, an indication of the volatile content of the rock.

Problem 4.3. The Fe-index is defined based on total iron expressed as FeO. Show that weight percent FeO = 0.9 multiplied by weight percent Fe_2O_3.

Problem 4.4. Construct templates for Fe-index, MALI, and ASI variation diagrams.

a. The boundary between ferroan and magnesian fields is described by Frost and Frost (2008): Fe-index + 0.46 = $0.005SiO_2$; it is applicable for 48–75 percent SiO_2.

b. The boundaries on the MALI diagram are as follows (Frost et al., 2001); they are applicable for 50–75 percent SiO_2:

alkali–alkali-calcic:

$$Na_2O + K_2O\text{-}CaO = -41.86 + 1.112*wt\%SiO_2 - 0.00572\ wt\%SiO_2{}^2$$

alkali-calcic–calc-alkalic:

$$Na_2O + K_2O\text{-}CaO = -44.72 + 1.094*wt\%SiO_2 - 0.00527\ wt\%SiO_2{}^2$$

calc-alkalic–calcic:

$$Na_2O + K_2O\text{-}CaO = -45.36 + 1.0043*wt\%SiO_2 - 0.00427\ wt\%SiO_2{}^2$$

c. ASI is the molecular ratio Al/(Ca + Na + K) (Shand, 1943). Derive the equation that converts weight percent oxide to molecular ratio of these ions.

Problem 4.5. Plot the analyses from Shasta volcano (below) on Fe-index, MALI, and ASI diagrams. Describe these andesites and dacites based on where they plot on these diagrams (oxides in weight percent).

	SiO_2	TiO_2	Al_2O_3	FeO^{tot}	MnO	MgO	CaO	Na_2O	K_2O	P_2O_5	LOI	Sum
82–91b	63.32	0.57	16.73	3.749	0.07	3.7	6.02	4.09	1.11	0.17	0.26	99.69
82–96	62.96	0.58	17.01	4.033	0.08	3.44	5.93	4.19	1.2	0.16	0.25	99.73
82–97	63.38	0.59	16.62	4.268	0.07	3.03	5.19	4.00	1.56	0.14	0.47	99.23
83–45	62.98	0.52	17.51	3.877	0.08	2.64	5.52	4.93	1.27	0.16	0.98	99.78
83–54	63.8	0.59	16.14	4.102	0.08	3.11	5.31	4.59	1.57	0.13	0.77	99.69
83–55	62.44	0.63	16.36	4.479	0.08	3.52	5.89	4.32	1.41	0.14	0.61	99.48
97–4	62.21	0.61	17.03	4.374	0.08	3.45	5.95	4.16	1.26	0.18	0.01	99.78
97–6	61.51	0.65	16.81	4.185	0.08	3.57	6.06	4.10	1.23	0.23	0.60	98.87
99–12A	56.9	0.36	17.1	5.382	0.11	5.9	8.82	3.03	0.74	0.13	0.40	99.07
99–12B	55.7	0.54	16.8	6.561	0.14	6.08	9.23	2.88	0.48	0.13	0.19	99.27
99–13	61.5	0.60	17.0	4.392	0.08	3.47	5.97	3.95	1.24	0.23	0.20	98.92
99–14	61.7	0.60	17.0	4.293	0.08	3.45	5.92	4.00	1.23	0.23	0.24	98.98
99–16	61.5	0.65	16.7	4.158	0.08	3.53	6.02	3.98	1.24	0.27	0.56	98.59
99–17	62.4	0.57	17.0	3.879	0.07	3.21	5.86	4.06	1.20	0.22	0.43	98.90
99–18	61.4	0.65	16.8	4.158	0.08	3.53	6.07	3.98	1.23	0.27	0.45	98.63

FeO^{tot} is total iron expressed as FeO.

Data from Grove et al. (2005).

Problem 4.6. Assume the partition coefficient, D, for Sr between plagioclase and melt is 2. If the initial Sr concentration of a melt is 200 ppm, what will the concentration of the melt be after plagioclase crystallizes, leaving 50 percent of the melt remaining? Do the calculation using Equation 4.6 and check by solving the problem graphically using Figure 4.13.

Problem 4.7. Are the REEs compatible or incompatible in plagioclase? Answer by inspection of Figure 4.15. (Recall that anorthosite is a rock composed of 90–100 percent plagioclase.)

Further Reading

Cox, K. G., Bell, J. D., and Pankhurst, R. J., 1979, *The Interpretation of Igneous Rocks*. Allen and Unwin, London.

Frost, B. R., Arculus, R.J., Barnes, C. G., Collins, W. J., Ellis, D. J., and Frost, C. D., 2001, A geochemical classification of granitic rock suites. *Journal of Petrology*, 42, 2033–48.

Frost, B. R. and Frost, C. D., 2008, A geochemical classification for feldspathic rocks. *Journal of Petrology*, 49, 1955–69.

Rollinson, H. R., 1993, *Using Geochemical Data: Evaluation, Presentation, Interpretation*. Longman/Wiley, Harlow, NY.

Chapter 5

Application of Stable and Radiogenic Isotopes in Petrology

5.1 Introduction

Isotope geochemistry is an exciting and dynamic field that plays an increasingly important role in petrology. Isotope geochemistry is used to determine the age of minerals and rocks, to identify the sources of various components that compose them, and to describe the processes that form and alter minerals and rocks. Petrologists and geochemists continue to develop new isotopic systems and to use them to constrain igneous and metamorphic processes in innovative ways. This chapter introduces the basic principles of isotope geochemistry with suggestions for further reading provided at the end of the chapter.

Elements are characterized by the number of protons in the nucleus; for example, carbon always contains six protons. The number of neutrons in the nucleus of a particular element can vary; for example, although oxygen always has eight protons, it may have eight, nine, or ten neutrons. As a result of the variation in neutrons, an oxygen atom may have the atomic weight of 16, 17, or 18. Neutron variants of a single element are called **isotopes**, and they can be either stable or radioactive. ^{16}O, ^{17}O, and ^{18}O are stable isotopes of oxygen. A radioactive isotope undergoes decay and produces another isotope. For example, ^{14}C is a radioactive isotope with six protons and eight neutrons. ^{14}C is referred to as the *parent* isotope, and it decays to form ^{14}N, the *daughter* isotope.

5.2 Stable-Isotope Geochemistry

For elements of low atomic mass, the differences in the mass of their isotopes are large. For example, the mass of the heavier of the two stable isotopes of hydrogen, ^2H (also known as deuterium, D), is twice that of the lighter isotope, ^1H. In another example, the mass difference between two of the isotopes of oxygen, ^{18}O and ^{16}O, exceeds 11 percent. These mass differences lead to differences in chemical behavior and result in readily measured variations in the isotopic ratios of light stable elements. Stable isotopic ratios of many elements, most commonly oxygen, sulfur, and hydrogen, are used widely for a number of applications in igneous and metamorphic petrology. These include:

- Stable-isotope **geothermometry**, the determination of the temperatures of igneous and metamorphic processes, which takes advantage of the fact that differences in chemical behavior of stable isotopes are temperature-dependent.

- Stable-isotope tracers, which make use of natural variations in isotopic compositions to infer processes and sources. Petrologic applications include discerning magmatic processes of differentiation and assimilation, and identification of source rocks that may have been partially melted and incorporated into magmas, and of water–rock interaction. Biologically mediated processes also can produce large variations in stable isotopic compositions of elements that have multiple oxidation states, including carbon, sulfur, and iron. The involvement of biotic and abiotic processes in the precipitation of minerals, including ores, may be discriminated based on the stable-isotope ratios of these elements.

As analytical techniques improve, differences in the chemical behaviors of isotopes have been measured in elements of increasingly higher atomic mass, including Si, Cr, Fe, Cu, Sr, Mo, Ba, and Hg, among many others. Measurement of stable-isotope fractionations in these elements has led to exciting new discoveries and descriptions of geologic processes that were previously difficult to identify and study (Teng et al., 2017).

Natural variations in stable-isotope ratios are expressed relative to a standard material. For example, isotopic ratios of hydrogen (D/H, where D = deuterium, ^2H and H denotes ^1H) and oxygen isotopic ratios ($^{18}O/^{16}O$) are reported relative to standard mean ocean water (SMOW):

$$\delta D = \left[\frac{\left(\frac{D}{H}\right)_{sample} - \left(\frac{D}{H}\right)_{SMOW}}{\left(\frac{D}{H}\right)_{SMOW}} \right] \times 10^3 \tag{5.1}$$

$$\delta^{18}O = \left[\frac{\left(\frac{^{18}O}{^{16}O}\right)_{sample} - \left(\frac{^{18}O}{^{16}O}\right)_{SMOW}}{\left(\frac{^{18}O}{^{16}O}\right)_{SMOW}} \right] \times 10^3 \tag{5.2}$$

The differences in isotope ratio of sample and standard are multiplied by 1000 because variations in the ratios are typically in the range of parts per thousand.

The differences in chemical behavior of isotopes of an element are referred to as **isotopic fractionation**. Isotopic fractionation can arise from both kinetic and equilibrium effects. Kinetic effects include diffusion, in which the lighter isotope diffuses faster than the heavier one. Fractional crystallization is an example of such a process. Equilibrium effects include isotope exchange reactions between minerals, in which the heavier isotope occupies the sites with stronger bonds. For example, the $\delta^{18}O$ of minerals containing strong Si–O bonds, such as quartz, are higher than the $\delta^{18}O$ of coexisting minerals with weaker bonds, such as Fe–O in magnetite (Figure 5.1). In decreasing order, the $\delta^{18}O$ of coexisting minerals in igneous and metamorphic rocks is quartz–feldspar–muscovite–biotite–magnetite and feldspar–pyroxene–hornblende–ilmenite–magnetite (Taylor, 1968).

To quantify processes that cause isotopic fractionation, it is helpful to define a fractionation factor, α. This constant represents the factor by which the isotopic ratio will change during a chemical reaction or a physical process. It is defined as:

$$\alpha_{X-Y} = \frac{R_X}{R_Y} \tag{5.3}$$

where R_X and R_Y are the isotopic ratios of two phases, X and Y. In general, isotope fractionation effects are small, such that α differs by no more than a few percent from a value of 1.00.

Because the isotopic compositions of phases are expressed using δ notation, the following relationship is used to relate the difference in δ between two phases, $\delta_X - \delta_Y$, to the fractionation factor, α. The difference, $\delta_X - \delta_Y$, may be expressed as Δ_{X-Y}. The relationship between Δ and α is:

$$\Delta_{X-Y} = \delta_X - \delta_Y \approx 10^3 \ln \alpha \tag{5.4}$$

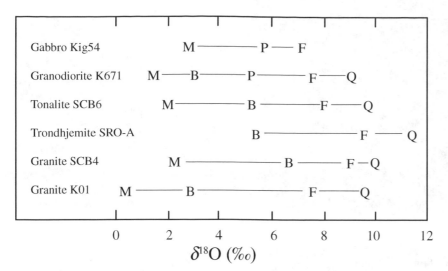

Figure 5.1 Oxygen isotopic compositions of coexisting minerals in various rock types, illustrating the relative $\delta^{18}O$ of rock-forming minerals (in decreasing order): quartz (Q), feldspar (F), pyroxene (P) and biotite (B), and magnetite (M). The absolute $\delta^{18}O$ of the coexisting minerals depends on a number of factors including $\delta^{18}O$ of the parental magma, temperature of crystallization, and subsolidus isotopic re-equilibration. Data from Bottinga and Javoy (1975).

These basic expressions are used to identify and quantify temperatures and petrologic processes affecting igneous and metamorphic rocks, as summarized below.

5.2.1 Stable-Isotope Geothermometry

Fractionation is temperature-dependent, such that processes at lower temperatures result in greater isotopic fractionations than in the same processes at higher temperatures. This means that the fractionation factor varies as the inverse square of temperature (T):

$$\Delta_{X-Y} \approx 10^3 \ln \alpha_{X-Y} = A + \frac{B}{T^2} \tag{5.5}$$

where A and B are constants.

Because of this temperature dependence, the differences in δ of pairs of minerals can be used as a geothermometer. An example is shown in Figure 5.2, which plots differences in $\delta^{18}O$ for mineral pairs as a function of temperature. For some mineral pairs, such as between the silicate pairs quartz–albite and anorthite–olivine, the difference is relatively small. For others, such as between the silicate and oxide pairs quartz–magnetite and albite-magnetite, the difference is larger. For all pairs, the difference in δ values of the mineral pairs decreases with increasing temperature. Assuming that equilibrium between the phases was achieved and the phases did not subsequently re-equilibrate, then the temperature of equilibration can be determined from Δ_{X-Y}. This is an example of an ion-exchange geothermometer, which involves the distribution of two ions, in this case ^{18}O and ^{16}O, between two phases. Other types of geothermometers are described in Chapter 18 (Section 18.3.1).

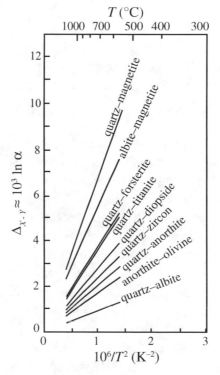

Figure 5.2 Calculated oxygen isotope fractionation for various mineral pairs as a function of temperature. Data from Chiba et al. (1989), King et al. (2001), and Trail (2009).

5.2.2 Stable-Isotope Tracers of Magmatic Processes

Stable isotopic ratios of rocks and minerals can serve as tracers of magmatic processes, including fractional crystallization, crustal assimilation, and water–rock interaction. Just as behavior of trace elements during fractional crystallization is described by a Rayleigh equation (see Equation 4.6), the change in δ of the crystallizing solids as they are continuously removed

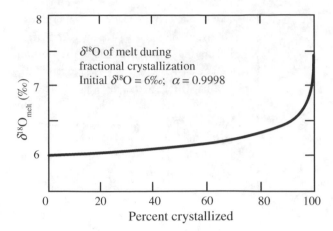

Figure 5.3 Change in $\delta^{18}O$ of residual melt due to fractional crystallization for magma with initial $\delta^{18}O$ of 6‰ and α of 0.9998. Note the minimal increase in $\delta^{18}O$ throughout most of the crystallization process.

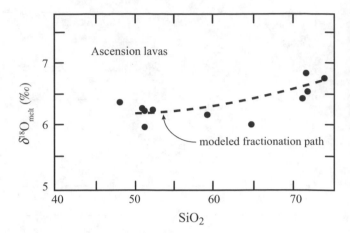

Figure 5.4 $\delta^{18}O$ as a function of SiO_2 content for alkali basalt to rhyolite lavas from Ascension Island. The variation in these lavas is attributed to fractional crystallization, a process that produced less than 1‰ difference between the mafic and felsic rocks (Taylor and Sheppard, 1986).

from the remaining melt is also described by a Rayleigh equation:

$$\Delta = \delta_{melt} - \delta_{initial\ melt} = 1000\left(f^{\alpha-1} - 1\right) \qquad (5.6)$$

During this kinetic process, the lighter isotope preferentially enters the crystals, leaving a melt enriched in the heavier isotope. In theory, with extreme degrees of fractional crystallization, this process can produce a residual, highly fractionated melt with a very high δ. In natural systems, such extreme fractionations are not observed because the fractionation factors between rock-forming minerals and melt are small, typically less than 2‰. For example, consider a melt with initial $\delta^{18}O$ of 6‰ and α of 0.9998 between crystals and melt, with the melt being enriched in ^{18}O relative to the crystals. As fractional crystallization proceeds, the $\delta^{18}O$ of the melt changes very little until almost all the melt is solidified but then increases dramatically (Figure 5.3). Fractional crystallization is thought to account for the diversity of lavas from Ascension Island, which range from alkali basalt to rhyolite. The slight increase in $\delta^{18}O$ with increasing SiO_2 content of the lavas has been described by a Rayleigh fractionation model, involving crystallization plagioclase, pyroxene, olivine, and magnetite (Figure 5.4; Sheppard and Harris, 1985). The limited variation in $\delta^{18}O$ observed in this and other igneous rock suites formed by fractional crystallization led Taylor and Sheppard (1986) to postulate that igneous rock suites that exhibit oxygen isotope variations greater than 1‰ must have been affected by other processes. Therefore, larger isotope variations in igneous rock suites are important indicators of open-system interactions, such as

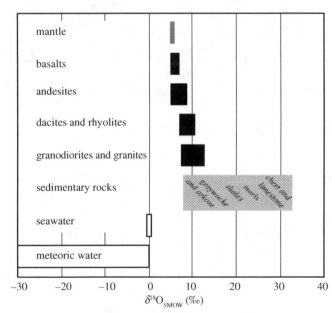

Figure 5.5 Comparison of $\delta^{18}O$ for the mantle, various igneous and sedimentary rocks, seawater, and meteoric water. Data from Taylor (1968), Taylor and Sheppard (1986), Hoefs (1987), and Valley et al. (1998).

assimilation of sedimentary rocks or interaction with hydrothermal fluids.

Sedimentary rocks tend to have $\delta^{18}O$ considerably higher than the mantle and most igneous rocks (Figure 5.5), and thus assimilation of sedimentary rocks tends to increase the $\delta^{18}O$ of the magma. The Earth's mantle has $\delta^{18}O$ of 5.7‰, yet many intermediate and felsic igneous rocks have significantly higher values

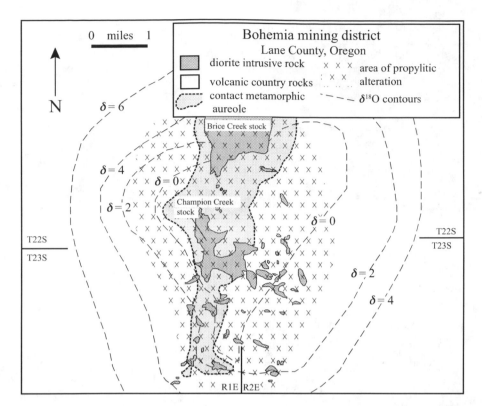

0 miles 1

N

Bohemia mining district
Lane County, Oregon

diorite intrusive rock

volcanic country rocks

contact metamorphic aureole

x x x area of propylitic
x x   alteration

- - - $\delta^{18}O$ contours

$\delta = 6$

Brice Creek stock

$\delta = 4$

$\delta = 0$

$\delta = 2$

Champion Creek stock

$\delta = 0$

T22S
T23S

$\delta = 2$

T22S
T23S

$\delta = 4$

R1E R2E

Figure 5.6 Contours of $\delta^{18}O$ for rocks of the Bohemia mining district, Oregon (Taylor, 1974). The bulls-eye pattern with lowest $\delta^{18}O$ in the center, where alteration and gold mineralization occur, is attributed to circulation of meteoric water and water–rock interaction that accompanied precipitation of ore from solution.

(Figure 5.5). Given that fractional crystallization will elevate the $\delta^{18}O$ of the residual melt by only around 1‰, interaction of magmas with ^{18}O-rich sedimentary rocks appears to have been a very important process in the formation of granitic continental crust.

Rocks that have $\delta^{18}O$ lower than typical mantle values are likely to have interacted with water: seawater has $\delta^{18}O = 0$, and meteoric waters even lower. When seawater or meteoric water interacts with rock, the $\delta^{18}O$ of the rock is lowered. Thus, stable-isotope compositions are sensitive indicators of water–rock interaction and have successfully identified hydrothermal systems around igneous intrusions. In some localities the hydrothermal fluids also carry metals in solution. One such example is the Bohemia mining district in the western Cascade Range in Oregon, where $US 1 million worth of gold was mined between 1870 and 1940. The diorite intrusive rocks are depleted in ^{18}O, as are the surrounding basaltic country rocks. The rocks with low $\delta^{18}O$ contain chlorite, epidote, calcite, and other minerals, consistent with water–rock interaction. Gold mineralization is concentrated in the area with the lowest $\delta^{18}O$ (Figure 5.6; Taylor, 1974). Sulfur isotope compositions also can be used to identify hydrothermal interaction and formation of sulfide ore deposits (Hoefs, 1987). Because stable-isotope ratios can identify the role of various fluids in the formation of ores, stable-isotope geochemistry is an important tool in economic geology.

5.3 Radiogenic Isotope Geochemistry

In recent years, an increasing number of radioactive parent–daughter pairs have been applied to igneous and metamorphic petrology. Applications are of two main types:

- **Geochronology**, which establishes timescales and absolute ages of igneous and metamorphic processes, and

- Isotopic petrogenesis, which identifies magma sources and traces the evolution of magmatic and metamorphic systems.

Some of the most common and geologically useful radioactive decay schemes are listed in Table 5.1. Most radioactive isotopes listed in this table decay directly to a single stable daughter, such as the decay of ^{87}Rb to ^{87}Sr or decay of ^{147}Sm to ^{143}Nd. Unusually, ^{40}K has two daughter isotopes, ^{40}Ca and ^{40}Ar, with 88.3 percent of decays yielding ^{40}Ca and 11.7 percent of decays yielding ^{40}Ar. Some radioactive parents decay through one or more intermediate radioactive daughters before a stable daughter is produced. Examples include ^{232}Th, ^{235}U, and ^{238}U, which decay via a series of short-lived intermediate radioactive daughters to stable isotopes of ^{208}Pb, ^{207}Pb, and ^{206}Pb, respectively. The decay rates of the radioactive parent isotopes vary, and are expressed as λ. The differences in λ relate to differences in half-life, that is, the time it takes for half of the parent isotopes to decay. All the radioactive parent–daughter pairs listed on Table 4.3

Table 5.1 *Radioactive isotopes and their daughter products commonly used in petrology*

Parent isotope	Daughter isotope	λ (y^{-1})	Half-life (billion years)	Ratio measured
^{40}K	^{40}Ca, ^{40}Ar	5.543×10^{-10}	1.25	^{40}Ar/^{36}Ar
^{87}Rb	^{87}Sr	1.42×10^{-11}	48.8	^{87}Sr/^{86}Sr
^{147}Sm	^{143}Nd	6.54×10^{-12}	106	^{143}Nd/^{144}Nd
^{176}Lu	^{176}Hf	1.867×10^{-11}	36	^{176}Hf/^{177}Hf
^{187}Re	^{187}Os	1.64×10^{-11}	42.3	^{187}Os/^{188}Os
^{232}Th	^{208}Pb	4.948×10^{-11}	14	^{208}Pb/^{204}Pb
^{235}U	^{207}Pb	9.8485×10^{-10}	0.707	^{207}Pb/^{204}Pb
^{238}U	^{206}Pb	1.55125×10^{-10}	4.47	^{206}Pb/^{204}Pb

are long-lived, suitable for dating events in the million- to billion-year age range. To constrain the timing of shorter and more recent geologic events such as crystallization of magma during ascent and eruption, it is useful to measure shorter-lived radioactive isotopes and their daughter products, such as intermediate daughters of U and Th.

The expression describing radioactive decay is:

$$N = N_0 e^{-\lambda t} \tag{5.7}$$

where N_0 is the original number of parent atoms, N is the number of atoms of parent isotope remaining after time t, and λ is the decay constant. Because the decay of one parent atom N creates one daughter atom D^* (where $*$ indicates the atom was created by radioactive decay), then:

$$D^* = N_0 - N \tag{5.8}$$

Rearranging Equation 5.7 gives:

$$N e^{\lambda t} = N_0 \tag{5.9}$$

and substituting into Equation 5.8 gives:

$$D^* = N e^{\lambda t} - N = N(e^{\lambda t} - 1) \tag{5.10}$$

Because, in general, there are atoms of the daughter isotope present at $t = 0$, a more general form of Equation 5.10 is:

$$D = D_0 + N(e^{\lambda t} - 1) \tag{5.11}$$

Equation 5.11 can be written for any specific decay system. For example, for the Rb–Sr decay pair, the equation becomes:

$$^{87}\text{Sr} = {}^{87}\text{Sr}_0 + {}^{87}\text{Rb}(e^{\lambda t} - 1) \tag{5.12}$$

Because it is analytically more convenient to measure isotopic ratios rather than absolute abundances, the expression can be written relative to a stable isotope of the daughter element. (The isotope ratio conventionally measured for each system is given in Table 5.1.) For the Rb–Sr system this is by convention ^{86}Sr:

$$\frac{^{87}\text{Sr}}{^{86}\text{Sr}} = \left(\frac{^{87}\text{Sr}}{^{86}\text{Sr}}\right)_0 + \frac{^{87}\text{Rb}}{^{86}\text{Sr}}(e^{\lambda t} - 1) \tag{5.13}$$

Generalizing, we can write the basic age equation as:

$$R = R_0 + R_{P/D}(e^{\lambda t} - 1) \tag{5.14}$$

where R is the ratio of daughter isotope to stable isotope of daughter element (such as ^{87}Sr/^{86}Sr) and $R_{P/D}$ is the ratio of parent isotope to stable isotope of daughter element (such as ^{87}Rb/^{86}Sr).

Both R and $R_{P/D}$ are the ratios measured today, at time t.

5.3.1 Geochronology

The basic age equation (5.14) includes two unknowns: age (t) and initial ratio (R_0). There are several approaches to solving for t and obtaining the date of a geologic event. First, it may be possible to identify a radioactive decay system where no atoms of the daughter isotope are present initially. In such cases, R_0 is zero and the only unknown is t. This can be assumed for the K–Ar decay pair. Second, if two samples have the same age, t, and initial daughter ratio, R_0, then the two unknowns can be solved using two equations. This approach is called the **isochron** method. Third, in the U–Pb system, the two parent–daughter pairs ^{238}U–^{206}Pb and ^{235}U–^{207}Pb provide two independent chronometers. The two age equations constructed for each decay scheme can be combined to leave a single unknown, t. Each of these cases is described briefly below.

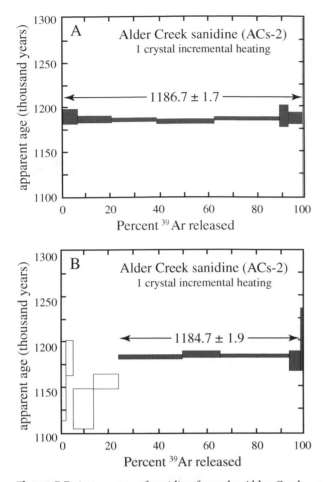

Figure 5.7 Age spectra of sanidine from the Alder Creek rhyolite at Cobb Mountain, Sonoma County, California. (A) Sanidine crystal with concordant ages in all incremental heating steps. (B) Sanidine crystal that yields younger ages at the beginning of the experiment but preserves the crystallization age in subsequent steps. Data from Jicha et al. (2016).

The K–Ar System

The K–Ar system is unusual in that the daughter product, Ar, is a noble gas that is not chemically bound in mineral lattices. Therefore, minerals should have little to no initial Ar and age (t) can be calculated from Ar and K present in the sample. In reality, samples in contact with the atmosphere absorb small but finite amounts of Ar on their surfaces. Fortunately, the Ar isotopic composition of the atmosphere is uniform, with $^{40}Ar/^{36}Ar = 296.16$. By measuring the amount of ^{36}Ar in the sample, the proportion of ^{40}Ar from atmospheric sources can be determined and subtracted, leaving the radiogenic ^{40}Ar produced by decay of ^{40}K to be entered in the age equation.

Because it is a gas, Ar may diffuse and be lost from samples, either during initial cooling or upon reheating.

Ar loss will result in ages younger than the crystallization age. An innovation that can identify Ar loss and recover the original age is ^{40}Ar–^{39}Ar dating. In this technique, samples are irradiated with fast neutrons in a reactor, and ^{39}K is converted to ^{39}Ar. The age of the sample can be determined from:

$$\frac{^{40}Ar^*}{^{39}Ar} = C\frac{^{40}K\left(e^{\lambda t}-1\right)}{^{39}K} \tag{5.15}$$

where C is a constant that relates the amount of ^{39}Ar produced in the reactor to the amount of ^{39}K present in the sample. Irradiated samples can be incrementally heated, and an age determined for each increment of argon gas released. Samples without Ar loss yield consistent ages across all heating steps (Figure 5.7A). Samples that have lost Ar from rims or fractures yield low $^{40}Ar/^{39}Ar$ during initial heating steps but higher $^{40}Ar/^{39}Ar$ during higher-temperature steps when more retentive Ar is released (Figure 5.7B). In this way, incremental heating can recover the crystallization age from the higher-temperature steps. In some samples, the time of episodic Ar loss also can be identified from the initial steps.

The Isochron Method of Age Determination

The basic age equation (5.14) has the form $y = b + mx$. This is an equation of a line, where y is R, b is R_0, x is $R_{P/D}$, and m is ($e^{\lambda t} - 1$). The age, t, is related to the slope of the line, which is known as an *isochron*. Although two analyses will define a line and thus an age, three or more collinear analyses are needed to evaluate whether all samples had the same initial ratio R_0 and the system remained closed, that is, no parent or daughter was lost or gained from the system, since time $t = 0$. Isochrons may be constructed for many isotopic systems, including Rb–Sr, Sm–Nd, Lu–Hf, Re–Os, and U–Pb; an example of the Sm–Nd system is shown on Figure 5.8. U–Pb data may be plotted with the axes $^{207}Pb/^{204}Pb$ versus $^{206}Pb/^{204}Pb$. In this way the plot shows data from both U–Pb decay chains on a single diagram. Linear arrays on this plot are also isochrons because the slope is proportional to age.

The U–Pb System

Because both isotopes of uranium are radioactive and follow different decay chains to different daughter isotopes of Pb, uranium-bearing minerals contain two independent "clocks," which in combination provide a robust dating system. U–Pb isotopic data are commonly plotted on a concordia diagram that shows the daughter/ parent ratios $^{207}Pb/^{235}U$ on the x-axis and $^{206}Pb/^{238}U$ on the y-axis. Young rocks and minerals will have accumulated few daughter isotopes, so will have low daughter/ parent ratios near 0 (Figure 5.9). With time, radioactive decay causes the ratio of each to increase along a curve called a **concordia** curve. Samples that have remained a closed system to U and Pb since formation plot "on concordia." If an analysis plots off the concordia curve, then that is an indication that the sample experienced open-system behavior. An important advantage of the two U–Pb decay systems is that if loss or gain of Pb or U occurred in a single episode, then the original age and time of disturbance can be recovered. This is shown on Figure 5.9, where a straight line drawn through the analyses, including a concordant point, intersects concordia at the original age. The lower intercept gives the time of episodic loss or gain of Pb or U.

Uranium-bearing accessory minerals, particularly zircon, but also monazite, titanite, and apatite, are commonly used for U–Pb dating. Dissolution of single grains, pre-treated by "chemical abrasion," commonly results in precise, concordant analyses (CA–TIMS

method; Mattinson, 2005). Microbeam analysis using secondary ion mass spectrometry (SIMS; Ireland and Williams, 2003) or laser ablation–inductively coupled plasma mass spectrometry (LA–ICP–MS; Woodhead

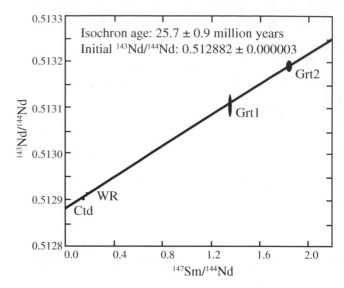

Figure 5.8 Sm–Nd isochron constructed from two garnet fractions, chloritoid (Ctd) and whole rock (WR) from a sample of garnet–chloritoid mica schist from the Tauern Window, eastern Alps (Favaro et al., 2015). The 25.7 ± 0.9 Ma age is interpreted to represent the age of garnet growth during peak temperature amphibolite-facies metamorphism during the Alpine orogeny.

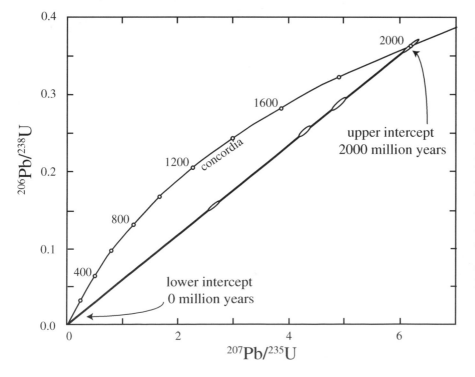

Figure 5.9 U–Pb concordia diagram, which shows data from both the $^{235}U-^{207}Pb$ and $^{238}U-^{206}Pb$ decay series. The $^{206}Pb/^{238}U$ and $^{207}Pb/^{235}U$ daughter/parent ratios increase with time along the concordia curve. One of the four analyses falls on the concordia curve, indicating that it has not lost or gained U or Pb since the sample formed 2000 million years ago. The analyses define a chord with an upper intercept of 2000 million years, the crystallization age of the sample. The lower intercept represents the time of open-system behavior when Pb was lost, which in this example is recent.

et al., 2016) provides the ability to date small spots, several microns to a few tens of microns in diameter. Multiple analyses on a single grain may illuminate its history by revealing features such as older cores, overgrowths, and altered areas.

A number of other geochronometers are based on the radioactivity of uranium. One of these, the (U–Th)/He thermochronometer, involves measuring helium, one of the decay products of uranium and its intermediate daughter isotopes. A noble gas, helium diffuses from radioactive minerals at relatively low temperatures. The (U–Th)/He method takes advantage of this property to document cooling histories. Another method, fission-track dating, makes use of the spontaneous fission of uranium. The number of damage tracks produced by uranium fission fragments is a function of uranium content and age. Finally, uranium-series dating focuses on the short-lived intermediate radioactive daughters of uranium, which have half-lives ranging from a few hundred thousand years to fractions of a second. The decay of intermediate daughters to their daughter products provides timescales of geologic processes such as generation of melt, magma residence times, and crystallization histories of minerals in plutonic and volcanic rocks.

5.3.2 Isotopic Petrogenesis

Another important application of isotope geology is in igneous petrogenesis, the study of magmatic sources. Isotopic compositions of heavy elements that have one or more radiogenic isotope, including Sr, Nd, Hf, and Pb, vary in rocks and minerals. The variations in $^{87}Sr/^{86}Sr$, $^{143}Nd/^{144}Nd$, $^{176}Hf/^{177}Hf$, $^{208}Pb/^{204}Pb$, $^{207}Pb/^{204}Pb$, and $^{206}Pb/^{204}Pb$ ratios are partly a function of age of the rock: the amounts of radiogenic daughter isotopes, ^{87}Sr, ^{143}Nd, ^{176}Hf, and ^{208}Pb, ^{207}Pb, and ^{206}Pb, increase with time as their parent isotopes decay, whereas the abundance of the non-radiogenic isotopes ^{86}Sr, ^{144}Nd, ^{177}Hf, and ^{204}Pb stays the same. However, the rate at which these radiogenic daughters are produced depends not just on time, but also on the parent/daughter ratio of the rock. Consider ^{87}Sr, which forms by decay of ^{87}Rb. The granitic parts of the continental crust contain a higher Rb/Sr ratio than the mantle. Over geologic timescales, this difference in abundance ratio has produced a mantle with a low $^{87}Sr/^{86}Sr$ ratio of around 0.703, whereas the present-day

upper continental crust contains rocks that have an average $^{87}Sr/^{86}Sr$ ratio of around 0.715. The isotopes of Sr (and other high-mass elements) have relatively small mass differences so they do not fractionate measurably during geologic processes. When a partial melt forms from a source region, the melt has exactly the same $^{87}Sr/^{86}Sr$ ratio as its source region. Therefore, a magma with a $^{87}Sr/^{86}Sr$ ratio of 0.720 cannot have formed by partially melting the mantle, but instead was generated somewhere in the crust. Measurements of potential source rocks narrow down the possible crustal sources. Note that the isotopic composition of a magma is also unaffected by crystallization: the growing crystals acquire the same $^{87}Sr/^{86}Sr$ ratio present in the magma. Isotopic tracers can "see through" the process of fractional crystallization and consequently they retain information about the magma source(s).

Table 5.2 lists the estimated abundances and parent/daughter ratios for the Rb–Sr, Sm–Nd, Lu–Hf, and U–Pb systems in average mantle, continental crust, upper crust, and lower crust. As discussed above, the high Rb/Sr ratio of the upper crust leads to a high $^{87}Sr/^{86}Sr$ ratio as ^{87}Rb decays to ^{87}Sr, whereas the low Rb/Sr ratio of the mantle results in a lower present-day $^{87}Sr/^{86}Sr$ ratio. In addition, because the Rb/Sr ratios of rocks in the upper crust are distinctly higher than those of rocks in the lower crust, the Rb–Sr system can be used to distinguish lower crustal source rocks from upper crustal source rocks. For the Sm–Nd and Lu–Hf systems, the parent/daughter ratio is higher in the mantle than in the crust – the opposite of the Rb/Sr ratio, which is higher in the crust. Therefore, with time, the $^{143}Nd/^{144}Nd$ and $^{176}Hf/^{177}Hf$ ratios become higher in the mantle than in the crust. The upper and lower crust have similar Sm/Nd and Lu/Hf ratios, making these systems best able to distinguish mantle sources from crustal ones but less useful in identifying various magma sources within the continental crust. The U/Pb ratios of major Earth reservoirs vary differently from the other radiogenic systems: the U/Pb ratio is highest in the upper crust and lowest in the lower crust, with the mantle and average continental crust falling in between. These different parent/daughter ratios in potential magma source regions lead to distinctive $^{87}Sr/^{86}Sr$, $^{143}Nd/^{144}Nd$, $^{176}Hf/^{177}Hf$, $^{208}Pb/^{204}Pb$, $^{207}Pb/^{204}Pb$, and $^{206}Pb/^{204}Pb$ ratios. Use of several

Table 5.2 *Abundances and ratios of radiogenic parent–daughter pairs in major Earth reservoirs*

	Average mantle	Average continental crust	Average upper crust	Average lower crust
Rb	0.605 ppm	69 ppm	84 ppm	56 ppm
Sr	22 ppm	285 ppm	320 ppm	308 ppm
Rb/Sr	**0.0275**	**0.242**	**0.263**	**0.182**
Sm	0.4347 ppm	4.84 ppm	4.7 ppm	4.65 ppm
Nd	1.341 ppm	27.4 ppm	27 ppm	25 ppm
Sm/Nd	**0.324**	**0.177**	**0.174**	**0.186**
Lu	0.07083 ppm	0.35	0.31 ppm	0.38 ppm
Hf	0.3014 ppm	4.71	5.3 ppm	4.2 ppm
Lu/Hf	**0.235**	**0.074**	**0.058**	**0.090**
U	0.0229 ppm	1.2 ppm	2.7 ppm	0.86 ppm
Pb	0.185 ppm	15 ppm	17 ppm	13 ppm
U/Pb	**0.124**	**0.080**	**0.158**	**0.066**

Abundance data from Palme and O'Neill (2014) and Rudnick and Gao (2014).

isotopic systems in combination can be successful in identifying the sources that contributed to the magma.

An early demonstration of the usefulness of radiogenic isotope ratios for magma source identification was provided by Kistler and Peterman (1973). They determined the $^{87}Sr/^{86}Sr$ isotopic compositions of granites from Mesozoic batholiths of California (Figure 5.10). Plotted in Figure 5.10 are contours drawn from the $^{87}Sr/^{86}Sr$ ratios of the granitic rocks at the time they crystallized. This is referred to as the *initial* $^{87}Sr/^{86}Sr$ because the subsequent decays of ^{87}Rb to produce additional ^{87}Sr after the rock solidified has been determined and subtracted out. Kistler and Peterman's data show the initial $^{87}Sr/^{86}Sr$ ratio of the granitic rocks is a function of geographic location, and that the ratios increase to the south and east. To the west of the contour defining $^{87}Sr/^{86}Sr = 0.704$ is an area composed of relatively young, mainly mantle-derived volcanic rocks and volcanogenic

sediment. To the east of the $^{87}Sr/^{86}Sr = 0.706$ contour are Precambrian to Triassic carbonates, shale, and sandstones. The $^{87}Sr/^{86}Sr$ ratio of the Mesozoic granites shows how they inherited their Sr isotopic compositions from the compositional differences in the crust they intruded. Kistler and Peterman (1978) later used the $^{87}Sr/^{86}Sr = 0.706$ contour to approximate the edge of Precambrian crust in the western United States. An Nd isotopic study of Mesozoic and Tertiary granite in the same area showed a similarly dramatic change in the Nd isotopic composition of granites, intruded on opposite sides of the inferred edge of the Precambrian basement (Farmer and DePaolo, 1983). Lackey et al. (2008) showed that oxygen isotopes, too, are effective tracers of crustal province boundaries and different magma sources. They analyzed the $\delta^{18}O$ values of bulk rocks, quartz, and zircon from samples of the Sierra Nevada batholith and used these data, combined with those from other isotope systems in prior studies, to better define the locations and

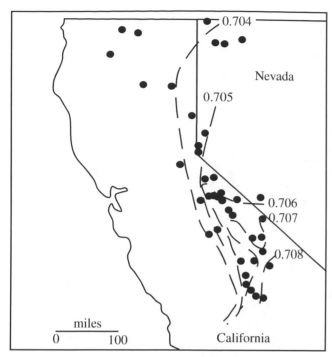

Figure 5.10 Contour diagram showing the regional variation in initial $^{87}Sr/^{86}Sr$ ratios of Mesozoic granitic rocks in central California. Solid dots indicate locations of analyzed samples. From Kistler and Peterman (1973).

compositions of different belts of rocks that were incorporated into the batholith.

A more recent example illustrates how radiogenic isotope systems can be used to establish the formation of continental crust early in Earth history. Lu–Hf isotopic data can be obtained by LA–ICP–MS on the same areas of zircon for which U–Pb isotopic data are collected using LA–ICP–MS or SIMS. Variations in measured $^{176}Hf/^{177}Hf$ ratios are small, so ratios are normalized in a fashion analogous to the δ notation used for stable isotopes (see Equations 5.1 and 5.2 above). Hf isotopic ratios are commonly normalized to chondritic meteorites, which have a composition similar to the Earth as a whole. The normalized ratio, denoted as ε_{Hf}, is defined as:

$$\varepsilon_{Hf} = \left[\frac{\left(\frac{^{176}Hf}{^{177}Hf}\right)_{sample} - \left(\frac{^{176}Hf}{^{177}Hf}\right)_{chondrites}}{\left(\frac{^{176}Hf}{^{177}Hf}\right)_{chondrites}} \right] \times 10^4 \qquad (5.16)$$

On a diagram of ε_{Hf} as a function of time, the chondritic reservoir (CHUR) is by definition always 0 (Figure 5.11). Because Hf is more incompatible than Lu, the Lu/Hf of the mantle is higher than in the crust. With time, the decay of ^{176}Lu produces $^{176}Hf/^{177}Hf$ ratios that are higher in the mantle than in the crust, leading to mantle compositions that have increasingly more positive ε_{Hf} with time. On the other hand, the lower Lu/Hf of the crust means that radiogenic Hf accumulates slowly such that $^{176}Hf/^{177}Hf$ ratios increase only slowly with time and ε_{Hf} of rocks and minerals with long residence in the crust are negative (Figure 5.11). Lu–Hf data from ancient zircon give the $^{176}Hf/^{177}Hf$ of the magma from which the mineral grains crystallized. The oldest dated zircon grains are found in Archean sandstone in the Jack Hills, western Australia. The least disturbed of these zircon grains have negative ε_{Hf}, as do zircon grains from the Acasta gneiss from the Slave craton in northern Canada and the oldest zircon grains in 3.4-billion-year-old gneiss from the Wyoming craton (Figure 5.11). Together, these negative ε_{Hf} values suggest that the zircon crystallized from magma formed by melting pre-existing crust of Hadean age (>4 billion years). Zircon from younger Wyoming craton gneisses, 2.8–3.6 billion years old, have ε_{Hf} close to 0. Because the crust containing the 3.8-billion-year-old zircon will evolve to increasingly negative ε_{Hf} values by 2.8–3.6 billion years ago, if it were partially melted those melts would have ε_{Hf} values lower than the values measured in the 2.8–3.6-billion-year-old gneiss. One possibility is that the 2.8–3.6-billion-year-old zircon grains crystallized from different magma sources, derived from younger crust. Alternatively, partial melts of the ancient crust mixed with melts from a higher ε_{Hf} source, such as contemporary depleted mantle (Figure 5.11). These data indicate that the process of forming continental crust began very early in Earth history, and also that new crust formation continued throughout the Archean Eon (2.5–4 billion years). Studies like this, which combine geochronology with radiogenic isotope ratio tracers of petrologic processes, belong to an emerging field called **petrochronology** (Engi et al., 2017).

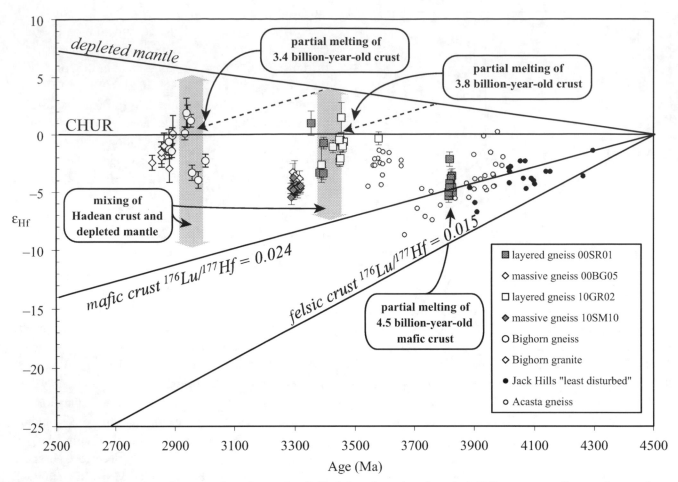

Figure 5.11 ε_{Hf} of zircon as a function of U–Pb age. Small filled dots show data from Jack Hills zircon, small open dots are from Acasta gneiss zircon, and other symbols are from gneisses from the Wyoming craton. The Jack Hills, Acasta, and oldest Wyoming zircon grains have negative ε_{Hf} indicating that they crystallized from partially melted Hadean-age crust. Younger, 3.4- and 2.9-billion-year-old Wyoming zircon crystallized from partial melts of younger crust, or mixtures of Hadean crust and depleted mantle. Modified from Frost et al. (2017).

Summary

- Isotope geochemistry is an increasingly important part of petrology. Both stable and radiogenic isotope data provide fundamental information about petrologic processes.

- Variations in the ratios of stable isotopes are the result of differences in chemical behavior. These differences are temperature-dependent, allowing stable isotope ratios to be used as geothermometers.

- Stable isotopes also are used to identify petrologic processes and the components incorporated into magmas. Fractional crystallization can produce variations in $\delta^{18}O$ of igneous rock suites of no more than 1‰. Differentiated members of igneous suites with higher ratios may have assimilated sedimentary rocks. Suites with $\delta^{18}O$ lower than mantle values ($\delta^{18}O = 5.7 \pm 0.6‰$) may have been affected by water–rock interaction.

- The decay of radioactive parent isotopes to daughter isotopes in rocks and minerals provides the basis for geochronology. Many different parent–daughter pairs are used to obtain ages of geologic events such as crystallization or metamorphism. The U–Pb system is especially powerful because two isotopes of uranium decay along distinct paths to two different isotopes of lead. This "double clock" enables open-system behavior to be detected and the original age recovered.

- Isotopic data also are used in igneous petrogenesis, the study of magma sources. Differences in parent/daughter ratios in different Earth materials, such as mantle reservoirs and various different compositions and ages of continental crust, lead with time to distinct isotopic compositions. When these materials are incorporated into magma by partial melting or assimilation, their isotopic signatures are transferred and serve as fingerprints of the magma sources.

Questions and Problems

Problem 5.1. $\delta^{18}O$ of quartz in a sample is measured to be 8.8‰, the $\delta^{18}O$ of albite is 7.9‰, and the $\delta^{18}O$ of magnetite is 2.0‰.

a. Using the relationships shown in Figure 5.2, estimate the temperature of equilibration of quartz, albite, and magnetite in this rock. What temperatures are given by the quartz–magnetite and the albite–magnetite thermometers?

b. The uncertainty on these measurements is 0.1‰. Do the two thermometers agree within error?

Problem 5.2. The following are Rb–Sr data for samples from the Sherman batholith, Laramie Mountains, Wyoming (Peterman et al., 1968):

Sample	$^{87}Rb/^{86}Sr$	$^{87}Sr/^{86}Sr$
W-5-15	1.125	0.7260
3-0-8	1.43	0.7326
W-2-7	1.72	0.7377
C235.1	2.21	0.7463
C-28.1	4.94	0.8058
81-46	12.5	0.9484
31-42a	15.1	0.9968
E-33-4	3.26	0.7676
E-33-4 (microcline)	4.38	0.7904

a. Plot these data on an isochron diagram. Evaluate the data. Is there evidence of open-system behavior? Why or why not?

b. Using a simple linear regression, determine the isochron age of these samples.

Problem 5.3. The Deccan Traps are a large continental flood basalt province associated with break-up of supercontinent Gondwana (see Chapter 8), and are temporally coincident with a bolide impact and mass extinctions at the Cretaceous–Paleogene boundary. Sample BOR14–1 (Renne et al., 2015) gave the following results:

Cumulative ^{39}Ar (%)	^{40}Ar*/^{39}Ar
0.753869	12.43485
3.612976	11.54639
12.957177	13.69519
21.394366	14.18945
29.759875	14.44622
38.991559	14.71509
43.151224	14.59891
48.824354	14.55679
55.074985	14.59244
60.546261	14.62277
65.481993	14.64721
70.145764	14.68477
73.828319	14.60791
76.122566	14.50252
78.008935	14.63861
80.854178	14.63752
85.456862	14.59683
90.445468	14.74240
94.535282	14.73469
96.467929	14.53266
100.000000	14.70886

Use Equation 5.15 to calculate ^{40}Ar*/^{39}Ar dates for each fraction (C^{40}K/^{39}K $= 391.853$) and plot them versus the percentage of ^{39}Ar released. Do three or more consecutive steps define ages that overlap within error and carry $>50\%$ of the total ^{39}Ar released? If so, what is the approximate age of this basalt based on those steps?

Problem 5.4. The continental crust is hypothesized to have formed by partial melting of the mantle repeatedly over geologic time, and by the ascent and crystallization of those partial melts at shallower depths in the crust.

a. Assuming that for mantle melting the bulk-distribution coefficient of Sr is greater than one and the bulk-distribution coefficient of Rb is less than one, suggest how the Rb/Sr ratio of the mantle from which partial melts were extracted (the depleted mantle) has changed during that time.

b. As a result, how will the ^{87}Sr/^{86}Sr ratio of the depleted mantle differ from the continental crust?

c. Is your reasoning compatible with the observed variations in ^{87}Sr/^{86}Sr ratio of granites in California shown in Figure 5.10? Explain.

Problem 5.5. The value of Δ(zircon–garnet) has been shown to be 0.0‰ at magmatic temperatures, that is, both minerals will have the same δ^{18}O if they crystallize from the same magma. Below are oxygen isotope data for garnet and zircon from Cretaceous garnet-bearing granites of the Idaho Batholith, USA, from King and Valley (2001). The $\delta^{18}O$ are not identical for garnet and zircon, which the authors interpreted to suggest crustal assimilation following zircon crystallization and prior to garnet crystallization. They identified Proterozoic Belt–Purcell Supergroup metasedimentary rocks ($\delta^{18}O = 17\permil$) as a likely contaminant. Your task is to determine the amount of assimilation that would be needed in this scenario.

a. Calculate Δ(garnet–zircon) for each sample.

b. Assuming that the concentration of oxygen is approximately the same in zircon, garnet, and metasedimentary rock, then the fraction, F, of sediment in the magma can be calculated using a simple mixing equation:

$$F*\delta^{18}O_{component\ 1} + \left(1 - F\right)*\delta^{18}O_{component\ 2} = \delta^{18}O_{mixture}$$

Calculate the amount of metasedimentary rock needed to raise the $\delta^{18}O$ from the value for zircon to the value for garnet for each sample.

c. Granites that have assimilated sedimentary rocks tend to have lower Ca, Na, and Sr and higher K and Rb. Does this general observation support your calculations?

d. Is the amount of assimilation you calculated a minimum or a maximum amount that may have been incorporated into these Idaho Batholith granites? Explain your reasoning.

Sample	$\delta^{18}O$ zircon	$\delta^{18}O$ garnet	SiO_2 (wt%)	CaO (wt%)	Na_2O (wt%)	Sr (ppm)	K_2O (wt%)	Rb (ppm)
98IB-31	5.78	6.02	73.0	3.38	5.08	722	0.94	6
98IB-33	6.33	6.45	72.9	3.24	5.46	893	0.84	2
98IB-7	6.93	7.69	71.4	1.96	4.22	507	3.66	112
98IB-40	7.41	8.28	71.0	2.96	4.81	599	2.38	72

Further Reading

Kohn, M. J., Engi, M., and Lanari, P., eds., 2017, Petrochronology: Methods and applications. *Reviews in Mineralogy and Geochemistry*, 83.

Schmitt, A. K., 2011, Uranium series accessory crystal dating of magmatic processes. *Annual Review of Earth and Planetary Sciences*, 39, 321–49.

Teng, F-Z, Watkins, J. M., and Daughas, N., eds., 2017, Non-traditional stable isotopes. *Reviews in Mineralogy and Geochemistry*, 82.

White, W. M., 2015, *Isotope Geochemistry*. Wiley, Chichester.

Chapter 6

Basalts and Mantle Structure

6.1 Introduction

Basalts are the most common rock type on the surface of Earth. The oceanic crust, which covers more than 70 percent of the surface of Earth, is composed of basalt and its intrusive equivalent, gabbro. Basalts dominate the rocks on oceanic islands and are also widespread on the continents. One of the major petrologic discoveries in the twentieth century was that basalts are partial melts of the mantle (Green and Ringwood, 1969). With this insight, basalts became more than simply interesting volcanic rocks: they took on significance as probes of the mantle. The chemistry of basalts, including their major- and trace-element compositions as well as their isotope geochemistry, provides direct evidence about the nature and composition of the mantle that is difficult to obtain by other means. This chapter describes the petrology of basalts, the structure and composition of the mantle from which they are derived, and the various processes by which the mantle may partially melt to form basaltic magmas.

6.2 Basalt Petrology

6.2.1 Classification

Because basalts are typically fine-grained to glassy rocks, the most common classification is based upon normative (as opposed to modal) mineralogy. One of the best ways to visualize basalt chemistry is by use of the basalt tetrahedron (Yoder and Tilley, 1962) (Figure 6.1). The basalt tetrahedron has the apices of normative augite, quartz, nepheline, and olivine. Normative albite plots one-third of the way from nepheline toward quartz and normative "hypersthene"[1] plots midway between olivine and quartz. Because hypersthene never coexists with nepheline, the olivine–albite–augite plane is always present. This plane is called the *plane of critical silica undersaturation* and separates hypersthene-normative bulk compositions from nepheline-normative bulk compositions. The nepheline-normative basalts are called **alkali basalts**; the hypersthene-normative basalts are called **tholeiites**. Because the magnesium-rich olivine found in basalts never coexists with quartz, the hypersthene–albite–augite plane is also important to basalt petrology. This plane is called the *plane of silica saturation* and it separates quartz-saturated tholeiites (i.e. *quartz tholeiites*) from olivine-saturated tholeiites (i.e. *olivine tholeiites*).

6.2.2 Chemistry and Petrography

The normative differences between the two basalt types reflect their subtle differences in chemistry. As their name implies, alkali basalts are richer in alkalis (Na_2O and K_2O) than are tholeiites (Table 6.1). As such, they plot in the alkalic or alkali-calcic fields of Peacock (1931), whereas tholeiites are typically calcic or calc-alkalic. Importantly, alkali basalts have slightly lower silica contents than tholeiites (46–48 percent compared to 48–52 percent). Because alkali feldspars contain more silica than calcic feldspars (cf. albite – $NaAlSi_3O_8$ and anorthite – $CaAl_2Si_2O_8$), crystallization of an alkali feldspar will deplete silica from a melt more effectively than will crystallization of a calcic feldspar. Thus, the combination of relatively high alkalis and low silica explains why alkali basalts are nepheline-normative. In addition to the differences in major-element abundances, alkali basalts also tend to be richer in incompatible elements than tholeiites. As we noted in Chapter 4, incompatible elements are elements that are incompatible with crystallizing silicates but compatible with the melt. These elements, such as rare-earth

Table 6.1 *Average major-element compositions (in weight percent) of tholeiites and alkali basalts (Nockolds et al., 1978)*

	Tholeiite basalt[a]	Alkali basalt[b]
SiO_2	50.83	46.19
TiO_2	2.03	2.54
Al_2O_3	14.07	15.02
Fe_2O_3	2.88	2.70
FeO	9.06	9.01
MnO	0.18	0.17
MgO	6.34	9.05
CaO	10.42	10.82
Na_2O	2.23	2.78
K_2O	0.82	0.89
P_2O_5	0.23	0.38
Total	99.09	99.55

[a] Average of 137 tholeiitic basalts (including diabase).
[b] Average of 45 alkali basalts.

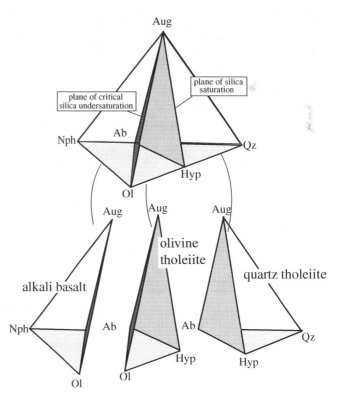

Figure 6.1 The basalt tetrahedron showing the differences in normative composition between alkali basalt, olivine tholeiite, and quartz tholeiite. After Yoder and Tilley (1962).

elements and TiO_2 and P_2O_5 (in the absence of ilmenite or apatite crystallization), will be concentrated in the magma as it crystallizes or are likely to be the first elements to enter a melt when melting begins.

The chemical differences between tholeiites and alkali basalts are reflected in the following petrographic differences. Tholeiites typically contain olivine only as a phenocryst. The olivine commonly shows signs of resorption or reaction to pigeonite or orthopyroxene. Groundmass phases include pyroxenes and plagioclase. The augite in tholeiites is typically colorless, indicating that it is poor in ferric iron and titanium. Some quartz tholeiites may contain groundmass quartz or vesicles lined with a silica mineral, although in many quartz tholeiites the excess silica will be hidden in its glassy matrix. Alkali basalts contain olivine as both phenocrysts and groundmass. The augite tends to be pleochroic because it contains small amounts of ferric iron and titanium. There may be a late-stage alkali feldspar, often anorthoclase, in the groundmass. In most alkali basalts the normative nepheline is hidden in the residual glass; however, if the basalt is very alkalic, nepheline may appear in the groundmass. Such a basalt is called a *basanite*.

As depicted in Figure 2.11, in the system Nph–silica, albite is a thermal barrier. Melts on the silica side of the albite composition evolve to a quartz-bearing eutectic, whereas melts on the nepheline side evolve to a nepheline-bearing eutectic. This behavior extends to more complex silicate systems as well. Those melts that lie to the nepheline side of the olivine–albite–augite plane in Figure 6.1 (i.e. alkali basalts) differentiate toward nepheline-saturation, whereas those melts on the hypersthene side (i.e. tholeiites) differentiate toward hypersthene- or quartz-saturation. As a result, alkali basalts and tholeiites follow very different differentiation paths. During differentiation, alkali basalts evolve to form nepheline-bearing rocks, such as phonolites or their plutonic equivalents, nepheline syenites (i.e. the rocks on the lower half of the IUGS diagrams shown in Figures 1.1 and 1.4). Tholeiites, in contrast, evolve toward silica saturation, forming residual magma with trachyitic or rhyolitic composition.

6.3 Melt Generation from the Mantle

6.3.1 Mantle Composition

Because the mantle cannot be directly sampled, petrologists deduce its composition indirectly. The proxy evidence includes:

Evidence from mantle-derived melts. The compositions of partial melts derived from the mantle, particularly mid-ocean ridge basalts and ocean island basalts, place important constraints on the composition of the mantle.

The composition of rocks of mantle origin. Samples of rock that formed in the mantle can be found at Earth's surface and give important indications of the rocks that compose the upper mantle. Mantle rocks occur as xenoliths in basalts or kimberlites, as well as in **ophiolites** (discussed further in Chapter 7), which represent pieces of the upper mantle and oceanic crust that have been thrust onto the continents.

The composition of chondritic meteorites. Chondritic meteorites have a similar composition to the bulk composition of Earth. The composition of the mantle can be estimated by taking the chondrite composition and subtracting those elements thought to make up Earth's core and crust.

Geophysical evidence. The geophysical properties of the mantle, in particular its density and seismic velocity, allow geologists to construct a fairly robust picture of mantle structure and place some constraints on composition.

These various kinds of evidence suggest the mantle has the composition of lherzolite: a peridotite dominated by olivine that contains both orthopyroxene and clinopyroxene (see Figure 1.3). An aluminous mineral is also present: either plagioclase, spinel, or garnet (Figure 6.2). Depth is the primary control determining which aluminum-bearing mineral is present; plagioclase forms at the shallowest levels, whereas garnet forms at greatest depth.

6.3.2 Crust and Mantle Structure

Geophysical evidence indicates that the outer 660 kilometers of Earth consist of the following major layers (Figure 6.3):

Oceanic or continental crust. Oceanic crust is between 7 km and 10 km thick, and continental crust is up to 70 km thick. The base of the crust is defined by the Moho, the seismic discontinuity across which S-wave velocity increases from around 3.5 km s^{-1} in the crust to 4.5 km s^{-1} in the mantle.

Lithospheric mantle. Lithospheric mantle is the upper portion of the mantle that deforms brittlely. It is

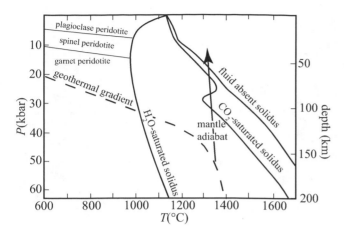

Figure 6.2 Diagram showing how adiabatic decompression of the mantle (arrow) can lead to melting even if the mantle is dry. Modified from Philpotts and Ague (2009).

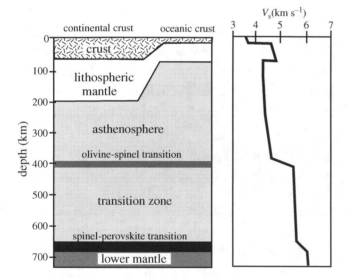

Figure 6.3 The major layers of the crust and mantle, along with characteristic S-wave velocities (V_s) for each layer.

defined by relatively high S-wave velocities of around 4.5–5 km s^{-1}. The thickness of oceanic lithosphere increases with age and is up to 80 km thick. The thickness of continental lithosphere varies as well but is generally 150–200 km.

Asthenosphere. The asthenosphere extends from the base of the lithosphere to the top of the transition zone at 410 km depth. It is a relatively weak zone that deforms by creep. S-wave velocities are lower than in the lithospheric mantle and may be attenuated in part because of the presence of a partial melt.

Transition zone. Between 410 km and 660 km depth, seismic wave velocities increase in a series of steps.

The velocity steps are related to increasing mantle density, which in turn correspond to changes in mineral structure in response to increasing pressure. Together, the lithospheric mantle, asthenosphere, and transition zone compose the *upper mantle.*

Lower mantle. The lower mantle extends from the base of the transition zone to the outer core at 2900 km depth.

6.3.3 Mechanisms for Partial Melting of the Mantle

The normal temperatures encountered at increasing depth in the mantle (the dashed line indicating the geothermal gradient in Figure 6.2) are always below the solidus for fluid-absent lherzolite. Under ordinary circumstances, therefore, the mantle is solid. However, a number of phenomena can generate mantle melting. First, the normal geothermal gradient could be perturbed, so that it is locally hot enough to melt. This may occur beneath ocean islands at "hot spots," such as Hawaii.

Second, the temperature at which melting begins could be lowered by addition of a component to dry lherzolite. The addition of H_2O- and CO_2-bearing fluids to peridotite lowers the solidus significantly (Figure 6.2). This means that the addition of such fluids can lower the solidus enough that melt can be produced from lherzolite at the temperatures and pressures thought typical of the normal mantle thermal regime (Figure 6.2). Because subduction carries water-rich fluids along with oceanic crust into the mantle, this process is likely an important mechanism for adding fluids that depress the mantle solidus and trigger partial melting.

A third mechanism that may produce melting is decompression of ascending mantle. Although the mantle is very viscous, with a viscosity of around 10^{-21} Pa s (King, 1995), mantle flow enables important geologic processes, such as isostatic rebound in response to erosion and deglaciation. A petrologically important process is the upward flow of mantle beneath mid-ocean spreading centers. As the mantle deep beneath the ridge rises, it begins to melt by decompression melting. The temperature gradient is approximately adiabatic; that is, no heat is transferred in or out of the mass under consideration. For mantle materials, the adiabatic gradient is around 0.3 °C km^{-1}, which means that mantle rising adiabatically does not cool appreciably as it ascends. By comparison, the gradient for the melting point of anhydrous mantle is much steeper (Figure 6.2). This relationship is shown by the arrow in Figure 6.2, which indicates the pressure–temperature path of a rising mantle diapir that originally lay on the mantle geotherm. As the diapir is decompressed adiabatically, it will melt when it crosses the fluid-absent solidus at around 50 km depth, even if the mantle lacked any fluid component such as CO_2 or H_2O. (If the mantle contains CO_2 or H_2O, then melting begins at greater depth.) Initially, the melt remains on the grain boundaries of the mantle peridotite, but as ascent continues and melting progresses the melt begins to migrate into channels and then upward into the overlying oceanic crust. Because mid-ocean ridges are located above up-going limbs of convection cells, **decompression melting** is particularly important at mid-ocean spreading centers.

6.3.4 The Process of Mantle Melting

The composition of partial melts of the mantle and their residual solids can be illustrated in the simplified system Di–Fo–En (Figure 6.4). Lherzolite plots near the center of this ternary system. On heating, the first melt to form is of eutectic composition (point X). As melting proceeds, the residual solids will become progressively more olivine-rich (gray arrows in Figure 6.4). After about 25 percent melting, the diopside (as well as most of the aluminous phase, spinel, or garnet) will have been completely incorporated into the melt, and the residue will consist only of olivine and orthopyroxene (i.e. the rock will be a harzburgite). If melting proceeds further, the melt composition will become increasingly enriched in the orthopyroxene component (black arrow in Figure 6.4), while the residua becomes enriched in olivine.

Production of basaltic melts therefore leaves the mantle enriched in olivine. This leads to the common terminology applied to peridotites. A *fertile* lherzolite (a lherzolite from which a basaltic melt can be extracted) contains abundant green (i.e. Al_2O_3-rich) spinel as well as clinopyroxene that may be rich in minor components such as TiO_2 and Na_2O. In a *depleted* lherzolite (a lherzolite from which a partial melt has been extracted), the spinel is chromium-rich and less abundant than in fertile lherzolite, and the clinopyroxene has lower Na_2O and

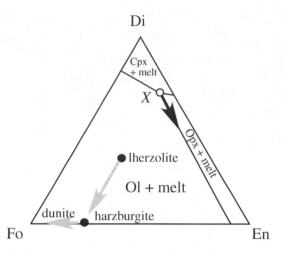

Figure 6.4 Simplified phase diagram for the system Di–Fo–En at about 20 kbar. Black arrow shows the path followed by the melt during melting of lherzolite; gray arrows show the path followed by the residua.

TiO_2 contents. Extremely depleted mantle rocks include harzburgite (a peridotite with little or no clinopyroxene) or dunite, which contains olivine with only minor amounts of orthopyroxene and clinopyroxene.

6.3.5 Origin of Tholeiitic versus Alkali Basalts

As noted at the beginning of this chapter, there are two main basalt types, tholeiitic and alkalic. It is natural to ask how melting of the mantle can produce basalts of varying compositions. Possible explanations include:

1. *The alkalic and tholeiitic basalts come from two different sources with different compositions.* As discussed in Chapters 9 and 10, mantle-derived, alkalic magmas display a wide range of compositions, from hyperpotassic magmas, such as those of the Roman province of Italy, to highly sodic magmas, such as those from the East African Rift. It is appealing to invoke a heterogeneous mantle to explain this broad compositional spectrum. If the extreme compositional range observed in alkaline rocks does reflect a heterogeneous mantle, then the same heterogeneity may also explain lesser compositional differences, such as those that distinguish tholeiites and alkali basalts.

2. *Both alkali basalts and tholeiites come from the same kind of source but they represent melting at different*

pressures or different degrees of partial melting. Most alkali basalts differ only slightly in composition from tholeiites, so a substantial difference in source composition is not required. Furthermore, in some places, the basalt types vary during the eruptive history of a volcanic center. On Hawaii, for example, the early eruptions were mixed tholeiites and alkali basalts, the main shield building stage was dominated by tholeiitic basalt, and the post-shield stage lavas were alkali basalt. To explain this, many petrologists argue that alkali basalts and tholeiitic basalts come from a single mantle source. Evidence supporting this argument derives from the fact that augite in the mantle is the major source of Na_2O, K_2O, TiO_2, and other incompatible elements, enriched in alkali basalts. As noted earlier, augite is the first silicate depleted from melting of lherzolite. If Na_2O is extracted preferentially from this pyroxene during early stages of partial melting of the mantle, then the first melts are alkaline. As melting proceeds, the magmas become progressively more calcic, approaching tholeiite in composition.

Another possible explanation for the origin of the different basalt compositions relates to the pressure of melting. Figure 6.5 shows the effect of pressure on the olivine–orthopyroxene–clinopyroxene–plagioclase eutectic as projected from an aluminous phase (plagioclase, spinel, or garnet). This diagram, a *pseudoternary* diagram, is read in a similar way to Figure 2.16 as long as the projected phase – plagioclase (or spinel or garnet at higher pressure) – is always present. Figure 6.5 shows that increasing pressure moves the eutectic away from silica toward olivine, meaning melts generated at high pressure will likely have less silica than those produced at lower pressure. Thus, alkali basalts may be generated from the same mantle as tholeiites either by lower degrees of partial melting, or at higher pressure, or both.

6.4 Environments where Magmas are Generated

Igneous activity observed today is confined to relatively few tectonic environments:

Constructive plate margins. These are divergent plate boundaries, such as mid-ocean ridges and back-arc spreading centers, where mantle upwells, decompression melting occurs, and magma is emplaced into the rift. Magmatism in this environment is described in Chapter 7.

Destructive plate margins. These are convergent plate boundaries that are either ocean–ocean or ocean–continent collision zones. In these collisions, water subducted into the mantle via the downgoing, hydrated plate induces melting in the downgoing plate or overlying mantle. The resulting magmatism forms oceanic and continental arcs discussed in Chapter 8.

Oceanic intraplate regions. These manifest as islands and sea-floor plateaus that decorate the ocean floor and that were probably caused by hot-spot magmatism. Oceanic intraplate magmatism is described in Chapter 7.

Continental intraplate regions. Within-plate continental magmatism produces igneous rocks that manifest a substantial range in composition because the magmas form by a number of processes and because the rocks that partially melt are compositionally varied, reflecting a range of mantle and continental sources. Continental intraplate volcanism and plutonism are the subjects of Chapters 9 and 10, respectively.

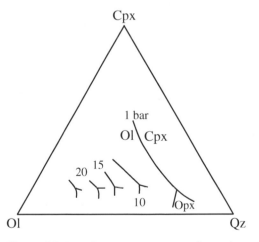

Figure 6.5 Pseudoternary projection from plagioclase on to the olivine–clinopyroxene–quartz plane showing how location of the basalt eutectic changes with increasing pressure. Modified from Elthon (1989).

Summary

- Basalts are classified as alkali basalts, quartz tholeiites, and olivine tholeiites based on their normative mineralogy.

- Alkali basalts evolve toward nepheline saturation and form phonolites and nepheline syenites. Tholeiites evolve toward silica saturation and form trachytes and rhyolites.

- The mantle is composed of an upper lithospheric mantle, which overlies the asthenosphere and transition zone, and a lower mantle.

- The temperatures and pressures in the mantle encountered along a typical geothermal gradient are always below the solidus for dry melting of the mantle, so the mantle is normally solid.

- Partial melting of the mantle produces basaltic magmas. Melts are generated by perturbing the normal geothermal gradient to raise the temperature of the mantle, lowering the melting point by adding H_2O or CO_2 or other components, or by bringing the mantle to shallower depths and producing melt by decompression.

- Alkali basalts and tholeiites could come from different mantle sources, or form from the same mantle source by different degrees of melting or melting at different pressures.

- The tectonic environments that generate the greatest volume of magma are at constructive plate margins, both at mid-ocean ridges and in back-arc spreading centers (Table 6.2). Subduction zones are the second most voluminous sites of magmatism, followed by oceanic intraplate regions where ocean islands and plateaus are formed. Lesser volumes of magma form within continental plates, but this tectonic setting creates the largest variety of igneous rock compositions.

Table 6.2 *Global rates (km^3 per year) of Cenozoic magmatism*

Location	Volcanic rocks	Plutonic rocks
Constructive plate boundaries	3.0	18.0
Destructive plate boundaries	0.4–0.6	2.5–8.0
Continental intraplate regions	0.03–0.1	0.1–1.5
Oceanic intraplate regions	0.3–0.4	1.5–2.0
Global total	3.7–4.1	22.1–29.5

Sources: Crisp (1983) and McBirney (1993).

Questions and Problems

Problem 6.1. Describe three mechanisms by which the mantle may partially melt. Give the tectonic environments in which each of these mechanisms may operate.

Problem 6.2. What are the mineralogical and chemical differences between alkalic and tholeiitic basalt? What rock types represent the extreme differentiates of each?

Problem 6.3. What rocks that can be collected at Earth's surface provide the best information about the composition of the mantle? Explain your answer.

Problem 6.4. Use the figure below (from Frost and Frost, 2008) to relate the FSSI index (Chapter 4) to the basalt tetrahedron (Figure 6.1). Give the range of FSSI for alkali basalt, olivine tholeiite, and quartz tholeiite.

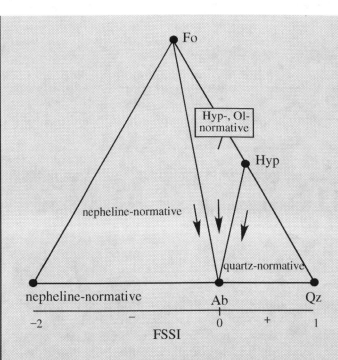

Problem 6.5. Examine Figure 6.4.

a. What are the proportions of Di, Fo, and En in the lherzolite indicated by the dot?

b. Given the composition of lherzolite indicated, what percent melt of composition X will be generated, and what percent residual harzburgite will remain?

c. What are the proportions of Di, Fo, and En in the residual harzburgite?

d. From these mineral proportions, determine whether mantle-derived basalt melt will be more or less siliceous than residual depleted mantle. Explain.

Further Reading

Frost, D. J., 2008, The upper mantle and transition zone. *Elements*, 4, 171–6.

Langmuir, C. H. and Forsyth, D. W., 2007, Mantle melting beneath mid-ocean ridges. *Oceanography*, 20, 78–89.

McKenzie, D. and Bickle, M. J., 1988, The volume and composition of melt generated by the extension of the lithosphere. *Journal of Petrology*, 29, 625–79.

Philpotts, A. R. and Ague, J. J., 2009, *Principles of Igneous and Metamorphic Petrology*, 2nd edn. Cambridge University Press, Cambridge, Chapter 23: Origin of rocks.

Wilson, M., 1989, *Igneous Petrogenesis: A Global Tectonic Approach*. Unwin Hyman, London, Chapter 3: Partial melting processes in the Earth's upper mantle.

Note

1 Hypersthene, an orthopyroxene containing both iron and magnesium, is a mineral name that is no longer in use. It remains in the normative mineral calculation and is used to define the plane of silica saturation in basalts.

Chapter 7

Oceanic Magmatism

7.1 Introduction

Because it is covered by kilometers of water, oceanic crust was long inaccessible to direct observation by geologists. Today, however, our knowledge of the ocean floor comes from the study of fragments of the ocean floor that have been thrust onto the land, called ophiolites, and from ship-based geophysical and geological studies that burgeoned during World War II. Important information comes from core retrieved by ship-based deep-sea drilling, which began in 1961 and continues today (see Box 7.1). Together, these investigations provided the foundation that underpins our understanding of oceanic magmatism. This chapter first discusses the structure and stratigraphy of ophiolites and to what extent they provide models that help us understand the oceanic crust. A description of advances achieved by recent research of the ocean floor, based on geophysical studies and ocean drilling, follows. Finally, this chapter describes the magmatic suites that compose ocean islands and oceanic plateaus.

BOX 7.1 | OCEAN DRILLING

Before ocean drilling programs commenced, the only information scientists had about the composition of the ocean floor was obtained by dredging. The US National Science Foundation began the deep-sea drilling program in 1961 with project MoHole, which aimed to drill through the oceanic crust to the mantle. Five holes were drilled off the Pacific coast of Mexico; the deepest was 183 m. The project proved that deep drilling could be done in the ocean, but the goal of drilling to the mantle was never reached and remains an elusive goal even today. Project MoHole was followed by the Deep Sea Drilling Project (DSDP), which began in 1966. In 1975, France, Germany, Japan, the former Soviet Union, and the United Kingdom joined in funding the drilling program. DSDP ran from 1968 until 1983, using the research vessel *Glomar Challenger*. The *Glomar Challenger* was retired in 1983 and the drilling program resumed in 1985 as the Ocean Drilling Program (ODP). In 2003, it changed its name to the Integrated Ocean Drilling Program (IODP), and in 2013 its name was changed to the International Ocean Discovery Program to encompass research conducted by IODP beyond drilling alone. IODP uses the US research vessel *Joides Resolution* and the Japanese vessel *Chikyu* along with their associated manned and unmanned submersible vehicles. Currently, the drilling program is supported by 26 countries, including the United States, the European Union, Japan, China, India, Australia, and New Zealand.

One of the major scientific themes of the IODP is to study the petrology of the oceanic crust, to understand the geochemical and geodynamic processes involved in the solid Earth system. The scientific value of ocean drilling became apparent within the first years of drilling. The first cores substantiated the young age of the oceanic crust and the dynamics of sea-floor spreading. These geologic observations and processes now underpin discussions of plate tectonics. Ocean drilling verified that the primary transfer of energy and material from the deep Earth to the surface occurs via sea-floor spreading and the creation of oceanic crust at mid-ocean ridges, as well as by upwelling magmas that form ocean islands, ocean plateaus, and island arcs. Further drilling documented that sea-floor spreading involves not only magmatic addition to the crust but, locally, may include tectonic denudation as well. As a result of sea-floor tectonics, geologists now recognize that a considerable area of the ocean floor is underlain by serpentinized mantle peridotite. Ocean drilling has enabled descriptions of the kinds of reactions involved in the alteration of the sea floor, including serpentinization; these reactions have proven critical to modeling the geochemistry of ocean water. Recent findings suggest that mid-ocean ridge basalts interact extensively with the mantle through which they move, producing hybrid troctolites whose existence was previously unexpected. In total, these observations help petrologists understand the evolution of mantle-derived basaltic magmas and the formation of oceanic crust.

Box Figure 7.1 Rainbow over the drilling rig of the *Joides Resolution*, 30° N on the Mid-Atlantic Ridge.

7.2 Petrology and Structure of the Oceanic Crust

7.2.1 Ophiolites as a Model of the Oceanic Crust

Geologists have long recognized that an association of peridotite (in many places hydrated to serpentinite), gabbro, basalt, and deep-water chert are exposed in many places around the world (Map 7.1). In some localities, these rocks form a complete stratigraphic section, but in many places one or more of these rock types exist within fault-bounded tectonic slices. As early as the 1820s, this association was called an *ophiolite*, but before the advent of plate tectonics, the significance of these rocks was cryptic. Geologists attending the September 1972 Penrose Conference defined the stratigraphy of a typical ophiolite, shown in Figure 7.1 (Anonymous, 1972). Implicit in the definition is the assumption that ophiolites are fragments of oceanic crust thrust onto the continents, and thus the stratigraphy described at the Penrose Conference represents an idealized cross-section of the oceanic crust.

The uppermost layer in an ophiolite is composed of deep-water sediments, mostly pelagic mud, although chert may be common in some places. The thickness of this layer depends on the age of the crust. On juvenile oceanic crust there are no sediments; the thickness of the sediment layer generally increases with age.

			Thickness (km)
sediments		Layer 1	ca. 0.5
pillow lavas		Layer 2	1.7
sheeted dikes			1.8
		Layer 3	
gabbro			3.0
layered peridotite		Seismic Moho	
deformed peridotite		Petrologic Moho / Layer 4	

Figure 7.1 Petrologic profile for an ideal ophiolite (Anonymous, 1972).

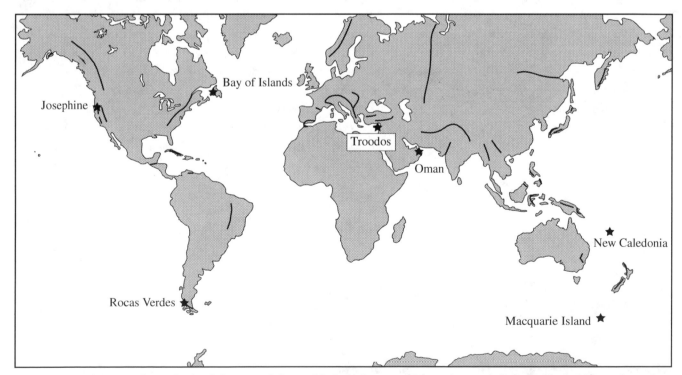

Map 7.1 Map showing select ophiolite belts around the world. Ophiolites occur along the trends indicated by bold lines. Stars show particularly well-known occurrences. Data from Irwin and Coleman (1974).

direction of spreading

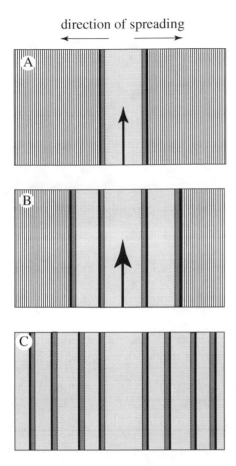

Figure 7.2 Diagram showing the development of a sheeted dike complex. (A) A basaltic dike is intruded into rifted oceanic crust and produces chilled margins. (B) Continued rifting splits the original basaltic dike and a new dike is intruded, forming chilled margins. (C) The result is a suite of dikes that have intruded other dikes. Dikes on the right side of the spreading center are chilled only on the right; those on the left are chilled only on the left.

A kilometer or so of *pillow basalts*, which represent lavas that were erupted directly onto the ocean floor, underlie the sediments. The pillow basalts grade into sheeted dikes, a horizon that may be over a kilometer thick.

Sheeted dikes are dikes that consistently chilled on one side only. They are interpreted to have been emplaced into a spreading center. As spreading continues, extension splits the dike, and the next dike is intruded into the core of a preceding dike (Figure 7.2). Over a long period of time, this process results in a horizon that is composed entirely of dikes (Figure 7.3).

A sheeted-dike complex is characterized by two features: (1) there is no evidence of a country rock into which the dikes were emplaced and (2) the dikes typically are chilled only on one side.

Stratigraphically, below the sheeted dikes lie several kilometers of gabbro. This layer is interpreted to have crystallized from intrusive bodies of basaltic magma that fed the overlying basalts. The top of the gabbro is directionless, but toward the bottom the gabbro may be layered or foliated. Below the gabbro lies layered peridotite, which is much denser than the overlying gabbro. The contact between peridotite and gabbro is the location of a distinct change in seismic velocity, marking the Moho. However, because the layered peridotites are interpreted to have formed as cumulates from the basaltic magma, they are unrelated to underlying mantle and actually represent an ultramafic portion of the crust. Below the layered peridotites lies a highly deformed peridotite, which is interpreted as mantle that has been depleted by partial melting during basalt genesis. Petrologically, this is true mantle, even though it is impossible to distinguish it seismically from the overlying cumulate peridotite.

7.2.2 Refinements of the Ophiolite Model

Nearly as soon as the Penrose ophiolite model was proposed, geologists began to debate whether or not the model describes a true picture of the oceanic crust (Miyashiro, 1975; Moores, 1982). It quickly became evident that ophiolites form in diverse tectonic environments, and not all of them reflect the idealized ocean-floor stratigraphy described by the Penrose model. Some ophiolites, such as the Troodos ophiolite in Cyprus, contain basalts more closely related compositionally to arc basalts than to mid-ocean ridge basalts (Miyashiro, 1973) and evidently formed above newly initiated subduction zones. These are called **suprasubduction-zone ophiolites** (Pearce et al., 1984b). Observations suggest that ophiolites form in a wide range of tectonic environments and thus resist a simplified, "one-size-fits-all" model. In addition to forming above subduction zones, ophiolites may form by back-arc spreading, as did the Rocas Verdes ophiolite in Chile (Stern and de Wit, 2003), and at the contact between a back-arc and an arc, as did the Bay of Islands ophiolite in Canada (Kurth-Velz et al., 2004). Others formed at mid-oceanic spreading centers, as described by the Penrose model,

chilled margin

Figure 7.3 Sheeted dikes from the Oman ophiolite. Dikes are all chilled on the right side (see inset), indicating that at the time they formed the spreading center was to the left. Average width of dike is about a meter. Photo courtesy of Benoit Ildefonse.

including the Macquarie Island ophiolite in the South Pacific (Varne et al., 2000) and the Oman ophiolite on the Arabian Peninsula (Boudier and Nicolas, 2011). Map 7.1 shows the global distribution of these and other major ophiolites.

A second problem with the ophiolite model arose in the 1990s and 2000s when seismic surveys and deep-ocean drilling showed the stratigraphy of the oceanic crust is far more complex than the ophiolite model suggested. Geophysical studies revealed significant differences in spreading rates among oceanic ridges (Map 7.2) and that ridges with different spreading rates have different morphology (Figure 7.4), which translates into differences in crustal cross-section.

Fast-Spreading Centers

The East Pacific Rise is an example of a fast-spreading center (half-rate 6–7 cm yr^{-1}). Fast-spreading centers are characterized by a 2.5 to 3.0-km-wide zone of magma extrusion, which forms a smooth topographic high of around 200 m above mean ocean floor (Figure 7.4A). Flat lava plains made of ponded lava lakes and small volcanic hills, composed of sediment-free pillow lavas, occur along the ridge axis. There is either no axial valley or only one that is poorly developed.

Early conceptions of mid-ocean rifts envisioned magma chambers in spreading zones on the order of kilometers wide. Seismic studies, however, have shown that large magma bodies are not present in fast rifts, such as the East Pacific Rise (Sinton and Detrick, 1992), which means that any magma bodies that exist are likely to be small (on the order of hundreds of meters) and transitory. Melt is supplied by decompression melting of upwelling mantle. The melt collects along grain boundaries and then migrates into channels and upward into the crust (Figure 7.5). Most of the crust beneath a mid-ocean ridge likely consists of a zone of melt with abundant crystals (a **crystal mush**) that is surrounded by a larger zone of mostly crystallized magma with small pockets of melt (Figure 7.5). The crystallization of the melt as the rifting continues produces a crustal cross-section that is similar to the one described by the ophiolite model.

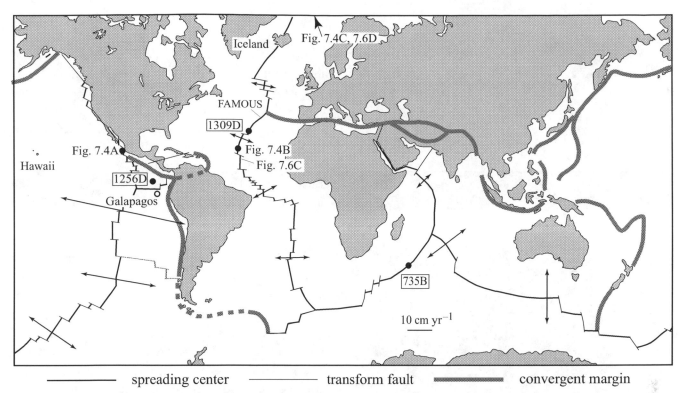

Map 7.2 Tectonic map of the ocean basins, showing mid-ocean ridges, convergent margins, transform faults, and areas discussed in the text. The length of the spreading rate vector arrows is proportional to the spreading rate. Numbers in boxes refer to ODP and IODP drill holes shown in Figures 7.6 and 7.7. Modified from Brown and Mussett (1981) with additional data from Dick et al. (2000), Teagle et al. (2011), and Blackman et al. (2006).

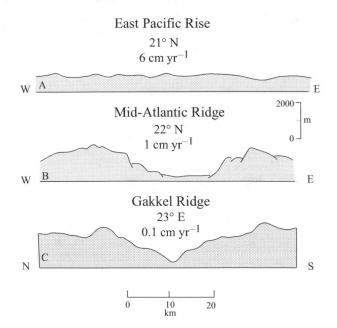

Figure 7.4 Morphology of (A) fast (East Pacific Rise), (B) slow (Mid-Atlantic Ridge), and (C) ultra-slow (Gakkel Ridge, Arctic Ocean) spreading centers. Data from Basaltic Volcanism Study Project (1981), and from Cochran (2008) by permission of Oxford University Press.

Slow- and Ultra-Slow-Spreading Centers

The Mid-Atlantic Ridge is a typical slow-spreading center (half-rate 1–2 cm yr^{-1}), and the Gakkel Ridge under the Arctic Ocean is an ultra-slow-spreading ridge (half-rate 0.1 cm yr^{-1}) (Figure 7.4B, C). The process of melt generation in these spreading centers is similar to that in fast-spreading centers (Figure 7.5) but, because the spreading rate is slower in these rifts, the volume of melt produced along the spreading center is lower than in fast-spreading ridges. As a result, some of the spreading has to be accommodated by extensional faulting. This produces a sea-floor topography that is distinct from that of the fast-spreading centers. Slow- and ultra-slow-spreading centers tend to have a well-defined axial valley. The slow-spreading center is characterized by a 25–30-km-wide axial valley, bounded by mountains. Within this broad valley is a second, well-defined inner valley, 3–9 km wide, where volcanic activity is concentrated (Figure 7.4B, C). Small volcanic hills occur within this inner valley, showing that volcanic activity is neither spatially nor temporally continuous.

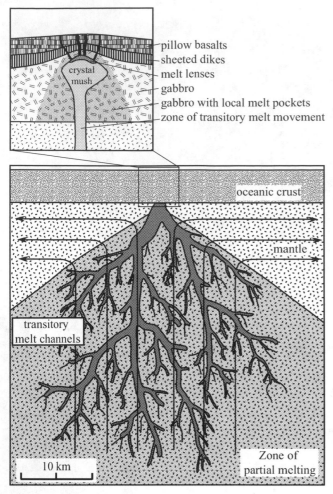

pillow basalts
sheeted dikes
melt lenses
gabbro
gabbro with local melt pockets
zone of transitory melt movement

crystal mush

oceanic crust

mantle

transitory melt channels

Zone of partial melting

10 km

Figure 7.5 Schematic diagram for the generation and emplacement of melts in a fast-spreading ridge. Spreading causes the mantle to flow upward beneath the ridge and to spread outward near the crust–mantle boundary (arrows). Decompression causes melting to occur along grain boundaries. The melt concentrates into melt channels and migrates upward into the crust. Inset shows the emplacement of melt into the base of the ridge, a surrounding mush zone, and a peripheral zone where the host gabbro contains dikes, pods, and small lenses of melt on grain interfaces. Diagram constructed from figures in Perfit et al. (1994) and Braun and Kelemen (2002).

The crust beneath the slow-spreading centers is more poorly layered and more heterogeneous than the ophiolite model predicts (Figure 7.6). Sub-crustal magma bodies beneath slow-spreading centers (Figure 7.6B, C, D) appear to be smaller and more widely spaced than those in fast-spreading centers (Figure 7.6A). The extensional faulting that is characteristic of slow- and ultra-slow-spreading centers may excise some of the units in the ophiolite stratigraphy. In some places, the basalt flows from the

spreading centers are in contact with gabbro; in others, the basalt is in contact with serpentinized peridotite (Figure 7.6 C, D). In many places in slow- and ultra-slow-spreading ridges, this extension has stripped the crust from the mantle, exposing serpentinized peridotite directly on the sea floor (Figure 7.6D).

Four decades of ocean drilling have revealed a great petrologic variability to the oceanic crust (see Box 7.1). Deep drill cores in fast-spreading crust, such as Integrated Ocean Drilling Program (IODP) hole 1256D, show relations similar to those that the ophiolite model predicts (Teagle et al., 2011) (Figure 7.7). However, drill holes into gabbroic crust exposed in slow-spreading ridges (ODP holes 735B and IODP hole 1309D) show relations that are much more complex (Dick et al., 2000; Blackman et al., 2011) (Figure 7.7). Core from hole 735B from the southwest Indian ridge contains mainly gabbro with minor amounts of oxide gabbro. The section is cut by several large, ductile shear zones. In contrast, core from hole 1309D from the Mid-Atlantic Ridge contains a complex series of gabbro and oxide gabbro interlayered with screens of peridotite. Magmatic differentiation produced the oxide gabbro retrieved from holes 735B and 1309D. As explained in Chapter 2, crystallization of olivine (and other iron–magnesium silicates) removed magnesium from a melt preferentially to iron (Figure 2.12B). Eventually, this saturated the melt in iron–titanium oxides (magnetite and ilmenite), producing the oxide gabbro. The presence of multiple oxide gabbro horizons in holes 735B and 1309D means the holes penetrated several discrete igneous bodies, each of which had differentiated to oxide gabbro. A significant amount of troctolite and olivine-rich troctolite is also present and these are inferred to have formed by reaction between the peridotitic mantle and the basalt emplaced into the spreading center (Blackman et al., 2011). Chromium and copper ores are also associated with ophiolites (Box 7.2).

In summary, the data obtained from recent drilling of the sea floor show that, although some or all of the components of an ophiolite may be present at any given locale, oceanic crust is immensely more complex than the ophiolite model implies. Fast-spreading centers produce crustal stratigraphy that is closest to the ophiolite model, but in slow-spreading areas faults truncate the stratigraphy and the gabbroic sequence consists of multiple injections of magma that interacted with the peridotite host rocks and differentiated in place.

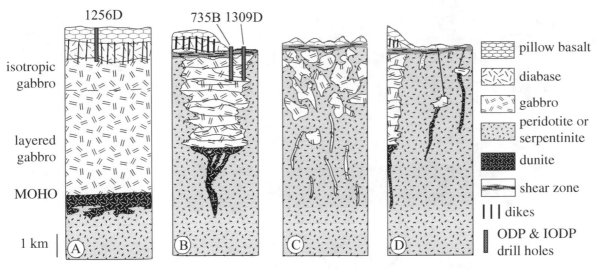

Figure 7.6 Cross-sections of oceanic crust. (A) The typical ophiolite model section as exemplified by the East Pacific Rise, penetrated by IODP hole 1256D. (B) A slow-spreading center, such as along the Mid-Atlantic and southwestern Indian ridge, penetrated by ODP hole 735B and IODP hole 1309D. (C) The cross-section inferred for the area at 23°N on the Mid-Atlantic Ridge. (D) An ultra-slow-spreading center, such as along the Gakkel Ridge in the Arctic Ocean. After Dick et al. (2006).

7.3 Petrography and Geochemistry of Oceanic Magmatism

Oceanic magmatism occurs in two distinct environments: at mid-ocean ridges and at off-ridge locations where ocean islands and oceanic plateaus are formed. In many instances the basalts erupted in each of these two environments are chemically and petrographically distinct. Volumetrically, the most important environment is along mid-ocean ridges where new oceanic crust continually forms as tectonic plates diverge (Map 7.2). The basalt erupted here is olivine and quartz-normative tholeiite; these basalts are referred to as **mid-ocean ridge basalts (MORBs)**. A significant volume of basaltic magma is also erupted from vents not located on ridges; this off-ridge magmatism occurs on ocean islands and oceanic plateaus. The rocks erupted off-ridge are referred to as **ocean island basalts (OIBs)**, and they include both tholeiitic and alkali basalts.

7.3.1 Mid-Ocean Ridge Basalt

The fine-grained groundmass of MORB reflects rapid cooling of magma, extruded into a cold submarine environment. Phenocryst assemblages in glassy basalts suggest that the first minerals to crystallize are olivine and spinel.

As the magma differentiates, plagioclase joins the crystallizing assemblage. Finally, a groundmass consisting of plagioclase + calcium-rich clinopyroxene (augite) + olivine forms. Olivine compositions are typically Fo_{65-91}. The spinel is magnesium- and chromium-rich and is frequently found as inclusions in olivine. Plagioclase is typically An_{88-40} (Grove and Bryan, 1983) and is commonly more calcium-rich in basalts that erupted on the Mid-Atlantic Ridge than those that erupted on the East Pacific Rise. The presence of more sodic plagioclase on the East Pacific Rise suggests that magmas differentiate to a greater extent at fast-spreading centers where larger magma bodies may be present (Hekinian, 1982).

Trace-element characteristics of MORB suggest that the type of mantle source rock from which partial melts are extracted is spinel or plagioclase lherzolite, rather than the high-pressure assemblage, garnet lherzolite. This mineralogy is consistent with geophysical studies of P- and S-wave attenuation, which suggest that the melting begins at depths of 60–80 km, and the melt segregates at about 20 km, to rise and feed shallow magma reservoirs. At minimum, 20 percent partial melting is required to produce the most MgO-rich MORB compositions (Wilson, 1989).

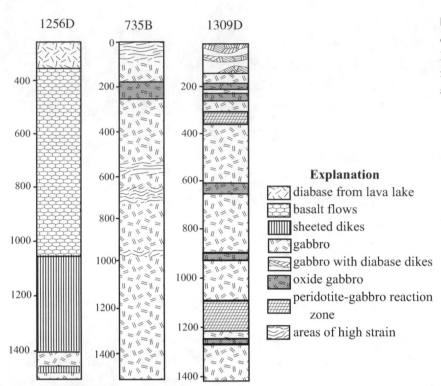

Figure 7.7 Cross-sections of the oceanic crust as obtained in several IODP drill holes. Drill hole locations shown on Map 7.2. Data from Dick et al. (2000), Teagle et al. (2011), and Blackman et al. (2006).

Explanation

- diabase from lava lake
- basalt flows
- sheeted dikes
- gabbro
- gabbro with diabase dikes
- oxide gabbro
- peridotite-gabbro reaction zone
- areas of high strain

MORB melts have a small but significant range in compositions, the origin of which is shown in Figure 7.8. This figure shows the system olivine–clinopyroxene–quartz–plagioclase as projected from anorthite to the olivine–clinopyroxene–quartz plane (see also Figure 6.5). Heavy lines on this pseudoternary projection are the locations of the olivine–clinopyroxene, olivine–orthopyroxene, and orthopyroxene–clinopyroxene cotectics in the presence of an aluminous phase at 1 bar, and 10, 15, 20, and 25 kbar. At 1 bar the aluminous phase is plagioclase, at 10, 15, and 20 kbar it is spinel, and at 25 kbar it is garnet. This diagram clearly illustrates that increasing pressure stabilizes clinopyroxene with respect to olivine and orthopyroxene. This occurs because, at increasing pressure, clinopyroxene is progressively enriched in sodium and aluminum.

As illustrated on Figure 7.5, partial melting of mantle to produce MORB takes place at a range of depths, and, hence, pressures. Consider upwelling dry mantle that begins melting at around 25 kbar. The first melt has a composition that lies on the olivine–orthopyroxene–clinopyroxene "eutectic" for 25 kbar (Figure 7.8). (The quotation marks recognize that in a pseudoternary projection, the eutectic does not lie in the plane of the diagram.) If this high-pressure melt moves to shallower crustal levels (i.e. to lower pressure), for example to a

pressure of 10 kbar, then the original melt will no longer lie on the "eutectic." Rather it will lie in the field of primary olivine crystallization. As olivine crystallizes, it drives the residual melt composition directly away from the olivine apex as the dashed grey arrow originating at olivine and passing through the "eutectic" composition shows. All melts derived by fractional crystallization of olivine from the 25-kbar "eutectic" lie along the dark portion of the arrow. Partial melting at a range of mantle depths and fractional crystallization of olivine during magma ascent will produce the observed compositional range of MORB glasses (gray field in Figure 7.8).

Figure 7.8 implies that the range in compositions of MORB results from a number of processes. First, magmas form by partial melting at various depths, producing parent magmas with a limited but varying composition. These magmas start to crystallize during ascent into the ridge center. As they ascend, magmas originating from different depths mix. These composite melts move to shallow magma chambers where their compositions are further changed by low-pressure differentiation. Finally, observations from IODP hole 1309D suggest the magmas interact with their mantle host rocks at relatively shallow crustal levels and that this process further affects the compositions of MORB (Blackman et al., 2011).

BOX 7.2 | ORE DEPOSITS IN OCEANIC CRUST

Most ophiolites lack economic mineral deposits, but two types of important ore deposits are found locally in ophiolites: chromite and copper-bearing, massive sulfide deposits.

Chromium. Concentrations of chromite are common in the deformed peridotite portion of an ophiolite (see Figure 7.1). In some localities the chromium concentration is rich enough to be ore grade. Because the peridotite and chromite bodies within an ophiolite have been subjected to ductile deformation within the mantle, the ore bodies have a discontinuous, pod-like shape. Hence, they are called *podiform chromite* deposits, to distinguish them from chromite deposits associated with layered mafic intrusions, which are called *stratiform chromite* deposits (see Box 10.1). Although the stratiform deposits are much larger than podiform deposits, podiform deposits contain chromite much richer in chromium, and are therefore valuable despite their smaller size. The chromite deposits of Kazakhstan and Turkey, the second and fourth largest chromium producers in the world, come from ophiolites. (South Africa and India are the first and third largest chromium producers.)

Copper. Associated with pillow basalts in some ophiolites are massive sulfide deposits consisting of pyrite, pyrrhotite, chalcopyrite, and sphalerite. How these deposits formed was a great mystery until the discovery of "black smokers" on the sea floor (Corliss et al., 1979). Geologists now recognize that the intrusion of gabbroic rocks at a ridge crest drives circulation of hydrothermal fluid through the overlying basalts. As these fluids react with the basaltic crust, they extract metals, mostly iron, copper, and zinc, as well as sulfur from the sulfides in the basalt. When these fluids are expelled into the sea, the sulfides precipitate as "black smoke," which cools and deposits as **volcanogenic massive sulfide (VMS) deposits**. The most famous ophiolite-hosted VMS deposits are associated with the Troodos ophiolite in Cyprus. Copper has been mined on Cyprus for 4000 years.

Box Figure 7.2 Copper ingot in the shape of an oxhide, from Zakros, Crete, from the Mediterranean Late Bronze Age. Artifact held by the Heraklion Archeological Museum, Crete. This Wikipedia and Wikimedia Commons image is from the user Chris 73 and is freely available at //commons.wikimedia.org/wiki/File:Minoan_copper_ingot_from_Zakros,_Crete .jpg under the creative commons cc-by-sa 3.0 license.

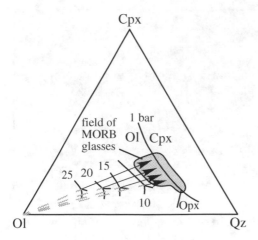

Figure 7.8 Pseudoternary projection from plagioclase onto the olivine–clinopyroxene–quartz plane, showing the composition range of MORB glasses. Also shown are the eutectics for aluminous peridotite at 1 bar and 10, 15, 20, and 25 kbar. See text for discussion. Modified after Elthon (1989).

These multiple processes result in a small but significant range in the major-element composition of erupted MORB. The composition of basalt liquids erupted on the sea floor can be determined from the composition of basalt glass. Compositions of basalt glasses from the Narrowgate region of the FAMOUS valley, Mid-Atlantic Ridge, plotted on Figure 7.9 show that, although SiO_2 contents are relatively constant, MgO varies between 7 percent and 9 percent. As the observed phenocrysts, particularly olivine and augite, crystallize, MgO in the remaining melt decreases. The decrease in Al_2O_3 and CaO with decreasing MgO is consistent with the crystallization of plagioclase and olivine. Crystallization of these phases alone would not result in the observed decrease in CaO; another calcium-rich phase, augite, is required to account for the observed compositional range of basalt glasses. Both FeO^{tot} and TiO_2 behave incompatibly in

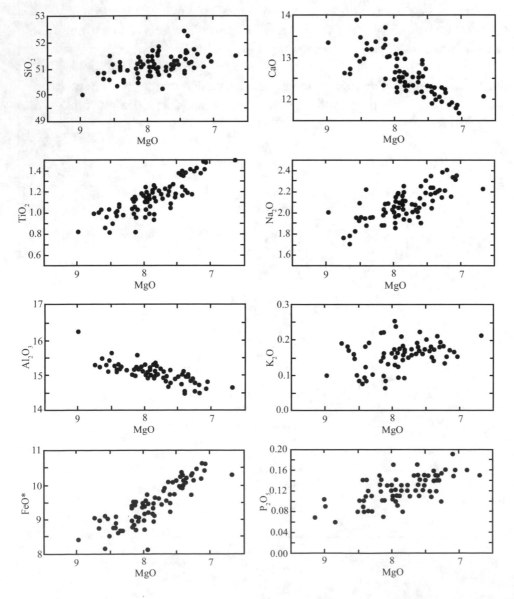

Figure 7.9 Compositions of basalt glasses from the Narrowgate region of the FAMOUS valley, Mid-Atlantic Ridge. Data from Stakes et al. (1984).

Table 7.1 *N-MORB and E-MORB compositions from the Mid-Atlantic Ridge, Iceland, and the Galapagos Islands*

	N-MORB	E-MORB	E-MORB
	Mid-Atlantic Ridge	Iceland	Galapagos Islands
SiO_2	51.15	49.64	47.42
TiO_2	1.12	2.90	2.50
Al_2O_3	15.09	13.37	15.47
FeO (total)	9.46	14.03	11.05
MgO	7.84	5.57	7.76
CaO	12.65	9.89	10.96
Na_2O	2.15	2.90	2.95
K_2O	0.13	0.50	0.51
P_2O_5	0.12	0.49	0.28
Total	98.90	99.28	99.71

Data for Mid-Atlantic Ridge are average values of 32 basalt glasses in the Narrowgate valley of the FAMOUS region (Stakes et al., 1984), data for Iceland are average values of 10 tholeiitic basalts (Carmichael, 1964), data for Galapagos are average values of 19 tholeiitic basalts (McBirney and Williams, 1969).

Table 7.2 *Sr and Nd isotopic compositions of mantle reservoirs identified at the FAMOUS area of the Mid-Atlantic Ridge, Iceland, and the Galapagos Islands*

Mantle reservoir	$^{87}Sr/^{86}Sr$	$^{143}Nd/^{144}Nd$
FAMOUS depleted mantle	0.7029	0.51318
Galapagos depleted mantle	0.7024	0.51317
Iceland depleted mantle	0.7027	0.51318
Galapagos plume-source mantle	0.7033	0.51289
Iceland plume-source mantle	0.7032	0.51300

Data for FAMOUS area from Frey et al. (1993), data for Galapagos from Harpp and White (2001), data for Iceland from Thirlwall et al. (2004).

MORB and increase with decreasing MgO, indicating that neither ilmenite nor Ti-magnetite were crystallizing during the limited fractionation of these rocks. This geochemical data set indicates that minor differentiation occurs during emplacement of MORB onto the sea floor. More extensive differentiation by crystal fractionation would not be expected, given the rapid cooling of the magma in contact with seawater.

There are two spreading centers where partial melting of the mantle has produced enough tholeiitic lava to produce islands that rise above sea level – Iceland and the Galapagos Islands. Iceland is located on the Mid-Atlantic Ridge, and the Galapagos Islands sit slightly off-axis of the Galapagos spreading center (Map 7.2). Both localities represent places where basalts formed by upwelling and decompression of mantle at a spreading center and are also affected by a complex interaction with a mantle plume or hot spot (Harpp et al., 2002; O'Connor et al., 2007). Whereas basalt generated at mid-ocean ridges is commonly considered "normal," or N-MORB, basalts at locations on or near hot spots exhibit higher concentrations of incompatible elements and may be referred to as "enriched," or E-MORB. Inspection of Table 7.1 reveals the lower silica and higher alkali contents of Iceland and Galapagos E-

MORB, and enrichment in incompatible elements such as titanium and phosphorous. N-MORB and E-MORB also have different radiogenic isotopic compositions. N-MORB has lower $^{87}Sr/^{86}Sr$ and higher $^{143}Nd/^{144}Nd$ ratios than E-MORB (Table 7.2). The isotopic composition of N-MORB is inherited from the isotopic composition of the shallow upper mantle from which it partially melts, a region that is depleted in incompatible elements relative to the mantle as a whole. The isotopic composition of E-MORB reflects involvement of other, deeper mantle reservoirs that supply mantle plumes. These reservoirs have higher $^{87}Sr/^{86}Sr$ and lower $^{143}Nd/^{144}Nd$ ratios than the "normal" mantle that supplies ocean ridges. Multiple mantle reservoirs with different isotopic compositions are required to explain all the variations in Sr, Nd, and Pb isotopic compositions of hot-spot basalts. Plume interaction with a spreading center, as for Galapagos and Iceland, produces a range in isotopic compositions of basalt that reflect different proportions of plume and depleted mantle sources.

Although dominated by basalt, both Iceland and the Galapagos contain volcanoes that have erupted lavas with a wide range of compositions, from basalt to rhyolite (Carmichael, 1964; McBirney and Williams, 1969). Lavas from both centers show similar differentiation trends (Figure 7.10). The magma undergoes extensive differentiation, and a large change in $Fe^{tot}/(Fe^{tot} + MgO)$ ratio coincides with a minimal change in silica content. This trend results from the crystallization of olivine and pyroxenes without participation of iron–titanium oxides. Such differentiation enriches the magma in iron and leads to the formation of ferrobasalt, a basalt with more than 13 percent FeO^{tot} and less than

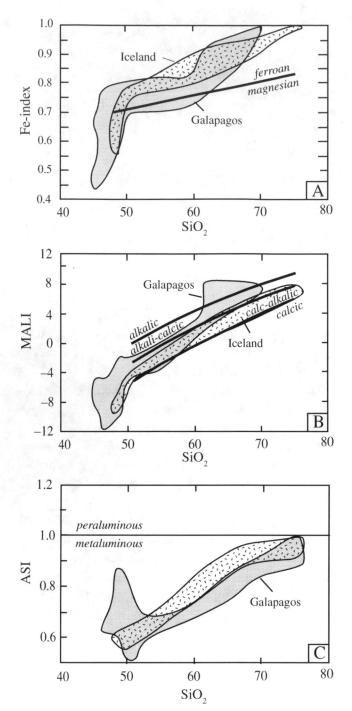

Figure 7.10 Geochemical trends of volcanoes on the Galapagos Islands and Iceland, both of which straddle oceanic ridges. (A) Fe-index versus SiO_2. (B) MALI versus SiO_2. (C) ASI versus SiO_2. Data from McBirney and Williams (1969) and Carmichael (1964).

6 percent MgO (McBirney and Williams, 1969). As such, the rhyolites formed by differentiation of these basalts are ferron, metaluminous, and calc-alkalic, in Iceland, to alkali-calcic and alkalic, in the Galapagos (Figure 7.10).

7.3.2 Off-Ridge Magmatism

Off-ridge magmatism falls into three broad categories:

1. *Seamounts.* These submarine volcanic structures either never grow enough to breach sea level or, if they do, they are eroded and subside. In the tropics, they may be capped with coral reef deposits, but as the volcano subsides only a guyot (a flat-topped seamount that lies just below sea level) remains. Seamounts are most abundant in the Pacific, where they number between 22 000 and 55 000 (Batiza, 1982). Many seamounts appear along fracture zones, which may have provided conduits for the magma. Others form linear chains, which show a progressive age relationship, and which suggest a genetic relationship with oceanic island volcanoes.

2. *Oceanic island volcanoes.* These immense volcanoes rise up 10 000 m above the ocean floor and have dimensions greater than the largest mountains on the continents. Usually an ocean island volcano has several centers, suggesting the focus of magmatic activity migrates with time. Ocean island volcanoes may be single islands, or in fast-spreading oceans like the Pacific, they may form linear chains. A good example is the Hawaii–Emperor chain, a chain of seamounts and subaerial oceanic islands that stretches from Hawaii nearly to Kamchatka. In this chain, the oldest volcanoes lie at the northwest end, and the active, but still submarine, volcano of Loihi is situated at the east end. Locations of some ocean islands are shown in Map 7.3.

3. *Oceanic plateaus.* Oceanic plateaus are topographic highs within ocean basins that have an area of several hundred square kilomters and rise an average of a thousand meters above the ocean floor. Large, well-studied oceanic plateaus include the Ontong Java plateau in the western Pacific Ocean and the Kerguelen plateau in the southern Indian Ocean (Map 7.3). Oceanic plateaus lie on oceanic crust that may have been thickened to between 10 km and 35 km thick. Many form at or near mid-ocean spreading centers, and appear to have formed by immense, short-lived eruptions of tholeiitic basalt (Kerr, 2004). Because of their thickness, they are not easily subducted and, instead, fragments of oceanic plateaus may accrete to continental margins. For example, the Ontong Java plateau collided with the Solomon Islands (Neal et al., 1977; Petterson et al., 1997), and the Caribbean plateau collided with northwestern South America (Kerr et al., 1997). Oceanic plateaus are similar to continental flood basalts (see Chapter 9) in that both

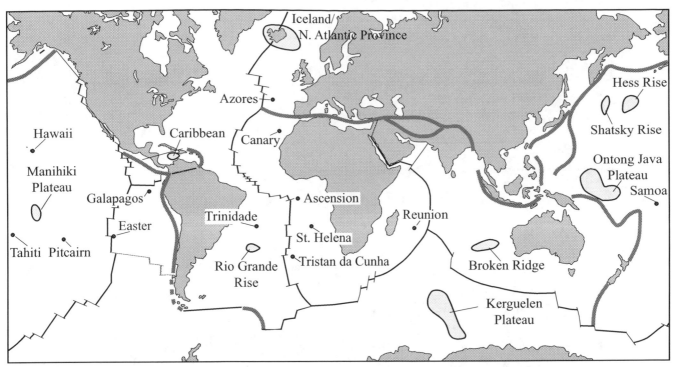

Map 7.3 Map showing the location of oceanic islands (points) and oceanic plateau (shaded). Plateau locations are after Coffin and Eldholm (1992).

involve large, rapid outpourings of basalt. Both are referred to as **large igneous provinces**, sometimes abbreviated to "LIPs" (Coffin and Eldholm, 1992).

A plume or hot-spot model appears to explain many of the intraplate volcanic features of the ocean floor, especially many oceanic island volcanoes and plateaus. Plumes may originate from a thermal boundary layer at the core–mantle boundary, or from the base of the upper mantle at a depth of 670 km. The rising plume of solid material undergoes decompression melting as it shallows. The composition of the basalt magma depends on the depth and extent of melting and the composition of the mantle diapir. A short-lived voluminous eruption of this basalt may form an oceanic plateau, whereas a plume that produces magma over a longer period of time will build an oceanic island volcano. As plate motion carries the overlying oceanic crust across the plume, the site of volcanic activity shifts to that part of the ocean floor that is directly above the plume. In this way, a chain of hot-spot volcanoes develops across the oceanic crust at the pace of plate motion.

Hawaii: An Example of an Oceanic Island Volcano

The Hawaii–Emperor island chain is over 6000 km long. The easternmost island, Hawaii, is the only volcanically active island, although to its east the submarine volcanic center of Loihi is developing (Map 7.4). Five overlapping shield volcanoes built the island of Hawaii, of which only Kilauea and Mauna Loa are active. Mauna Loa, at 4170 m above sea level, and Mauna Kea, at 4205 m, have the highest relief above base level of any mountain on Earth (>10 000 m), which indicates the huge amounts of magma involved in their formation. The focus of volcanism on Hawaii moves 8 cm a year, which is essentially identical to the movement rate of the Pacific plate. A long-lived, more-or-less stationary hot spot appears to explain the spatial relationships of Hawaiian volcanism very well.

Structure. The Hawaiian islands are associated with a large-amplitude, free-air gravity anomaly, which can be explained in terms of a huge volcanic load that down-warps the oceanic lithosphere by as much as 5 km. The crust beneath Hawaii is 15–20 km thick, as opposed to 5–6 km in the adjacent Pacific, and the lithosphere thickness is estimated to be around 90 km (Forsyth, 1977). The distribution of magma-related earthquake hypocenters has been used to construct the three-dimensional layout of the magma chamber and feeder conduit for the active volcanic center of Kilauea. It appears that melting occurs below 60 km depth, and this magma is transported up through narrow conduits to shallow magma chambers (Wright, 1984; Clague and Sherrod, 2014). The main magma storage area is a zone, 3–20 km^3 in volume, which extends from 3 km to 5 km in depth. As magma fills the magma reservoirs the summit of the volcano inflates, as is

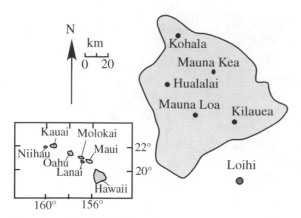

Map 7.4 The island of Hawaii is a compound volcanic edifice formed by five overlapping shield volcanoes.

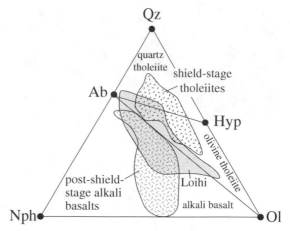

Figure 7.11 Comparison of the compositions of early Hawaiian basalts (as exemplified by Loihi), the main shield-building-stage tholeiites, and the post-shield-stage alkali basalts projected onto the normative Qz–Nph–Ol pseudoternary phase diagram. Qz = quartz, Nph = nepheline, Ol = olivine, Ab = albite, Hyp = hypersthene. Data from Muir and Tilley (1957), MacDonald (1968), Frey and Clague (1983), Hawkins and Melchior (1983), Wilkinson and Hensel (1988), Fodor et al. (1992), West et al. (1992), and Ren et al. (2009).

measured by tilt meters, global positioning system (GPS), and interferometric synthetic-aperture radar (InSAR) (Poland et al., 2014). After a major eruption, it takes a period of months for the Kilauea magma chamber to refill (Ryan et al., 1981; Dzurisin et al., 1984).

Evolution of an Oceanic Island Volcano. Nearly all oceanic island volcanoes exhibit a complex pattern of igneous evolution, although the pattern is not the same for all ocean islands. At Hawaii, four characteristic stages can be identified (Clague and Sherrod, 2014). The first, the pre-shield stage, is seen at Loihi, the seamount, south of Hawaii, constructed of alkali basalt and tholeiite (Figure 7.11). The second, the shield-building stage, consists of large volumes of tholeiitic basalt (Figure 7.11). As volcanism wanes, tephra deposits from scoria cones and domes become more abundant and the eruptive rocks more alkalic in composition (Figure 7.11). The volume of magma produced during this post-shield stage is only about 1 percent of the total production of the volcano. After a long period of dormancy and erosion, a rejuvenated stage of volcanism begins. Again, the eruptions are explosive, and the magmas produced are highly alkaline and silica-poor. The rock types produced are alkali basalts and nepheline-bearing basalts. The amount of time between stages varies from volcano to volcano and can be between 100 000 years and 2 million years. On the island of Hawaii, Mauna Loa and Kilauea are in the shield-building stage, while two older volcanoes (Mauna Kea, Hualalai) are in the post-shield stage. Kohala's youngest post-shield lavas are around 0.12 Ma but the volcano has no rejuvenated products. Hazards associated with ocean island volcanism are rarely life-threatening (Box 7.3).

Petrography and Geochemistry. Hawaii is a little unusual compared to other oceanic island volcanoes in the large volume of tholeiite and small volume of alkaline

rock it has produced. Most other hot spots, such as the Azores, St. Helena, and Tristan da Cunha, have a larger proportion of alkaline rocks. The petrography of oceanic island tholeiites is similar to MORB, although ocean island tholeiites may have orthopyroxene as well as olivine, spinel, clinopyroxene, and iron–titanium oxides. In terms of major-element chemistry, these tholeiites have higher K_2O and TiO_2 relative to MORB, but lower Al_2O_3 contents. Alkali basalts commonly contain ultramafic xenoliths, whereas ultramafic xenoliths are rare in tholeiites.

Magma Sources for Hawaii Volcanism. Neodymium, strontium, and lead isotopic compositions from Hawaiian volcanic centers vary significantly, and the variations correlate with geographic location. These inter-volcano isotopic differences strongly suggest that no single mantle source of magma can produce all the rock types found there. The Hawaiian plume appears to be heterogeneous, bringing up various mantle components that are identified by their different isotopic compositions. Most plume components are thought to originate from the lower mantle, where different types of ancient lithosphere that was subducted into the mantle are stored. A depleted upper-mantle component also is identified in Hawaiian magmas. This material may be incorporated as the upwelling plume interacts with the upper-mantle asthenosphere through which it passes (Clague and

BOX 7.3 | VOLCANIC HAZARDS FROM OCEANIC VOLCANISM

Despite the fact that oceanic volcanism is by far the most voluminous magmatism on Earth, it is associated with comparatively low volcanic hazard. This is partly because most eruptions take place under the ocean where eruptions are isolated from the atmosphere by a kilometer or more of seawater. Equally important, oceanic magmas are dominated by highly fluid basalt, which generally lacks the explosive nature of arc magmatism. However, oceanic magmatism is not without its hazards. The huge volcanic edifaces that compose ocean islands are prone to infrequent but giant landslides. Over 15 submarine landslides, among the largest known on Earth, have affected the Hawaiian islands and caused both loss of land and catastrophic tsunami, recorded as far away as Australia. Oceanic island volcanism can be disruptive, as any air traveler to Europe in the spring of 2011 could tell you. The eruption of the Eyjafjallajökull volcano in Iceland ejected a plume of ash that interrupted air travel to western Europe for weeks. This ash plume was small compared to the fissure eruption of Laki, also on Iceland, in 1783. Laki is the largest fissure eruption in history, and the volcanic gases emitted by the eruption killed around 80 percent of the livestock on the island, as well as approximately a quarter of the population because of famine and fluorine poisoning.

Box Figure 7.3 Lava from Kilauea engulfing the Kalapana Gardens Subdivision in 1990, causing residents to abandon their community as it was consumed by the slow-moving flows. Photo from United States Geological Survey Hawaii Volcano Observatory, http://hvo.wr.usgs.gov/kilauea/history/1990Kalapana/.

The islanders living on Hawaii are accustomed to the volcanic hazards of the island, which they attribute to activity of the goddess Pele. Volcanic eruptions and associated destruction of human structures are part of Hawaii's history. In one of Pele's repeated tricks, slow-moving lava flows from Kilauea's East Rift overran homes and other buildings in 1955 and 1960, and from 1983 to 2014 destroyed the town of Kalapana Gardens. Over a period of eight months in 1990, 214 homes were destroyed in that area. Lava flows erupting from East Rift fissures in May 2018 engulfed 700 homes in the eastern part of the Big Island within the first two months of renewed activity. Although property damage is considerable, no lives have been lost. One of the owners of a house engulfed in lava in 2012 explained that he stayed until the last minute because "it is very easy to outrun a lava flow."

Sherrod, 2014). Current research focuses on the identification of the end-member components, their chemical and isotopic characteristics, and how these different components are located spatially in the mantle source region beneath Hawaii.

Ontong Java: An Example of an Oceanic Plateau

The Ontong Java plateau is the largest known oceanic plateau, with a volume of 60 million km^3 occupying an area of 2.0 million km^2 of Pacific sea floor (Fitton et al., 2004). It is composed of tholeiitic basalt, most of which was erupted in a major episode at 122 Ma. The Ontong Java plateau crust averages 30–35 km thick and the plateau sits 2–3 km above the surrounding sea floor today. To generate such a large quantity of basalt over such a short interval, petrologists call upon decompression melting of

an anomalously hot ($T > 1500$ °C) mantle diapir (Fitton et al., 2004).

The Ontong Java plateau is composed of two major basaltic units: the Kwaimbaita-type and the Kroenke-type. The Kroenke-type basalts overlie the Kwaimbaita-type basalts in two ODP drill holes. The two basalt types are isotopically indistinguishable and have similar trace-element compositions, suggesting that they came from a similar source (Fitton and Godard, 2004). The Kwaimbaita-type basalts have moderate amounts of MgO and are compositionally similar to the MORB glasses from the FAMOUS region. Like MORBs, the major phenocrysts consist of olivine, augite, and plagioclase. In contrast, the Kroenke-type basalts are enriched in MgO (Figure 7.12) and contain only olivine and minor chromite phenocrysts (Sano and Yamashita, 2004).

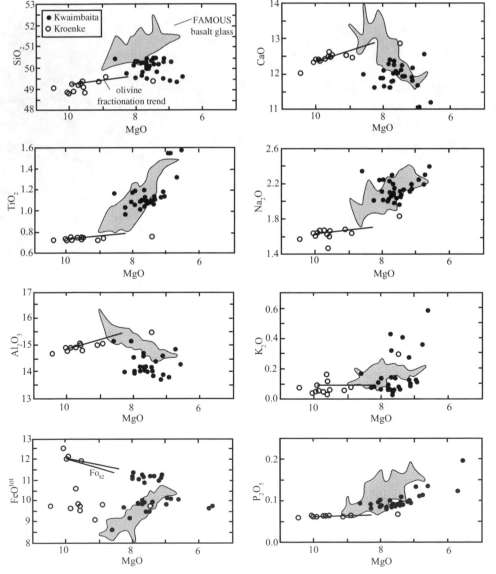

Figure 7.12 Major-element composition of the Kwaimbaita and Kroenke basalts from the Ontong Java plateau compared to basaltic glass from the FAMOUS region of the Mid-Atlantic Ridge (shaded fields, data from Figure 7.9). Line shows trend for olivine fractionation, suggesting that the Kwaimbaita basalts could have been derived from the Kroenke basalts by olivine fractionation. Data from Fitton and Godard (2004).

Petrologists propose that the Kwaimbaita-type basalts could have been derived from basalts similar to the Kroenke-type. Just as olivine fractionation plays a major role in the compositional variation of MORBs (see Figure 7.9), it may explain the relationship of the Kwaimbaita-type basalts to the Kroenke-type basalts.

Extensive partial melting of more than 30 percent in a hot and voluminous rising mantle diapir would have produced a high-MgO basalt like the Kroenke-type basalts. Olivine fractionation, following the trends shown on Figure 7.12, would produce basalts of a typical tholeiitic composition, similar to MORB (Fitton et al., 2004).

Summary

- Ophiolites are commonly used as a simplified model for oceanic crust. The stratigraphy of an ophiolite from top to bottom is: sea-floor sediment, pillow basalt, sheeted dikes, gabbro, cumulate peridotite, and deformed peridotite.

- Ophiolites may form in back-arc spreading centers, in suprasubduction-zone settings, or in mid-ocean spreading centers.

- Oceanic crust is more complex than the ophiolite model suggests. Parts of the crust may be excised by faults, and the gabbroic horizon may involve multiple injections of melt that may interact with mantle host peridotite and evolve toward oxide gabbro.

- Ocean islands and oceanic plateaus form above mantle plumes ("hot spots"). Basaltic magma forms by decompression melting within rising mantle diapirs. The composition of the basalt magma varies depending on the depth and extent of partial melting and the composition of the mantle involved.

- Tholeiites are found in MORB, ocean islands, and oceanic plateaus.

- Alkali basalts are found on many ocean islands, and on some of these islands they have evolved to extremely alkaline magmas.

Questions and Problems

Problem 7.1. Compare and contrast fast mid-ocean spreading centers with slow- and ultra-slow-spreading centers. Include in your comparison:

- Topography and dimensions of the ridge;

- Stratigraphy and structure of the rocks produced;

- Differences in plagioclase compositions; and

- Magma volumes and differentiation histories.

Problem 7.2. How can hot-spot tracks be used to determine plate motion? What assumptions are involved?

Problem 7.3. What types of basalt are found at mid-ocean ridges? At oceanic islands? What might account for any differences?

Problem 7.4. Plot data for MORB glasses from the FAMOUS area and basalts from Galapagos on Fe-index, MALI, and ASI diagrams. How do the two groups of basalts differ? How might these differences relate to the presence or absence of mantle-plume activity? (Data for this problem can be downloaded from the book's website at www.cambridge.org/frost.)

Further Reading

Basaltic Volcanism Study Project, 1981, *Basaltic Volcanism on the Terrestrial Planets*. Pergamon Press, New York, NY, sections 1.2.5, 1.2.6, 6.2.1, and 6.2.2.

Dilek, Y. and Furnes, H., 2014, Ophiolites and their origins. *Elements*, 10, 93–100.

Kerr, A., 2004, Oceanic plateaus. In *Treatise on Geochemistry*, eds. H. D. Holland and K. K. Turekian. Elsevier, Amsterdam, vol. 3, pp. 537–65.

Kerr, A. C., 2005, Oceanic LIPs: The kiss of death. *Elements*, 1, 289–92.

Saunders, A. D., 2005, Large igneous provinces: Origin and environmental consequences. *Elements*, 1, 259–64.

Wignall, P., 2005, The link between large igneous provinces and mass extinctions. *Elements*, 1, 293–7.

Winter, J. D., 2010, *Principles of Igneous and Metamorphic Petrology*, 2nd edn. Prentice Hall, New York, NY, Chapter 13: Mid-ocean ridge volcanism and Chapter 14: Oceanic intraplate volcanism.

Chapter 8

Convergent-Margin Magmatism

8.1 Introduction

Subduction produces some of the major topographic features on Earth and consumes large amounts of oceanic crust each year. At modern rates, it would take only 160 million years to subduct an area equal to that of the entire surface of Earth. Regardless of whether the overriding plate is oceanic or continental, convergent plate margins share many of the same characteristics. These include a deep (6000–11 000 m deep) oceanic trench marking the plate boundary, chains of volcanoes on the overriding plate located about 100–200 km inboard from the trench, and a dipping zone of seismicity called the **Benioff zone**, which includes shallow, intermediate, and deep-focus earthquakes. The occurrence of a zone of earthquakes associated with oceanic trenches was first noted by Hugo Benioff (1949) but the significance of these earthquakes was poorly understood for two decades. The development of the plate tectonic theory in the 1960s and early 1970s led to the realization that the Benioff zones mark the plane of descent of the oceanic lithosphere into the mantle in subduction zones. The volcanoes and plutonic rocks above the subduction zones are constructed of magmas that range from basalt to rhyolite, with andesite the dominant composition. Volcanism in this tectonic setting is frequently highly explosive. This, coupled with the fact that large population centers are located in the shadow of many of these volcanoes, makes study of arc magmatism important for hazard prediction. The volcanic and plutonic rocks formed at convergent margins are also relevant to the study of the growth and evolution of the continental crust. Insofar as convergent-margin magmas transfer material from the mantle to the crust, they represent a mechanism by which continents form and grow. The volume and composition of crust formed at subduction zones is therefore of considerable interest to geologists who study the formation and development of continental crust over Earth's history.

This chapter introduces the main features of oceanic and continental arc magmatism using well-studied examples of arc volcanic and plutonic complexes. The petrography and geochemistry of island and continental arcs provide important clues to the petrogenesis of arc magmas, although the details of the process remain incompletely understood.

8.2 Oceanic and Continental Arcs

Convergent-margin magmatism can be divided into two groups, even though the overall tectonic setting is similar in each. First, *island arc magmatism* generates melts in response to subduction beneath an upper plate composed of oceanic crust. The second is *continental arc magmatism*, wherein subduction zone magmas ascend through and interact with continental crust. Although many of the magmatic processes are the same in both environments, the resulting magma suites have somewhat different compositions that relate to whether the magmas encountered continental crust during their ascent into and through the overriding plate.

8.2.1 Island Arc Magmatism

Subduction of one oceanic plate beneath another has produced currently active oceanic island arcs, including the Aleutian, Kurile, Marianas, Ryuku, Sunda–Banda, Solomon, New Hebrides, and Tonga–Kermadec arcs of the western Pacific; the Lesser Antilles of the Caribbean; and Scotia in the southern Atlantic Ocean (Map 8.1). In these young island arcs the plutonic rocks are rarely exposed, so that most of the petrologic information available comes from extrusive materials. In some island arc terrains, erosion has exposed their plutonic roots. A good example is Tobago in the West Indies, where the crust has tilted to expose a cross-section that includes arc basement and plutonic rocks underlying a volcanic sequence (Frost and Snoke, 1989; Snoke et al., 2001).

8.2.2 Continental Arc Magmatism

Whereas oceanic island arcs are formed by the subduction of one oceanic plate beneath another, continental-margin magmatism results from the more complex tectonic environment in which the overriding plate is continental. Magmas generated in this environment today occur along the west coast of North and South America, Japan, New Zealand, and along the Aegean Sea (Map 8.1). The eruptions of both continental and island arc volcanoes represent a significant geologic hazard (Box 8.1).

Not only are volcanic edifices common in continental arcs, some arcs have been eroded deeply enough to expose granitic batholiths that formed beneath the volcanoes. These batholiths represent some of the most voluminous granitic intrusions in the world. Late Mesozoic subduction along the western margin of North and South America produced several large granitic batholiths, including the

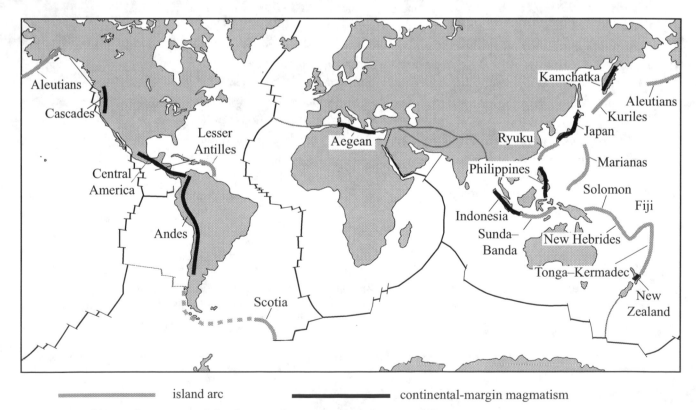

Map 8.1 World map showing active island arcs and continental-margin magmatism.

BOX 8.1 | VOLCANIC HAZARDS

There are approximately 600 active arc volcanoes in the world. When these volcanoes erupt they may cause loss of life and damage to property by a number of different processes. Some destruction is a direct result of the eruption: ash falls, pyroclastic flows, debris avalanches, explosions, and emission of volcanic gases and acid rain all may cause damage and death. Other hazards are indirect: the volcanic eruptions may trigger earthquakes, tsunamis, and post-eruption famine. Fortunately, volcanic eruptions are relatively infrequent events and they involve less economic loss and human casualties than other natural hazards such as floods, hurricanes, and earthquakes. Nevertheless, there have been 32 eruptions since 1000 CE that killed more than 300 people. Moreover, approximately 10 percent of the world's people live near potentially dangerous volcanoes (Tilling, 1989).

The **Volcanic Explosivity Index (VEI)** was developed to provide an estimate of the magnitude of volcanic eruptions. It is a logarithmic index based upon the volume of material ejected by an eruption (Newhall and Self, 1982). The scale ranges from 0 for small, non-explosive eruptions of lava to 8 for huge, paroxysmal eruptions of pyroclastic material and injection of significant amounts of ash into the stratosphere (Box Table 8.1). As one might expect, the larger the eruption the less frequent it is likely to be. Mason et al. (2004) identified five eruptions with a VEI of 7 or greater in the last 10 000 years, the most recent being Tambora in 1815. There has been no eruption with a VEI of 8 in that time. The entrainment of ash from Tambora into the stratosphere caused the following summer to be much cooler than usual. In the past 2 million years, only six eruptions with VEIs of 8 or greater have occurred, and they were cataclysmic. Eruptions on this order of magnitude may have caused "volcanic winters," with significant effects on life on the planet.

Box Table 8.1 *Volcanic Explosivity Index*

VEI	Volume	Plume height	Frequency	Example	Death toll
0	$<10^4$ m^3	<100 m	Constant	Kilauea	4 since 1900
1	$>10^4$ m^3	100–1000 m	Daily	Nyiragongo (2010)	245
2	$>10^6$ m^3	1–5 km	Weekly	Galeras (1993)	9
3	$>10^7$ m^3	3–15 km	Few months	Nevado del Ruiz (1985)	23 000
4	>0.1 km^3	10–25 km	$\leq$yr	Eyjafjallajökull (2010)	0
5	>1 km^3	20–35 km	$\leq$10 yr	Mount Saint Helens (1980)	57
6	>10 km^3	>30 km	$\leq$100 yr	Mt. Pinatubo (1990)	700
7	>100 km^3	>40 km	$\leq$1000 yr	Tambora (1815)	92 000
8	>1000 km^3	>50 km	$\leq$10 000 yr	Toba (70 000 yr BP)	Unknown

Coast Range batholith of western Canada, the Sierra Nevada batholith of California, and the Peninsular Ranges batholith of California and Baja California. The magmatic arc along the western margin of North America has been more deeply eroded than the corresponding arc in South America: huge andesite volcanoes still cap much of the Andes. These large batholiths, known as Cordilleran batholiths after the extensive mountain chains along western North and South America, are not single intrusions but rather composite bodies made up of numerous plutons that

---

BOX 8.2 | PORPHYRY COPPER DEPOSITS

A major economic characteristic of arc magmatism is the occurrence of porphyry copper deposits. These deposits supply nearly three-fourths of the world's copper, half of the molybdenum, and around one-fifth of the gold (Sillitoe, 2010). Porphyry copper deposits are associated with shallow intrusions in arc settings. Typical examples include the Bingham mine in Utah, El Teniente in Chile, and the Ok Tedi Mine in Papua New Guinea. There is no consistent relationship between porphyry deposits and the composition of the host rocks, which may range from diorite to granodiorite in calc-alkalic suites and from diorite to syenite in more alkalic suites (McMillan and Panteleyev, 1988). Instead, the common feature of the host plutons for porphyry copper deposits is their shallow level of emplacement. The shallow emplacement level means that, during crystallization of the igneous pluton, the aqueous fluids exsolved from the magma undergo a large volume increase (see Figure 3.4). This causes hydrofracturing of the rocks within, above, and around the intrusion, allowing circulation of fluids through the rock. Because transition metals such as copper, molybdenum, and gold and elements such as chlorine and sulfur behave incompatibly in silicate melts, these become enriched during the fractionation of the melt and will, in turn, fractionate into the exsolving fluids. The metals, which are dissolved in the fluid as chloride complexes, are transported through the fractured rocks and precipitated as the fluids become neutralized by reaction with the country rock (see Chapter 17).

---

range in composition. Arc batholiths are important as a source of metals, including copper, molybdenum, and gold (Box 8.2).

8.2.3 Structure of Island and Continental Arcs

Cross-sections through island and continental arcs can be divided into four regions: the trench, fore-arc, arc, and back-arc. Arcs are marked by a distinctive negative gravity anomaly over the trench that is paired with a positive anomaly over the fore-arc (Figure 8.1). The negative gravity anomaly near the trench is caused by the presence of relatively light, water-saturated sediments in the fore-arc, which are scraped off the downgoing plate to form an **accretionary prism**, whereas the positive gravity anomaly over the fore-arc reflects the presence of cold, dense subducted lithosphere beneath the arc. The down-going slab in subduction zones is marked by the Benioff zone earthquakes, whose epicenters lie along the top of its plunging surface. The trench and fore-arc are marked by low heat flow, whereas the arc and back-arc are characterized by high heat flow. The low heat flow over the fore-arc is produced by the cold slab that lies beneath it. The high heat flow over the arc and back-arc is caused by the heat carried to high crustal levels by hot magma.

8.2.4 Examples of Island and Continental Arcs

Island Arc Volcano: Seguam, Aleutian Islands, Alaska

Seguam, an island in the middle of the Aleutian chain, is dominated by Pyre Peak, the highest of the young volcanic edifices on the island (Figure 8.2). In this portion of the chain, modern volcanoes are built atop Eocene to Miocene arc crust (Singer et al., 1992). The subaerial lavas of Seguam consist of three major eruptive phases (Map 8.2). The oldest of these three units is the Turf Point Formation, which consists of flows that range from 1.07 Ma to 0.07 Ma. Overlying this, the Finch Cove Formation ranges in age from 0.08 Ma to 0.03 Ma. The deposits from the youngest eruptive phase compose the Holocene Pyre Peak Formation, which occupies the western half of the island. The most recent volcanic activity on Seguam occurred in May 1993.

The geochemical studies of Seguam were conducted on the Turf Point Formation where the wave-cut cliffs on the island shore expose a cross-section through the eruptive sequence. The Turf Point Formation consists of about 70 percent basalt, 15 percent andesite, and 15 percent dacite. Similar abundances of these rocks are found in both the Finch Cove and Pyre Peak Formations.

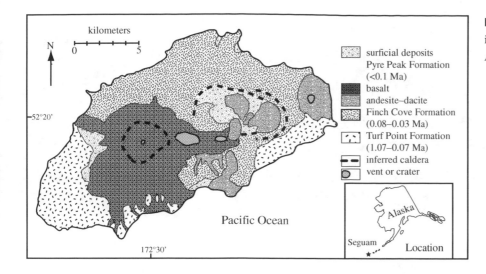

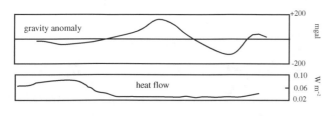

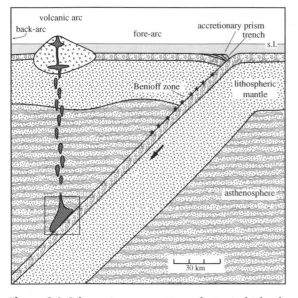

Figure 8.1 Schematic cross-section of a typical island arc. The graphs above show the gravity and heat flow profiles across the arc. Stars indicate locations of earthquake epicenters. s.l. = sea level. Gravity anomaly generalized from Watts and Talwani (1974). Heat flow generalized from Manga et al. (2012).

Figure 8.2 Photo of Pyre Peak, Seguam island, Alaska. Note characteristic steep-sided cone of the composite volcano breached by a summit caldera. Photo by Brad Singer.

Island Arc Plutonic Complex: Tobago, West Indies

The island of Tobago lies at the northeast corner of the South American shelf in the southern Caribbean. The island preserves a crustal cross-section through a 105–103-Ma (Albian) oceanic island arc (Snoke et al., 2001). The arc is built on older, metamorphosed, and deformed Cretaceous arc rocks, referred to as the North Coast Schist (Map 8.3). Both plutonic rocks and volcanic rocks of the Albian arc are exposed. The plutonic rocks of the oceanic arc include ultramafic rocks, gabbro and diorite, and a small volume of tonalite. The ultramafic rocks include dunite, wehrlite, olivine clinopyroxenite, and hornblendite. The gabbroic rocks include olivine melagabbro, hornblende gabbro, and gabbronorite. Mineralogical layering in the gabbro unit has been interpreted to result from crystal accumulation, and texturally the ultramafic and gabbroic rocks appear to be cumulate rocks.

Parts of the plutonic complex intruded and contact metamorphosed the volcanic rocks of the Tobago plutonic–volcanic complex, which include volcaniclastic

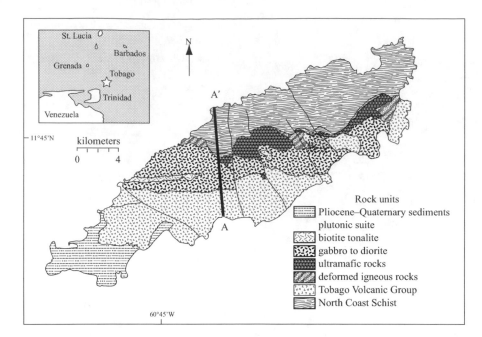

Map 8.3 Geologic map of Tobago, West Indies, showing a cross-section through the Cretaceous oceanic arc complex that exposes both plutonic and volcanic rocks. From Frost and Snoke (1989).

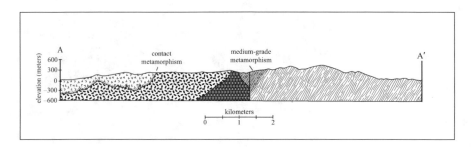

breccias and lava flows. Volcanogenic mudstone and sandstone are also found within the volcanic sequence. Both plutonic and volcanic rocks are cross-cut by a suite of mafic dikes interpreted by Frost and Snoke (1989) as similar in composition to the basaltic parent magmas from which the Tobago plutonic and volcanic rocks formed.

Continental Arc Volcano: Mount Saint Helens, Washington

Before its explosive eruption on May 18, 1980, Mount Saint Helens, located in southern Washington State, was a stratovolcano rivaling Mount Fuji for its symmetry (Figure 8.3). The geologic map of Mount Saint Helens before the eruption displays many of the features distinctive of continental arc volcanoes (Map 8.4). One characteristic is that the Mount Saint Helens volcano erupted basalt, andesite, and dacite lavas. The basalt flows extended up to 20 km from the vent (Map 8.4). The fluid basalt formed lava tubes and flowed through a thick forest of large trees, typical of the Pacific Northwest rain

forest, solidifying around the tree trunks. After the trees rotted away, the flow was littered with huge holes, several feet in diameter, where the trees used to stand. The andesite flows were much more viscous than the basalt flows. Most of the andesite froze on the slopes of the mountain, although some flowed about 8 km from the vent. Dacite volcanism formed domes, which seldom exceed 1 km in diameter. In addition to forming domes, the dacite volcanism and, to a lesser extent, the andesite volcanism produced pyroclastic flows that flowed down the mountain and into the valleys of the Toutle and Lewis rivers, which drain the north and south sides of the volcano, respectively.

Even before a magnitude 4.1 earthquake on March 15, 1980 marked the awakening of Mount Saint Helens, the volcano was recognized as the most active volcano in the Cascades arc, which extends from northern California to southern British Columbia. Geologists had determined that over the past 40 000 years Mount Saint Helens had erupted lavas, emplaced dacitic domes, produced pyroclastic flows, and emitted ash (Figure 8.4). Several events

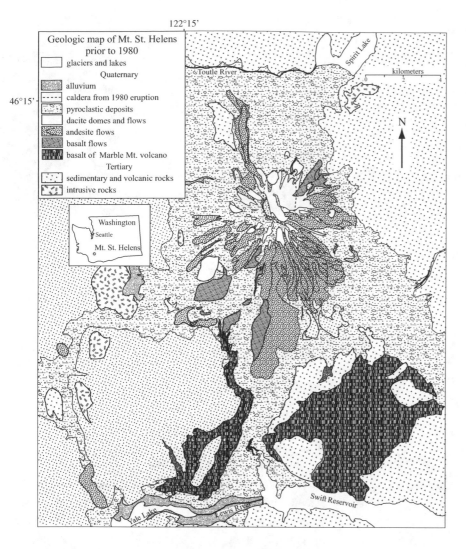

122°15'

Geologic map of Mt. St. Helens
prior to 1980

glaciers and lakes
Quaternary
alluvium
caldera from 1980 eruption
pyroclastic deposits
dacite domes and flows
andesite flows
basalt flows
basalt of Marble Mt. volcano
Tertiary
sedimentary and volcanic rocks
intrusive rocks

46°15'

Washington
Seattle
Mt. St. Helens

Map 8.4 Geologic map of Mount Saint Helens, Cascade Range, Washington, prior to the 1980 eruption. Heavy dashed line shows the extent of the caldera produced by that eruption. Inset shows location in the state of Washington. Modified after Hopson (2008).

Figure 8.3 Mount Saint Helens prior to the eruption of May 1980.

had occurred within the last 200 years. Because of extensive petrologic study and an intensive monitoring program, the 1980 eruption of Mount Saint Helens was predicted successfully, minimizing loss of life (Box 8.3).

Continental Arc (Cordilleran) Batholith: The Tuolumne Intrusive Suite

A classic example of a Cordilleran batholith is the Tuolumne intrusive suite in the Sierra Nevada batholith, which is spectacularly exposed in Yosemite National Park in California (Figure 8.5). The Tuolumne intrusive suite makes up only a small portion of the immense Sierra Nevada batholith, which is over 500 km long by 50–80 km wide and is composed of up to 200 separate plutons. The majority of exposed rock in the Tuolumne intrusive suite is granodiorite, rather than true granite (Map 8.5). Many other Cordilleran batholiths are similarly dominated by granodiorite. Another important feature of the Tuolumne intrusive suite is that it is composite, meaning that it formed from multiple intrusive episodes. Zircon U–Pb geochronology reveals that the batholith was emplaced over a period of 10 million years. Detailed field mapping indicates that the batholith was constructed

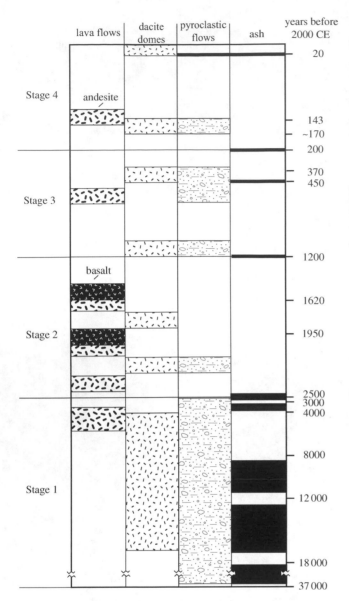

Figure 8.4 Diagram showing the geologic history of Mount Saint Helens volcano. Modified after Hopson (2008).

Figure 8.5 Granodiorite of the Tuolumne intrusive suite in Yosemite National Park, California. Photo by Arthur W. Snoke.

incrementally from a series of ephemeral magma chambers composed of melt and crystals (Figure 8.6). The occurrence of abundant xenoliths along the margins of the individual intrusions indicates that large amounts (up to around a third) of the earlier intrusions were consumed by stoping or by partial melting during the intrusion of later magmas (Paterson et al., 2016).

8.3 Petrographic Characteristics of Island and Continental Arc Rocks

8.3.1 Petrography of Island Arc Rocks

The Turf Point Formation lavas on Seguam are a calcic island arc suite that span a compositional range from basalt to rhyodacite. This suite consists of an anhydrous phenocryst assemblage of plagioclase, olivine, clinopyroxene, orthopyroxene, magnetite, and rare ilmenite. The proportions of the phenocryst phases vary between basalts, basaltic andesites, andesites, and dacites, but even the dacites and rhyodacites lack hydrous minerals such as hornblende and biotite (Singer et al., 1992).

The calcic lavas of Seguam differ from those of other Aleutian volcanoes and from most of the volcanoes of the Pacific and Lesser Antilles arcs, which tend to be calc-alkalic. Many calc-alkalic island arc suites are dominated by voluminous, two-pyroxene andesites. Such rocks are typically phenocryst-rich, containing an assemblage of orthopyroxene, augite, and plagioclase. Plagioclase usually is andesine and commonly shows complex zoning. Olivine phenocrysts may occur, but they are rarely in equilibrium with the pyroxenes and may be xenocrysts from the mantle. Some andesites may contain phenocrysts of hornblende in addition to, or even to the exclusion of, the pyroxenes. Plagioclase tends to be stubby (see Figure 1.7C). In addition to andesite, calcic and calc-alkalic island arcs commonly erupt small volumes of evolved dacite. Like the andesites, dacitic lavas from these suites are phenocryst-rich. The key phenocryst for dacite is quartz, which occurs with complexly zoned andesine. Ferromagnesian minerals may include pyroxenes, hornblende, cummingtonite, or biotite.

Alkali-calcic to alkalic, high-potassium series island arc rocks are around 50 percent basalt, 40 percent andesite, and 10 percent dacite. They are distinguished from calc-alkalic suites by a higher abundance of biotite.

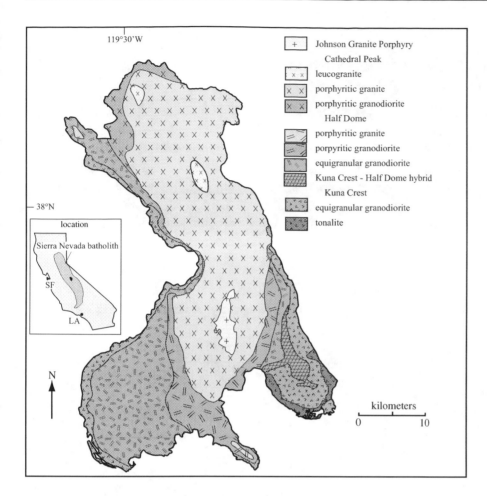

Map legend:

+ Johnson Granite Porphyry
Cathedral Peak
x x leucogranite
x x porphyritic granite
x x porphyritic granodiorite
Half Dome
porphyritic granite
porpyritic granodiorite
equigranular granodiorite
Kuna Crest - Half Dome hybrid
Kuna Crest
equigranular granodiorite
tonalite

Map 8.5 Geologic map of the Tuolumne intrusive suite from Memeti et al. (2010). Inset shows the location of the intrusive suite within the Sierra Nevada batholith, and the location of the Sierra Nevada batholith in California.

Alkalic island arc suites occur in Fiji and Sunda in the Pacific, and in Grenada in the southern Lesser Antilles (Map 8.1).

The plutonic rocks associated with modern island arc volcanoes are rarely exposed, but older complexes, such as Tobago, indicate that the plutonic roots of island arc volcanoes are composed of ultramafic to gabbroic cumulate rocks, with lesser amounts of diorite and tonalite (Frost and Snoke, 1989). Some mafic layered intrusions appear to represent the roots of arc volcanoes. The Duke Island mafic layered intrusion in southeastern Alaska, described in Chapter 10, is interpreted as a sub-arc intrusion. Other examples include the Protero-zoic Mullen Creek and Lake Owen complexes of south-eastern Wyoming, which are composed of mafic and ultramafic cyclic units that suggest repeated intrusion of mafic magma into a sub-arc magma chamber (Premo and Loucks, 2000). Other plutonic arc rocks, such as the Jurassic Smartville complex of northern California, are dominated by unzoned gabbros or zoned plutons composed of olivine gabbro in the core with quartz

diorite rims (Beard and Day, 1988). A common feature in all these examples of island arc plutons is the preser-vation of both cumulate and differentiated rocks in the complexes.

8.3.2 Petrography of Continental Arc Rocks

Magma series erupted along continental margins are compositionally similar to those erupted in island arcs, except that arcs erupting through continents have a greater abundance of silica-rich rock types. Much of this felsic material occurs as pyroclastic flow deposits. Continental-margin magmatism contains significant amounts of andesite and dacite, which are petrologically similar to rocks of those compositions that erupted in island arcs. Continental-margin magmatic arcs may also contain rhyolite, which may be distinguished from dacite by the presence of phenocrystic sanidine.

Plutonic rocks are more commonly exposed in continental-margin magmatic arcs than in island arcs. Rock types are generally gabbro, diorite, tonalite, granodiorite, and granite. The minerals characteristic of

BOX 8.3 | **VOLCANIC HAZARDS ASSOCIATED WITH CONVERGENT MARGIN MAGMATISM: EXPLOSIVE VOLCANISM AND PYROCLASTIC DEPOSITS**

The intermediate to felsic magmas of magmatic arcs are characterized by high water contents and high viscosity (see Chapter 3). These magmas lead to explosive volcanism, in contrast to the relatively quiescent basaltic volcanism of oceanic magmas. Explosive magmatism produces pyroclastic deposits, which may form by multiple mechanisms. One mode of origin is collapse of a growing lava dome. Growing lava domes are unstable, and commonly break up to form landslides. If the melt is close to water saturation at the time the landslide forms, sudden decompression of the underlying magma can lead to explosion, which triggers an avalanche of hot blocks, ash, and gas. Transported individual blocks can reach tens of meters in diameter. The pyroclastic flow deposits of Mount Saint Helens associated with the climactic eruption on May 18, 1980 formed by this mechanism (Christiansen and Peterson, 1981).

Pyroclastic flow deposits also may form by collapse of a vertical eruption column. Magma is disrupted by expanding gases within the conduit and discharges pumice, ash, and gases as a mixture into the air.

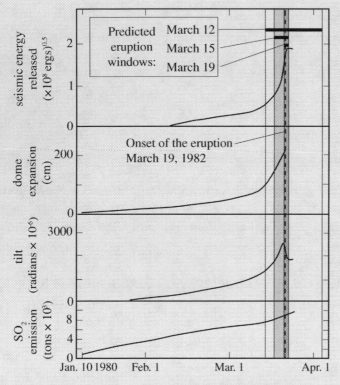

Box Figure 8.3 Volcanic monitoring capabilities improved substantially following the May 1980 eruption of Mount Saint Helens. Evidence of potential new volcanic activity at Mount Saint Helens became increasingly compelling through the early months of 1982, when seismic activity, dome expansion, inflation of the edifice, and SO_2 emissions increased. Geologists issued a series of warnings of a volcanic eruption with successively narrowing predictive windows, all of which accurately predicted the March 19 event. Modified from Swanson et al. (1985).

A turbulent jet structure is formed, mixing cold air into the sides of the column. This air is heated by the ash, expanding the column and lowering its density. The mixture of ash and gas is almost always denser than the surrounding atmosphere when first discharged. In addition, the amount of kinetic energy available from gas expansion is only enough to discharge the mixture to a few kilometers height at most. To produce a column 30–50 km high, as is typical for continental arc volcanoes, the discharged mixture must absorb and heat enough air to lower the density of the mixture to less than that of the atmosphere. The mixture is then buoyant and rises as a convective plume, like smoke from a forest fire or an exhaust pipe. In columns where the whole or part of the mixture is still denser than the atmosphere by the time all kinetic energy is lost, collapse occurs and forms a density current of ash and hot gases. Because gases have very low viscosity and density, the particles settle rapidly to form a dense avalanche or pyroclastic flow. These flows can travel several tens of kilometers at velocities up 10 to 300 m s^{-1}. The July 22, 1980 eruptions of Mount Saint Helens included pyroclastic flows of this origin (Christiansen and Peterson, 1981).

Arc volcanoes, including Mount Saint Helens, have produced massive eruptions (VEI scale 5 and 6; see Box 8.1), many of which have resulted in extreme loss of life. Relatively recent eruptions on this scale include Krakatoa in 1883, Katmai in 1911, Mount Pinatubo in 1990, and Volcán de Fuego in 2018. Because volcanic eruptions can have disastrous consequences for people living near the volcano, petrologists and volcanologists seek to understand the plumbing of volcanoes, to characterize and mitigate volcano hazards. Because Mount Saint Helens was so close to population, geologists studied the development of the volcanic edifice attentively. Researchers established seismographs around the volcano to sense the movement of magma, and tiltmeters around the edifice to recognize how the mountain inflated as the magma moved into the shallower plumbing. These measurements provided the basis for a series of successively narrowing predictive windows that proved remarkably accurate (Swanson et al., 1985; Tilling, 1989; Box Figure 8.3).

The information obtained from the eruption of Mount Saint Helens became indispensable in predicting the eruption of Mount Pinatubo 10 years later. Scientists from the Philippine Institute of Volcanology and Seismology, working closely with those from the United States Geological Survey, were able to monitor emissions of volcanic gases and to use earthquake data to follow the movement of magma into the volcano from a depth of 32 km. They used this information to predict the eruption far enough in advance that people living on the volcano's slopes were able to evacuate before the eruption. These efforts saved an estimated 5000 lives and at least $250 million in property (Newhall et al., 1997). These results indicate that in the future, geologists will be able to predict volcanic eruptions with sufficient accuracy to minimize the effects of eruptions of arc volcanoes threatening cities from Seattle to Tokyo. Hazard forecasts could give the millions of people who live in the shadows of these beautiful mountains enough time to avoid the immediate effects of volcanic eruptions.

these rocks are plagioclase, alkali feldspar, quartz, amphibole, biotite, magnetite, and ilmenite. Pyroxenes, both clinopyroxene and orthopyroxene, may be found locally. Titanite and apatite are common accessory minerals, even in the mafic rocks, whereas allanite is common in highly differentiated granites. As with most slowly cooled rocks, there is evidence of subsolidus growth of minerals such as biotite, amphibole, and chlorite due to the interaction of solid rocks with high-temperature hydrothermal fluids. Grain boundaries also may be altered by post-crystallization reactions, resulting in a sutured texture.

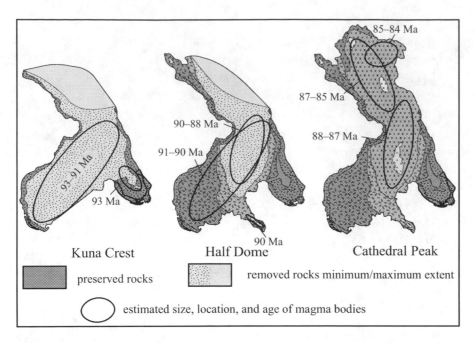

Figure 8.6 Diagram showing the minimum (stippled) and maximum (shaded) areas of the Kuna Crest, Half Dome, and Cathedral Peak intrusions, resulting in the present shape of the Tuolumene intrusive suite. Black ellipses show estimated age, and size, of magma chambers in each of the units. Modified after Paterson et al. (2016).

Pyroxene is dominantly augite, sometimes joined by orthopyroxene in intermediate-composition plutonic rocks. In many diorites, clinopyroxene is rimmed by hornblende. Amphibole is mainly hornblende and, unlike in the volcanic rocks where pyroxenes are common, hornblende is abundant and often euhedral in plutonic rocks, reflecting the fact that the plutonic rocks crystallized with higher water activity than the volcanic rocks. The color of the hornblende ranges from green to brown and tends to correlate with increasing amounts of TiO_2. Biotite is a common mafic mineral in granitic rocks. Plagioclase is the major rock-forming mineral in nearly all plutonic rocks. It is often complexly zoned (see Figure 1.8C), and myrmekite is common. Orthoclase is the most common alkali feldspar, whereas microcline is found in the most differentiated rocks and forms under volatile-rich conditions. Exsolution textures are common, and magmatic fluid may be the cause of perthite coarsening (Parsons, 1978). Granophyric intergrowths are characteristic of the most differentiated granitic rocks, which form from the most volatile-rich magmas. Granophyres may form under conditions of supercooling, either by reduction in temperature as the magma rises in the crust, or by sudden loss of volatiles from the system (Vernon, 2004). Magnetite and ilmenite are the major opaque oxides.

8.4 Geochemical Characteristics of Convergent-Margin Magma Series

8.4.1 Comparison of Oceanic and Arc Differentiation Trends

The tholeiitic basalts from island arc environments are similar in terms of major-element composition to oceanic basalts but they follow a different differentiation trend. This is best illustrated on a plot of the iron-enrichment index (Fe-index) versus SiO_2 (Figure 8.7). During differentiation, tholeiitic melts from oceanic environments become enriched in iron relative to magnesium. The ferromagnesian silicates in the lavas become increasingly enriched in their iron end members as differentiation progresses. Only late in the differentiation history of oceanic tholeiites do the melts undergo silica enrichment (see data for the Galapagos and Iceland in Figure 7.10). In contrast, most arc magma suites show strong enrichment in silica with increasing differentiation, but the ferromagnesian silicates show only moderate increases in the Fe/(Fe+Mg) ratio. In terms of the classification discussed in Chapter 4, oceanic magmas are typically ferroan, whereas those from magmatic arcs are magnesian.

These different trends are caused by the fact that the oceanic tholeiitic magmas tend to be more reducing than the arc magmas, and as a result iron–titanium oxides crystallize relatively late in tholeiites. The crystallization of

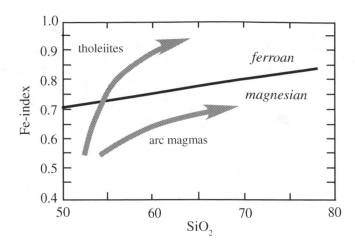

Figure 8.7 Plot of $FeO^{tot}/(FeO^{tot} + MgO)$ showing the different differentiation trends for arc magmas and oceanic tholeiites.

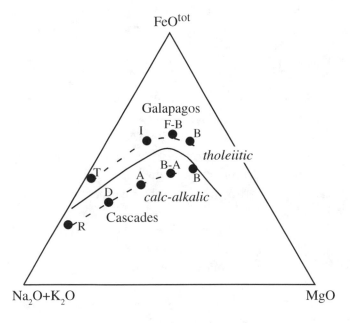

Figure 8.8 Alkalis–FeO^{tot}–MgO (AFM) diagram comparing the differentiation trends of tholeiitic rocks (as indicated by the lavas of the Galapagos) and calc-alkalic rocks (as indicated by the lavas of the Cascades). A = andesite, B = basalt, B-A = basaltic andesite, D = dacite, F-B = ferrobasalt, I = icelandite, R = rhyolite, T = trachyte. From Irvine and Baragar (1971) and McBirney and Williams (1969).

olivine extracts magnesium preferentially to iron and this causes the melts to become iron-enriched, a feature noted earlier in the formation of oxide gabbros in oceanic crust. In contrast, in arc magmas, iron–titanium oxides crystallize relatively early, inhibiting the iron enrichment of the residual magma and causing the evolving magmas to become enriched in silica (Frost and Lindsley, 1992).

The two differentiation trends are often shown in a ternary diagram, plotting $(Na_2O + K_2O)$–$FeO^{(tot)}$–MgO (Figure 8.8). On such a diagram differentiation drives the oceanic magmas (magmas from the Galapagos as an example) toward the iron apex before alkalis become enriched. In contrast the magmatic arc magmas (often called "calc-alkalic," *sensu lato*), as exemplified by magmas of Mount Saint Helens and other Cascade volcanoes, evolve toward the alkali apex without increases in FeO.

8.4.2 Comparison of Island and Continental Arc Magma Series

Although island and continental arc magma series both tend to be magnesian, there are some important differences in alkali content. This is shown in Figure 8.9, where fields for Seguam, Tobago, Mount Saint Helens, and Tuolumne all follow magnesian differentiation trends and are metaluminous, becoming peraluminous with higher silica contents (Figure 8.9A, C), but Seguam and Tobago are calcic whereas Mount Saint Helens and Tuolumne are calc-alkalic (Figure 8.9B). Arc volcanoes also may be characterized according to their trends on a plot of K_2O versus SiO_2 (Figure 8.10). A few arc volcanoes, such as St. Kitts, are poor in K_2O. These magmas are referred to in Figure 8.10 as the low-potassium (low-K) series and are calcic by the modified alkali–lime index (MALI) classification. Some arcs, such as those of Papua New Guinea, are potassic and define a high-K series. These are alkali-calcic by the MALI classification. Most arc volcanoes, including Mount Saint Helens and the more differentiated lavas of Seguam, fall into the medium-K series. As noted in Figure 8.10, some researchers have called this medium-K group the "calc-alkalic" series even though Seguam is calcic (Figure 8.9). Volcanic rocks of this composition are so common in arcs that the term "calc-alkalic" commonly is used to describe island arc magmas in general. However, because the term has a strictly defined geochemical connotation and because some arc suites are calcic and some are alkali-calcic, it is not advisable to use the word "calc-alkalic" loosely to refer to island arc magmatism.

There are several explanations for the variations in K_2O in arc lavas. One is based on the observation that in some arcs there is a spatial pattern along the axis of the arc. In the Lesser Antilles, for example, the lavas vary from low-K magmas at St. Kitts at the

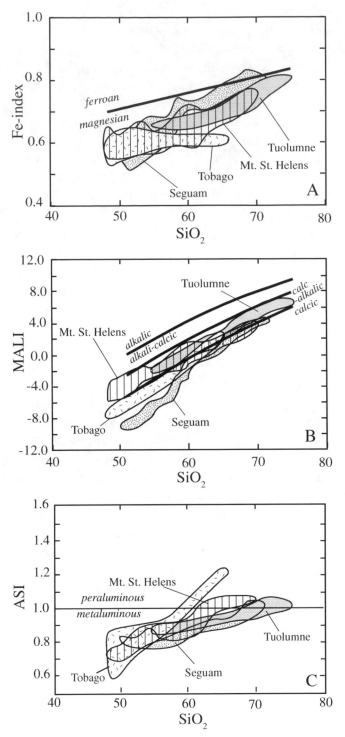

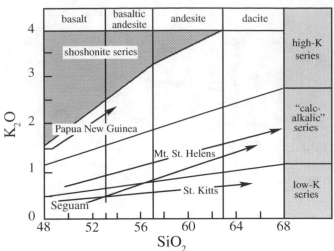

Figure 8.10 Plot of weight percent SiO_2 versus weight percent K_2O showing the composition variation of volcanic rocks from various island arcs. Data from Brown et al. (1977), Johnson et al. (1978), Halliday et al. (1983), Singer et al. (1992), and Smith and Leeman (1993).

correlates with the amount of sediment being eroded off the South American continent, which increases southward toward South America. This sediment, which is a rich source of K_2O, is carried into the trench. A portion of the sediment is subducted and carried to depth where it can be incorporated into arc magmas.

Another pattern observed in some arcs, including the Quaternary High Cascade volcanoes of Washington and Oregon and many Quaternary island arcs of the circum-Pacific, is a correlation between K_2O content and a volcano's distance from the subduction zone (Dickinson, 1970). The alkali content of volcanoes in an island arc tends to increase with increasing depth to the Benioff zone. Volcanoes erupting close to the trench tend to be calcic in terms of the modified alkali–lime index (MALI; see Chapter 4). Volcanoes erupting further away from a trench are calc-alkalic, whereas those that are erupted farthest from a trench are alkali-calcic. This change may be related to depth of melting, since K_2O increases in the magma with increasing pressure (Marsh and Carmichael, 1974).

In comparison to island arcs, calcic suites are uncommon along active continental margins. Instead, calc-alkalic suites are most typical. As with island arcs, the K_2O content of continental-margin volcanism increases inland from the coast, with alkalic suites occurring on the landward side of the volcanic front.

Figure 8.9 The chemical trends of island arc rocks, represented by Seguam and Tobago, compared to continental arc rocks from Mount Saint Helens and Tuolumne. See text for details. Data from Bateman and Chappell (1979), Halliday et al. (1983), Frost and Snoke (1989), Singer et al. (1992), and Smith and Leeman (1993).

northern end of the arc, to "calc-alkalic" magmas in the center of the island chain, to high-K lavas at the southern end at Grenada (Brown et al., 1977). This variation

8.4.3 Comparison of Oceanic and Continental Arc (Cordilleran) Plutonic Complexes

As the Tobago plutonic complex documents (Map 8.3), island arc plutonic rocks are dominated by ultramafic and gabbroic cumulates, with only small volumes of tonalite. By contrast, continental arc plutons are composed mainly of granodiorite with subsidiary granite and quartz diorite. On the QAP diagram, the oceanic arc plutons like Tobago define a slightly different differentiation trend from the continental arc batholiths such as Tuolumne (Figure 8.11). Tuolumne follows a differentiation trend of quartz diorite to granodiorite, with minor early gabbro or diorite and late-forming granite (the trend marked C-A). Oceanic arc plutons like Tobago have lower abundance of alkalis; they are less likely to contain granodiorite or granite and typically follow a trend from gabbro and diorite to quartz diorite to tonalite (marked C in Figure 8.11).

The divergent differentiation trends of oceanic and continental arc plutonic complexes illustrated on the QAP diagram are also evident on plots of Fe-index and MALI (Figure 8.12). Although both Tobago and Tuolumne are magnesian, Tobago is dominated by calcic rocks and Tuolumne by calc-alkalic ones. The ultramafic and gabbroic cumulates exposed on Tobago account for the lower silica contents of that island arc's plutonic complex compared to the continental arc rocks that extend to more siliceous compositions.

Also shown in Figure 8.12 is the compositional range of the Sierra Nevada batholith, of which Tuolumne is a small part. The granitic rocks of the batholith are dominantly magnesian, although a few of the most siliceous granites are ferroan. The granitic rocks define a large field on a MALI diagram that covers the range from calcic to alkalic (Figure 8.12B). This broad range in composition correlates to geographic location. Plutons in the western part of the Sierra Nevada tend to be calcic, those in the core Sierra Nevada are calc-alkalic, and those lying in the eastern Sierra Nevada are alkali-calcic or alkalic. As was observed for arc volcanic rocks, this change in composition may reflect a shift to more alkalic compositions farther inboard from the subduction zone. In addition, the inboard plutons may have incorporated more continental crust, which also is indicated by their increase in initial Sr isotopic composition from west to east (see Figure 5.10).

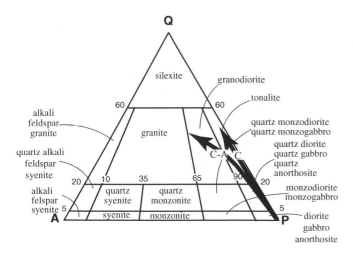

Figure 8.11 QAP diagram showing the compositional ranges of rocks found in calcic (C) and calc-alkalic (C-A) batholiths (after Frost and Frost, 2008).

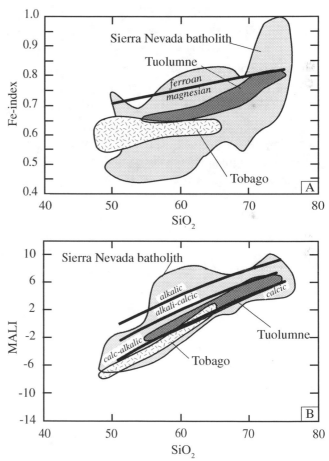

Figure 8.12 The chemical trends, Fe-index (A) and MALI (B), of plutonic island and continental arc rocks as represented by Tobago and Tuolumne. Also shown is the field for the Sierra Nevada batholith as a whole. Data from Bateman and Chappell (1979), Halliday et al. (1983), Singer et al. (1992), Smith and Leeman (1993), and Frost et al. (2001).

8.4.4 Geochemical and Isotopic Identification of Contrasting Processes Forming Seguam and Mount Saint Helens

Although both Seguam and Mount Saint Helens have erupted basalt, andesite, and dacite, the two rock suites developed by very different processes. Geochemical studies (Halliday et al., 1983; Smith and Leeman, 1993) strongly suggest the range of rock compositions from Mount Saint Helens formed from mixing two melts: a mantle-derived basaltic melt and a crustally derived dacite melt. In contrast, Singer et al. (1992) concluded that Seguam formed by closed-system fractional crystallization.

Evidence for these different processes is seen in geochemical variation diagrams (Figure 8.13A, B). The plots of TiO_2 and FeO^{tot} abundances from Seguam define a distinct inflection at 60 percent SiO_2, with TiO_2 increasing with increasing silica up to this inflection point. The inflection indicates the appearance of iron–titanium oxides (ilmenite or titanomagnetite) on the liquidus, after which point iron and titanium are removed from the melt by incorporation into crystallizing oxides. In contrast, the TiO_2 and FeO concentrations in rocks from Mount Saint Helens form straight trends on variation diagrams, a feature explained by mixing basaltic and dacitic magma. This hypothesis is supported by Sr isotopic data (Figure 8.13C). As described in Chapter 5, in a closed system, the $^{87}Sr/^{86}Sr$ ratio will increase over time as ^{87}Rb decays to ^{87}Sr. The decay rate for ^{87}Rb is slow enough that the $^{87}Sr/^{86}Sr$ ratio for a magma varies little over the lifetime of a volcano. Figure 8.13C shows that the $^{87}Sr/^{86}Sr$ ratio of lavas from Seguam remains constant regardless of the silica content of the rock, a pattern consistent with closed-system fractionation of a magma with a particular $^{87}Sr/^{86}Sr$ ratio. By contrast, the $^{87}Sr/^{86}Sr$ ratio for rocks from Mount Saint Helens increases with increasing SiO_2. This pattern would be expected if a dacitic melt with a relatively high $^{87}Sr/^{86}Sr$ ratio mixed with a basaltic melt with a low $^{87}Sr/^{86}Sr$ signature.

This evidence led Halliday et al. (1983) and Smith and Leeman (1993) to conclude that the various rock types on Mount Saint Helens formed from mixed magmas. Intrusion of hot basaltic magma melted mafic country rock to produce a dacitic melt. The two initial melts, the crustal dacite melt and the basalt mantle-derived melt, then mixed in various proportions to produce the andesites. On the other hand, Singer et al. (1992)

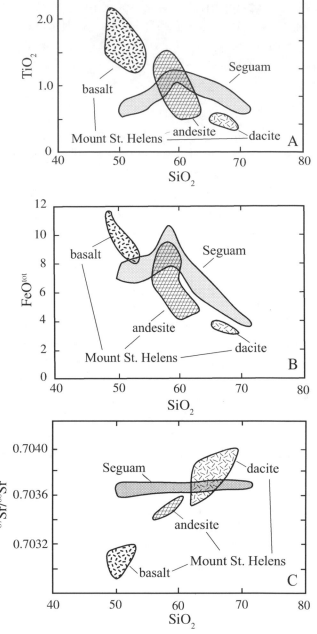

Figure 8.13 Variation diagrams showing the differences in trends exhibited by Seguam and Mount Saint Helens volcanic rocks, which reflect different processes of magmatic differentiation. See text for details. Data from Singer et al. (1992), Halliday et al. (1983), and Smith and Leeman (1993).

concluded that Seguam formed from mafic melt that differentiated in a closed system to form the more evolved rocks, including the andesites, dacites, and rhyodacites.

8.5 Magma Generation at Convergent Margins

The sources of arc magmas are the subject of long and continuing debate. Compared to oceanic magmas at ridges or within oceanic plates, a more diverse combination of materials may be involved or, in some way, influence the magmas in arc settings (see Figure 8.1):

1. *Seawater* is the ultimate source for most of the H_2O that appears in island arc magmas. It is trapped in pore spaces in sea-floor sediment, gets incorporated into oceanic crust during metamorphism, and is subducted along with the oceanic crust.

2. *Oceanic crust* consists of basalt, gabbro, and sea-floor sediments including clays, carbonates, and chert. Oceanic crust may be incorporated into arc magmas in two environments. Subducted oceanic crust could melt at depth, or the oceanic crust on which the arc volcano is built could melt as magmas pass through it.

3. *The mantle wedge above the subducted slab* consists of two components. One is the mantle lithosphere, 40–70 km thick, which is depleted in some constituents because mid-ocean ridge basalt (MORB) melts have already been extracted. These depleted rocks are unlikely to melt readily. The other component is the asthenospheric mantle, which consists of peridotite, likely to be more fertile than the lithosphere mantle, and which could partially melt if the asthenosphere was to interact with fluids or melts discharged by the subduction zone.

4. *Continental crust.* Unlike island arc magmas, magmas formed by subduction along continental margins must pass into and through continental crust. Many of the compositional differences between continental-margin magmas and island arc magmas can be attributed to the fact that continental-margin magmas must pass through a ~50-km-thick section of continental crust.

8.5.1 Primary Arc Magma-Forming Processes

As was described in Section 7.2, the subducting oceanic plate, around 5–7 km thick, is composed of a sedimentary veneer of varying thickness, basalt and gabbro of the oceanic crust, and underlying ultramafic mantle lithosphere. These rocks will have been altered and hydrated by interaction with seawater on the ocean floor, prior to subduction, a process described in more detail in Section 19.5. As the plate enters the trench, some of the oceanic sediment may be mechanically eroded to form an accretionary prism. The remainder is subducted along with the oceanic crust and underlying mantle lithosphere. Deformation of the downgoing plate produces a complex zone of blocks and slabs of subducted rock mixed with overlying mantle along the slab–mantle interface (Figure 8.14). The subducting plate also undergoes metamorphic reactions in response to increased temperature and pressure, and a series of fluids and melts are released. Initially, these are water-rich, as hydrous minerals in the altered ocean crust break down. Although the subducting ocean crust loses most of its initial water content before it reaches depths of around 60 km, dehydration reactions continue to sub-arc depths. In addition to hydrous fluids, the subducting plate also may partially melt. Fluid-saturated partial melting is especially likely in subducted warm, young oceanic crust. Fluid-absent melting of amphibole and biotite also takes place at sub-arc pressures and temperatures. The distinction between fluid and melt breaks down at pressures over 50–60 kbar, where a continuum exists between fluids and melts and a single supercritical liquid phase is present (Schmidt and Poli, 2014).

Interaction of these slab-derived fluids and melts with the overlying mantle wedge may promote partial melting of fertile mantle peridotite. As this magma rises through the mantle wedge, additional partial melt may form by decompression melting. Together, these processes produce magma of basaltic composition. This basaltic magma ascends to the base of the crust of the overriding plate by buoyancy-driven upwelling.

8.5.2 Evolution of Arc Magmas During Ascent Through the Crust

Arc magmas undergo evolution throughout their ascent through the crust, differentiating and assimilating wall rocks at all levels, from the base of the crust to shallow, sub-volcanic depths (Figure 8.15). The incremental evolution of arc magma during its passage through the crust starts at the crust–mantle transition, where the basalt accumulates and starts to differentiate. This zone has been referred to as a deep-crustal **MASH zone** (Hildreth

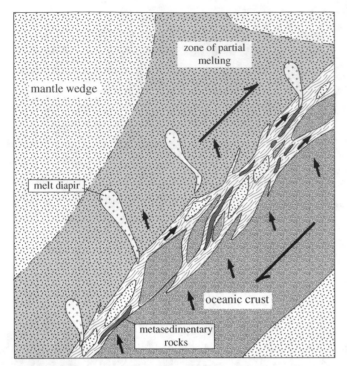

Figure 8.14 Schematic depiction of the slab–mantle interface beneath an arc. Blocks and slabs of subducted sediment, oceanic crust, and mantle peridotite occupy the zone between the downgoing oceanic plate and the overlying mantle wedge. Fluids and partial melts generated below, within, and above this zone rise into the mantle wedge where they induce partial melting of fertile peridotite. This magma rises through the mantle toward the base of the overlying crust. Modified from Schmidt and Poli (2014) and Bebout (2014).

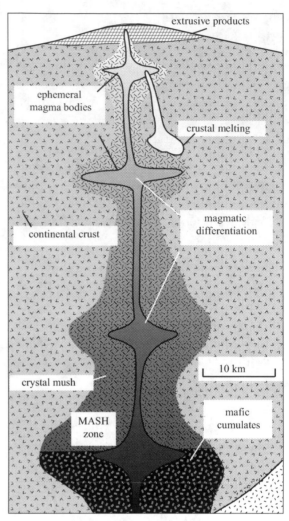

Figure 8.15 Schematic illustration of the processes accompanying magma ascent and differentiation through continental arc crust, from a deep crustal MASH zone to shallow, differentiated magma reservoirs and erupted products.

and Moorbath, 1988), where MASH stands for **m**elting, **a**ssimilation, **s**torage, and **h**omogenization. MASH zones are complex: the intrusion of hot basalt provides heat for partial melting and assimilation of lower crust. Fractional crystallization produces ultramafic cumulates. These dense cumulates may delaminate and founder or be sheared back into the mantle. Magmas within the MASH zone may include differentiated magma, new influxes of less differentiated magma, and partial melts of lower crustal wall rocks, all present together and variably homogenized. Buoyant melt may segregate from crystals, leave the MASH zone, and rise higher in the crust.

Although some magmas traverse the crust with little evidence of interruption, the ascent of many magmas involves stops and starts. Magma may accumulate in the middle and shallow crust at variable depths and multiple times, wherever there are changes in density or rigidity. When magma stalls and differentiates it can produce crystal mush, a system of crystals and melt in which the crystals form a continuous framework of touching grains, through which melt is distributed (Cashman, 2017). An active arc system may supply basalt to the base of a series of vertically stacked and differentiating accumulations of magma and crystal mush. This process can form large felsic continental arc batholiths like the Tuolumne intrusive suite, which appears to be assembled from repeated intrusion and solidification of bodies of crystal mush (see Section 8.2.4).

When magma flux rates are high, melt-rich magma chambers may form at the top of this crustal-scale vertical magmatic system (Figure 8.15; Cashman, 2017). Magma brought to this level will be composed of melt, volatiles, and crystals. The crystals typically include crystals newly formed by differentiation as well as older crystals formed at deeper levels. Some magma chambers appear to form within a few hundred meters of the surface. The existence of a crater or caldera at the summit of volcanoes is evidence of shallow magma chambers. They are formed by eruption of large volumes of magma and the subsequent foundering of the overlying rock into the space previously occupied by the magma.

Explosive volcanic eruptions may be triggered by exsolution of volatiles close to the surface, as described in Section 3.2.1. Eruptions may be presaged by changes in gas emissions from volcanic edifaces, increasingly shallow volcanic tremors, and the deformation of the ground surface that is associated with the movement of magma at shallow depths. Not all shallow magma systems erupt. In some cases, volatiles evolved from the magma may remain trapped beneath the surface where they drive hydrothermal systems or form ore deposits (Boxes 8.2, 11.1).

Summary

- Oceanic and continental arcs form on an overriding oceanic or continental plate, respectively, above a subducting oceanic plate.

- Magmas from arcs typically follow a magnesian differentiation trend in contrast to the ferroan trend exemplified by oceanic tholeiites.

- Island arc magmas may be calcic, calc-alkalic, or alkalic. The K_2O content correlates spatially either with position along the arc (as in the Lesser Antilles), or with increasing distance from the subduction zone (as in the Sierra Nevada batholith).

- The common volcanic rock types associated with arcs are basalts, andesites, dacites, and minor rhyolites. Magmatic arcs on continental margins tend to produce magmas that are more siliceous than arcs in oceanic environments.

- Arc volcanism is commonly associated with explosive eruptions, which form pyroclastic deposits. These pose considerable hazards to human and other life.

- Oceanic arc plutonic complexes typically contain ultramafic and gabbroic cumulates as well as differentiated rocks such as tonalites.

- The plutonic roots of continental arcs are exposed as Cordilleran granitic batholiths, composed mainly of granodiorite. Cordilleran batholiths tend to be magnesian, calcic to calc-alkalic, and dominantly metaluminous.

- A combination of sources may contribute to arc magmas, including the oceanic crust of the downgoing slab and its carapace of subducted sediment, the overlying mantle wedge, and the oceanic or continental crust of the overriding plate. Partial melts of subducted oceanic crust and sediment, fluids derived by dehydration reactions in the downgoing plate, and partial melting of fertile peridotite in the mantle wedge produce primary basaltic arc magma. This magma is modified as it passes through the crust and undergoes differentiation and assimilation in a series of complex magma storage regions before erupting or solidifying. Differences in the thickness and composition of this crust may account for the petrologic, geochemical, and isotopic differences observed in oceanic and continental arc magmas.

Questions and Problems

Problem 8.1. Construct a table comparing the petrography and geochemical composition of island arc magmas to mid-ocean ridge magmas and to ocean island magmas.

Problem 8.2. What are the essential differences between island and continental arc magmatic suites? What are possible causes of those differences?

Problem 8.3. Describe the sequence of rock types erupted during stage 2 of Mount Saint Helens shown in Figure 8.4. What magma chamber processes could have produced this sequence of eruptive products?

Problem 8.4. Are volcanic and plutonic suites from continental arcs identical, except that one was erupted on Earth's surface and the other crystallized at depth? Why or why not?

Problem 8.5. U–Pb ages of zircon, a common accessory mineral in granitic rocks, provide information about the construction of continental arc intrusions. Zircon populations may include xenocrysts (inherited from older wall rocks or melt source regions), antecrysts (recycled from earlier batches of magma), and autocrysts (grown in final melts) (Paterson et al., 2016). In the figure below are plotted ages of single zircon crystals from three samples of the Tuolumne intrusive suite reported by Memeti et al. (2010) and Paterson et al. (2016), along with thick lines indicating the crystallization ages of each lobe.

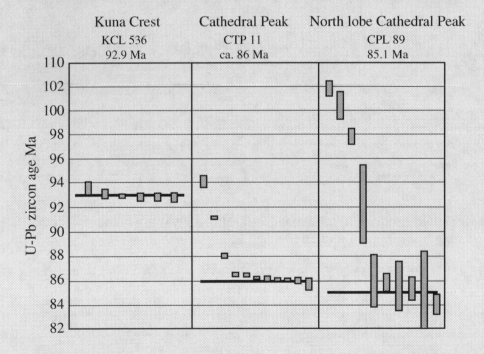

a. Identify each zircon analyzed as xenocryst, antecryst, or autocryst and explain your reasoning.

b. Write a paragraph explaining the zircon age populations in terms of the interpreted development of the Tuolumne intrusive suite shown on Figure 8.6.

Further Reading

Cashman, K. V., Sparks, R. S. J., and Blundy, J. D., 2017, Vertically extensive and unstable magmatic systems: A unified view of igneous processes. *Science*, 355, eaag3055, doi: 10.1126/science.aag3055.

Ducea, M. N., Paterson, S. R., and DeCelles, P. G., 2015, High-volume magmatic events in subduction systems. *Elements*, 11, 99–104.

Kelemen, P. B., Hanghoj, K., and Greene, A. R., 2014, One view of the geochemistry of subduction-related magmatic arcs, with an emphasis on primitive andesite and lower crust. In *Treatise on Geochemistry*, 2nd edn, eds. K. Turekian and H. Holland. Elsevier, Amsterdam, vol. 4, pp. 749–805.

Paterson, S. R. and Ducea, M. N., 2015, Arc magmatic tempos: Gathering the evidence. *Elements*, 11, 91–8.

Self, S. and Blake, S., 2008, Consequences of explosive supereruptions. *Elements*, 4, 41–6.

Wilson, C. J. N., 2002, Supereruptions and supervolcanoes: Processes and products. *Elements*, 4, 29–34.

Chapter 9

Intracontinental Volcanism

9.1 Introduction

Although continental intraplate magmatism produces only a small proportion of the igneous rocks erupted on Earth, it produces the largest variety of rock types. This diversity of intracontinental igneous rocks results from various combinations of mantle, continental lithosphere, and crustal sources. Different combinations and proportions of these sources, together with processes of magmatic differentiation, create a broad spectrum of igneous suites. Intracontinental magmatism is addressed in this and the following chapter. This chapter is concerned with intracontinental volcanism, whereas Chapter 10 describes plutonic rocks that formed in an intraplate environment. The discussion of volcanism precedes that of plutonism because the tectonic environment of young volcanic rocks is typically the most straightforward. The information assembled in this chapter is then used to address the origin of intracontinental plutonic rocks, the tectonic environments of which are not always definitive. It is important to keep in mind that in many instances intrusions are spatially related to coeval volcanic rocks, and hence the division of this topic into two chapters is somewhat arbitrary.

Intracontinental volcanism is typically associated with mantle plumes or rifts. Plume-related magmatism in intracontinental settings is analogous to that in oceanic environments: volcanic centers define tracks of progressively younger magmatism, approaching the present site of the plume. The plume track along the Snake River Plain in Idaho, USA, is comparable to the one that produced the Hawaii–Emperor island chain in the Pacific Ocean. Intracontinental volcanism associated with active continental rift zones includes the Basin and Range province and the Rio Grande rift of the southwestern United States, the Rhine graben of Europe, the East African rift zone (including the Red Sea), the Cameroon volcanic line of Africa, and the Baikal rift in Siberia (see Map 9.1). Of these, the East African rift system involves the greatest volumes of magma (500 000 km^3 compared to 12 000 km^3 in the Rio Grande rift). Continental rift zones form above areas of local lithospheric extension and are characterized by a central depression, uplifted flanks, and a thinning of the underlying crust. These structures are associated with high heat flow, broad zones of regional uplift, and magmatism. The rift zones are usually tens of kilometers wide and tens to a few hundred kilometers long. The Basin and Range province is unusual among continental rifts in that it extends hundreds of kilometers in both length and breadth. Some rift zones, such as the East African rift, represent continental crust undergoing the first stages of continental break-up. These may eventually evolve into Atlantic-type, rifted continental margins. Other rift zones are the products of "escape tectonics." In these zones, the continent outboard of a

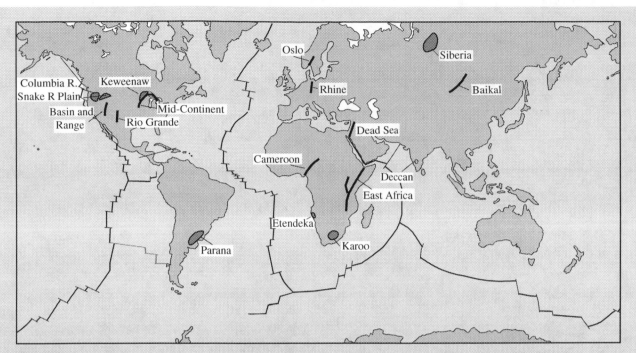

Map 9.1 Locations of major continental flood basalt provinces (shaded areas) and of major continental rifts (heavy lines).

continental collision extends in response to that collision. Examples include the Rhine graben, which formed in response to the Alpine collision, and the Baikal rift, which formed in response to the Himalayan collision.

It is difficult to generalize about the composition of lavas erupted in continental rifts. In some environments, rifting involves eruption of huge volumes of tholeiitic basalt as flood basalt, such as in the Rio Grande rift. In other areas, such as the East African rift, magmatism involves the formation of individual, point-source volcanoes that may range in composition from tholeiitic to alkaline. Most alkaline rocks are rich in sodium, rather than potassium, although potassic alkaline rocks are present as well. This chapter examines this wide range of primary magmas, their origin, and their differentiation to form a wide range of derivative magmas.

9.2 Continental Flood Basalt Provinces

Perhaps the most dramatic examples of continental volcanism are flood basalts. Large areas of the continents have been covered by great thicknesses of basaltic magma that appear to have been fed from fissures rather than from central vents. Some of these flood basalts, such as the Keweenawan basalts in the north-central United States, are clearly related to rifts; for others, the relationship is cryptic. Continental flood basalt provinces, together with oceanic plateau (see Chapter 7), are sometimes referred to as large igneous provinces (sometimes abbreviated to "LIPs"). The volume of lava produced in these provinces is far greater than any continental volcanism occurring today. Most continental flood basalts

are tholeiitic, and they appear to form in extensional environments. Although they form lava sequences that are 1–10 km thick, the surface upon which the lavas flowed never seems to have developed much relief. Apparently, the underlying rocks subsided to form a basin of about the same extent as the lavas that flowed out, so that mass was simply transferred from depth to the surface.

Map 9.1 shows the distribution of major continental flood basalt provinces in the world; the ages and dimensions of the provinces are given in Table 9.1. Many of the flood basalts erupted during continental break-up or during failed continental rifting. The Karoo, Paraná, and Deccan provinces evolved in tensional environments associated with the break-up of Gondwana during the Jurassic and Cretaceous periods. These provinces form a

Table 9.1 *Ages and dimensions of major continental flood basalt provinces*

Flood basalt	Location	Age (Ma)	Area (km²)	Thickness (m)	Reference
Snake River Plain	Idaho, USA	16–present	0.5×10^5	Up to 1200	1
Columbia River	WA, ID, OR, USA	16–6	2.0×10^5	>1500	2
Deccan Traps	India	65.5–66.3	$>5.0 \times 10^5$	>2000	3
Paraná Plateau	Brazil	134–135	1.5×10^6	1800	4
Karoo	Southern Africa	172–182	$>1.4 \times 10^5$	9000	5
Siberian Traps	Russia	253–227	$>15 \times 10^5$	3500	6
Keweenaw	Lake Superior, USA	1110–1084	$>1.0 \times 10^5$	12 000	7

1 = Perkins and Nash (2002); 2 = Brueseke et al. (2007); 3 = Schoene et al. (2015); 4 = Baksi (2018); 5 = Jourdan et al. (2007); 6 = Ivanov et al. (2005); 7 = Fairchild et al. (2017).

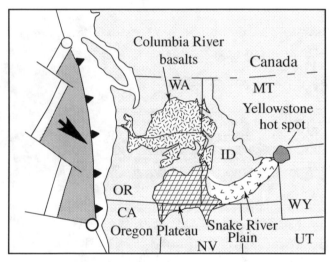

Map 9.2 Map showing the relation between the Columbia River basalts, the Oregon Plateau volcanic belt, the Snake River Plain, Yellowstone, and the direction of movement of the Juan de Fuca plate relative to North America.

band that parallels the Atlantic margin of Gondwana. A plume source is implicated in the origin of other occurrences of continental flood basalts. The Columbia River basalts and the lavas of the Snake River Plain and Yellowstone may be the product of a single plume (Camp, 1995). Mantle plumes are commonly associated with divergent plate margins; many ocean islands, including Iceland, the Galapagos, Bouvet, and Tristan da Cunha, are located on mid-ocean ridges. Continental flood basalt provinces may be related to rifting and to plume activity. For example, the Etendeka and Paraná flood basalt provinces were formed over the Tristan hot spot during the opening of the South Atlantic Ocean (Mohriak et al., 2002).

One potentially important aspect of the formation of large igneous provinces is their possible impact on Earth's environment. There is tantalizing evidence of temporal correlations between the formation of large igneous provinces and oceanic anoxia, rapid global warming, and mass extinctions. For example, the Siberian Traps erupted at the end of the Permian Period, coinciding with the mass extinction event at the end of the Paleozoic Era, and the Deccan Traps formed at the end of the Cretaceous Period, coinciding with the mass extinction event at the end of the Mesozoic Era. Beyond the temporal coincidence, the causal links between these global impacts and the outpouring of large quantities of basaltic magma are not well established, representing a fruitful area for future research (Wignall, 2005).

9.2.1 The Columbia Plateau–Snake River Plain Province

One of the youngest and best-preserved continental flood basalt provinces in the world is the Columbia Plateau–Snake River Plain province of Washington, Oregon, and Idaho. The basaltic volcanism composes part of a swath of Miocene and younger basaltic (and minor rhyolite) volcanism that extends across the northwestern United States, including the Columbia Plateau, the Oregon Plateau, and to the east, the Snake River Plain and Yellowstone (Map 9.2). Between 6 and 17 million years ago, approximately 200 000 km³ of relatively phenocryst-free basaltic lavas covered an area of about 200 000 km² in the Columbia Plateau. Since the Miocene Epoch, a much smaller volume of basaltic lava has been erupted in the Snake River Plain, where the last eruption took place only 2000 years ago.

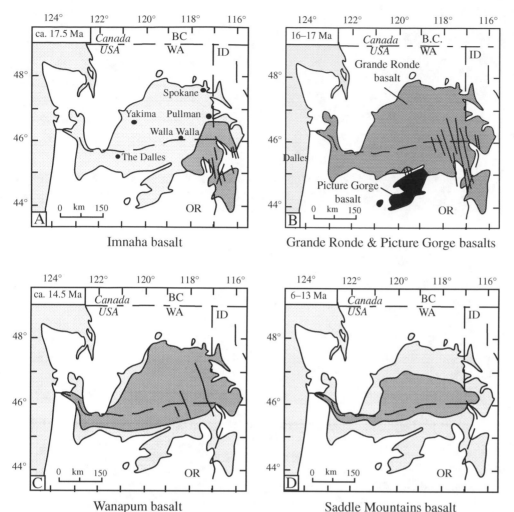

Imnaha basalt

Grande Ronde & Picture Gorge basalts

Wanapum basalt

Saddle Mountains basalt

Figure 9.1 The aerial extent (in dark shading) of the four units of the Columbia River basalts. Light shading gives the modern extent of the Columbia River basalts. The location of the fissures are marked with heavy lines. From Swanson et al. (1979).

The Columbia River basalts have been divided into four major units, which are, from oldest to youngest, the Imnaha, Grande Ronde, Wanapum, and Saddle Mountains basalts (Figure 9.1). The Picture Gorge basalt is a smaller unit coeval with the Grande Ronde basalt. The Grande Ronde flow is by far the largest by volume and was erupted over a narrow time interval of 0.42 million years, beginning at 15.79 Ma (Barry et al., 2010). This age is somewhat younger than the 16.5-Ma age for the Steens basalt, one of the major basalt flows of the southeastern Oregon Plateau, which is considered part of the Columbia River Basalt Group (Brueseke et al., 2007; Baksi, 2010). Like the Grande Ronde basalt, the Steens basalt was erupted over a very short time of 0.2 million years (Baksi, 2010). Recent work by Kasbohm and Schoene (2018) suggests that more than 95 percent of the Columbia River basalts erupted between 16.7 and 15.9 Ma. The origin of these flood basalts is the subject of ongoing debate. Some authors emphasize the location of the basalt province east of the Cascades arc in a back-arc setting and propose a magma source in the shallow mantle, akin to that supplying arcs. Alternatively, the fact that huge volumes of basalt were erupted over a very short period of time has led others to conclude that the eruption of the Steens basalt marked the onset of hot-spot volcanism in western North America (see discussion and reply in Hooper et al., 2007). Hooper et al. (2007) proposed that one edge of the plume died out over 6 million years ago, after the eruption of the Columbia River basalts, while the main portion of the plume produced the basaltic and rhyolitic volcanism of the Snake River Plain. The rhyolite volcanism of the Snake River Plain becomes younger to the east as the North American plate drifted westward over the head of the plume. The locus of the plume-related volcanism is now centered over Yellowstone and the eastern Snake River Plain.

Chemistry of the Columbia River Basalts

The Columbia River basalts show a differentiation trend typical of tholeiitic suites, with moderate increases in SiO_2 and considerable decreases in CaO with decreasing MgO (Figure 9.2). However, at any value of MgO there is a considerable range in SiO_2 among the eruptive suites. This suggests each suite experienced a different degree of crustal assimilation. The well-defined trend on the CaO versus MgO diagram is consistent with fractionation of clinopyroxene and plagioclase, which produces residual lavas poorer in both of those two oxides. The increase in K_2O with decreasing MgO indicates that potassium was behaving incompatibly; no potassium-bearing phase, such as K-feldspar, was crystallizing in these basalts. TiO_2 behaved incompatibly in the Imnaha, Grande Ronde, and Picture Gorge basalts, indicating that the titanium-bearing phase, such as Ti-magnetite or ilmenite, was not crystallizing. In both the Saddle Mountains and Wanapum basalts, the TiO_2 reaches a maximum at MgO ~ 5.0 percent, which indicates the composition where ilmenite or Ti-magnetite began to crystallize.

A key point indicated by the data shown in Figure 9.2 is that each formation in the Columbia River Basalt Group has a distinct geochemical signature. The oldest unit, the Imnaha basalt, has a relatively restricted compositional range, and SiO_2 and MgO compositions are near those of primitive mantle-derived basalts. Despite the huge volume of basalt erupted in the Grand Ronde basalts, these basalts have a rather restricted range in composition, which is richer in SiO_2 and poorer in MgO than the Imnaha basalt. Compared to the Grande Ronde, the time-correlative Picture Gorge is less evolved chemically and has lower SiO_2 and higher MgO (Figure 9.2). The smaller lava volumes of the Wanapum and Saddle Mountains formations were erupted with highly variable chemical compositions.

Two end-member models describe the formation of the Columbia River basalts: either each eruption represents a separate batch of mantle-derived magma, or the basalts formed from a single large batch of magma, which ponded and differentiated in a chamber at depth, from which eruptions tapped magma periodically. Since the lavas do not systematically become more evolved with decreasing eruptive age, the geochemical data best support the first alternative.

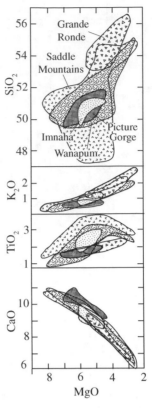

Figure 9.2 Major-element compositions (in weight percent) of the Columbia River Basalt Groups. From Hess (1989) and Hooper and Swanson (1990).

9.2.2 Petrography and Chemistry of Continental Flood Basalts

Flood basalt lavas typically are aphanitic and phenocrysts are scarce. When phenocrysts are present, plagioclase is the most abundant mineral; it is accompanied by augite, pigeonite, and Ti-magnetite, with lesser amounts of olivine. This mineral assemblage is indicative of shallow-level crystallization. There are some distinct differences between continental flood basalts and mid-ocean ridge basalt (MORB). In both rock types, plagioclase is the dominant phenocryst phase, but in MORB, olivine and magnesium–chromium spinel are very common, whereas in continental flood basalts, augite and sometimes pigeonite are the main ferromagnesian minerals.

In terms of major-element chemistry, continental flood basalts are similar to MORB. One important difference is the the iron-enrichment index (Fe-index), which is typically higher in continental flood basalts ($FeO^{tot}/[FeO^{tot} + MgO] = 0.7–0.8$; Hooper, 2000) than in MORB ($FeO^{tot}/[FeO^{tot} + MgO] = 0.5–0.6$; Stakes et al., 1984). Either continental flood basalts are generated from sources that have more iron than "normal" basalt (N-MORB) source mantle, such as enriched (E-MORB) mantle (see Figure 7.10), or magmatic differentiation lowered the magnesium content.

9.2.3 Models for the Generation of Continental Flood Basalts

The petrography, major- and trace-element chemistry, and isotope geochemistry suggest fundamental differences between continental flood basalts and oceanic basalts, some of which may be due to contamination of flood basalts by subcontinental lithospheric mantle and continental crust (Hooper et al., 2007).

Continental flood basalts contain a low-pressure phenocryst assemblage. Consequently, it is reasonable to assume that at least part of the differentiation that produces continental flood basalts takes place at shallow depths, and that gabbroic and ultramafic cumulates lie beneath the flood basalt provinces. Cox (1980) envisions a two-stage process: basaltic sills near the base of the crust that undergo partial crystallization. The residual magmas, which are now less MgO-rich, rise through a network of dikes toward the surface. The magmas may pond again near the surface and undergo further fractional crystallization to produce the low-pressure phenocryst assemblage.

Ascent through a system of dikes provides ample opportunity for crustal contamination. The extent of contamination will vary with the temperature of the magma, the flux of magma through the dikes, the width of the dikes, and the composition (and solidus temperature) of the crust through which they pass. Where flow in the dikes is turbulent, magma will erode its walls and become more contaminated than if flow were laminar. The Columbia River basalts appear to have been contaminated at two levels: first, at the base of the crust by the subcontinental lithosphere or lower crust; and second, at shallower crustal levels where assimilation of continental material occurred during magma ascent (Hooper et al., 2007).

9.3 Bimodal Volcanism

The Columbia River flood basalt province contains essentially no coeval rhyolites, but most flood basalt provinces and many tholeiitic volcanoes in rifts erupt rhyolite as well as basalt (Bryan et al., 2002). In fact, the association of basaltic rocks and rhyolite with a comparative lack of intermediate rocks is common enough in rifted areas to carry its own term – **bimodal volcanism**.

9.3.1 Bimodal Volcanism in the Yellowstone–Snake River Plain Province

A good example of bimodal volcanism is seen in the Snake River Plain–Yellowstone system, which, as noted earlier, appears to be located along a continuation of the plume track that may have formed the Columbia River basalt and Oregon Plateau (Map 9.3). The Yellowstone–Snake River Plain province erupted in a two-stage process. The first event involved rhyolitic caldera eruptions that migrated from west to east across Idaho during the past 16 million years (Perkins and Nash, 2002). This age progression apparently is caused by the westward migration of North America over a stationary hot spot. Indeed, the fact that the eastward migration of the calderas matches the rate of the westward movement of the North America plate is one of the strongest arguments that the Yellowstone–Snake River Plain volcanism is caused by a mantle hot spot. The famous thermal features of Yellowstone National Park are the result of the magmatic heat that remains after the most recent eruption in this series, a massive eruption that formed the Yellowstone Caldera,

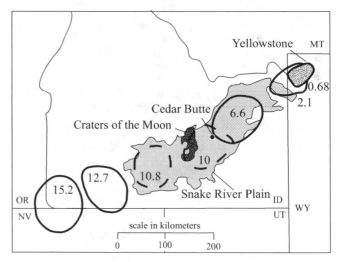

Map 9.3 A portion of the northwestern United States, showing the location of the Snake River Plain, Craters of the Moon volcanic rocks, Cedar Butte rhyolite volcano, the Yellowstone Caldera, and the calderas formed by the Yellowstone hot spot as it moved across Idaho, dashed where approximate. Numbers give the ages for the caldera eruptions in millions of years. Modified after Perkins and Nash (2002).

640 000 years ago. The second series of events involved the eruption of extensive flood basalts, which filled most of the calderas that were formed by the rhyolitic eruptions.

The eruptions of the Yellowstone Caldera produced voluminous deposits of fayalite-bearing, high-silica rhyolitic tuffs, which record some of the world's most catastrophic volcanic events (Box 9.1). The presence of fayalite in these tuffs indicates that the magma was iron-enriched. The remnants of similar iron-enriched, high-silica rhyolites, found on the margins of the Snake River Plain across southern Idaho and northern Nevada, were used to determine the locations and ages of the calderas that have long been buried by the basaltic Snake River Plain lavas.

The Snake River Plain volcanism is dominated by olivine tholeiite, but locally, including at Cedar Butte, the plain is dotted with small rhyolite volcanoes. Like the rhyolite from Yellowstone, the rhyolite from the small rhyolite volcanoes in the Snake River Plain is iron-enriched. Additionally, the rhyolite is isotopically indistinguishable from the basalts of the Snake River Plain, causing McCurry et al. (2008) to conclude that the rhyolite was produced by extreme differentiation of the Snake River Plain tholeiitic basalt. This association of rhyolite with basalt is typical of bimodal volcanism. In many rift environments, only basalt and rhyolite are found and lavas with intermediate silica contents are conspicuously missing. In the Snake River Plain, lavas with intermediate silica contents are found in the 400 000-year-old lava flows from Cedar Butte, which record a complete geochemical transition from basalt to rhyolite in the Snake River Plain magmatic system (see Figure 9.3).

9.3.2 Geochemistry of the Yellowstone–Snake River Plain Bimodal Suite

The differentiation trends of intracontinental, plume-related magmas (Figure 9.3) are distinctly different from those of arc magmas (Figure 8.9). The notable increase in the Fe-index with only a minor change in SiO_2 is probably an indication that magnetite and ilmenite didn't begin crystallizing until late in the fractionation history of the lavas. The lack of iron–titanium oxides in the initial crystallizing assemblage leads to the formation of rhyolites that are strongly enriched in iron over magnesium (Figure 9.3A). Similar to the Fe-index, the modified alkali–lime index (MALI) of Snake River Plain basalts shows a strong increase with only a minor change in SiO_2 before following a trend parallel to the alkalic–alkali-calcic boundary (Figure 9.3B). This trend is probably caused by fractionation of clinopyroxene at relatively high pressures, conditions that suppress plagioclase crys-

---

BOX 9.1 | VOLCANIC HAZARDS IN INTRACONTINENTAL ENVIRONMENTS

Arc volcanoes account for more eruption-related fatalities in the past thousand years than eruptions in any other tectonic setting (see Box 8.1). However, some of the largest volcanic eruptions derive from intracontinental volcanoes. The Yellowstone volcanic field has erupted catastrophically three times over the past 2.1 million years:

Eruption	Date	Volume of ejecta
Huckleberry Ridge	~2.1 Ma	2500 km^3
Mesa Falls	~1.3 Ma	280 km^3
Lava Creek	~0.64 Ma	1000 km^3

Each eruption produced extensive blankets of hot rhyolitic ash that covered much of western and central North America (Box Figure 9.1). Each eruption also resulted in the formation of a caldera when the ground collapsed above the partially emptied magma chamber.

Eruptions on this scale are rarest, but the hazards they pose are the most significant. Each eruption at Yellowstone probably lasted for days to weeks. The area closest to the volcanic centers was overrun by ash flows, and a larger area was affected by ash falls. Fine ash encircled the globe, cooling climate for several years. The United States Geological Survey estimates that the probability of another major, caldera-forming eruption at Yellowstone in our lifetimes is so small as to be below the threshold of calculation (Christiansen et al., 2007). However, the area remains geologically active, and the hydrothermal features – including geysers, hot springs, and fumaroles – draw millions of visitors to Yellowstone National Park each year.

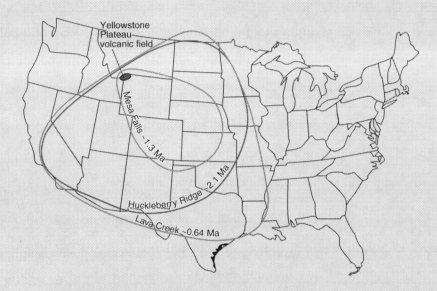

Box Figure 9.1 The extent of ash fall deposits from the three major caldera-forming eruptions of the Yellowstone Plateau volcanic field: the Huckleberry Ridge ash bed (~2.1 Ma), the Mesa Falls ash bed (~1.3 Ma), and the Lava Creek ash bed (~0.64 Ma). From Christiansen et al. (2007).

tallization (Frost and Frost, 2011). Crystallization of clinopyroxene extracts CaO from the melt without changing the K_2O or Na_2O contents, unlike crystallization of plagioclase.

The rhyolitic tuffs of Yellowstone lie at the silica-rich margins of the Snake River trends in Figure 9.3. The Yellowstone rhyolites, however, are a little less iron-enriched than the rhyolites of Cedar Butte: they border on being calc-alkalic instead of alkali-calcic. In addition, the Yellowstone rhyolites are slightly peraluminous instead of metaluminous (Figure 9.3C). Isotopic studies of the intracaldera Yellowstone rhyolitic rocks indicate that they formed dominantly by partial melting of tholeiitic basalt but may contain up to 15 percent crustal components (Hildreth et al., 1991). The compositional differences between the Yellowstone rhyolites and those of Cedar Butte can be explained by this crustal contamination. The continental crust in the area, as indicated by the gneisses exposed in the adjacent Wind River Range, is dominated by magnesian calc-alkalic granitic rocks (Frost et al., 1998). Melting of these granitic rocks likely produces mildly peraluminous magnesian calc-alkalic melts that, when added to the melts of Cedar Butte composition, could produce the magmas of Yellowstone.

9.3.3 Models for the Generation of Bimodal Volcanism

Models for the origin of bimodal volcanism must both identify the sources of the basaltic and rhyolitic magmas and also explain the near absence of intermediate

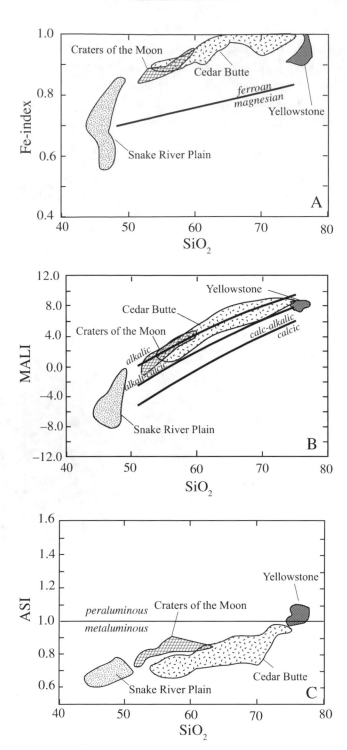

Figure 9.3 Geochemical trends in the volcanic rocks of the Yellowstone–Snake River Plain province. (A) Fe-index versus SiO₂. (B) MALI versus SiO₂ (a = alkalic, a-c = alkali-calcic, c-a = calc-alkalic, c = calcic). (C) AI versus FSSI. Data from McCurry et al. (2008), McCurry (unpublished data), and Hildreth et al. (1991).

composition lavas. The source of the basaltic components in bimodal suites is the least controversial aspect of the problem; a mantle source is almost certainly required. Mantle-derived tholeiitic basalt may be produced either when the asthenosphere upwells during rifting or when a mantle plume ascends. This tholeiitic basalt may pond at the base of the crust, where it would differentiate following a trend of iron enrichment, producing magmas enriched in FeO/(FeO + MgO). Contamination of the primary magma by subcontinental lithosphere and the continental crust may occur in bimodal associations, just as it does in flood basalts. Eruption of these magmas accounts for the basalt flows in bimodal associations.

The origin of the rhyolitic rocks in bimodal associations is more contentious. One possibility is that the basaltic magma stalled at depth continues to differentiate until producing andesitic to rhyolitic magmas. These magmas may assimilate continental crust as they evolve, and crustal contamination may be reflected in their isotopic compositions. The magmas become more siliceous and, accordingly, more buoyant, triggering ascent to shallower depths or eruption at the surface. This hypothesis is favored for the eastern Snake River Plain, where an unusual continuum of magma compositions is preserved in a few Quaternary volcanic centers (McCurry et al., 2008; Figure 9.3).

To explain the lack of intermediate compositions typical of bimodal associations, Frost and Frost (1997) proposed that partial remelting of earlier formed differentiates could produce granitic melts. These, too, could assimilate felsic crust prior to eruption on the surface. Hildreth et al. (1991) identified partial melts of Cenozoic basalt as a possible source of Yellowstone rhyolites and were able to quantify varying amounts of crustal assimilation using Sr, Nd, and Pb isotopic data.

A third mechanism to produce the rhyolitic magmas in bimodal associations is partial melting of felsic crust in response to heating by tholeiitic magma. This mode of origin accounts for the lack of intermediate rocks. However, as Christiansen and McCurry (2007) pointed out, rocks formed by crustal melting tend to be magnesian and calc-alkalic, whereas the silicic volcanic rocks in bimodal associations are typically ferroan and alkali-calcic to alkalic. Nevertheless, it seems likely that rhyolites formed in extensional environments associated with tholeiitic basalt may include a spectrum from those that are formed exclusively by fractionation or partial melting of basalt and its subsequent differentiates, to those that were produced largely or entirely by partial melting of felsic crust.

9.4 Alkaline Volcanism

Small volumes of alkaline volcanic and intrusive rocks occur in intracontinental settings (Map 9.4). Alkaline rocks are so named because they have relatively high abundances of sodium and potassium. Most commonly, alkaline rocks are unusually rich in sodium; however, potassic alkaline rocks do occur. As noted in Chapter 4, three broad classes of alkaline rocks exist: nepheline-bearing rocks that are metaluminous (i.e. there are enough alkalis to stabilize nepheline, but not enough to make sodic amphiboles and pyroxenes); nepheline-bearing rocks that are peralkaline (i.e. there are enough alkalis to form sodic pyroxenes and amphiboles); and quartz-bearing rocks that are peralkaline (see Figure 4.12). The rocks associated with alkaline magmatism range from mafic to felsic silicate rocks and carbonatites.

Alkali and Nepheline Basalts. These basalts typically contain plagioclase, olivine, and augite that is rich in titanium and sodium. The absence of orthopyroxene in the groundmass allows one to distinguish alkali basalt from tholeiite. Many alkali basalts do not contain nepheline as a modal phase; it is a normative component hiding in the glass phase. Nepheline is a groundmass phase in the nepheline basalts. These rocks include basanite, where olivine is present with nepheline, and tephrite, where olivine is absent.

Trachyte. The primary mineral in trachyte is alkali feldspar. Metaluminous trachyte may contain minor phenocryst phases, including calcic pyroxene, iron-rich olivine, hornblende, and iron–titanium oxides. In peralkaline trachyte, the groundmass phase is sodic pyroxene and sodic amphibole. Trachyte is close to silica-saturation, and so trachyte can be either quartz-bearing or nepheline-bearing. With increasing silica, quartz trachyte grades into rhyolite; with decreasing silica, trachyte grades into trachytic phonolite.

Phonolite. Phonolite is a felsic volcanic rock that contains nepheline and alkali feldspar. The groundmass minerals include alkali feldspar, pyroxene, and

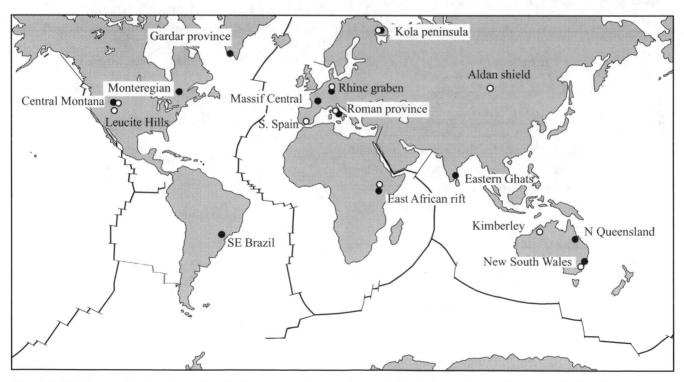

Map 9.4 World map showing the major alkaline magmatism provinces. Sodic alkaline rocks are shown in filled circles; potassic alkaline magmatism is shown in open circles.

amphibole. In peralkaline phonolites, the pyroxenes and amphiboles are sodium-rich (aegirine or riebeckite/arfvedsonite).

Peralkaline Rhyolites. Peralkaline rhyolites contain alkali feldspar and quartz as the main phenocryst phase. The presence of sodic pyroxenes and amphiboles distinguishes peralkaline from metaluminous rhyolites.

Carbonatites. Carbonatites contain at least 50 percent carbonate minerals, mostly calcite, although Ol Doinyo Lengai, a volcano of the East African rift in Tanzania, has erupted sodium carbonate. Other minerals include sodic pyroxenes and amphiboles, and magnesium-rich biotite. Carbonatites may form by unmixing of a carbonate-rich melt from phonolite or alkali basalt.

9.4.1 Sodic Alkaline Magmatism of the East African Rift

The East African rift, one of the great alkaline provinces in the world, extends 6500 km from the Red Sea to Mozambique (Map 9.5). Volcanoes in the rift, which have been erupted from the Miocene to Holocene, include compositions that range from tholeiitic to highly alkaline basalts. Most of the volcanoes in the East African rift have erupted sodic lavas, although a few, such as Nyiragongo, are potassic. Three volcanoes from the East African rift are described to provide examples of the range in compositions present and some of the different differentiation processes capable of producing those compositions. One is the Quaternary Boina volcano from the Afar region, which consists of mildly alkaline olivine tholeiite, ferrobasalt, trachyte, and peralkaline rhyolite (Barberi et al., 1975). The Nyambeni range consists of multiple volcanic centers in Kenya that have been active from the Pliocene to the Holocene. These have erupted a series of lavas, ranging from highly alkaline basalts (basanite and tephrite), to phonolite with minor peralkaline phonolite (Brotzu et al., 1983). Mount Suswa is a Quaternary volcano in Kenya that has erupted peralkaline trachytes and phonolites; basaltic rocks are absent (Nash et al., 1969).

Trends on the Fe-Index and MALI Diagrams

During differentiation, the basaltic rocks of the Boina volcano show a strong increase in the iron ratio with only a minor change in SiO_2 content (Figure 9.4A), defining a trend very similar to rocks from the Snake River Plain

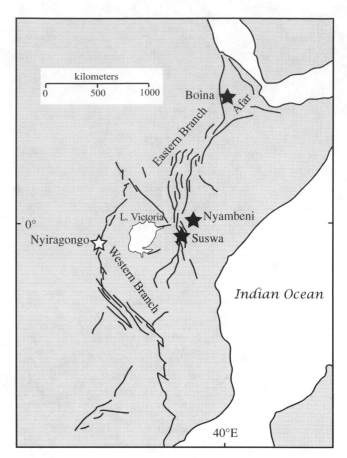

Map 9.5 Map of the East African rift showing the locations of the Boina, Nyambeni, and Suswa volcanoes, all of which are sodic alkaline rocks (filled stars), and Nyiragongo, which is a potassic alkaline volcano (unfilled star).

(compare Figure 9.4A to Figure 9.3A). This early iron enrichment probably indicates that substantial differentiation took place before magnetite began to crystallize. Nyambeni shows an iron-enrichment trend similar to Boina, but because the primary magma was undersaturated with respect to SiO_2 differentiation never led to significant silica enrichment. The rocks from Suswa are restricted to an iron-rich composition. Because iron-rich melts are never derived directly from the mantle, the melts from Suswa must have formed from differentiation of a parent melt that was not erupted.

On a MALI diagram (Figure 9.4B), the lavas of Boina form a narrow trend that initially increases until the rocks lie in the alkalic field and then plateau at high SiO_2. The increase at low SiO_2 can be explained by substantial fractionation of augite, which extracts CaO from the melt, while the plateau at high silica may be produced by the

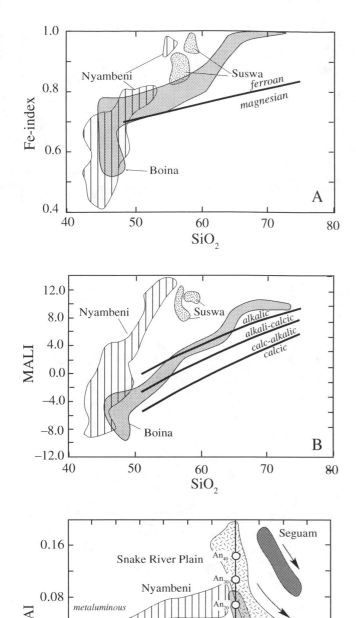

Figure 9.4 Geochemical trends for the Boina, Nyambeni, and Suswa volcanoes. (A) Fe-index versus SiO_2. (B) MALI versus SiO_2. (C) AI versus FSSI. Trends for the Snake River Plain and Seguam are shown for comparison. Data from Barberi et al. (1975), Brotzu et al. (1983), Nash et al. (1969), Singer et al. (1992), and from references cited in Figure 9.3.

fractionation of aegirine and sodic amphibole, which extract Na_2O from the melt. Rocks from Nyambeni form a broad field with a very steep increase in relative MALI with minor change in silica.

Trends on the AI–FSSI Diagram

On a plot of alkalinity index (AI) versus FSSI (feldspathoid silica-saturation index) (see Figure 4.12), compositional data from each of the three volcanoes define separate fields that radiate from FSSI = 0.0, which represents the albite thermal barrier (Figure 9.4C; cf. Figure 2.11). As noted in Chapter 2, fractional crystallization of magmas slightly to the right of this barrier can move the composition of a melt farther to the right (toward higher silica contents). In the same way, a magma slightly to the left of this barrier may differentiate to compositions farther to the left (toward lower silica contents), but in neither case can the melt composition cross this barrier. The rocks from Boina become increasingly silica-rich during differentiation (as indicated by the arrow in Figure 9.4C), but also become more alkaline. In contrast, the rocks from Nyambeni reflect a decrease in silica with differentiation. The rocks from Suswa define two groups, each of which show increasing alkalinity and decreasing silica abundance with increasing differentiation. These trends of increasing alkalinity with increasing differentiation are not unique to alkaline rocks, as indicated by the fact that the lavas from the Snake River Plain and Seguam show similar trends. These trends are a manifestation of the crystallization of plagioclase, because, as indicated in Figure 2.12A, crystallization of plagioclase will extract calcium from a melt, enriching it in alkalis. In a calcium-bearing melt that is sodium-rich, CaO will preferentially enter plagioclase, leaving the pyroxene enriched in Na_2O. This effect, called the "plagioclase effect" by Bowen (1945), means crystallization of plagioclase can drive a melt that was originally metaluminous into the peralkaline field. The positions of An_{10}, An_{20}, An_{30}, and An_{40} are shown in Figure 9.5C to show how extraction of plagioclase of these compositions could have caused the trends observed (although the differentiation was certainly more complex than that).

Figure 9.4C shows that a suite of rocks derived from a basaltic parent will evolve to become peralkaline through fractional crystallization, depending on the relative abundances of CaO and Na_2O in the parent rock. Rock suites such as Seguam, which are relatively calcic, do not become peralkaline during differentiation. In contrast, those that are sodium-rich may become peralkaline if significant amounts of plagioclase fractionates.

Implications for the Evolution of Boina, Nyambeni, and Suswa Volcanoes

As noted in Chapter 4, geochemical variation diagrams such as those in Figure 9.4 cannot prove that a suite of rocks evolved through fractional crystallization, but the fact that the rocks from Boina and Nyambeni form tight trends strongly supports this contention. The fact that Sr isotopes are similar for the basaltic and rhyolitic rocks of Boina and that the trace elements behave in a manner consistent with fractional crystallization (Barberi et al., 1975) is strong corroboration that the suite of rocks erupted by Boina evolved through this process. Similarly, the well-defined trends on major- and trace-element variation diagrams in the Nyambeni lavas suggest derivation through fractional crystallization (Brotzu et al., 1983). The same cannot be said for the rocks from Suswa because they occupy two separate fields in the discrimination diagrams, particularly in Figure 9.4C. Based on whole-rock and trace-element chemistry, Nash et al. (1969) argued that Suswa volcano erupted melts from different sources at different times, although each of the individual batches of melt may have evolved toward increasing peralkalinity and decreasing silica activity.

9.4.2 Potassic Alkaline Volcanism

Although most alkaline rocks are sodic (i.e. have $Na_2O > K_2O$), a number are potassic. In some localities, such as in the East African rift, potassic alkaline magmas are clearly rift-related. In other localities, such as the Roman province, the tectonic environment is cryptic at best. In this section, Mount Vesuvius from the Roman province exemplifies potassic alkaline volcanism. Another group of mafic potassic alkaline rocks, kimberlites and lamproites, forms small bodies (often a kilometer or less across) that are important as hosts of diamonds. This section also discusses the origins of kimberlites and lamproites and the diamonds they contain.

Mount Vesuvius and the Roman Province

Mount Vesuvius is the southern-most volcano in the Roman province, a group of potassic alkaline volcanoes

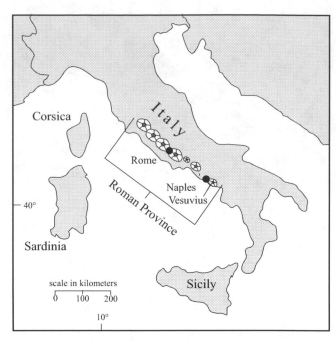

Map 9.6 Geologic sketch of the Roman province in Italy, showing the location of major eruptive centers. Locations of volcanoes from Geological Survey of Italy (1960).

that extends for more than 500 km along the west coast of Italy (Map 9.6). Mount Vesuvius is a recent cone (less than 19 000 years old), built in the caldera of Mount Somma, an older volcano, active from 39 000–19 000 years ago. Long periods of quiescence, punctuated by explosive eruptions, characterize the activity of Vesuvius (Di Renzo et al., 2007). One of these eruptions buried Pompeii in 79 CE, an event witnessed by Pliny the Younger. His letters on the eruption of Vesuvius (*Epistulae* VI.16 and VI.20) are certainly some of the earliest geologic descriptions of a volcanic eruption. Because of Pliny's letters, eruptions such as those of Vesuvius are called "plinian."

Over the last 19 000 years, Vesuvius has erupted about 50 km^3 of magma with wide-ranging composition. Trachyte from Vesuvius may be either nepheline- or quartz-normative, whereas phonotephrite and phonolite are both nepheline- and leucite-normative (Di Renzo et al., 2007). The rocks from Mount Vesuvius do not form any clear trends on chemical discrimination diagrams (Figure 9.5), which is not surprising considering that both quartz-saturated and nepheline-saturated rocks are involved. Lack of a simple trend indicates the evolution of the rocks from Vesuvius did not involve simple fractional crystallization. In fact, Di

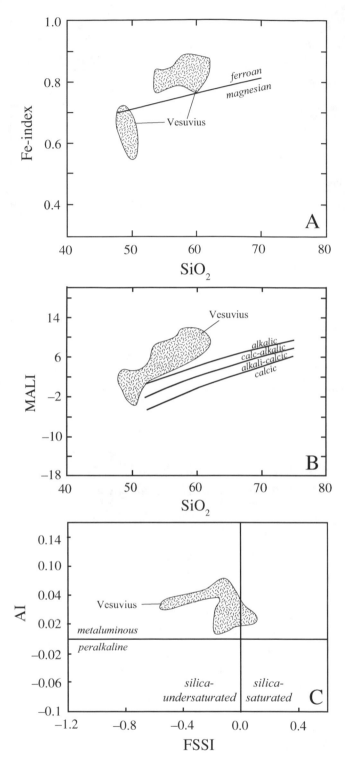

Figure 9.5 Geochemical trends for the rocks from Mount Vesuvius. (A) Fe-index versus SiO_2. (B) MALI versus SiO_2. (C) AI versus FSSI. Data from Di Renzo et al. (2007).

Renzo et al. (2007) postulate that the melting of a heterogeneous source area, crustal contamination, and magma mixing could all have been involved in the evolution of the lavas from Vesuvius.

Kimberlites and Lamproites

Kimberlites and lamproites occur almost exclusively within continental plates, typically erupting through Precambrian crust (Dawson, 1967; Mitchell and Bergman, 1991) (Map 9.7). Kimberlites and lamproites have similar major-element compositions and the two rock types are often grouped together (Mitchell and Bergman, 1991). Both are potassic ultramafic rocks that contain olivine and phlogopite. The tectonic environment that produces such rocks is often enigmatic and may be different for different occurrences. These potassic mafic rocks are interesting because (1) they may bear xenocrystic diamond, and (2) they have exotic mineralogies (and names) that give petrologists important information about the mantle.

Kimberlites are products of continental intraplate magmatism, confined to regions of the crust underlain by ancient cratons; no occurrences have been described from oceanic environments or young fold belts. Kimberlites don't appear to be associated with continental rifts, although in some cases their location can be correlated with zones of weakness in the crustal lithosphere. Kimberlites cluster in small volcanic diatremes, sills, and dikes. Some kimberlite provinces have experienced multiple episodes of kimberlite magmatism – southern Africa, Wyoming, and Russia in particular.

Kimberlites appear to originate at depths of 100–200 km and ascend into the crust as a kimberlite magma. At relatively shallow depths (maybe around 2 km), they exsolve a CO_2-rich fluid and the fluid-rich magma erupts explosively to the surface (Dawson, 1967; Mitchell, 1986). This rapid emplacement results in brecciation of the host rock as well as fragmentation of the conduit up which the melt moves. As a result, kimberlites consist of a fragmented and altered groundmass, containing olivine, phlogopite, diopside, apatite, spinel, and ilmenite. Many of these minerals have been partially or completely altered to serpentine and carbonate. Kimberlites also often include lithic fragments and xenocrysts from crustal and mantle rocks through which they ascend. The diamonds commonly present in kimberlites are usually found within garnet peridotites or eclogite xenoliths; rarely, they are found as xenocrysts within the kimberlite itself.

Lamproites occur mainly as dikes, minor intrusions, and flows (Mitchell and Bergman, 1991). Like kimberlites, lamproites are associated with lineaments that

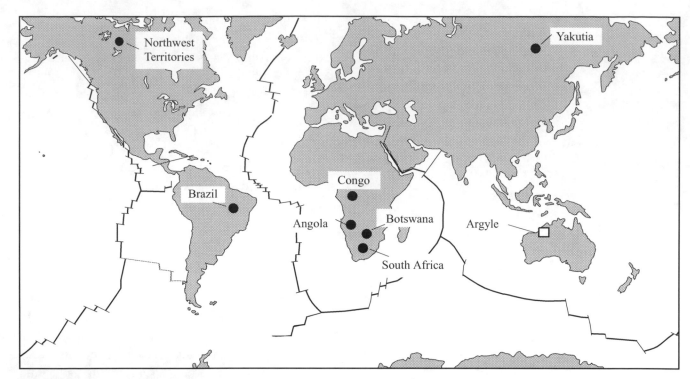

Map 9.7 Major diamond-producing regions of the world. Filled circles = diamonds sourced from kimberlite, open square = diamonds sourced from lamproite.

may reflect zones of lithospheric weakness. Some of the classic lamproite occurrences include the Leucite Hills, Wyoming; Smoky Buttes, Montana; and occurrences in Antarctica, southern Spain, and western Australia (Map 9.4). Most lamproites are Cenozoic, in contrast to kimberlites, which are exclusively pre-Cenozoic. Like kimberlites, lamproites may host diamonds.

Lamproites are characterized by major phases of phlogopite, olivine, diopside, sanidine, and leucite. In most rocks, olivine is partially or totally pseudomorphed by serpentine, iddingsite, carbonate, or quartz. Fresh leucite is also rare, commonly replaced by sanidine, analcite, quartz, zeolite, or carbonate. Lamproites may contain diamonds, but the diamonds tend to be much smaller than those found in kimberlites. One explanation for this size disparity is that lamproites are emplaced at a slower velocity than the explosive emplacement of kimberlite. Whereas kimberlites can carry the larger xenoliths and the larger diamond xenocrysts to crustal levels, in lamproites, those diamonds and larger xenoliths may sink into the mantle before arriving at shallow crustal levels.

9.5 Origin of the Chemical Diversity of Intracontinental Basaltic Magmas

The preceding sections illustrated how a wide variety of mantle-derived rocks of basaltic composition erupted in continental settings. These intracontinental mafic rocks range from voluminous tholeiitic flood basalts through alkali basalts to highly alkaline sodium-rich or potassium-rich mafic magmas. A reasonable question to ask is: What causes this extreme variability of magma compositions? As noted in Chapter 6, alkali basalts could form by small degrees of partial melting of a normal mantle. However, it is difficult to adopt this explanation for the origin of the highly alkaline basaltic rocks, such as those that formed the Nyambeni eruptive centers, or for potassic magmas such as those that feed the Roman province. These highly alkaline melts must have been derived from a mantle whose composition was different from the "normal" mantle that produces tholeiites.

This process by which the composition of the mantle has changed is called **metasomatism**, which technically describes any metamorphic process that changes the composition of a rock. Metasomatized mantle is increasingly recognized as a precursor for alkaline magmatism, both in the East African rift (Rosenthal et al., 2009) and in the Roman province (Bianchini et al., 2008; Nikogosian and van Bergen, 2010). Movement of sodium-bearing fluid through the mantle will produce hornblende-rich veins, whereas potassium-bearing fluids will produce phlogopite. Preferential melting of these veins will produce sodium-enriched and potassium-enriched alkaline magmas. Mantle xenoliths brought up in alkali basalts provide compelling evidence for metasomatism (Dawson and Smith, 1988; Meshesha et al., 2011). The mineralogical composition in these veins and their abundance in the mantle are likely to be irregularly distributed. Thus melting of a metasomatized mantle is likely to produce a wide variety of magmas. This explains why areas such as the East African rift and the Roman province erupt such a wide variety of lavas (Bianchini et al., 2008; Nikogosian and van Bergen, 2010).

Having established how highly alkaline magmatism can derive from a heterogeneous, metasomatized mantle, the next question concerns the source of the fluids that metasomatized the mantle. Because both Na_2O and K_2O are enriched in the crust over the mantle, one reasonable source is through subduction (Bianchini et al., 2008; Markl et al., 2010). If a subduction zone entrained substantial sea-floor sediment, dewatering of these rocks may produce a potassium-rich fluid that could metasomatize the overlying mantle. Alteration of basaltic rocks on the sea floor causes albite to form from calcic plagioclase. Subduction of these rocks provides a source for sodium-rich fluids that can affect the composition of the overlying lithosphere (Markl et al., 2010). It is important to emphasize that this type of metasomatism must have occurred before (in some areas, millions of years before) alkaline melts were generated. In most areas, the thermal event that produced the alkaline melts is in no way directly related to subduction.

The fact that metasomatized mantle that is enriched in K_2O or Na_2O melts to produce alkaline rocks may explain observations from the Roman province. The Roman province lies above a steep east-directed subduction zone (Bianchini et al., 2008), in an area where one would expect broadly calc-alkalic volcanism. Indeed, calc-alkalic magmas are found locally in the Roman province (Boria and Conticelli, 2007). One proposal to explain why the magmatism in the Roman province is highly potassic suggests that, unlike most arcs, the mantle under the Roman province had been previously invaded by potassium-rich fluids, derived from an earlier subduction event (Bianchini et al., 2008). Thus, although this book describes igneous rocks in the context of the tectonic environment where they most frequently occur, it is important to note that the factors controlling the composition of mantle melts are many and include mantle composition, degree of melting, and temperature and pressure conditions of melting. The tectonic environment usually controls the conditions of melting, but not the composition of the mantle that partially melts in these environments. For this reason, a variety of magma compositions may be produced by mantle melting in any given tectonic environment.

Summary

- Intracontinental volcanism includes continental flood basalts, bimodal associations of basalt and rhyolite, and alkaline rocks.

- Intracontinental volcanism is usually associated with mantle plumes or rifts.

- Hot-spot-related magmatism in intracontinental settings defines tracks of progressively younger volcanic centers, recording the movement of the plate over a plume.

- Rift-related volcanism may be associated with broad areas of extension, such as in the Basin and Range in southwestern United States, or in narrower rifts such as the East African rift.

- A wide range of lavas is erupted in continental rifts.

- Flood basalt provinces are dominated by large volumes of tholeiitic basalt.

- Bimodal associations are composed of tholeiitic basalt and ferroan rhyolite.

- Rift-related volcanism may range in composition from tholeiitic to alkaline.

- Kimberlites and lamproites are derived from unusual mafic, high-potassium magmas that may transport diamonds to the surface.

- The wide range of mantle-derived rocks of basaltic composition, from tholeiitic basalt, through alkali basalt, to alkaline sodium- or potassium-rich mafic magmas, may reflect:
  ○ different degrees of partial melting of the mantle, and/or
  ○ variations in the composition of the mantle sources.

- Alkaline rocks are commonly attributed to partial melting of mantle that is compositionally heterogeneous as a result of metasomatism.

Questions and Problems

Problem 9.1. Compare and contrast oceanic plateau and continental flood basalt provinces.

Problem 9.2. Calculate the approximate plate velocity for the North American plate from the age of volcanic centers along the Snake River Plain and Yellowstone, shown in Map 9.3. What is the direction of plate motion?

Problem 9.3. Provide at least two explanations for the absence of intermediate compositions of volcanic rocks found in bimodal volcanic suites.

Problem 9.4. What is metasomatism?

Problem 9.5. What are the similarities and differences between kimberlites and lamproites?

Problem 9.6. Given the size of the eruptions at Yellowstone (Box 9.1), determine their Volcanic Explosivity Index (VEI; Box 8.1).

Further Reading

Eldholm, O. and Coffin, M. F., 2000, Large igneous provinces and plate tectonics. In *The History and Dynamics of Global Plate Motions*, eds. M. A. Richards, R. G. Gordon, and R. D. van der Hilst. Geophysical Monograph 121, American Geophysical Union, Washington, DC, pp. 309–26.

Mitchell, R. H., 1995, *Kimberlites, Orangeites, and Related Rocks*. Plenum, New York, NY.

Pierce, K. L. and Morgan, L. A., 1992, The track of the Yellowstone hot spot: Volcanism, faulting, and uplift. In *Regional Geology of Eastern Idaho and Western Wyoming*, eds. P. K. Link, M. A. Kuntz, and L. B. Platt. Geological Society of America Memoir 179, pp. 1–52.

Sørensen, H., 1974, Introduction. In *The Alkaline Rocks*, ed. H. Sørensen. John Wiley & Sons, New York, NY, pp. 1–11.

Sparks, R. S. J., 2013, Kimberlite volcanism. *Annual Reviews in Earth and Planetary Sciences*, 41, 497–528.

Chapter 10

Intracontinental Plutonism

10.1 Introduction

The previous chapter described the wide variety of volcanic rocks produced in intracontinental settings through the participation of different source magmas, which have differentiated through fractional crystallization, crustal assimilation, or magma mixing. Intracontinental magmatism also produces a variety of plutonic rocks, which are compositionally distinct from igneous rocks found in arc environments, including layered mafic intrusions, Archean and massif anorthosites, alkaline intrusions, and ferroan granites. Some of these plutonic rocks are associated with the coeval volcanic rocks described in the previous chapter, which allow geologists to ascribe a tectonic environment to those plutons. Many intracontinental plutons, however, are not spatially and temporally related to volcanic rocks, so their tectonic environment must be determined by inference.

This chapter discusses the textures, occurrence, and compositions of four categories of plutonic rocks found in an intracontinental environment. Layered mafic intrusions are described first. These bodies, many of which exhibit spectacular layering, are also important sources for chromium, nickel, cobalt, and platinum-group elements (Box 10.1). The second group of rocks are anorthosites, rocks composed of at least 90 percent plagioclase, which occur in two associations. Archean anorthosites occur as dikes and sills, and as their name suggests, they are most common in the Archean rock record. Massif anorthosites are batholith-sized bodies of anorthosite that are relatively abundant in the Proterozoic Eon, but Phanerozoic examples are also known. Associated with massif anorthosite is the third group of intrusions, ferroan granites, also introduced in this chapter. However, ferroan granites are as likely to occur where anorthosite is absent. As the name implies, ferroan granites are iron-rich and contain more potassium feldspar than plagioclase, giving them a pink color. The final group of intracontinental plutonic rocks is alkaline intrusions. These intrusions are uncommon but contain unusual minerals and host scarce metals, in high demand for personal electronics and clean-energy technologies (see Box 10.3).

BOX 10.1 | ORE DEPOSITS ASSOCIATED WITH LAYERED MAFIC INTRUSIONS

Chromium. Layered mafic intrusions host important deposits of chromium, nickel, cobalt, and **platinum-group elements (PGEs)**: platinum, palladium, ruthenium, rhodium, osmium, iridium. The partition coefficient, D, for chromium between chromite and melt is on the order of 1500, which means the chromium content of chromite is about 1500 times that of the melt. Chromium deposits are found in the ultramafic portions of a layered mafic intrusion where chromite-rich layers may be more than 5 m thick. These deposits are thought to result from the injection of a new magma into a magma body. The interaction of cumulates or interstitial melt in the intrusion with the new melt may drive the system into the primary field of chromite crystallization (Duke, 1988; Mathez and Kinzler, 2017), producing chromite-rich layers. In some LMIs, these chromite deposits extend for kilometers and are called stratiform chromite deposits to distinguish them from podiform chromite deposits found in ophiolites (see Box 7.2). The biggest chromium deposit in the world is the Bushveld intrusion in South Africa; it supplies more than 70 percent of the world's chromium (see photo of chromite layers in Figure 10.1C). Chromium deposits from the base of the Stillwater intrusion in Montana were mined during World War II, but are not mined for chromium today, in part because the chromite in the Stillwater is not as pure as that in the Bushveld.

Platinum-Group Elements. In the process of differentiation, the basaltic magma in the magma chamber of a layered intrusion becomes saturated in sulfur, causing a separate sulfide melt to form along with the silicate melt. The sulfide melt is mostly composed of FeS, but economically important elements such as copper, nickel, cobalt, and the PGEs are strongly fractionated into sulfide melts. If, during the formation of a layered mafic intrusion, the sulfide melt was in communication with a large volume of silicate melt, it might sequester enough of these elements to form an ore deposit, particularly if the sulfide melt was concentrated into a limited horizon. Layered mafic intrusions that contain important PGE ore bodies include the Bushveld, Great Dyke, and Stillwater intrusions, all of which are being mined for PGEs today.

Nickel. The partition coefficient for nickel between olivine and melt is around 4, which means crystallization of olivine will deplete the melt of nickel. Therefore, nickel sulfide ore bodies are only found in intrusions where sulfur saturation occurred before significant olivine crystallized. Intrusions in which this occurred include Sudbury, which, being silica-saturated, never crystallized olivine, and Duluth, which attained early sulfur saturation because of assimilation of sulfur-rich pelitic country rocks.

Iron, Titanium, and Vanadium. Ferric iron, titanium, and vanadium behave as incompatible elements during early crystallization of mafic igneous intrusions until magnetite and ilmenite begin crystallizing. In the Bushveld intrusion, cumulate magnetite began crystallizing at the base of the upper zone, and the upper zone contains more than 20 layers, up to 10 m thick, which are rich in vanadium-bearing titaniferous magnetite. Magnetite concentrations are also found in irregular plugs and pegmatitic bodies within portions of the main zone (Willemse, 1969). These magnetite-rich areas are economic mainly because of their high concentrations of vanadium, and they make South Africa the world leader in the production of this metal.

10.2 Layered Mafic Intrusions

Layered mafic intrusions (sometimes referred to as LMIs) are some of the most important types of igneous bodies; the locations of some of the best-studied layered mafic intrusions are shown in Map 10.1 and their size and age are given in Table 10.1. Not only do these intrusions preserve spectacular igneous textures and structures, they provide petrologists with critical insights into the differentiation processes of mafic melts. In addition, some layered mafic intrusions host chromium, nickel, and platinum deposits. Although most layered mafic intrusions are Precambrian in age (Table 10.1), layered mafic intrusions have formed throughout Earth's history. A distinctive feature of layered mafic intrusions is their igneous layering (Figure 10.1). Layering is also found in oceanic gabbros and in gabbroic and ultramafic intrusions associated with arc volcanoes, but large-scale, persistent layering is the hallmark of layered mafic intrusions.

Igneous layering is postulated to form primarily by accumulation of minerals on the floor of an igneous intrusion. Usually, the magma from which it forms is mafic, because mafic magmas are fluid enough to allow minerals to sink during the crystallization of the melt. Layering, however, is not unknown in felsic intrusions, and in both compositions, layering forms on all scales. Kilometer-scale layering is governed by the appearance or disappearance of

Table 10.1 *Major layered mafic intrusions*

Name	Location	Age (Ma)	Area (km²)	Reference
Bushveld	South Africa	2059–2054	66 000	1
Dufek	Antarctica	183	50 000	2
Duluth	Minnesota, USA	1099	4700	3
Stillwater	Montana, USA	2712–2709	4400	4
Muskox	Northwest Territories, Canada	1269	3500	5
Great Dyke	Zimbabwe	2575	3300	6
Sudbury	Ontario, Canada	1849	1300	
Kiglapait	Labrador, Canada	1305	250	7
Skaergaard	Greenland	56	100	8
Duke Island	Alaska, USA	108–111	~100	9
Rum	Scotland	60	~100	10

1 = Rajesh et al. (2013); 2 = Burgess et al. (2015); 3 = Paces and Miller (1993); 4 = Wall et al. (2018); 5 = Mackie et al. (2009); 6 = Oberthür et al. (2002); 7 = Yu and Morse (1992); 8 = Hirschmann et al. (1997); 9 = Saleeby (1992); 10 = Hamilton et al. (1998).

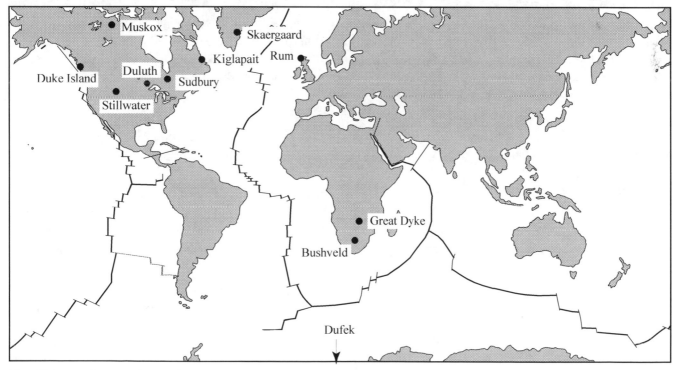

Map 10.1 Distribution of major layered mafic intrusions around the world.

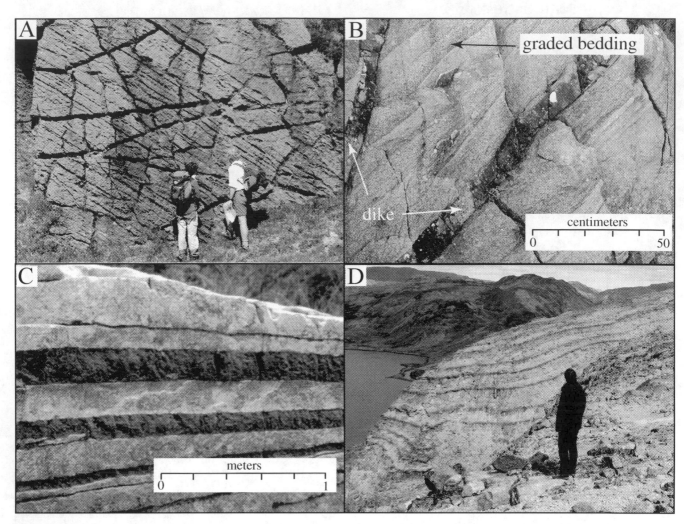

Figure 10.1 Photos of layering in igneous rocks. (A) Layering in feldspathic peridotite. Relatively resistant layers are richer in plagioclase than the less-resistant layers. Rum intrusion, Scotland. Photo courtesy of Michael Cheadle. (B) Layering in gabbros from the Rum intrusion. Layering is based on modal variation of olivine, augite, and plagioclase. Photo by Michael Cheadle. (C) Layering of chromite in anorthosite, Bushveld intrusion, South Africa. Photo by Michael Cheadle. (D) Layering in nepheline syenites, Ilimaussaq intrusion, southern Greenland.

key minerals, such as olivine or plagioclase, during the crystallization of an igneous body. The base of many layered intrusions is marked by ultramafic horizons, rich in olivine, pyroxene, and locally chromite (Figure 10.1A). The first appearance of cumulus plagioclase typically marks the transition from ultramafic to gabbroic composition and forms a horizon that often can be traced across the entire intrusion.

In addition to the kilometer-scale layers, most mafic layered intrusions contain layering on the meter scale. Many processes have been postulated for the formation of these small-scale layers. They could be caused by the injection of new magma into a magma chamber, which causes a new mineral, such as olivine or chromite, to crystallize and sink to the floor of the chamber. Such a process could have formed the Frost pyroxenite, a narrow layer that is a major marker, extending for more than 35 km in the Dufek intrusion in Antarctica (Figure 10.2). Small-scale layering may also be caused by density currents containing crystals that crystallized on the relatively cool walls of the magma chamber and then slid onto the floor of the chamber. These crystals could accumulate as graded layers (Figure 10.1B), similar to graded beds formed by turbidity currents in sedimentary environments. Such a process may also have produced the unusual chromite layers in anorthosite (Figure 10.1C),

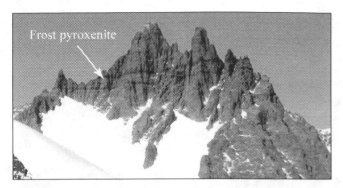

Figure 10.2 Photograph of the Dufek intrusion exposed on 1840-m Neuberg Peak in Antarctica. The Frost pyroxenite is a major marker in the intrusion and extends for about 35 km along strike. Photo by Michael Cheadle.

found in the Bushveld intrusion in South Africa (Voordouw et al., 2009). Another process forming layered rocks is postulated from the Ilimaussaq intrusion (Figure 10.1D), where periodic eruptions and magmachamber degassing could have produced layering (Pfaff et al., 2008). Because high water pressure depresses the crystallization temperature of a silicate melt (Figure 3.3), each time fluid was released the crystallization temperature increased and a small amount of melt crystallized. The denser minerals (the dark bands) in Figure 10.1D sank to the bottom of the chamber before the less-dense feldspars and feldspathoids. As crystallization of minerals proceeded, the fluids in the remaining melt increased, inhibiting further crystallization until another degassing event occurred. Understanding the processes by which igneous layering forms is the subject of ongoing research because these processes provide insights into the chemical evolution of intrusions over time.

10.2.1 The Bushveld Intrusion

The largest and arguably the most important layered mafic intrusion in the world is the 2.06-Ga Bushveld intrusion in South Africa (Map 10.1). The Bushveld intrusion, which contains a stratigraphy nearly 8000 m thick, is layered on scales ranging from less than a meter to hundreds of meters. The intrusion forms a complex bowl shape in which the shallower portions of the intrusion are exposed as one moves inward from the margins. The intrusion is capped by a series of granitic rocks and felsic volcanic rocks, which are the youngest units in the intrusive complex. The Bushveld is considered to have formed from a mafic magma that differentiated in place, although in detail the crystallization history is complex.

Locally, the base of the intrusion contains a narrow zone of fine-grained gabbroic rock, known as the *marginal zone* (Map 10.2), which is interpreted to represent the chilled magma that produced the intrusion. Above this is the *lower zone*, a sequence of peridotite and pyroxenite, more than a kilometer thick. These rocks formed by the accumulation of early crystallizing phases, including olivine, chromite, orthopyroxene, and clinopyroxene. Plagioclase, where present in the lower zone, occurs in the interstices of the ferromagnesian minerals. Above the lower zone are three thick gabbroic horizons: the *critical zone*, marked by the appearance of plagioclase; the *main zone*, marked by the disappearance of chromite; and the *upper zone*, marked by the appearance of cumulus magnetite.

10.2.2 Mineralogical Variation in LMIs

Because the rocks from layered mafic intrusions usually consist of varying proportions of cumulus and postcumulus grains, whole-rock analyses of these rocks do not retain direct information about melt compositions. For example, a rock from the base of an intrusion with 40 percent chromite and 60 percent olivine could contain around 16 percent Cr_2O_3. There is no melt that has anywhere near that chromium content; analyses with this composition would merely show that chromium has been concentrated by accumulation of chromite. To understand the chemical variation present within layered mafic intrusion, therefore, it is better to use mineral chemistry, because the composition of olivine, for example, that formed at any point in the crystallization of a melt will be the same regardless of whether the rock contains nearly 100 percent cumulus olivine or only a few percent postcumulus olivine.

Figure 10.3 shows that, at increasing stratigraphic positions in the Bushveld intrusion, the plagioclase becomes increasingly more sodic and the ferromagnesian silicates become more iron-rich (Atkins, 1969). These are exactly the kind of chemical changes one would expect of olivine and plagioclase formed by fractional crystallization. As Figure 2.12 demonstrated, olivine is always higher in Mg/(Mg + Fe) and plagioclase is always higher in Ca/(Ca + Na) than the melt from which they crystallize. Fractional crystallization of olivine and plagioclase will therefore preferentially extract MgO and CaO from a melt, leaving the residual melt relatively enriched in FeO and Na_2O.

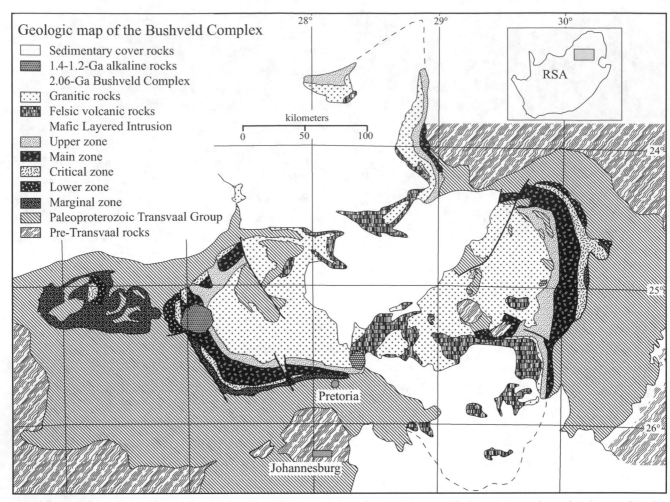

Map 10.2 Geologic map of the Bushveld Complex after Vermaak and Von Gruenewaldt (1981). Inset shows location within the Republic of South Africa. Age data after Cawthorn et al. (2012) and Scoates and Friedman (2008).

The mineralogy of most layered mafic intrusions is broadly similar to the Bushveld in that the intrusions record a differentiation trend with ferromagnesian minerals becoming enriched in iron relative to magnesium and plagioclase becoming more sodic at progressively higher stratigraphic levels. In detail, however, each pluton is distinct. Some of these differences reflect differences in silica abundance. Like the Bushveld, the middle zones of the Skaergaard intrusion in Greenland contain no olivine; low-calcium pyroxene is present instead (Wager and Brown, 1967). By contrast, the Kiglapait intrusion in Canada contains no orthopyroxene and olivine occurs throughout (Morse, 1980); the Dufek intrusion in Antarctica contains no olivine or quartz, only pyroxenes (Himmelberg and Ford, 1976), and Sudbury in Canada is quartz-saturated throughout (Naldrett et al., 1970). In most of the large layered intrusions, magnetite appears only in the upper portions of the stratigraphy, but in a

few intrusions, as exemplified by the Duke Island intrusion, magnetite appears in the ultramafic horizons (Irvine, 1974). As a result of magnetite extraction, the Duke Island intrusion, located in Alaska, does not show the extreme iron-enrichment characteristic of the Bushveld. Intrusions like the Duke Island intrusion are therefore considered of arc affinity, rather than rift-associated. The early appearance of magnetite in these intrusions inhibits iron enrichment during differentiation and produces magnesian rocks (cf. Figure 8.7).

10.2.3 Granitic Rocks Associated with LMIs

The uppermost portions of some layered mafic intrusions contain granites and granophyres (granites containing fine-grained intergrowths of quartz and feldspar) that appear to derive from extreme fractional crystallization of the basaltic parent melt. These granitic rocks are markedly anhydrous, which means biotite and

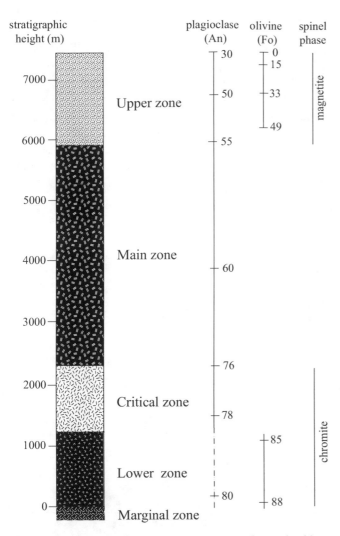

Figure 10.3 Stratigraphic cross-section across the Bushveld Complex showing the variation in olivine and plagioclase compositions. After Atkins (1969).

hornblende are rare, and they commonly contain fayalite, indicating extreme iron enrichment. Examples include the granophyres of the Skaergaard intrusion (Wager and Brown, 1967), and the expansive, iron-rich granitic rocks that cap the Bushveld intrusion (Map 10.2; Vantongeren et al., 2010).

10.2.4 Tectonic Environments of LMIs

The major tectonic environments established for the formation of layered mafic intrusions include (1) rifting environments, (2) deep portions of volcanoes, and (3) astroblemes.

Rifting Environments. Some layered mafic intrusions are clearly associated with continental rifts. Perhaps the best example is the Duluth intrusion, which intruded the coeval flood basalts of the Keweenaw rift (Green and Fitz,

1993). Another example is the Dufek intrusion, which was emplaced in the Jurassic Period, when Gondwana began to rift apart.

Other LMIs are not associated with continental extension. The Great Dyke of Zimbabwe is a distinctly tabular body, more than 500 km long. Although dikes are commonly interpreted to indicate rifting, the Great Dyke was emplaced at a time when there is no known local rifting event (Oberthür et al., 2002). Other large mafic intrusions, including the Bushveld and Stillwater intrusions, contain magma emplaced on a scale similar to that of flood basalts, but there is no clear geologic evidence that either formed in a rifting environment.

Plutons Beneath Volcanoes. Some smaller layered intrusions, including Skaergaard, Rum (in Scotland), and Duke Island, probably formed within a magma chamber beneath a volcano. Both Skaergaard and Rum represent basaltic volcano-plutonic complexes associated with the Tertiary opening of the Atlantic Ocean, and therefore are rift-related. The Duke Island intrusion, which lacks iron-enriched silicates and contains early magnetite crystallization, is of calc-alkalic affinity and probably formed at the base of an arc volcano. Frost and Lindsley (1992) list several other mafic intrusions that likely formed beneath arc volcanoes. Although technically these calc-alkalic intrusions form in the magmatic arc-type environments discussed in Chapter 8, they are included here because their structure and layering is similar to that of other layered mafic intrusions.

Astroblemes. There is very strong evidence that the Sudbury intrusion, which hosts the world's largest nickel deposit, formed as the result of a meteorite impact, approximately 1.85 billion years ago (Faggart and Basu, 1985). Evidence for this includes the fact that the area beneath the intrusion is highly brecciated and includes shatter cones – rocks deformed into cone-shaped, nested shears – a feature found only in meteorite impacts. Furthermore, the intrusion is blanketed by fallback breccia that was deposited following the impact. Finally, Sudbury is the only quartz-bearing layered mafic intrusion (Naldrett et al., 1970), suggesting its parent melt formed from a mixture of mafic, mantle-derived melt and felsic, crustal melt. Because Sudbury has been successfully explained as an astrobleme, some petrologists argue that the Bushveld and Stillwater layered mafic intrusions, of uncertain origin, may also have formed by a similar process. It is important to note that neither of these contains the overwhelming evidence seen in the Sudbury intrusion for an extraterrestrial trigger for the magmatism.

10.3 Anorthosites and Related Rocks

Anorthosite is a rock composed of 90 percent or more plagioclase. Anorthosite layers, up to several meters thick, are found in most layered mafic intrusions (Figure 10.1C). In addition, anorthosite occurs in two types of intrusions that may occur on a batholithic scale. These intrusions, which are compositionally and texturally distinct, are called *Archean* and *massif anorthosites* (Map 10.3; Tables 10.2 and 10.3).

10.3.1 Archean Anorthosites

Archean anorthosites consist of clusters of plagioclase megacrysts (~$An_{85\pm5}$) that may be 1–5 cm in diameter, surrounded by a matrix of mafic material of basaltic composition (Figure 10.4). They generally form small dike- and sill-like bodies, but the largest crop out over areas of more than 100 km^2 (Ashwal, 2010). Perhaps the best-studied Archean anorthosite is the Fiskenæsset

intrusion in west Greenland (Myers, 1976; Polat et al., 2011). Although the intrusion has been intensely deformed and metamorphosed to amphibolite (and, locally, granulite) facies, studies in low-strain zones where cumulate textures survive have allowed geologists to construct a reasonable stratigraphy for the intrusion (Figure 10.5). The lower portions consist of mafic and

Table 10.2 *Differences between Archean and massif anorthosites*

Anorthosite type	Archean	Massif
Age	Entirely Archean	Mostly Proterozoic
Plagioclase composition	ca. An_{85}	ca. An_{50}
Size	Small, generally <100 km^2; largest is 2200 km^2	Large, largest is 17 000 km^2
Shape	Sill-like	Domal
Associated rocks	Gabbro, peridotite, pyroxenite	Monzonite, syenite

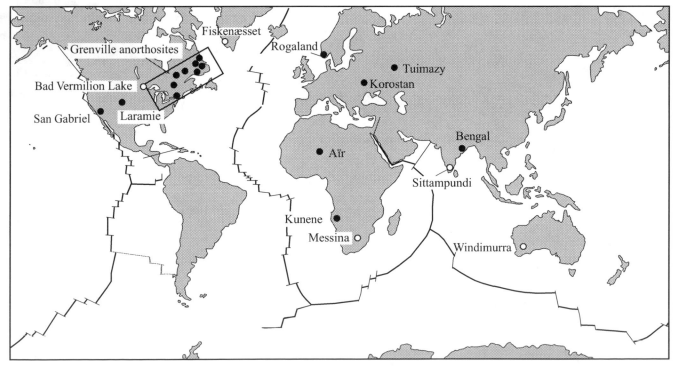

Map 10.3 Worldwide distribution of massif anorthosites (filled circles) and Archean anorthosites (open circles). The box outlines an area encompassing the Grenville anorthosites; the names of the individual Grenville anorthosites are given in Map 10.4.

Table 10.3 *Major anorthosite complexes*

Pluton	Location	Age (Ma)	Area (km²)	Reference
Archean anorthosites				
Windimurra	Western Australia	2813	2250	1
Fiskenæsset	Greenland	2970	500	1, 2
Bad Vermilion Lake	Ontario, Canada	~2750	100	1
Messina	South Africa	3345	~100	1, 3
Sittampundi	India	~2935	~30	1, 4
Massif anorthosites				
Lac Saint-Jean	Quebec	1150	17 000	1
Kunene	Angola	1385	18 000	1, 5
Harp Lake	Labrador, Canada	1460	10 000	1
Nain	Labrador, Canada	1305	7000	1
Lac Allard	Quebec, Canada	1060	5500	1
Marcy	New York, USA	1155	3000	6
Mealy Mountains	Labrador, Canada	1632	2600	1
Morin	Quebec, Canada	1155	2500	1
Korosten	Ukraine	1790	2185	1
Michikamau	Labrador, Canada	1450	2000	1
Tuimazy	Russia	2570	1800	1
Laramie	Wyoming, USA	1435	680	1, 7
Rogaland	Norway	930	580	8
Air	Niger	430	475	1
Labrieville	Quebec, Canada	1010	250	9
San Gabriel	California, USA	1125	250	1
Bengal	India	1550	250	1

1, Ashwal (2010); 2, Polat et al. (2011); 3, Zeh et al. (2010); 4, Bhaskar Rao et al. (1996); 5, Mayer et al. (2004); 6, McLelland et al. (2004); 7, Frost et al. (2010); 8, Schärer et al. (1996); 9, Owens et al. (1994).

Figure 10.4 Photo showing agglomeration of coarse plagioclase grains in leucogabbro, a texture common in Archean anorthosites. The pencil in the upper right is 15 cm long. Fiskenæsset intrusion, west Greenland. Photo by John Myers.

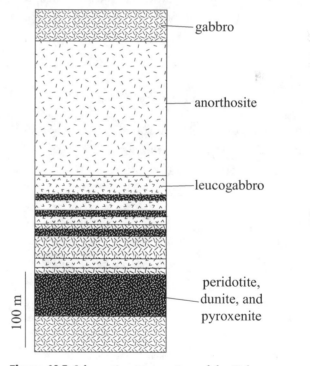

Figure 10.5 Schematic cross-section of the Fiskenæsset intrusion. Modified after Myers (1976).

ultramafic rocks which become less abundant stratigraphically upwards, while leucogabbro and anorthosite become more abundant. The primary melt probably resembled that supplying modern arc magmas, as indicated by the presence of magnetite in the peridotites and primary hornblende in the gabbro (Polat et al., 2011) (attributes shared with the Duke Island intrusion; Section

10.2.2). Archean anorthosites are assumed to have formed by shallow-level intrusion of dikes and sills of basaltic magma, carrying a slurry of plagioclase crystals. Subsequent differentiation formed the ultramafic horizons (Polat et al., 2011) whereas the plagioclase floated to the top of the intrusion. The exact tectonic environment is uncertain because, as the name implies, Archean anorthosites are restricted to Archean shields; no Phanerozoic example has yet been discovered.

10.3.2 Massif Anorthosites

Massif anorthosite intrusions are batholith-sized plutons composed almost entirely of plagioclase. These differ from Archean anorthosites in that they are composed almost entirely of relatively sodic plagioclase (An_{40-50}) and lack ultramafic units. These anorthosite bodies have been emplaced as domed intrusions that may cover an area of thousands of kilometers (see Table 10.2). To distinguish them from Archean anorthosites, this kind of anorthosite intrusion is called *massif anorthosite*. Ore deposits associated with massif anorthosite include iron–titanium oxides, nickel, copper, and cobalt (Box 10.2).

Massif anorthosite plutons also are different from layered mafic intrusions in several ways. First, massif anorthosite intrusions are dominated by anorthosite and rarely contain rocks with less than 90 percent plagioclase. In addition to lacking ultramafic units, gabbroic rocks are not abundant. Second, although some massif anorthosite intrusions are strongly layered, such as the Laramie anorthosite complex (Scoates et al., 2010), many are not. Finally, massif anorthosites are mostly restricted to the Proterozoic Eon, with only a few rare occurrences in the Archean or Phanerozoic eons (Table 10.3).

The Laramie Anorthosite Complex. In North America, Proterozoic massif anorthosites are exposed in a band extending from northeastern Canada to the southwestern United States. They are part of the mid-Proterozoic igneous event that led to the emplacement of granitic rocks

---

BOX 10.2 | ORE DEPOSITS ASSOCIATED WITH ANORTHOSITE COMPLEXES

Iron–Titanium Oxides. Because the floor of anorthosite complexes is rarely exposed, massif anorthosite intrusions lack the chromium and platinum-group ore deposits commonly associated with layered mafic intrusions. However, anorthosite massifs do contain concentrations of iron–titanium oxide ore, similar to the concentrations of magnetite found in the upper levels of layered mafic intrusions (see Box 10.1). These ore bodies are associated with late leucogabbroic rocks where they form concentrations near the floor of the leucogabbro, or as dike-like bodies intruding the underlying anorthosite. The iron–titanium oxide bodies in anorthosite are poorer in vanadium and richer in titanium than similar bodies associated with LMIs. The Tellnes deposit in south Norway, which is associated with the Rogaland anorthosite, is the largest hard-rock titanium deposit in the world.

Nickel, Copper, and Cobalt. Geologists had assumed that, like chromium and PGE deposits, base-metal[1] deposits associated with anorthosites were left behind, deep in the crust, during the initial crystallization of an anorthosite intrusion. That assumption was proved wrong by the 1992 discovery of giant nickel deposits at Voisey's Bay. The Voisey's Bay deposit occurs in an early troctolite intrusion within the Nain anorthosite suite. Evidently, before olivine crystallization could deplete nickel from the melt, assimilation of country rock drove the melt toward sulfide saturation (Amelin et al., 2000). Because of its density the sulfide melt sank and concentrated in constrictions of the magma conduit, forming one of the largest nickel deposits in the world.

[1] Base metals, which include iron, nickel, copper, and lead, are so called because they are inexpensive, in contrast to precious metals such as silver and gold.

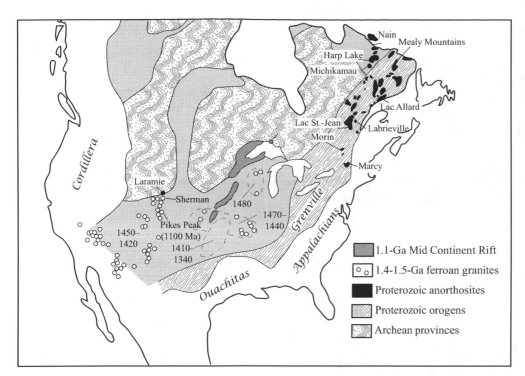

Map 10.4 Geologic map showing the Archean and Proterozoic provinces of North America and the distribution of Proterozioc massif anorthosites and ferroan granites. Age ranges of ferroan rocks given in millions of years.

across all of the southwestern and central United States and extending into northeastern Canada and Fennoscandia (Map 10.4). Unlike many of the Grenville anorthosite complexes, the Laramie anorthosite complex in southeastern Wyoming was not subsequently metamorphosed, and therefore preserves the relationships between massif anorthosite and coeval granitic intrusions. The Laramie anorthosite complex, which crops out over an area of 800 km^2 in the southern Laramie Range, was emplaced across the Cheyenne belt, a major crustal suture that separates Archean rocks of the Wyoming Province to the north from Proterozoic rocks to the south (Map 10.4).

The Laramie anorthosite complex consists of three anorthositic intrusions, three syenitic to monzonitic intrusions, and a number of smaller intrusions of leucogabbro and ferrodiorite (Map 10.5; Frost et al., 2010; Lindsley et al., 2010; Scoates et al., 2010). The earliest intrusion, the Poe Mountain anorthosite, has an average plagioclase composition of An$_{45}$. Plagioclase in the Chugwater anorthosite averages around An$_{55}$, and plagioclase from the Snow Creek anorthosite ranges from An$_{45}$ to An$_{55}$. Each of these three anorthositic plutons forms a broad dome in which stratigraphically deeper layers of the intrusion are encountered as one moves inward from the margin of the intrusion. For each

intrusion only a portion of the dome is preserved; the rest has been truncated by later intrusions or by ca. 65-Ma Laramide faulting, which produced the uplift in which the rocks are exposed. The rocks of the Laramie anorthosite complex display a complete range in textures, from clearly magmatic textures that are similar to those seen in layered mafic intrusions, to the highly deformed and recrystallized textures that record deformation related to emplacement, typical of most massif anorthosites.

At high structural levels in the Poe Mountain anorthosite, the anorthosite and associated leucogabbro exhibit igneous layering, manifested by differing abundances of plagioclase and pyroxene (Figure 10.6). The rocks have typical igneous textures with tabular plagioclase and interstitial pyroxene and olivine (cf. Figure 1.7A). At lower structural levels the rocks become more plagioclase-rich and more deformed (Scoates et al., 2010). Intergrown equant grains replace the tabular plagioclase. Undeformed igneous textures are less common in the Chugwater anorthosite, even at higher structural levels, and are rare in the Snow Creek anorthosite.

Two later monzonitic plutons, the Sybille monzosyenite and the Maloin Ranch pluton, were intruded along the flanks of the anorthosite domes (Kolker et al., 1991; Scoates et al., 1996). A third

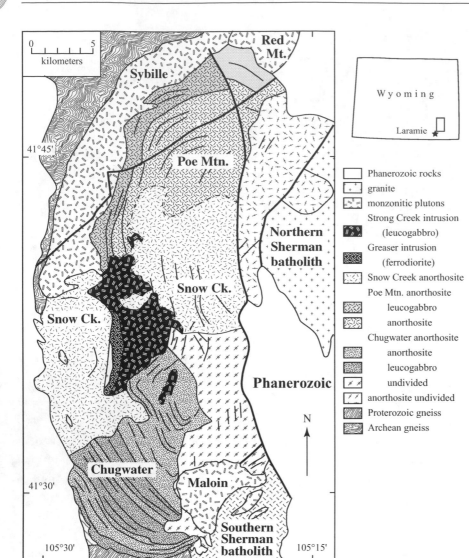

Map 10.5 Geologic map of the Laramie anorthosite complex. Inset shows its location within Wyoming, USA. Modified after Frost et al. (2010).

monzonitic body, the Red Mountain pluton (Anderson et al., 2003), intruded the Sybille monzosyenite in the northeastern part of the complex. The Sybille, Maloin, and Red Mountain plutons are inferred to have formed from residual, anorthositic liquids, with minor to moderate crustal contamination (Kolker et al., 1991; Scoates et al., 1996; Anderson et al., 2003).

Origin of Massif Anorthosite. Two observations from experimental work provide insight into understanding the origin of massif anorthosites. One, as pressure increases, plagioclase in equilibrium with augite becomes more sodic. Two, high-pressure conditions restrict the crystallization field for plagioclase. These observations have led to the conclusion that the Archean anorthosites, with calcic plagioclase, probably crystallized at shallow levels in the crust, and that the relatively sodic plagioclase in the massif anorthosite formed at deeper

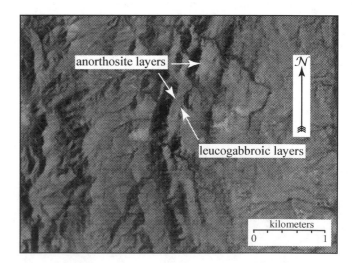

Figure 10.6 Google Earth image showing layering in the Poe Mountain anorthosite, Wyoming, USA. Light-colored anorthosite layers tend to weather in relief, whereas darker leucogabbroic layers tend to erode to form low topography.

crustal levels. The most widely accepted model for the formation of massif anorthosite involves a two-stage process (Longhi and Ashwal, 1985). First, basaltic magma, probably tholeiitic in composition, ponded near the base of the crust. Olivine and aluminum-rich pyroxene crystallized in this deep-level magma chamber, just as they do in layered mafic intrusions. This stage of crystallization is recorded in high-aluminum pyroxenes, which occur as xenocrysts and enclaves in many anorthosites around the world (Charlier et al., 2010). The second stage involved the injection of the residual, feldspathic magma, produced by crystallization in the deep magma chamber, into a second, mid-crustal-level magma chamber, leaving the ultramafic cumulates behind. Plagioclase crystals accumulated in this second chamber, trapping variable amounts of interstitial melt and resulting in thick sequences of layered anorthositic cumulates. These relatively buoyant plagioclase-rich cumulates and interstitial melt tended to rise diapirically, producing large-scale domal structures. Deformation was accommodated by varying degrees of recrystallization of plagioclase and may have led to local mobilization of interstitial melt. The interstitial melt and residual melt produced during extensive fractionation of plagioclase form the minor iron–titanium oxide, ferrodiorite, and gabbroic dikes that intrude most of the massif anorthosite plutons. The monzonitic rocks may represent a portion of residual liquids produced during crystallization of the anorthositic cumulates, which were then variably contaminated by crustal wall rocks (Scoates et al., 1996).

10.3.3 Lunar Anorthosites

Few of us will ever hold a lunar anorthosite in our hands, but these rocks are visible every night when the Moon is out. Lunar anorthosite makes up most of the light regions of the Moon, known as the lunar highlands, whereas the dark areas are made up of basalts. The anorthosites consist of extremely calcium-rich plagioclase (An_{94-99}), pigeonite, and augite (Taylor, 2009). Lunar anorthosites are very old (4.456 Ga; Norman et al., 2003) and are inferred to have formed from a magma ocean that encircled the Moon in the early stages of its formation. Plagioclase floated to the surface of the residual basaltic melt, forming the anorthosite. Evidence for this process is recorded in the rare-earth element patterns of these rocks (see Figure 4.15).

10.4 Ferroan Granites

Anorthosite massifs are commonly associated, both in time and space, with distinctive, iron-rich granites. Many massif anorthosite complexes include plutons of these granites, such as in the Laramie anorthosite complex (Map 10.5), but there are also large batholiths of iron-rich granites that are completely devoid of anorthosite. These ferroan granites are distinct from granites found in Cordilleran batholiths. Not only are they more iron-rich, they are also more potassic and usually contain much more potassium feldspar than plagioclase, giving them a distinctly pink color. Many of the ferroan granites are anhydrous, containing fayalite and clinopyroxene rather than biotite or hornblende. These granites were originally called "anorogenic granites" (Loiselle and Wones, 1979) because they were not associated with compressive structures or subduction.

Subsequently, a greater variety of tectonic environments have been proposed for the formation of these rocks and the term has been shortened to "A-type granites." A-type granites have a broad range of geochemical characteristics yet are united by their iron-rich compositions (Frost and Frost, 2013). Because the differences in composition reflect different geologic environments of formation, we have proposed that instead of "A-type" they be called *ferroan granites* (Frost and Frost, 2011). Ferroan granites are more reduced than other granite types. This is indicated mineralogically by the common presence of fayalite, the fact that magnetite is rare or absent in these rocks, and that ilmenite is the main iron–titanium oxide. Iron is present mainly as Fe^{2+} dissolved into the silicates. Common emplacement temperatures are 900 °C or higher, well above the water-saturated solidus for granite. The major fluid species was probably CO_2 rather than H_2O.

10.4.1 The Pikes Peak Batholith

A classic example of ferroan granite is the Pikes Peak batholith, a 1.1-Ga pluton in central Colorado (Map 10.6). As is typical of ferroan granites, the Pikes Peak batholith is not zoned. Rather, it is dominated by rather coarse-grained, pink granite. Relatively mafic rocks, ranging in composition from gabbro to monzonite or syenite, are present locally but they are not distributed around the margins of the plutons as is common in zoned Cordilleran plutons (cf. Map 8.5). Although feldspar fabrics in the Pikes Peak batholith indicate that it was emplaced though three intrusive centers, the

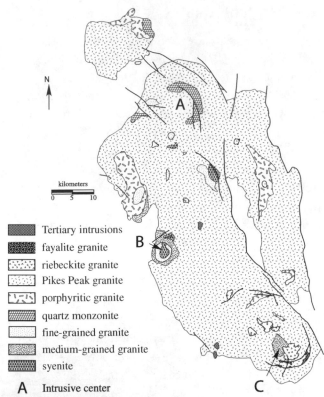

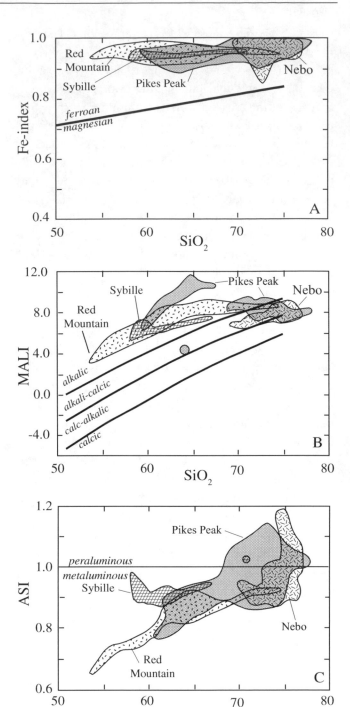

Map 10.6 Geologic map of the Pikes Peak batholith. The batholith includes three intrusive centers, labeled A, B, and C. Modified after Scott et al. (1978), Bryant et al. (1981), and Hutchinson (1976).

Figure 10.7 Geochemical characteristics of ferroan granites associated with mafic intrusions and anorthosites. (A) Iron-enrichment index (Fe-index) versus silica. (B) Modified alkali–lime index (MALI) versus silica (a = alkalic, a-c = alkali-calcic, c-a = calc-alkalic, c = calcic). (C) ASI vs. SiO₂. Data from Anderson et al. (2003), Barker et al. (1975), Kleeman and Twist (1989), Smith et al. (1999), and Scoates et al. (1996).

overwhelming impression one gets from Map 10.6, and from ferroan granites in the field, is their large-scale homogeneity.

10.4.2 Composition of Ferroan Granites

Ferroan granites, as exemplified by Pikes Peak, are compositionally similar to rhyolites derived from differentiation of basalt (Figure 10.7; see Cedar Butte in Figure 9.3). They are ferroan, and alkalic to alkali-calcic. Also shown in Figure 10.7 are the Nebo granite, which caps the Bushveld intrusion, and the Red Mountain and Sybille plutons, which are associated with the Laramie anorthosite complex. Figures 10.7 and 9.4 show that the granitic rocks generated by extreme differentiation of basaltic rock tend to be ferroan, alkalic, and metaluminous. However, portions of both the Nebo and the Pikes Peak granite are peraluminous. This leads to the question: What makes a granitic melt peraluminous?

It is clear that crystallization of olivine, pyroxene, and iron–titanium oxides, none of which contain Al_3O_3, will increase alumina in the residual melt, which is why the aluminium-saturation index (ASI) for most suites tends to increase with increasing silica. However, since all feldspars have an ASI = 1.0, it is impossible for fractional crystallization of feldspar to drive melt compositions to peraluminous compositions. The closer the melts get to ASI = 1.0, the more feldspar crystallizes. Fractional

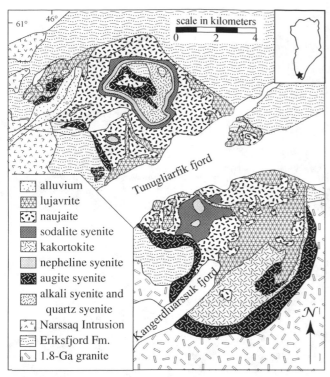

Map 10.7 Geologic map of the Ilimaussaq intrusion. Inset shows its location in Greenland. After Ferguson (1964), © GEUS, Geological Surveys for Denmark and Greenland.

The legend for the map reads:

- alluvium
- lujavrite
- naujaite
- sodalite syenite
- kakortokite
- nepheline syenite
- augite syenite
- alkali syenite and quartz syenite
- Narssaq Intrusion
- Eriksfjord Fm.
- 1.8-Ga granite

crystallization of augite or hornblende can push the melt to weakly peraluminous compositions (Zen, 1986). Experimental petrology shows that peraluminous melts can also be produced by melting rocks that contain minerals with an ASI of >1.0. The classic example is muscovite, which has an ASI of 2.5 to 3.0 and which melts at low temperatures (see Figure 4.11B). Thus, partial melting of a pelitic rock containing muscovite will produce peraluminous melts. Nominally, biotite [K(Fe, Mg)$_3$Si$_3$AlO$_{10}$(OH)$_2$] has an ASI of 1.0; this is the consequence of its formula, which has an equal number of potassium and aluminum atoms. However, most natural biotites have excess aluminum substituting for magnesium, iron, and silicon by the substitution $Al_2 \rightleftharpoons (Mg,Fe)Si$. For this reason, partial melting of tonalitic biotite granites and gneisses can also produce peraluminous melts. Because the excess alumina for these melts resides in biotite, and biotite is one of the first minerals to melt, initial melts generated from tonalitic gneisses will be peraluminous; the ASI of these melts will decrease with increasing degree of melting (Frost and Frost, 2011).

The extremely high FeO/(FeO + MgO) ratio of these granites indicates that, like the Yellowstone rhyolites, they are largely derived from residual melts, produced by partial melting or differentiation of basalt. However, those portions of the Nemo and Pikes Peak granites that are weakly peraluminous may have assimilated a certain amount of crustal melt. This is not surprising considering that the high magmatic temperatures from which these granites were derived almost certainly melted adjacent wall rock.

10.5 Alkaline Complexes

Alkaline intrusions make up an extremely small volume compared to other rocks described in this chapter; they contain minerals uncommon in any other environment, and they are graced with rock names obscure even to most geologists. As a result, the study of alkaline complexes has traditionally been an aspect of petrology with few devotees. However, the proliferation of cell phones, computers, flat-panel televisions, and a variety of clean-energy technologies has changed this picture entirely. These devices and technologies contain permanent magnets, lighting phosphors, and other components that use REEs, which has led to an increased demand for these and other relatively scarce elements (Box 10.3). REEs are concentrated in alkaline intrusions and, as a result, economic geologists now study alkaline intrusions as sources for these metals. Alkaline intrusions are found in ancient rifts, such as in the Gardar province, as well as in areas where rifting is not obvious, as in the Kola peninsula (Map 9.4).

10.5.1 Geology of the Ilimaussaq Intrusion

The Ilimaussaq intrusion serves as a good example of an alkaline pluton for two reasons. First, its tectonic context is clear and both intrusive and extrusive rocks compose the complex: it is intruded into coeval rift-related sediments and alkaline lava flows (the Eriksfjord formation in Map 10.7). Second, it is the most alkaline intrusion in the world and also the host of one of the largest known deposits of rare metals. However, there are hurdles to understanding the crystallization of the Ilimaussaq intrusion, namely, the bizarre minerals that comprise the intrusion and the even more bizarre names to the rocks. In addition to containing common minerals any geologist should know, including alkali feldspar, sodalite, nepheline, and aegirine, the Ilimaussaq contains two rock-forming minerals rarely found except in alkaline rocks: eudialyite and arfvedsonite. Eudialyite is a complex sodium–zirconium silicate that forms in sodium-rich rocks in place

BOX 10.3 | ENRICHMENT OF RARE METALS IN ALKALINE PLUTONS

Because of their use in a number of "green" energy technologies, including wind turbines, electric vehicles, photovoltaic cells, and fluorescent lighting, there is an increasing demand for REEs and other scarce metals, including indium, gallium, tellurium, lithium, and yttrium. Alkaline intrusions are a major source for many of these metals. For example, the Ilimaussaq intrusion hosts one of the richest REE deposits in the world, where the lujavrite contains more than 1 percent total REEs (Bailey et al., 2001). Alkaline rocks are enriched in these rare metals by the same process that causes layered mafic intrusions to be enriched in chromium, nickel, copper, and PGEs – igneous fractionation. PGEs, chromium, and nickel have extremely large partition coefficients between mineral and melt (or between silicate melt and sulfide melt) and concentrate early in the fractionation history of an igneous system, such as in layered mafic intrusions. In contrast, elements such as the REEs have extremely low partition coefficients between mineral and melt. These elements concentrate by orders of magnitude in the residual melt if sufficient fractionation has taken place (see Figure 4.13).

Extreme enrichments in rare metals in an igneous system are the result of two features. First, the bulk distribution (D) for the rare metals between melt and the crystallizing phase must remain low throughout the crystallization history of a rock. Second, igneous fractionation must proceed until only a very small fraction of melt remains. In most igneous rocks, zircon concentrates REEs, particularly the heavy REEs, so once zircon begins to crystallize out of a melt, the bulk D between melt and the crystallizing phases becomes greater than one. At this point these heavy metals behave compatibly and their concentration in the residual melt decreases with increasing fractional crystallization. This effect is observed in the Red Mountain pluton, where zircon saturation occurred at around 60 percent SiO_2 and fractional crystallization after this point depleted zirconium (as well at REEs, hafnium, and tantalum) from the melt (see Figure 4.14). Zircon has high solubility in alkaline melts (Linnen and Keppler, 2002), which inhibits crystallization of zircon in alkaline melts. If zircon does not crystallize, then the bulk D for REEs remains near zero, allowing them to be enriched in the residual melt. Eventually, the zirconium concentration becomes so high that eudialyite, the sodium–zirconium silicate, crystallizes instead of zircon.

The other feature that enriches REEs in alkaline rocks is that alkaline melts continue crystallizing (and hence differentiating) down to very low temperatures. Most granitic magmas solidify at around 650–700 °C (see Figure 3.3). However, the alkaline melt at Ilimaussaq may have remained molten to temperatures well below 600 °C (Markl, 2001). Alkaline rocks are rich in elements such as lithium, fluorine, sodium, beryllium, and boron that are known to flux silicate melts to relatively low temperature. The reason alkaline rocks crystallize to such low temperatures is complex and is related to the sodium-rich composition of the melt. Because highly sodic rocks contain more Na_2O than feldspars can accommodate, the extra Na_2O must be taken up by amphiboles (riebeckite or arfvedsonite) and pyroxenes (aegirine), in which $NaFe^{3+}$ substitutes for $MgFe^{2+}$. Iron in the parent magma is likely to be ferrous, as in the augite syenite in the Ilimaussaq intrusion. The excess oxygen necessary to convert the ferrous iron to ferric iron likely sources from H_2O and CO_2 bound in the melt structure (Markl et al., 2010). The reaction is:

$$8\,\underset{\text{melt}}{FeO} + 2\,H_2O + CO_2 = 4\,\underset{\text{melt}}{Fe_2O_3} + \underset{\text{fluid}}{CH_4}$$

The depletion of water from the melt to make ferric iron and methane means that, unlike typical granitic magmas, the alkaline magmas never reach water saturation; they become saturated in CH_4-rich fluids instead. Indeed, CH_4-rich fluid inclusions are common in alkaline rocks (Konnerup-Madsen and Rose-Hansen, 1982). Because water never evolves during fractionation of alkaline intrusions, elements such as lithium, fluorine, sodium, beryllium, and boron remain in the melt rather than being fractionated into the fluid. This allows alkaline melts to fractionate down to lower temperature, thus increasing the REE content of the residual liquid.

Box Figure 10.3 REE ore from the Bear Lodge alkaline igneous complex, Wyoming. The sample is composed of >50 percent calcite (Cal). The primary REE minerals have been replaced with dendritic ancylite [$SrCe(CO_3)_2(OH)H_2O$], fibrous aggregates of synchysite [$Ca(Ce,La,Nd,Y,Gd)(CO_3)_2F$], and parisite, with interstitial hematite, strontianite, barite, celestine, calcite, and sulfides. Galena, pyrite, sphalerite (Sp), phlogopite, and moderately to strongly altered potassium feldspar (Kfs) are disseminated throughout the sample. Photo courtesy of Danielle Olinger and Rare Earth Resources, Ltd.

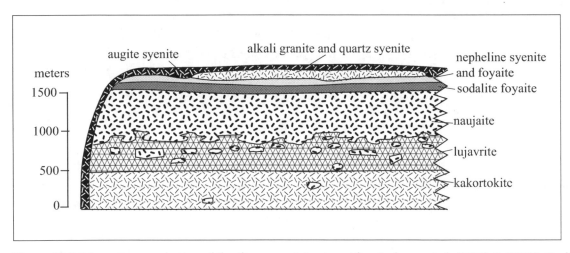

Figure 10.8 Schematic cross-section of the Ilimaussaq intrusion. After Andersen et al. (1981), © GEUS, Geological Surveys for Denmark and Greenland.

of zircon. Arfvedsonite is a highly sodic amphibole with the formula $Na_3Fe^{2+}_4Fe^{3+}Si_8O_{22}(OH)_2$. In this amphibole, sodium fills both the calcium site (as in riebeckite) and the vacant site (as in hornblende). The charge balance is maintained by ferric iron substituting for ferrous iron.

Although some of the rock units in the Ilimaussaq intrusion are named using mineralogical naming terminology (see Figure 1.1), the three most important units do not, and are nearly unpronounceable to an English speaker unfamiliar with the Nordic alphabet (in which

J is pronounced "ya"). These three terms are based more on textural features than mineralogy. Kakortokite is a layered nepheline syenite containing nepheline, alkali feldspar, arfvedsonite, and eudialyite. Naujaite is a sodalite syenite that consists of euhedral crystals of sodalite surrounded by crystals of alkali feldspar, arfvedsonite, and eudialyite. Lujavrite is an arfvedsonite- or aegirine-rich nepheline syenite.

The Ilimaussaq intrusion is nearly completely rimmed by augite syenite (Map 10.7 and Figure 10.8), which is a

Figure 10.9 Photo of the Ilimaussaq intrusion taken on the southeast shore of Kangerdluarssuk fjord, looking northeast. Kakortokite, composed of cumulus minerals on the floor of the complex, is in the foreground; naujaite is the sodalite-rich floating cumulates in the background, and lujavrite is the horizon sandwiched between the two.

syenite that is neither quartz- nor nepheline-bearing. This grades inward to nepheline syenites and various other K-feldspar-rich syenites. The main portion of the pluton consists of the crystallization sequence kakortokite, naujaite, and lujavrite. These rocks comprise the original

magma chamber for the intrusion. During the crystallization of the pluton, dense, crystallizing phases, eudialyite and arfvedsonite, accumulated on the floor of the chamber, interlayered with alkali feldspar, and formed the layered kakortokites (see Figure 10.9). At the same time, sodalite grains, which were less dense than the melt, floated to the top of the chamber, where they were cemented together with eudialyite, arfvedsonite, and alkali feldspar, forming the naujite. From time to time, blocks of the naujite broke off the roof of the intrusion and fell into the residual melt and sometimes were incorporated into the kakortokites. The residual melt eventually crystallized to make the lujavrite.

In addition to the aforementioned differentiation of the Ilimaussaq intrusion by fractional crystallization processes, at least two other magmatic processes were involved in forming the complex. First, limited assimilation of the lower crust is implicated in the formation of the peralkaline granite. Second, evidence that the rocks of the complex interacted with late-stage gaseous and aqueous fluids is preserved in late-stage pegmatites and hydrothermal veins (Marks et al., 2004). These fluids are implicated in the enrichments in uranium, thorium, REEs, and beryllium found in Ilimaussaq and other alkaline plutons (see Box 10.3).

Summary

Intraplate magmatism involves a complex assortment of magma types, both extrusive and intrusive. The wide variety of intraplate magmatism is caused by different compositions of mafic magmas that can be produced during mantle melting, the extent to which they interact with the crust, and the various depths at which they differentiate. The formation of this diversity of igneous rocks is summarized in Figure 10.10 and Figure 10.11.

- Partial melts of the mantle are implicated in the formation of intraplate magmatism. As shown in Figure 10.10, the melts generated from the mantle may range from tholeiite to highly alkaline in composition, depending on the composition of the mantle melted and the depth and extent of melting. Abundant melting produces tholeiitic melts, which may be emplaced as flood basalts or may pond at relatively shallow levels to produce a layered mafic intrusion (Figure 10.11A).

- Differentiation of the mafic melts in LMIs or anorthosite complexes will lead to the production of syenite, which could undergo further differentiation or crustal assimilation to produce ferroan granite (Figure 10.11A).

- Ponding of the melt at the base of the crust may produce feldspar-rich, crystal-laden magmas that ascend to form anorthositic intrusions (Figure 10.11B). The felsic differentiates of these basalts may be emplaced into the upper crust to produce ferroan granites (Figure 10.11B).

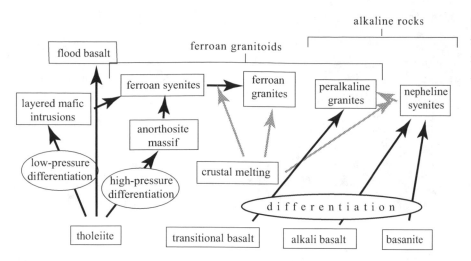

Figure 10.10 Flow chart showing how the range of rock types found in intracontinental magmatism are related. Modified after Frost and Frost (2011).

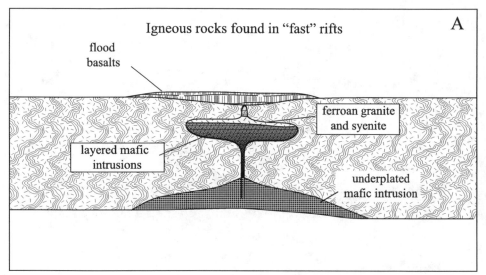

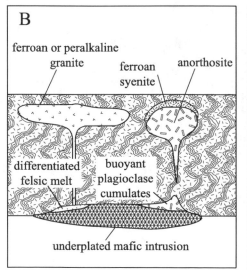

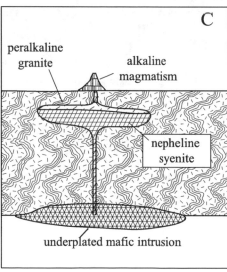

Figure 10.11 Diagrams showing the igneous rocks found in continental rifts. (A) Fast rifts produce flood basalts and their plutonic equivalent, layered mafic intrusions. (B) Slow rifts with tholeiitic magmatism produce underplated mafic magma. Floating plagioclase cumulates may be emplaced as solid diapirs as anorthosite and felsic differentiates could produce ferroan or peralkaline granites. (C) Slow rifting with alkali basalt magmatism will result in the formation of nepheline syenite. Crustal assimilation could produce peralkaline granite.

- Differentiation of diverse parent magmas at various depths in the crust also is a factor in the diversity of intraplate magmatism (Figure 10.10).

- Transitional basalts, those that are sodium-rich but not necessarily alkaline, may differentiate to produce peralkaline granites and alkali basalts, or basanites will differentiate to produce nepheline syenite (Figure 10.11C).

- Throughout the differentiation process the mantle melts may interact with crustal melts (Figure 10.10). Small degrees of melting of crustal rocks may produce ferroan granite. Alternatively, assimilation of crustal melts by syenitic melts produced by differentiation of tholeiites may also make ferroan granite (Figure 10.11A). Mixing of crustal melts with nepheline syenite melts can produce peralkaline granites (Figure 10.11C).

Questions and Problems

Problem 10.1. Summarize the tectonic environments in which layered mafic intrusions may form and give an example of each.

Problem 10.2.

a. Referring to Figure 2.17, identify several different sequences of layering that might develop in layered mafic intrusions by the crystallization of a basaltic parent. (Note that the figure shows one such sequence starting with basalt liquid composition Y.)

b. Compare these to the stratigraphic section through the Fiskenæsset intrusion shown in Figure 10.5.

   i. Can the layering in the Fiskenæsset intrusion be explained by any of the crystallization sequences you identified in 2a?

   ii. Under what circumstances might a layer of dunite be followed by a layer of pyroxenite?

   iii. How might the pure anorthosite layer be formed?

Problem 10.3. What are the differences between Archean and massif anorthosites?

Problem 10.4. How do ferroan granites like the Pikes Peak granite (Map 10.6, Figure 10.7) differ from continental arc plutons such as the Tuolumne intrusion (Figure 8.9 and Map 8.5)?

Further Reading

Ashwal, L. D., 2010, The temporality of anorthosites. *Canadian Mineralogist*, 48, 711–28.

Frost, C. D. and Frost, B. R., 2011, On ferroan (A-type) granites: Their compositional variability and modes of origin. *Journal of Petrology*, 52, 39–53.

Frost, C. D., Frost, B. R., and Beard, J. S., 2016, On silica-rich granitoids and their eruptive equivalents. *American Mineralogist*, 101, 1268–84.

O'Driscoll, B. and VanTongeren, J. A., 2017, Layered intrusions: From petrologic paradigms to precious metal repositories. *Elements*, 13, 383–9.

Saunders, A. D., 2005, Large igneous provinces: Origin and environmental consequences. *Elements*, 1, 259–63.

Winter, J. D., 2010, *Principles of Igneous and Metamorphic Petrology*, 2nd edn. Prentice Hall, New York, NY, Chapter 12: Layered mafic intrusions.

Chapter 11

Interpretation of Granitic Rocks

11.1 Introduction

Granitic rocks are the most abundant rocks in continental crust and geologists have developed many ways to classify them. This text has relied heavily on the geochemical classification of Frost et al. (2001), but it is important that petrology students understand other classification schemes are in use. This chapter begins with a summary of other classification schemes for granitic rocks: (1) the mineralogical classification, (2) a classification based on the presence or absence of magnetite, and (3) an "alphabet" classification, based on the inferred origin of the granites.

The chapter continues with a discussion of granitic rocks from continental collisions. They are compositionally distinct from the Cordilleran and ferroan granites discussed in previous chapters. A summary of the four geochemical types of granitic rocks, noting the compositional characteristics of each type and showing how the compositional changes are manifested in the mineralogy of the granitic rocks concludes the chapter.

11.2 Classification of Granitic Rocks

11.2.1 Mineralogical Classification

The mineralogical classification discussed in Chapter 1.1, which is the simplest of the classifications for granitic rocks, is based on modal proportions of quartz, plagioclase, and alkali feldspars. The advantage of this classification is that it can be readily applied in the field and it is simple to use. The major drawback is that it ignores compositional variations apart from those that affect the feldspar compositions. Thus mafic and felsic rocks may plot in the same field even if they have significantly different chemical compositions. Furthermore, the classification cannot address the presence or absence of minor phases, such as muscovite, which may convey significant petrologic information.

11.2.2 Classification Based on Opaque Oxides

A second classification scheme for granites is based on the presence of magnetite. Most granites contain magnetite but some do not. This has led to the classification of granites in terms of the occurrence or absence of magnetite (Ishihara, 1977). Magnetite contains Fe^{3+}; therefore magnetite will be present in relatively oxidized rocks and absent in relatively reduced ones. *Magnetite granites* are relatively oxidized granites that contain magnetite and ilmenite as the major oxides, whereas *ilmenite granites* are relatively reduced and contain ilmenite as the only iron–titanium oxide. One reason that granites may lack magnetite relates to the composition of the magma source. Partial melting of pelitic rocks produces magma that crystallizes peraluminous, ilmenite granites. Since pelitic rocks contain graphite, and graphite is a reducing agent, magnetite will not be stable in these magmas. The advantage of this classification scheme is that it is simple; one can classify granites by measuring their magnetic susceptibility. In fact, aeromagnetic surveys map the distribution of magnetite and ilmenite granites over a wide terrain. The disadvantage is that the classification scheme telescopes the whole range of granite compositions into two categories and hence overlooks the chemical complexity of granitic rocks.

11.2.3 Alphabetic Classification

Chappell and White (1974) developed an alternate classification scheme that emphasizes the origin of granites.

For obvious reasons it is sometimes referred to as the alphabetic classification. The major types of granites in this classification are:

I-type (I = igneous): Metaluminous granites that are typically magnetite-bearing. I-type granites are inferred to be produced by differentiation of andesite or partial melting of an igneous source.

S-type (S = sedimentary): These peraluminous granites are typically magnetite-free. They are inferred to be produced by partial melting of pelitic rocks. Hence S-type granites are assumed to come from a sedimentary source.

A-type (A = anorogenic): These are granites not associated with an obvious penetrative contractional orogeny. They are compositionally distinct from I-type granites, being almost exclusively ferroan and higher in potassium, rare-earth element (REEs), and zirconium. They are inferred to be produced by partial melting or fractional crystallization of mafic rocks.

Even though this classification scheme is widely used, many petrologists look upon it with disfavor because it is genetically dependent. It assumes that petrologists know what rocks melted to make the granite. However, it turns out that peraluminous granites don't only come from melting of sedimentary rocks; they can form from small degrees of melting of a tonalite (Skjerlie and Johnston, 1993; Patiño Douce, 1997). Similarly, melting of an immature greywacke can form granite that is compositionally "I-type." An ideal classification scheme depends only on the features seen in the rock, not on inferred origin. For these reasons the alphabetic classification of granites is slowly falling out of favor.

11.2.4 Geochemical Classification

A number of major-element indices of differentiation were presented in Section 4.4, and these have been used in Chapters 8, 9, and 10 to identify the processes by which magmas evolve. However, these indices are also useful in classifying granitic rocks. The classification scheme of Frost et al. (2001) and Frost and Frost (2008) relies on four geochemical indices to classify granitic rocks. The indices, which are reviewed later in this chapter, are based on the major-element abundances in the rock analyses and include:

Iron-enrichment index (Fe-index) ($FeO + 0.9Fe_2O_3$)/($FeO + 0.9Fe_2O_3 + MgO$): This index measures the iron–magnesium ratio of the ferromagnesian silicates. *Ferroan* rocks have either undergone extensive fractional crystallization of olivine and pyroxene and minor magnetite fractionation or formed by minor melting of crustal rocks. *Magnesian* rocks have undergone early crystallization of magnetite, which suppressed iron enrichment.

Modified alkali–lime index (MALI) ($Na_2O + K_2O - CaO$): This index monitors the compositions of the feldspars in the rock. This is modified after the alkali–lime index of Peacock (1931) and involves four classes. In order of increasing alkalinity these are *calcic, calc-alkalic, alkali-calcic,* and *alkalic.*

Aluminum-saturation index (ASI) [molecular Al/(Ca – 1.67P + Na + K)]: This index compares the amount of aluminum, calcium, sodium, and potassium in the rock to the amounts needed to make feldspars. Phosphorous is included because small amounts of apatite are present in rocks and the calcium in apatite is not available for incorporation into feldspars. This index determines whether a rock is *metaluminous,* in which case it has more calcium, sodium, and potassium than feldspars consume, or *peraluminous,* in which case the rock has excess aluminum. Peraluminous rocks contain aluminous minerals including muscovite, garnet, sillimanite, and cordierite.

Alkalinity index (AI) [molecular Al – (Na + K)]: This index determines the balance between aluminum and alkalis (sodium and potassium). A rock that contains excess aluminum may be metaluminous or peraluminous (see the ASI index). A rock with excess sodium and potassium is *peralkaline* and will contain sodic pyroxene or amphibole.

Using these indices, Frost et al. (2001) described 14 chemical varieties of granitic rocks, which fall into four distinct tectonic environments. Chapters 8 and 10 examined granitic rocks formed in two of these tectonic environments. Cordilleran-type granites form in arc environments and are characterized by magnesian compositions, dictated by early crystallization of magnetite. In contrast, ferroan granites are characteristic of rifting environments, where they form by partial melting or extreme differentiation of basaltic magma.

This chapter describes the other two types of granite, both of which occur during continental collision. *Peraluminous leucogranites* may form as tectonic decompression brings hot rocks to shallow crustal levels. After collision, the base of the thickened crust may delaminate as the lower portion of it sinks into the mantle. Mantle lithosphere rising into the space vacated by delaminated crust may melt, causing the formation of "post-tectonic" or *Caledonian granites.* Granites are associated with a number of different ores (Box 11.1).

11.3 Peraluminous Leucogranites

Many orogenic belts contain small plutons of leucogranite, emplaced during the height of deformation. These granites are silica-rich and are composed mostly of quartz and feldspars (and hence are *leuco*cratic) with only a small amount of muscovite, garnet, or biotite. These rocks are distinctly peraluminous and hence are termed *peraluminous leucogranites.* Some peraluminous leucogranites may also contain sillimanite, cordierite, or tourmaline as important phases. Peraluminous leucogranites form by decompression melting, in a manner analogous to the process that forms basaltic melts at a spreading center. In a mountain belt where the crust has been tectonically thickened, deeply buried rocks may get much hotter than is necessary to melt water-saturated granite (Figure 11.1). However, because high-grade metamorphism leaves rocks H_2O undersaturated, little melt, if any, is produced when the water-saturated granite melting curve is crossed. However, above the temperature of the granite solidus, several **dehydration melting** reactions can produce substantial amounts of granitic melt. If deep, relatively hot rocks are brought to shallower levels by tectonic activity, either by thrusting or extension, then the rocks will experience rapid decompression (heavy arrow in Figure 11.1), causing the rocks to undergo one or two important melt-producing reactions: muscovite + plagioclase + quartz → sillimanite + K-feldspar + melt, and biotite + plagioclase + sillimanite + quartz → K-feldspar + garnet + melt. In crustal rocks, decompression melting, involving rocks that contain muscovite, plagioclase, sillimanite, biotite, and quartz, will take place at much lower temperatures than melting of the mantle at a mid-ocean ridge.

BOX 11.1 | ORE DEPOSITS AND GRANITES

Granitic rocks commonly are enriched in incompatible elements, including metals that form important ore deposits. The ore deposits found in a granite pluton are closely tied to their composition and, to a lesser extent, the tectonic environment of the intrusion. For example, porphyry copper deposits (discussed in Box 8.2) occur in subduction-related settings. Other ore deposits hosted in granites yield gold, tin, and tungsten.

Gold. Gold deposits in granite include Bonanza-type epithermal deposits. These deposits, as exemplified by the high-grade vein deposits associated with the Comstock Lode in Nevada, are deposited by low-temperature hydrothermal fluids associated with shallow-level granitic plutons. (The term *epithermal* means veins formed at 50–200 °C.) During circulation of hydrothermal fluids, gold is leached from the country rock (commonly rhyolitic or dacitic lavas) and is deposited when the fluids boil. The Comstock Lode produced 258 tons of gold out of a vein system that extends for more than three miles. It formed as the result of granitic magmatism that occurred during Miocene rifting of western North America (Vikre, 1989). It is likely that the geothermal processes that formed this deposit are similar to those in Yellowstone today (Box Figure 11.1). It is important to note that epithermal deposits are not restricted

Box Figure 11.1 Old Faithful Geyser, Yellowstone National Park, Wyoming. The hydrothermal system, driven by Yellowstone magmatism, is analogous to those that form epithermal gold deposits. It was likely a thermal environment like this that led to the formation of the Comstock Lode.

(continued)

BOX 11.1 | (CONT.)

to any given plate tectonic environment but can occur in both rifting and arc environments. A good example of an epithermal deposit currently in the process of forming is the Volcan Galeras, in Colombia, which Goff et al. (1994) calculate emits 500 to 5000 tons of SO_2 and 0.5 kilograms of gold per day.

Tin. Tin is present in biotite and muscovite as a trace element, in abundances up to 1000 ppm (Eugster, 1985). Because tin is present as a minor component in magnetite and titanite, it will be disseminated rather than building up in the magma to ore concentrations, unless those minerals have not crystallized. Both magnetite and titanite are destabilized by reducing conditions. Thus, tin granites are typically peraluminous granites, formed by melting of aluminous metasedimentary rocks, which are reducing because of the presence of graphite. Tin granites can form in arc settings if assimilation of oxidized continental crust is not involved. They may also form in rift settings (Eugster, 1985). Tin granites are not widespread, but instead concentrate in certain belts, including the Cornwall area of Great Britain, Bolivia, and Malaysia. Although cassiterite, the major ore for tin, may crystallize out as a magmatic mineral in granites, most of the ore zones for tin granites represent reaction zones between the fluids that evolved late during the crystallization of the granite and the host rock.

Tungsten. Tungsten behaves geochemically like tin. Tungsten deposits, like tin deposits, are found in granites formed from melting of tungsten-bearing sedimentary rocks (Li, 1988). Although some tungsten deposits occur as veins within granites, the richest deposits occur in **skarns**, reaction zones between granite and surrounding carbonates. Major tungsten districts include the McKenzie Mountains on the Yukon–Northwest Territory boundary of Canada (Rassmussen et al., 2011) and in southern China (Li, 1988).

11.3.1 Himalayan Leucogranites

The best-known example of peraluminous leucogranite comes from the Himalaya, where small leucogranite bodies were emplaced into the highest-grade portion of the area during peak metamorphism (Figure 11.2). Peraluminous leucogranites are common in other regions of the world and include the Harney Peak granite in the Black Hills of South Dakota (Nabelek et al., 1992) and the leucogranites of Brittany in France (Strong and Hanmer, 1981). One of the best studied of the Himalayan leucogranites is the Manaslu granite (Le Fort, 1981) (Map 11.1). We note three distinctive features of the Manaslu granite. First, compared to the continental arc batholiths, it occupies a relatively small volume. Second, the body has a sheet-like form, oriented parallel to the structural grain of the Himalaya. Third, the granite is compositionally homogeneous. Unlike continental arc batholiths, it contains no granodiorite or more mafic components. The absence of these components is consistent with the theory that these leucogranites are a product of crustal melting due to tectonic thickening, and that no mantle heat source, and hence no mafic melts, are involved (Figure 11.3A).

11.3.2 Geochemistry of Peraluminous Leucogranites

Because of their restricted high silica contents (above 70 percent SiO_2), peraluminous leucogranites plot in a small field on MALI and the Fe-index plots (Figure 11.4). They span the range from magnesian to ferroan and from calcic to alkalic. Part of this variation occurs because the melt is dominated by silica and aluminum. This means MALI and iron-enrichment parameters are very sensitive to small changes in the abundances of Na_2O, K_2O, CaO, FeO, and MgO. The change from ferroan to magnesian may reflect changes in degree of melting from the source. Because iron is fractionated into the melt over magnesium, initial melts will tend to be ferroan, whereas progressive melting will enrich the melt in magnesium over

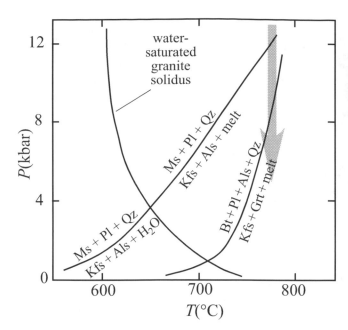

Figure 11.1 Pressure–temperature diagram, showing the relation between the water-saturated granite solidus and two common dehydration melting curves for pelitic rocks. Als = aluminosilicate (andalusite, kyanite, or sillimanite depending upon pressure–temperature conditions). Heavy grey arrow shows a possible pressure–temperature path followed in continental crust that has been tectonically thickened and then rapidly decompressed by extension. From Clarke (1992).

Figure 11.2 South face of Bhagirtahi III, showing 1500-m-thick tourmaline garnet and muscovite biotite leucogranite sill, Garhwal Himalaya, North India. The high-grade layered gneisses below contain kyanite and/or sillimanite, whereas the rocks above are low-grade schists, locally andalusite-bearing. Photo by Mike Searle.

iron. The differences in MALI appear to reflect differences in water pressure in the source region. In areas of low water pressure, melting is restricted to the micas, and the resulting melt is relatively enriched in K_2O over CaO and Na_2O. Higher water pressure allows more plagioclase to melt along with the mica, resulting in more calcic melts. Thus, early melts should be relatively calcic and, as water is depleted from the source, the later melts should become increasingly alkalic. In addition to crustal thickening, the insulating effect of overriding plate, radioactive heat production within the crust, and shear strain heating near large crustal shear zones may contribute to the higher temperatures required for dehydration melting (Strong and Hanmer, 1981; Patiño Douce et al., 1990).

11.4 Caledonian Granites

Early geologists studying the geology of Great Britain recognized a suite of granitic rocks emplaced late in the history of the Caledonian orogeny. The Caledonian orogeny involved the collision of two continents, Europe and North America, during the Ordovician to Early Devonian periods. After subduction ceased, a number of relatively small granitic plutons intruded the metamorphosed Precambrian rocks of northwestern Scotland (Map 11.2). These granites have been termed "post-orogenic" because they were emplaced late in the Caledonian orogeny, around 20 million years after Laurentia (North America) collided with Baltica (Europe) (Atherton and Ghani, 2002). The term Caledonian granite is preferred in this text because this description carries no tectonic implications.

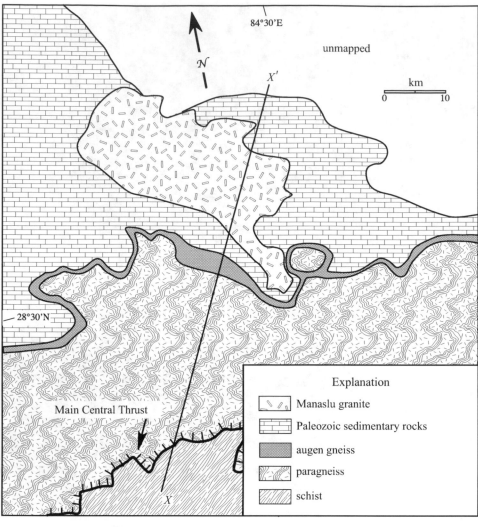

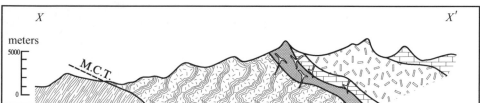

Map 11.1 Geologic map of the Manaslu granite after Colchen et al. (1980). Inset at bottom gives cross-section along the line X–X', modified from Pecher (1989).

unmapped

84°30'E

N

X'

km
0   10

28°30'N

Main Central Thrust

X

Explanation

Manaslu granite

Paleozoic sedimentary rocks

augen gneiss

paragneiss

schist

X   X'

meters
5000

M.C.T.

0

11.4.1 The Etive Granite

An example of Caledonian granite is the 415 million-year-old Etive granite of Scotland (Map 11.3; Oliver et al., 2008). The map of the Etive granite shows several features characteristic of many Caledonian granites. First, like the Tuolumne pluton (Map 8.5), it is zoned, with mafic units located on the margin and more evolved rocks in the center. Second, it locally intrudes the volcanic rocks exposed in Glen Coe that were erupted as part of the Etive magmatic event.

Although portions of the Sierra Nevada batholith contain roof pendants of coeval volcanic rocks, most plutons in the Sierra Nevada, including Tuolumne, intrude other plutonic rocks; their coeval volcanic rocks have eroded away. The presence of volcanic rocks associated with the Caledonian granites is a manifestation of the fact that the Caledonian granites tend to form small, isolated plutons (Map 11.2) rather than coalescing into the giant batholiths, like the Cordilleran granites. Third, the Etive granite contains

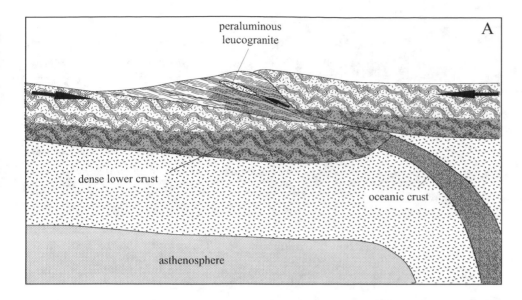

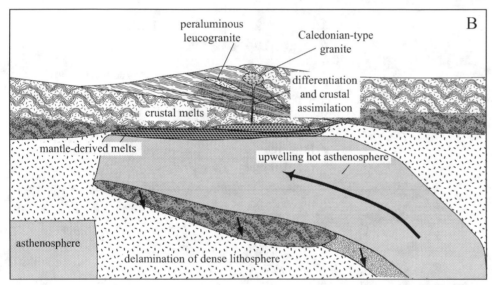

Figure 11.3 Schematic diagram showing the formation of granitic rocks in continental collisions. (A) Leucogranites are formed by decompression melting during continental collisions, either on the upper plate of large thrust sheets (as shown in the figure) or on the lower plate of large extensional faults. (B) Delamination of the high-pressure roots of a continental collision produces Caledonian-type granites.

a substantial volume of true granite, unlike the Tuolumne, which is dominantly granodiorite (Map 11.3).

11.4.2 Geochemistry and Origin of Caledonian Granites

Chemically, the Etive granite is distinct from the Cordilleran granites (Figure 11.5). Whereas both are magnesian, the Etive granite is more potassic. On the MALI diagram (Figure 11.5B), the Etive granite is mainly alkali-calcic, rather than calc-

alkalic or calcic. Possible equivalents to these post-collisional granites related to the Caledonian orogeny include Cenozoic volcanic bodies found in the Tibetan Plateau (Arnaud et al., 1992). Continent–continent collision zones, including the Caledonides and Himalaya, produce dramatically thickened crust.

Caledonian granites like the Etive are thought to form through the process of **crustal delamination** (Miles et al., 2016). In areas of continental collision, the base of the

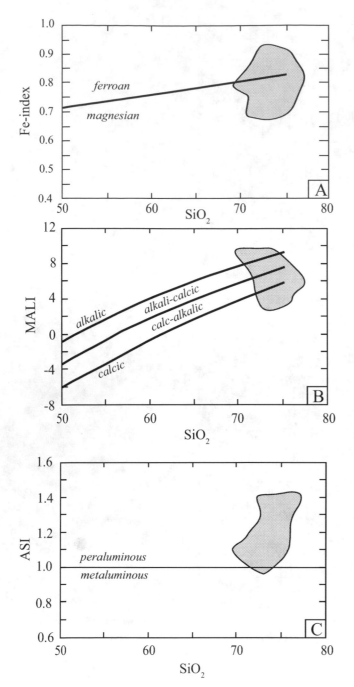

Figure 11.4 Chemical characteristics of peraluminous leucogranites. (A) Fe-index versus SiO_2; (B) MALI versus SiO_2; (C) ASI versus silica (Frost et al., 2001).

thickened lithosphere is metamorphosed at very high pressures, producing very dense rocks (Figure 11.3A). These rocks are denser than the mantle beneath them and therefore may delaminate and sink into the mantle (Figure 11.3B) (Bird, 1979). Hot asthenospheric mantle moves upward to replace the delaminated lithosphere

and undergoes decompression melting. These melts intrude, heat, and partially melt the crust, ultimately producing the plutons we see as Caledonian granites. This model could explain why the Caledonian granites occupy a smaller volume than do the Cordilleran batholiths. The Sierra Nevada formed above a subduction zone in a magmatic event that was active for more than 80 million years. During this time, individual plutons were continuously intruded, eventually forming a coalescing mass. In contrast, the delamination episode following the Caledonian orogeny was relatively short-lived, and hence there may have been insufficient time to produce the interlocking plutons overlying continental magmatic arcs.

11.5 Review of the Four Main Granite Types

As noted previously, petrologists recognize four distinct environments in which granites are generated. In order of decreasing volume of magma produced these include the magmatic arc (or Cordilleran-type) granites, intercontinental rift (or ferroan granites), post-collisional (or Caledonian-type) granites, and leucogranites (or Himalayan-type granites), formed from continental collisions (Table 11.1). For the most part, these granite groups are geochemically distinct (Figure 11.6), which allows geologists to assign a tectonic context to ancient granitic rocks even in a situation where other geologic evidence is lacking. Cordilleran granitic rocks tend to be magnesian and calcic to calc-alkalic. They are magnesian because of early magnetite crystallization and relatively calcic since they are intruded into the roots of an arc where the crustal rocks are basaltic or andesitic, and hence poor in K-feldspar. Ferroan granites are iron-rich because they contain significant portions of melt that has undergone extensive fractional crystallization in reducing conditions. They range from alkalic to calc-alkalic depending on how much crustal material has been assimilated. Caledonian granites are magnesian and alkali-calcic to alkalic. Their relatively high potassium content probably reflects deep, basaltic sources for these granitic rocks because, as was argued in Chapter 8, deeper melting produces higher potassium content. Finally, peraluminous leucogranites are silica-rich, peraluminous, and their geochemistry

Table 11.1 *The four major granitic rock types in order of decreasing relative abundance*

Granite type	Tectonic environment	Composition type	Other names
Cordilleran	Magmatic arcs	Magnesian calc-alkalic and calcic granitic rocks	Calc-alkalic, I-type
Ferroan	Intercontinental rifts	Ferroan alkali-calcic to calc-alkalic granitic rocks	Anorogenic or A-type granites
Caledonian	Post-collisional orogens	Magnesian alkali-calcic granitic rocks	Post-orogenic
Himalayan	Continent–continent collision zones	Peraluminous leucogranites	S-type

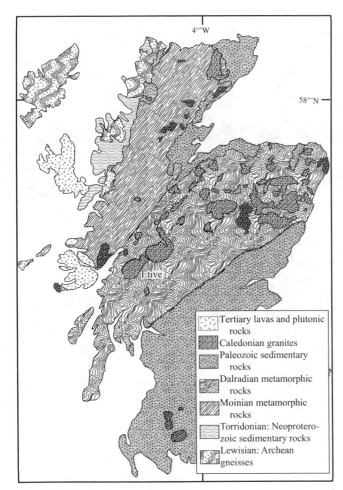

Map 11.2 Geologic map of Scotland, showing the locations of the Caledonian granites.

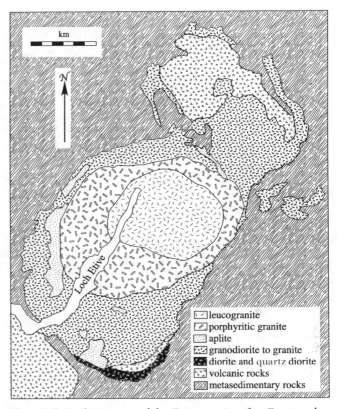

Map 11.3 Geologic map of the Etive granite after Frost and O'Nions (1985).

ranges widely in both the Fe-index and the MALI (Section 11.3).

Although the compositional trends outlined here are generally applicable, there is quite an overlap in the fields of the various geochemical families. This is particularly evident at high silica contents where a significant population of Caledonian and Cordilleran granites are ferroan and a significant population of the ferroan granites are calc-alkalic. A good explanation for this overlap is that the silica-rich magmas for all groups probably contain a significant amount of crustal-derived melt, as indicated by the fact that many Cordilleran, ferroan, and Caledonian granitic rocks with high silica are peraluminous. Mixing of crustal melts into the

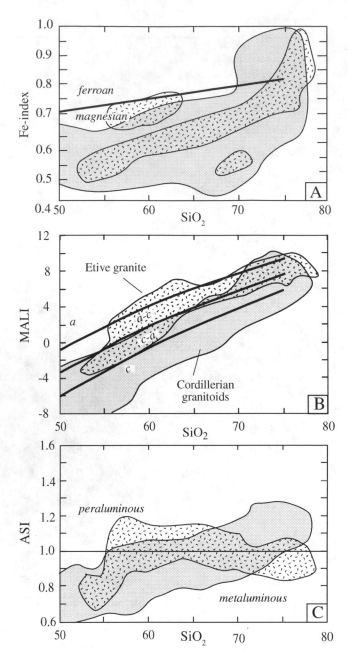

Figure 11.5 Comparison of the chemical characteristics of Caledonian-type Etive granite (stippled area) with Cordilleran granites (shaded area). (A) Fe-index versus SiO_2; (B) MALI versus SiO_2 (a = alkalic, a-c = alkali-calcic, c-a = calc-alkalic, c = calcic); (C) ASA versus SiO_2. (Data for Etive from Anderson, 1937; Batchelor, 1987; data for Cordilleran granites from Frost et al., 2001).

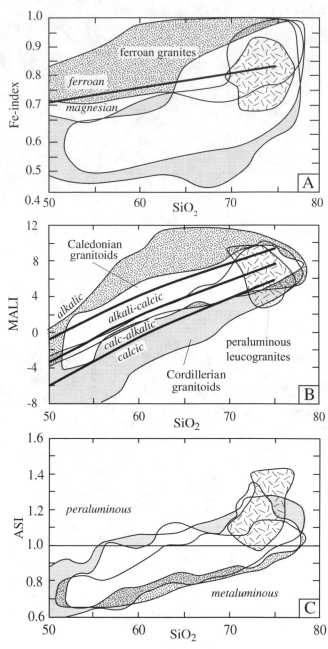

Figure 11.6 Comparison of the geochemical characteristics of peraluminous leucogranites, Caledonian granites, ferroan granites, and Cordilleran granites. (A) Fe-index versus SiO_2; (B) MALI versus SiO_2; (C) ASI versus SiO_2 (Frost et al., 2001).

fractionation trend of Cordilleran, ferroan, or Caledonian magmas will produce the overlapping compositions shown in Figure 11.6. Some late-stage, fluid-rich, siliceous magmas crystallize to form pegmatites that contain strategic metals and industrial minerals (Box 11.2).

Because of their geochemical differences, these four granite types follow distinct trends on a QAP diagram (Figure 11.7). As noted in Chapter 8, Cordilleran batholiths, being relatively rich in CaO, follow trends characterized by an increase in quartz and only moderate increases in potassium feldspar

BOX 11.2 | GRANITE PEGMATITES

Pegmatites are coarse-grained dikes or segregations that usually have a granitic composition. Most form from fluid-rich granitic magma, enriched in elements that are not incorporated into rock-forming minerals during crystallization of a pluton. A small fraction of the world's pegmatites contain uncommon minerals, including gem-quality beryl, tourmaline, topaz, spodumene (Li-pyroxene), and spessartine. Pegmatites are also important sources of industrial minerals and strategic metals.

Some of the most spectacular gemstones occur in cavities in the coarse interior of pegmatites. Crystallization of feldspars and quartz from granitic magma enriches the residual melt in elements that are excluded from those minerals. Water and other volatiles may exsolve from the melt as crystallization continues or as the magma ascends to lower pressure. Minerals growing in the volatile-rich cavities may develop into beautifully euhedral, gem-quality crystals (Simmons et al., 2012).

The feldspars, quartz, micas, and Li-pyroxene in pegmatites are important industrial minerals. Most feldspar is used for ceramics and glass manufacture, but it is also used as a filler in paint and plastic and as an abrasive in household cleansers. Quartz is important in the manufacture of all types of glass. High-purity quartz is used in a number of industrial applications, supplying a high thermal stability filler. In the past, mica from pegmatites was used as window material, particularly in stoves and ovens. Now it is used in coatings, lubricants, and fillers. Spodumene (Box Figure 11.2) is used in ceramics, and is also an important source of lithium, a strategic metal that is in growing demand for batteries and pharmaceuticals (Glover et al., 2012).

Box Figure 11.2 Gigantic crystals of spodumene are exposed in the walls of the Etta Mine, Keystone, South Dakota. The molds of two large spodumene crystals in this photo are more than 10 m long and their thickness approaches 1 m. The mine was opened in 1883 as a mica mine, and spodumene was mined from 1898 to 1959. This photo appeared in a 1916 USGS report by Schaller (1916). USGS photo library image file: /htmllib/btch368/btch368j/btch368z/ swt00066.jpg.

(continued)

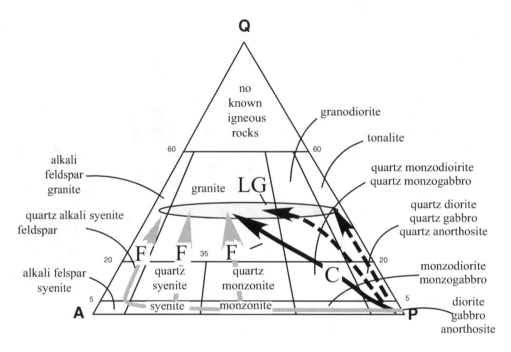

Figure 11.7 QAP diagram showing the differentiation path followed by various granite types. Dashed lines = granites of Cordilleran batholiths, C = Caledonian granites, F = ferroan granites, and LG = peraluminous leucogranites.

(dashed lines in Figure 11.7). The dominant rock type in these intrusions is tonalite or granodiorite. Himalayan-type granites (i.e. peraluminous leucogranites) have high quartz content but show a range in feldspar composition, from nearly pure albite (i.e. leucotonalite) to nearly pure microcline (i.e. alkali feldspar granite) (shaded field in Figure 11.7). Because Caledonian plutons are more alkalic than Cordilleran plutons, they contain granitic rocks that follow a more potassic path during differentiation (path C in Figure 11.7). The end product is abundant granite rather than granodiorite, and the less-differentiated portions may contain quartz monzodiorite and monzogabbro rather than quartz diorite and gabbro. Ferroan granites may follow a range of paths, depending on the alkali content of the parent basalt from which they differentiated and the extent of crustal assimilation (gray arrows in Figure 11.7; Frost and Frost, 2013). A tholeiitic host trends from monzonite to granite, whereas progressively more alkalic hosts follow paths involving syenite or even alkali feldspar syenite.

Summary

· ·

- Granites may be classified according to a number of schemes:
  - The mineralogical classification is based on the proportion of quartz, alkali feldspar, and plagioclase
  - The classification based on opaque oxides evaluates the presence or absence of magnetite and reflects the oxidation state of the magma
  - The alphabetic classification emphasizes the inferred origin of the granite, whether from igneous sources (I-type), sedimentary sources (S-type), or formed in an anorogenic environment (A-type)
  - The major-element geochemical classification discriminates groups of granites based on four indices, including the Fe-index, modified alkali–lime index, aluminum-saturation index, and alkalinity index

- Peraluminous leucogranites are high-silica granites formed by partial melting of continental crust, commonly in continent–continent collision zones. Granitic melt is produced by dehydration melting of metamorphosed crustal rocks brought to high temperature by processes that may include crustal thickening, radioactive heating, and frictional heating. Coeval mafic rocks are not associated with peraluminous leucogranites.

- Caledonian granites are alkali-calcic, magnesian granites that post-date collisional events. They may be related to crustal delamination. Ascending mantle may melt by decompression and interact with the crust to form these relatively small, compositionally zoned plutons and associated volcanic rocks.

- The compositional differences between peraluminous leucogranites, Caledonian granites, ferroan granites, and Cordilleran granites reflect their magma sources and subsequent differentiation.

Questions and Problems

· ·

Problem 11.1. What is the difference between "granite" and "granitic rock"? Do any of the four types of granitic rock suites described in this chapter consist solely of granite?

Problem 11.2. Explain the trend to more peraluminous compositions with increasing silica shown in Figure 11.6.

Problem 11.3. What tectonic environment may produce granites with the greatest involvement of continental crust? Which may produce granites with the least involvement of continental crust? Explain.

Problem 11.4. Using the maps of granitic intrusive complexes in Chapters 10 and 11 as a guide, what tectonic environments appear to produce the largest volumes of granitic rocks? Why is this so?

Further Reading

· ·

Barbarin, B., 1999, A review of the relationships between granitoid types, their origins and their geodynamic environments. *Lithos*, 46, 605–26.

Clarke, D. B., 1992, *Granitoid Rocks*. Chapman and Hall, London.

Coleman, D. S., Mills, R. D., and Zimmerer, M. J., 2016, The pace of plutonism. *Elements*, 12, 97–102.

London, D. and Kontak, D. J., 2012, Granitic pegmatites: Scientific wonders and economic bonazas. *Elements*, 8, 257–62.

Lundstrom, C. C. and Glazner, A. F., 2016, Silicic magmatism and the volcanic–plutonic connection. *Elements*, 12, 91–6.

Pitcher, W. S., 1997, *The Nature and Origin of Granite*. Chapman and Hall, London.

Whitney, J. A., 1988, The origin of granite: The role and source of water in the evolution of granitic magmas. *Geological Society of America Bulletin*, 100, 1886–97.

Chapter 12

Introduction to Metamorphic Petrology

12.1 Introduction

Metamorphic petrology studies rocks that recrystallized in the solid state within Earth's crust or, rarely, within the upper mantle. In most instances, recrystallization of rocks within the crust involves deformation, because deformation enhances mineral reactions by introducing strain energy and by opening pathways for fluid movement. As such, metamorphic petrology overlaps with the field of structural geology. Many ore deposits form from hydrothermal fluids at temperatures well below those of igneous activity. Consequently, the study of the reactions between these ore fluids and the country rocks can be included within the realm of metamorphic petrology.

As with igneous petrology, a knowledge of mineralogy is critical to the study of metamorphic petrology.[1] Metamorphic petrologists use the minerals present in a metamorphic rock to determine what its parent was (i.e. the **protolith**) and to estimate the conditions of metamorphism. Unlike igneous petrology, metamorphic petrology does not rely heavily on whole-rock geochemistry. Whereas variations in chemistry may provide important information about the origin and evolution of igneous rocks, chemical variations in metamorphic rocks merely change the relative abundance of minerals in the rock. For example, a basaltic rock metamorphosed at moderate grades will have the assemblage hornblende–plagioclase, regardless of whether the parent rock was an alkali basalt, tholeiite, or andesite, but the relative proportions of minerals will vary with the composition of the rock.

This chapter introduces terms that are important to the discussion of metamorphism and metamorphic rocks. The next chapter (Chapter 13) approaches metamorphic rocks through the interpretation of metamorphic phase diagrams and petrogenetic grids. After these tools are introduced, Chapter 14 discusses the metamorphism of mafic rocks and metamorphic facies. Metamorphic facies, which establish the relative temperature and pressure conditions of metamorphism, were introduced long before petrologists had the ability to quantify temperatures and pressures from metamorphic mineral compositions. After this, the text describes how pressure, temperature, and fluid composition affect the metamorphism of major protoliths, starting with peridotites, which are chemically the simplest (Chapter 15). Subsequently, more complex systems are discussed, such as pelitic rocks (Chapter 16). Attention then turns to how mineral **assemblages**, especially those in calc-silicate rocks, monitor fluid composition (Chapter 17). Chapter 18 covers **thermobarometry** – the quantitative estimate of metamorphic conditions – and describes the pressure and temperature conditions that constitute metamorphism. Finally, Chapter 19 summarizes the tectonic environments where metamorphic and igneous rocks are found.

12.2 Scope of Metamorphism

Sedimentary rocks usually contain minerals such as clays and carbonates that contain substantial amounts of volatiles (H_2O or CO_2) incorporated in their structures. As a result, these volatile-bearing rocks are stable only in low-temperature environments. Should they be subjected to higher temperatures, be it by sedimentary or tectonic loading, or by intrusion of magmas, these low-temperature minerals will react by releasing all or part of their volatile constituents. In a similar manner, should igneous rocks, which typically are dominated by anhydrous minerals, interact with fluids after solidification from the melt, new minerals will form that are more stable in lower-temperature, hydrous conditions. If anhydrous igneous rocks undergo deformation at high temperatures, where the igneous minerals are still stable, the rocks will acquire a deformed texture and are considered metamorphic rocks, even if the mineralogy has not changed. Thus, in the core of any orogenic belt the original sedimentary and igneous rocks will have been deformed at high temperatures and pressures and will have mineral assemblages indicative of those pressure and temperature conditions. Understanding how these reactions occur and how the fabric and mineralogy of metamorphic rocks reflect the conditions under which the rocks formed is the field of metamorphic petrology.

To provide a simple definition, **metamorphism** is the recrystallization of a rock at conditions below those of the liquidus. At low temperatures, metamorphism merges with the processes known as **diagenesis**, which are the changes that occur during cementation of sedimentary rocks. Diagenesis involves reactions in the pore space of sedimentary rocks, with little or no participation of the clastic grains. For the sake of simplicity, the boundary between diagenesis and metamorphism can be described to lie at those conditions where the clastic grains significantly participate in the mineral reactions. At high temperatures, metamorphism overlaps with igneous processes at conditions where melts are generated. The process of partial melting of rocks is called *anatexis*. The melts generated by this process are true igneous rocks, whereas the residue left behind after extraction of the melt, a rock known as the **restite**, is a metamorphic rock. The rock is metamorphic because the minerals in a restite, although they may have been in equilibrium with a melt, did not crystallize directly from it.

12.3 Types of Metamorphism

Several types of metamorphism have been recognized out of the continuum of subsolidus reactions that occur in the crust. The most common types are regional, contact, burial, dynamic, and hydrothermal metamorphism. Impact metamorphism results from collision of meteorites with the Earth's surface.

12.3.1 Regional Metamorphism

Regional metamorphism is metamorphism on a regional extent. It is usually related to orogeny and accompanying deformation. A geologic traverse across the Alps, Himalaya, or the Canadian Rockies would reveal metamorphic rocks in the core of the ranges, where rocks were subjected to deformation and metamorphism at the time the mountains were forming. Similarly, a traverse inland from the coast of California would document rocks deformed and metamorphosed during accretion of sediments and ocean crust along the west coast of North America in the Mesozoic Era. The metamorphic terrain in these areas is typical of the scale of regional metamorphism.

12.3.2 Contact Metamorphism

Intrusion of magmas into the crust brings heat to mid- or upper-crustal levels. This heat contact metamorphoses the country rock adjacent to the intrusion. Contact metamorphism is localized to the area adjacent to the intrusion. At high crustal levels, rocks that have undergone contact metamorphism are not deformed, but in deep environments the rocks may be deformed and contact metamorphism may grade into regional metamorphism.

12.3.3 Burial Metamorphism

At increasing depth within a deep sedimentary basin, the temperature and pressure increase simply because of regional heat flow. The study of the effect of heating on the diagenesis of sedimentary rocks in deep basins is key to understanding the origin of petroleum, which is produced from organic debris at elevated temperatures. If the basin is deep enough the temperature will exceed that necessary for petroleum generation and the rock enters the realm of metamorphism. Areas that have undergone burial metamorphism typically have been subjected to the lowest grades of metamorphism and have not been deformed.

12.3.4 Dynamic Metamorphism

Most rocks in Earth's crust are not in equilibrium with the temperature and pressure conditions at which they now reside. Most granitic crust contains assemblages that equilibrated at temperatures of 600 °C or more. Although the rocks may now be at shallower depth and lower temperature, most areas will show no sign of reacting to lower-temperature minerals (i.e. undergoing **retrograde metamorphism**). Because the crust is generally dry, there commonly is no water available to flux reactions that produce low-temperature hydrous assemblages. However, during deformation of deep crustal rocks, hydrous fluids migrate into deformation zones and interact with the minerals in the protolith. The consequence is dynamic metamorphism, a localized zone of retrograde metamorphism that may extend for several meters on either side of a shear zone into the surrounding undeformed rocks.

12.3.5 Hydrothermal Metamorphism

In many geologic environments, the presence of magma near the surface of Earth leads to the circulation of hot water through the upper crust, which triggers hydrothermal (i.e. hot-water) metamorphism. The water reacts with the original rock-forming minerals, such as feldspars, pyroxenes, and amphiboles, to make micas and clays. Commonly, this type of metamorphism is associated with the deposition of sulfide ore minerals to make hydrothermal ore deposits. Good examples of areas where hydrothermal processes are taking place today are the mid-ocean ridges or the thermal features of Yellowstone National Park.

12.3.6 Impact Metamorphism

Impact metamorphism is a type of local metamorphism that occurs around meteorite impact sites (or astroblemes). This type of metamorphism is characterized by intense deformation, ultra-high pressures, and locally high temperatures. Key indicators of impact metamorphism are:

Shatter cones: These are collections of fractures that are nested in a conical shape with the apex of the cone pointing toward the point of impact (Dietz, 1959). They are only found in areas of intense deformation such as impact sites.

Shocked quartz: Quartz displaying deformation bands or deformation twins in thin section is a good indication of impact metamorphism (Chao, 1966). This is because quartz flows readily at temperatures above about 350 °C and the deformation structures would survive only if they formed in an environment where deformation was intense and high temperatures were transient.

Coesite and stishovite: These are high-pressure forms of silica. Coesite is monoclinic and forms at pressures above 25 kbar. It is known from some very high-pressure metamorphic rocks as well as from impact craters. Stishovite, which indicates pressures around 75 kbar, is tetragonal and is known only from impact sites. The occurrence of these minerals, often together, in shocked rocks, is a clear indication of an impact origin (Spray and Boonsue, 2018).

Diamond: The conversion of coal directly to diamond is observed in some impact sites (Shumilova et al., 2018).

12.4 Basic Goals of Metamorphic Petrology

In general, the mineralogy and fabric of a metamorphic rock will provide answers to three questions:

(1) *What was the protolith?* (i.e. What were the parent rocks?)

  Were they igneous or sedimentary rocks and what kind of igneous and sedimentary rocks were they?

(2) *What were the conditions of metamorphism?*

  At what temperature and pressure did the metamorphic rock crystallize?

Table 12.1 *Clearly sedimentary protoliths*

Rock type	Characteristics	Parent rock
Quartzite	90–100% quartz	Quartz arenite, chert
Quartzose schist	>60% quartz	Quartz sandstone
Pelitic schist	Abundant micas and quartz aluminum-rich silicates (garnet, Al_2SiO_5, staurolite), graphite	Shale or mudstone
Semi-pelite, psammite	Abundant micas and quartz, few aluminum silicates	Impure sandstones
Carbonate	Abundant calcite or dolomite calcium–magnesium silicates: tremolite, diopside	Limestone or dolomite
Iron formation	Magnetite, hematite, quartz iron-rich silicates, carbonates or sulfides	Iron formation
Manganese formation	Manganese-rich silicates, oxides, or carbonates	Manganese-rich chert

What was the composition of the fluid present at the time of crystallization?

Was the metamorphism **isochemical**? (i.e. did the rock change composition during metamorphism?) Were major chemical components added to or extracted from the rock in a process called metasomatism?

(3) *What was the structural history of the rock*?
Is the rock undeformed or was it subjected to **penetrative deformation**, deformation that affects the rock at all scales?

If it was deformed, did the deformation take place before, during, or after the metamorphic minerals crystallized?

The mineralogy gives clues to the nature of the protolith and is the major feature used to determine the conditions of metamorphism, whereas the textures and fabric of the rock give information relevant to answering all three of these questions.

12.5 Identification of Protolith

12.5.1 Rocks of Clearly Sedimentary Parentage

The first question a petrologist asks with regard to the protolith is whether the parent rock was sedimentary or igneous. In weakly metamorphosed rocks where the original igneous or sedimentary textures and structures of the rock are readily recognized, this question is easily answered. In high-grade rocks, where the original textures and structures commonly have been obliterated, the petrologist must infer whether the protolith was

sedimentary or igneous from the mineralogy of a rock. The distinction between an igneous or sedimentary protolith will be difficult in high-grade rocks if the original sediment was immature; in other words, if it was deposited in an active tectonic environment, where igneous minerals in a rock are disaggregated by mechanical weathering and are incorporated in sediments with little chemical weathering or change in composition. Such immature sediments, if subjected to high-grade metamorphism, may be chemically indistinguishable from a metamorphic equivalent of the igneous source rock.

In a mature sedimentary environment, however, intense weathering is an effective agent of chemical differentiation. The original minerals of the rock react to form clay, quartz, and iron oxides, while sodium and calcium and, to a lesser extent, magnesium ions go into solution. In a continental-margin environment, these weathering products are deposited as quartz sandstone (**psammite**), shale (**pelite**), carbonate (limestone and dolomite), and evaporites, depending on the sedimentary environment. These sedimentary rocks are so chemically distinctive that they can be recognized even after being subjected to the highest grades of metamorphism (Table 12.1).

The most characteristic feature of metamorphosed mature sediments is that they tend to be quartz-rich and feldspar-poor. Quartz contents range from 100 percent in quartzites to less than 50 percent in pelitic rocks; plagioclase and microcline are not abundant in either. As quartz abundance decreases, the abundances of micas, biotite, and muscovite, which represent the original clay components of the protolith, increase, leading to a gradation from quartzose, through psammitic, to pelitic

Table 12.2 *Igneous protoliths or protoliths of uncertain parentage*

Rock type	Characteristics	Parent rock
Igneous protoliths		
Ultramafic	Virtually no feldspars, abundant olivine, pyroxenes, amphiboles, or serpentine	Mantle peridotite, ultramafic cumulates
Alkaline	Presence of feldspathoids or sodic pyroxenes and sodic amphiboles	Alkaline lava, nepheline syenite
Protoliths of uncertain parentage		
Mafic	Plagioclase + quartz < mafic minerals (amphiboles or pyroxenes)	Basalt, gabbro, andesite, diorite, immature sandstone, calcareous shale
Quartzo-feldspathic	Quartz + feldspars > mafic minerals	Dacite, rhyolite, granitic rock, arkose, immature sandstones

protolith. The psammitic protolith can be recognized because it has moderately abundant quartz and micas but only minor feldspar. Psammitic rocks are often garnet-bearing, but the presence of other aluminous minerals, such as the aluminosilicates, staurolite, or cordierite, mark the rock as pelitic rather than psammitic.

Another important mineral in many sedimentary rocks is carbonate, mainly calcite and dolomite. Metamorphosed limestones, dolomites, and **marls** are distinguished by high CaO and MgO and low SiO_2 and FeO. At low and medium metamorphic grade, carbonate-bearing metasedimentary rocks are identified by the presence of calcite and dolomite. At high grade, however, metamorphic reactions may have removed the carbonates. In such rocks, a carbonate-rich protolith can be identified by calcium- and magnesium-bearing minerals such as wollastonite, diopside, forsterite, and tremolite. In marls, where the protolith contained abundant clay as well as carbonate, the calcareous identity of the protolith is marked by the presence of calcium-rich minerals such as grossular and epidote.

As with carbonates, other chemical sediments are compositionally distinct and their metamorphic equivalents are easily recognized. Evaporites are rarely preserved in the metamorphic record because the salts are easily dissolved in the metamorphic fluids. Iron formations are distinctive by the interlayering of quartz-rich horizons with zones richer in iron silicates, iron oxides, or iron sulfides. Manganese formations, which are rich in manganese oxides or pink manganese silicates, are metamorphosed equivalents of manganese cherts and, hence, record metamorphosed deep-water sediments.

12.5.2 Rocks of Clearly Igneous Parentage

Because most igneous rock compositions occur in both immature sediments and igneous rocks, there are few definitively igneous protoliths (Table 12.2). The most important of these select few is the peridotitic protolith. Olivine is easily hydrated to serpentine and serpentine does not persist in a detrital environment. True, there are some spectacular olivine beach sands in Hawaii, but in a sedimentary basin such sands would be hydrated to serpentine, which subsequently reacts with any detrital quartz to make talc. Thus, there is no immature sedimentary equivalent to the peridotitic protolith. The presence of olivine- and serpentine-rich rocks therefore indicates that the protolith was either a fragment of the mantle or cumulate peridotites from the base of a layered mafic intrusion. As with peridotites, alkaline rocks do not form immature sediments. Nepheline is more easily weathered than albite, so silica-deficient alkaline rocks are rarely present as sedimentary equivalents. Thus, the rare occurrences of alkaline gneisses may be interpreted as metamorphosed alkaline intrusions.

12.5.3 Rocks of Uncertain Parentage

Many sedimentary rocks and igneous rocks have similar geochemical compositions. When metamorphosed, it is difficult to distinguish a sedimentary from igneous protolith. Nevertheless, the metamorphic rocks can be categorized into two broad protoliths: mafic and quartzo-feldspathic (Table 12.2).

Mafic metamorphic rocks broadly classify as basaltic, although andesitic compositions may also be present. These rocks are often called metabasites and are identified by subequal amounts of mafic minerals (amphiboles or, more rarely, pyroxene) and plagioclase (or metamorphic equivalents). Rocks where feldspars and quartz are more abundant than amphiboles or pyroxene are called quartzo-feldspathic.

12.6 Determination of Metamorphic Conditions

In addition to preserving information about the protolith, the mineral assemblage of a rock also may record the conditions of metamorphism, including temperature, pressure, and fluid composition present at the time the rock formed. There are four levels of sophistication in this approach: two consider stability of minerals and their assemblages; and two others, facies analysis and thermobarometry, combine chemical and environmental variables to deduce metamorphic regime.

12.6.1 Stability Range of Single Minerals

Many metamorphic minerals, such as the aluminosilicate (Al_2SiO_5) polymorphs, have a limited range of pressure–temperature conditions over which they are stable. The occurrence of these minerals constrains the conditions at which the rock equilibrated.

12.6.2 Stability of Mineral Assemblages

Petrologists can get a more precise estimate of metamorphic conditions by determining the mineral association that was stable in the rock when metamorphism peaked. An association of minerals that is interpreted to have equilibrated during metamorphism is known as an assemblage. Careful study of a suite of rocks may reveal that two or more assemblages are related by a common chemical reaction. The trace of this reaction on a map (or in the field) is known as an **isograd**.

12.6.3 Metamorphic Facies

Metamorphic facies are defined by mineral assemblages that are repeatedly associated in rocks of varying age and in many places around the globe. Whereas a given mineral assemblage or reaction is applicable only to a specific rock composition, metamorphic facies include all possible protoliths. The facies designation developed into a coarse classification system for metamorphic conditions, widely used to describe regional metamorphism.

12.6.4 Thermobarometry

During the past century, technological advances in metamorphic petrology have enabled precise chemical measurements of metamorphic minerals. These data are incorporated into thermodynamic data sets to obtain quantitative estimates of temperature and pressure stability for many mineral assemblages. This field, thermobarometry, transformed metamorphic petrology from a qualitative description of relative stability conditions of mineral assemblages to an essential tool in the modern study of orogenesis.

12.7 Metamorphic Textures

To determine the metamorphic history of a rock, petrologists use metamorphic textures, as they are observed, both in hand sample and in thin section. In addition to providing key information about the deformation history of a rock, the texture of a metamorphic rock provides information about the protolith and, thus, the rock's metamorphic history. Geologists recognize two types of textures: primary textures, or features formed with the protolith; and secondary textures, those that formed during metamorphism or deformation.

12.7.1 Primary Textures

Primary textures formed during the deposition of sedimentary rocks or during the crystallization of igneous rocks. Some of these textural features survive to surprisingly high metamorphic grades, especially when the texture involves variations in bulk composition.

Sedimentary Textures

One of the most distinctive features of sedimentary rocks is bedding. Following metamorphism, bedding manifests as compositional layering (Figure 12.1A). Especially in a high-grade gneiss, compositional layering may be an important indication of a sedimentary protolith. This guideline is not infallible because a thick greywacke that originally was very poorly bedded may become metamorphosed to a quartzo-feldspathic gneiss without layers. Conversely, intense deformation of a homogeneous granite may produce a strongly layered gneiss because the deformation may be focused in high-strain zones, which become rich in mica. To determine if gneissic layering is of sedimentary origin, look for thin quartzose, calcareous, or pelitic layers, which are indicative of bedding planes from previously sedimentary rocks.

Many sedimentary textures are destroyed during metamorphism. Cross-bedding is often obliterated because the quartz in the rock recrystallizes at relatively low metamorphic grades. On the other hand, if fine hematite grains blanket the original cross-bedded surfaces, the hematite-decorated cross-bedding may survive to moderate metamorphic grades. Similarly, pebbles in a conglomerate may survive to moderate metamorphic grades, particularly if the pebbles were distinctly different in composition from the matrix (Figure 12.1B, C). The most robust sedimentary feature in metamorphism is graded bedding (Figure 12.1A) because this involves variations in bulk composition. The lower portion of the graded bed may be quartzose, but the thin, upper layer, which originally consisted of clay, will be pelitic. Because the clay-rich horizons are far more reactive during metamorphism, the clay-rich upper portion of a graded bed may become a relatively coarse-grained pelitic layer, whereas the underlying quartz-rich layer remains relatively fine-grained, producing **reverse graded bedding**.

Igneous Textures

One of the most important igneous textures that can survive a wide range of metamorphic conditions is tabular feldspar (Figure 12.2A). When crystallizing from a melt, both plagioclase and K-feldspar will grow as tabular plates that are elongated along the *c*-axis and flattened along the *b*-axis. A typical texture in gabbro is shown in Figure 1.8A, wherein tabular plagioclase is surrounded by irregularly shaped pyroxene. Tabular potassium feldspar also commonly occurs as large phenocrysts in granitic rocks. Tabular feldspars do not grow in metamorphic rocks, where surrounding crystals constrain their shape (Figure 12.3A). Thus, the existence of tabular feldspars in a metamorphic rock is clear evidence of an igneous precursor. Relict tabular plagioclase allows geologists to recognize an amphibolite as a metadiabase (Figure 12.2A) or metagabbro, and relict tabular plagioclase or K-feldspar may provide the key in identifying a quartzo-feldspathic gneiss as an **orthogneiss** (a gneiss derived from an igneous parent) rather than a **paragneiss** (a gneiss derived from a sedimentary parent).

Another igneous feature likely to survive to high grades of metamorphism is the pillow structures that form when basaltic lava is quickly chilled in a subaqueous environment (see Figure 1.9D). Pillows survive to high grades of

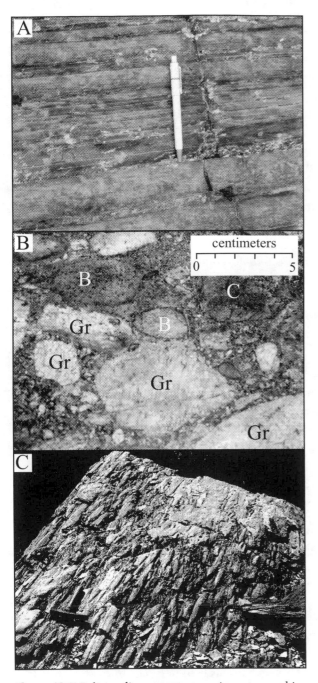

Figure 12.1 Relict sedimentary textures in metamorphic rocks. (A) Turbidite metamorphosed to greenschist facies shows relict graded bedding. Pencil indicates inverted stratigraphic-up direction. Yellowknife, Northwest Territories, Canada. (B) Conglomerate metamorphosed to greenschist facies, containing clasts indicated as Gr = metamorphosed granite and granitic gneiss, B = epidotized basalt, and C = chert. (C) Deformed pebble conglomerate from the Raft River Mountains, Utah. Photo by Arthur W. Snoke.

Figure 12.2 Relict igneous textures in metamorphic rocks. (A) Porphyritic diabase metamorphosed in amphibolite facies, showing relict plagioclase phenocrysts. From the Laramie Mountains, Wyoming. (B) Greenschist facies pillow basalts from the Smartville ophiolite, northern Sierra Nevada, California. Photo by Arthur W. Snoke. (C) Weakly metamorphosed basalt containing calcite-filled amygdules. Photo from United States Geological Survey photographic library image Bastin ES 112.

metamorphism because the quickly chilled rims of the pillows are very glassy and are more susceptible to alteration by seawater than the crystalline pillow core. This alteration, which may result in the leaching of alkalis or the oxidation of iron, produces rims compositionally distinct from cores. Because of this compositional difference, pillows can be recognized even in highly metamorphosed and deformed rocks (Figure 12.2B).

Another igneous feature that may survive to moderate grades of metamorphism are vesicles that form on the upper portions of lava flows. In low and medium grades of metamorphism these empty pockets fill with zeolite, calcite, epidote, or prehnite. These minerals form light-colored, rounded spots in a metabasalt called **amygdules** (Figure 12.2C). Amygdules, though important for determining the metamorphic assemblages in weakly metamorphosed rocks, seldom survive above medium grades of metamorphism.

12.7.2 Metamorphic Textures

Petrologists can use the texture of metamorphic rocks to determine if solid-phase recrystallization took place in a static environment or whether deformation was occurring when the rocks crystallized. Petrologists recognize two broad textures to metamorphic rocks: (1) static textures in which minerals grew without deformation or after deformation had ceased, and (2) tectonic textures in which deformation controlled the distribution and shape of minerals in the rock.

Static Textures

Static textures are characterized by equant and polygonal mineral grains that intersect at angles of 120° (Figure 12.3A). This angle, called the **dihedral angle**, is governed by the surface energy of the minerals, which, simply put, is how easily minerals form crystal faces. In a mineralogically simple rock, such as quartzite or dunite, where all grains have the same surface energy, the intersections are uniformly 120°; in polymineralogic rocks the intersections tend to deviate slightly from 120° because each mineral has slightly different surface energy. Some minerals, such as micas, which have strong cleavage, will not form the 120° grain boundaries because grain boundaries trend along the cleavage planes of the mineral.

Although euhedral feldspars are indicative of an igneous texture, euhedral crystals can also form in metamorphic rocks, where the texture is called **idioblastic**. In metamorphic rocks, coarse-grained minerals bounded by crystal

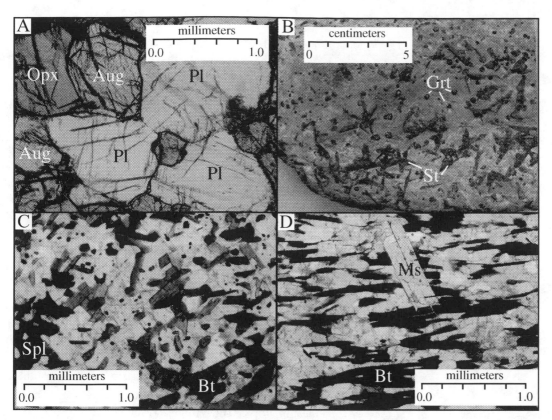

Figure 12.3 Textures in metamorphic rocks. (A) Photomicrograph in plane-polarized light (PPL) showing relations between orthopyroxene (Opx), augite (Aug), and plagioclase (Pl) in granulite from southwestern Montana. Note how grains meet at triple boundaries with angles approximating 120°, a texture distinctly different from igneous textures shown in Figure 1.8A, C. (B) Porphyroblasts of garnet (Grt) and staurolite (St) in mica schist from near Taos, New Mexico. (C) Photomicrograph in PPL, showing randomly oriented biotite (Bt) from cordierite–corundum–spinel (Spl) – plagioclase–K-feldspar hornfels adjacent to the Laramie anorthosite complex, Wyoming. Light-colored matrix consists of intergrown cordierite, plagioclase, and K-feldspar. Corundum is not shown. (D) Photomicrograph in PPL, showing biotite with strong preferred orientation from southern Ontario, Canada. Also present is muscovite (Ms), growing across the foliation defined by the biotite. Matrix consists of quartz and plagioclase.

faces are called **porphyroblasts** (Figure 12.3B). Porphyroblasts grow by reactions that consume the surrounding matrix and produce minerals, such as garnet, staurolite, and aluminosilicates, which tend to have low surface energies, which favor the formation of rational crystal faces.

Granoblastic textures, such as those shown in Figure 12.3A, are common in metamorphic rocks that have undergone heating without deformation, or where crystallization has occurred after deformation has ceased. A special type of granoblastic texture is found in a fine-grained rock, called **hornfels**, which occurs as fine-grained rocks in a contact aureole around an igneous intrusion. Usually hornfelses are so fine-grained that individual grains are indistinguishable in hand sample. However, the presence of an intimately intergrown crystal network is indicated by the fact that these fine-grained rocks often break with conchoidal fractures.

The term hornfels is entirely a texture term; it may be modified by a compositional adjective such as pelitic, mafic, or ultramafic.

Tectonic Textures

If a rock has deformed, the minerals align in a preferred orientation. During deformation, they tend to rotate, grow, or nucleate so that they are oriented in a position that most readily accommodates the strain. Thus, unlike in rocks with static textures, minerals in deformed rocks exhibit crystallographic **preferred orientation**. Figure 12.3C shows an example of a hornfels where the biotite has no preferred orientation. Biotite is a strongly pleochroic mineral, meaning that in transmitted light the biotite in a thin section shows a range of color intensity. The range of intensities exhibited by biotite in Figure 12.3C suggests that the elongate biotite grains have

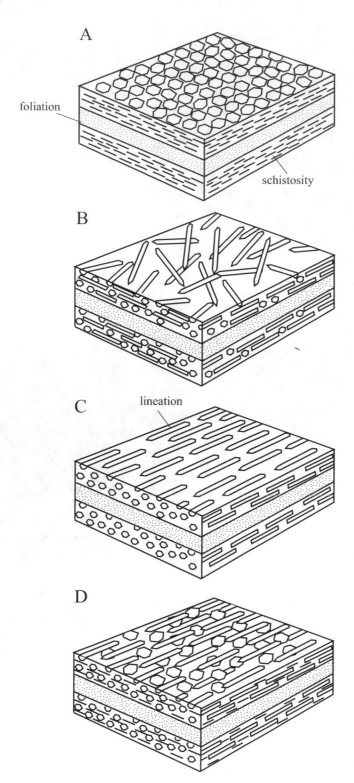

A

foliation

schistosity

B

C

lineation

D

Figure 12.4 Illustrations comparing the relationship between schistosity, foliation, and lineation in metamorphic rocks. (A) Rock with foliation and schistosity. (B) Rock with foliation and weak schistosity, marked by preferred orientation of tabular minerals. (C) Rock with foliation and lineation, but no schistosity. (D) Rock with foliation, schistosity, and lineation.

many different crystallographic orientations. This random distribution is in contrast to the schist shown in Figure 12.3D, where all the biotite grains are parallel and have the same intensity, indicating a strong preferred orientation in response to strain.

The distribution of minerals in a rock and their type of preferred orientation produces several macroscopic textural features. Deformed metamorphic rocks contain three types of fabric: **foliation**, **schistosity**, and **lineation** (Figure 12.4). *Foliation* is a term to describe any planar feature in a rock, including primary layering, such as bedding, or a secondary planar feature caused by deformation. *Schistosity* describes planar features defined by preferred orientation of platy or tabular minerals. *Lineation* is a linear feature that may form by preferred orientation of tabular minerals or by intersection of foliation planes. Figure 12.4A illustrates a rock that contains two foliations, a primary bedding plane and schistosity parallel to that bedding. The schistosity is caused by the strong preferred orientation of micas and is recognizable in hand sample because the rocks split along parallel planes, mirroring the parallel arrangement of platy mineral crystals. When mica defines schistosity, reflections off the mica cleavage identify the schistose surface. Rocks dominated by less platy minerals, such as feldspars, will not break along the plane of the preferred orientation; in such rocks the texture is called **gneissosity**. Foliation need not be parallel to the schistosity, as it is in Figure 12.4A; it may be oblique or several sets of foliation may intersect primary foliation at different angles.

Figure 12.4B illustrates a rock with foliation and weak schistosity, defined by an elongate mineral (an amphibole) that has its *c*-axes all lying in the same plane. The *c*-axes are not parallel so the rock does not display lineation. If the *c*-axes of the tabular minerals all lie in the same plane and are all pointed in the same direction, then the rock has a lineation (Figure 12.4C). The rock in Figure 12.4C has a foliation and a lineation, but not schistosity. Figure 12.1C provides an example of a lineation defined by relict pebbles, stretched into elongate shapes. Finally, if a rock contains both platy minerals, such as mica, and tabular minerals, such as sillimanite, it can develop foliation, schistosity, and lineation (Figure 12.4D). In this book these structural features are defined so a student can interpret hand samples of metamorphic rocks. The tectonic relations between foliation, schistosity, and lineation is treated more exhaustively in the field of structural geology.

Table 12.3 *Common rock names for metamorphic rocks*

Rock	Characteristics
Structural terms conventionally prefixed by mineral names	
Granofels	A relatively coarse-grained rock with equant grains
Hornfels	A massive, fine-grained rock that breaks with a conchoidal fracture
Slate	A very fine-grained rock with a perfect planar cleavage
Phyllite	A fine-grained rock with a silky luster on the cleavage surface
Schist	A strongly schistose rock
Gneiss	A foliated metamorphic rock dominated by feldspars
Rock names that imply a mineral assemblage and protolith	
Greenschist	A foliated mafic rock with actinolite as the major amphibole
Greenstone	A massive mafic rock with actinolite as the major amphibole
Amphibolite	A mafic rock with hornblende as the major amphibole
Blueschist	A mafic rock with sodic amphibole as the major amphibole
Granulite	A mafic rock containing orthopyroxene and clinopyroxene
Eclogite	A mafic rock containing clinopyroxene and garnet
Marble	A rock dominated by calcite and dolomite
Quartzite	A metamorphic rock dominated by quartz
Serpentine	An ultramafic rock containing serpentine as the major mineral

12.8 Naming a Metamorphic Rock

Compared to igneous petrology, naming metamorphic rocks is relatively straightforward. Most metamorphic rock names (Table 12.3) are prefixed by the minerals present. Geologists usually do not prefix the names "slate" or "phyllite" with minerals because these rocks are typically so fine-grained that detailed determination of their mineralogy is not practical without X-ray techniques.

In a metamorphic rock name, the minerals present are listed in the order of decreasing abundance (Fettes and Desomns, 2007). An example of such a rock name is quartz–plagioclase–biotite–muscovite–garnet–kyanite schist. If a very small amount of a mineral, say staurolite, was present in this rock, we could call it a staurolite-bearing quartz–plagioclase–biotite–muscovite–garnet–kyanite schist. This cumbersome scheme is often simplified by omitting some of the most common phases. For example, since most schists contain quartz, plagioclase, biotite, and muscovite, the rock named here can also be described as a staurolite-bearing garnet–kyanite schist.

Only a small population of metamorphic rock names exists, many of which apply to mafic rocks and carry an implication of protolith deduced from the metamorphosed assemblage (Table 12.3). These rock names need not be prefixed by any mineral, but the addition of a mineral prefix is often used, particularly if the prefix provides important information about the metamorphic conditions of the rock. These rock terms are discussed in detail in subsequent chapters that describe how mineral assemblages vary with increasing metamorphic grade for different protoliths.

Summary

- Metamorphism involves the crystallization or recrystallization of a rock in the solid state and occurs under a range of conditions, from diagenesis to the beginning of melting.

- The most common types of metamorphism are:

  ○ Regional metamorphism

  ○ Contact metamorphism

  ○ Burial metamorphism

  ○ Dynamic metamorphism

  ○ Hydrothermal metamorphism

- Metamorphic petrology endeavors to answer three fundamental questions about a metamorphic rock:
  - What were the conditions under which it formed?
  - What was the metamorphic protolith?
  - What do metamorphic textures imply about the structural history of the rock?
- Protoliths that are clearly sedimentary are:
  - Pelitic (metamorphosed shale)
  - Psammitic (metamorphosed sandstone)
  - Quartzose
  - Carbonate
- The only common clearly igneous protolith is *ultramafic*.
- Protoliths that may have either an igneous or sedimentary source include:
  - Mafic
  - Quartzo-feldspathic
- Metamorphic conditions can be inferred from:
  - The stability of a single mineral
  - The stability of mineral assemblages
  - Metamorphic facies
  - Thermobarometry
- Metamorphic textures encode evidence about whether a rock was metamorphosed in a static environment or whether it was deformed during metamorphism in a tectonic environment.
- Metamorphic textures also assist protolith determination, whether it is igneous or sedimentary.

Questions and Problems

Problem 12.1. What process marks the high-temperature limit of metamorphism? The lower-temperature and pressure limit of metamorphism? Which limit is easier to define?

Problem 12.2. List the minerals *in order of decreasing abundance* found in a staurolite-bearing quartz–plagioclase–biotite–muscovite–garnet–kyanite schist.

Problem 12.3. In the figure below, which rock (A or B) shows a lineation, and which shows a foliation?

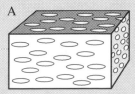
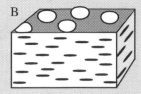

Problem 12.4. In the figure below, draw the texture of a rock that shows both a lineation and a foliation.

Problem 12.5. How does one distinguish rocks affected by contact metamorphism from those affected by regional metamorphism?

Problem 12.6. What information is used to determine the conditions of metamorphism?

Problem 12.7. What are the diagnostic minerals for the following protoliths?

a. pelitic

b. metasedimentary

c. mafic

d. peridotitic

Further Reading

Fettes, D. and Desmons, J., eds., 2007, *Metamorphic Rocks: A Classification and Glossary of Terms. Recommendations of the International Union of Geological Sciences Subcommission on the Systematics of Metamorphic Rocks.* Cambridge University Press, Cambridge.

Holness, M. B., Cesare, B., and Sawyer, E. W., 2011, Melted rocks under the microscope: Microstructures and their interpretation. *Elements*, 7, 253–60.

Jamtvelt, B., 2010, Metamorphism: From patterns to processes. *Elements*, 6, 149–52.

Jamtvelt, B. and Austerheim, H., 2010, Metamorphism, the role of fluids. *Elements*, 6, 153–8.

Vernon, R. H. and Clarke, G. L., 2008, *Principles of Metamorphic Petrology*. Cambridge University Press, Cambridge, Chapter 7: Parent rocks.

Winter, J. W., 2010, *An Introduction to Igneous and Metamorphic Petrology*, 2nd edn. Prentice Hall, New York, NY, Chapter 21: An introduction to metamorphism.

Yardley, B. W. D., 1989, *An Introduction to Metamorphic Petrology*. Longman, London, Chapter 1: The concept of metamorphism.

Note

1 A review of igneous and metamorphic minerals is found in the appendix.

Chapter 13

Interpretation of Metamorphic Phase Diagrams

13.1 Introduction

Phase diagrams are as important to metamorphic petrology as they are to igneous petrology because they give information about the stability fields of minerals with respect to pressure, temperature, and fluid composition. It is relatively straightforward to read the phase diagram for a one-component system, such as the alumino-silicate phase diagram (see Figure 2.1). The diagram readily indicates the stability field for sillimanite, for example. In more complex systems, rather than determining the stability of single phases, it is important to determine the stability fields for *assemblages*. Phase diagrams show the univariant reactions that relate minerals in a given system. Unfortunately, most mineral assemblages in rocks are not univariant and it is usually not intuitive to determine the divariant assemblages in a given system simply from the phase diagram. To make sense of divariant assemblages in a system, petrologists use **chemographic projections**. These are diagrams that graphically depict mineral compositions and, when used in combination with phase diagrams, help petrologists visualize what mineral assemblages occur in a given divariant field. This is important because the mineral assemblages one finds in a rock rarely are invariant; most assemblages are divariant or have a higher variance.

13.2 A Little History

The first study that tried to make sense of the mineral relations found in metamorphic rocks was that of George Barrow (1893, 1912), who recognized six metamorphic zones, based on the mineral assemblages found in pelitic schists of the Scottish Highlands. Barrow's classification worked well in pelitic rocks but had little relevance to other rock compositions. A few years later, Finnish petrologist Penti Eskola (1915, 1920) introduced the concept of metamorphic facies, which identified broader categories of mineral assemblages that described metamorphic conditions. At about the same time, the German school of petrologists, headed by Paul Niggli, established a classification of metamorphic rocks based on depth (epizone = shallow, low grade; mesozone = moderate depths, medium grade; katazone = deep level, high grade; Grubenmann and Niggli, 1924). Throughout the 1920s and 1930s, the German school held sway with the result that metamorphic facies did not gain acceptance until Francis Turner's classic 1948 (Turner, 1948) text on metamorphism. Today, the Germanic classification system is all but forgotten except for the term *epizonal*, which is usually applied to plutonic rocks, despite a proposal to revive it in a slightly different form by Winkler (1976).

Beginning in the late 1950s and continuing today, metamorphic petrologists have been concerned with determining the relative stabilities of mineral assemblages in various protoliths. The diagrams produced by such studies are known as **petrogenetic grids**. The grids are thermodynamically reasoned and corroborated by high-temperature, high-pressure experimental work. For pelitic rocks, the petrogenetic grids were established by students at Harvard University under J. B. Thompson in the 1950s and 1960s (see Albee, 1965). Grids for metaperidotites were developed in the 1960s and 1970s by B. W. Evans and V. Trommsdorff (see Evans, 1977). Grids for carbonate rocks were developed in the 1970s by many workers, including Metz and Trommsdorff (1968) and Skippen (1971).

Although petrologists have long known that metamorphic mineral assemblages preserve information about the composition of the attendant fluid phase, the quantitative use of mineral assemblages for this purpose didn't become important until the explosion of metamorphic petrology in the late 1960s and 1970s. At present, many studies estimate metamorphic fluid composition, considering this variable to be as important as pressure or temperature of formation. Among the fluid species that petrologists commonly evaluate in metamorphic rocks are H_2O, CO_2, sulfur, and oxygen.

The development of the electron microprobe in the mid 1960s revolutionized petrology. Not only did this instrument provide the data necessary to better constrain the petrogenetic grids in pressure–temperature (P–T) space and to determine the composition of the fluid phase, it also led to the development of the disciplines of geothermometry and **geobarometry** (these terms are often combined to form the name *thermobarometry*). These disciplines use thermodynamics and mineral equilibria to quantify the pressure and temperature conditions of crystallization. Although a detailed treatment is beyond the scope of this book, an introduction to thermobarometry is presented in Chapter 18. This text shows that for many bulk compositions careful use of petrogenetic grids can constrain metamorphic conditions nearly as well as thermobarometry. The evolution of petrogenetic grids and thermobarometry does not necessarily make metamorphic facies obsolete, but it does mean the facies designation is best used in a regional context or as a field term. When dealing with metamorphic rocks on a small scale, it is more helpful to determine conditions, if possible, from petrogenetic grids or thermobarometry.

The development of analytical instruments, including the Super High Resolution Ion Microprobe (SHRIMP) in the 1980s (Compston et al., 1984; Williams, 1998) and the laser ablation inductively coupled plasma mass spectrometry (LA–ICP-MS) in the 1990s (Storey et al., 2004), which allow for spatial analysis of minerals and determination of their age and chemical composition, led to the development of metamorphic geochronology – the use of accessory minerals to date metamorphic and deformation events. Geochronology is one of the active fields of research in metamorphism today (see, for example, Kohn, 2016) and promises to provide geologists closer constraints on the duration of tectonic events.

13.3 Use of Chemographic Projections

Sometimes it is possible to write a reaction between all the phases in a particular metamorphic rock. If so, the assemblage is univariant and the reaction traces a line on a phase diagram or an isograd on a geologic map. Most of the time, however, rocks contain assemblages of higher variance. It is important to find ways to treat multivariant assemblages, so that it is possible to visualize the pressure, temperature, or fluid compositions that control them. Divariant assemblages are easy to understand in a system with one component, such as the system Al_2SiO_5, for these assemblages consist of only one phase (see Figure 2.1). In systems with more components, however, it becomes more difficult to recognize divariant assemblages merely through inspection. In complex systems, chemographic projections make it possible to recognize divariant assemblages and to find their stability fields in a phase diagram. In chemographic projections, mineral compositions are expressed graphically and the graphical relationship between minerals determines whether or not they are in a reaction relationship.

13.3.1 Chemographic Projections in a Two-Component System

Consider a hypothetical two-component system consisting of the components a and b and the phases $V = a$, $W = a_2b$, $Y = ab_2$, and $Z = b$ (Figure 13.1; note the convention that components are written in lower case letters, whereas phases are written in capitals). With a little bit of work with a pencil and paper, you can show that this system contains the following univariant reactions:

$$2V + Z \rightleftharpoons W \qquad (13.1)$$

$$V + 2Z \rightleftharpoons Y \qquad (13.2)$$

$$3V + Y \rightleftharpoons 2W \qquad (13.3)$$

$$3Z + W \rightleftharpoons 2Y \qquad (13.4)$$

These reactions could potentially intersect at an invariant point, as in Figure 13.2. The diagram in Figure 13.2 is called a **topology**. It shows the arrangement of reactions around an invariant point but does not locate the reactions in P–T space. Because a topology is constructed to show the orientation of the reactions in P–T space, the stoichiometry of the individual reactions is not important. For this reason, stoichiometric coefficients are not usually presented on these diagrams.

Even with a quick glance at Figure 13.2 one can determine where three-phase assemblages such as W–V–Z are stable, since each three-phase assemblage is marked by one of the reactions listed previously. The problem comes when one has a divariant assemblage or suite of divariant assemblages. For example, where would the assemblages

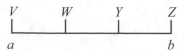

Figure 13.1 Chemography for the two-component system a–b with phases V, W, Y, Z. The topology for this system is shown in Figures 13.2 and 13.3.

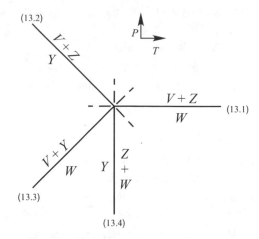

Figure 13.2 Topology for the hypothetical two-component system a–b shown in Figure 13.1.

WY and YV be stable? This question is answered by using chemographic diagrams for this system (i.e. Figure 13.1) to indicate assemblages that are stable in each of the divariant fields. Realize that each reaction is an expression of a change in chemography: two phases react to form one phase (or one phase breaks down to form two others; chemographically these reactions are equivalent). The chemographic expression of a reaction is the appearance (or disappearance) of the product phase on the segment of line that lies between the two reactant phases. For example, consider Equation 13.1 in Figure 13.2. On the high-pressure side of this reaction, phases $V + Z$ are stable; this means there is a line running from V to Z with no intervening phases. On the low-pressure side of this reaction, W has appeared. Chemographically this means that on this side of the reaction point, W lies between V and Z. Consequently, the stable two-phase assemblages in this field are VW and WZ. This is shown in Figure 13.3. Using this rule, it is possible to fill in all the rest of the divariant fields in Figure 13.3 with chemographic projections that indicate what two-phase assemblage is stable in each field. With this information added to Figure 13.3, it becomes straightforward to answer the question raised earlier about where WY and YV are stable.

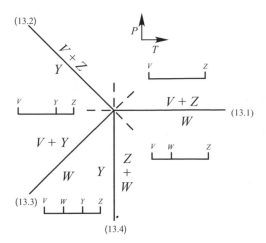

Figure 13.3 Topology for the hypothetical two-component system a–b shown in Figure 13.1, with divariant assemblages identified using chemographic projections.

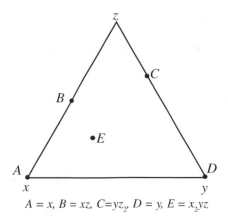

$$A = x, \; B = xz, \; C = yz_2, \; D = y, \; E = x_2yz$$

Figure 13.4 Chemographic diagram for a hypothetical three-component system x–y–z.

13.3.2 Chemographic Projections in a Three-Component System

Now consider a hypothetical three-phase system x, y, z, containing the phases: $A = x$, $B = xz$, $C = yz_2$, $D = y$, and $E = x_2yz$. Again, after a little work with a pencil and paper, it is possible to balance the following reactions:

$$3A + C \rightleftharpoons B + E \tag{13.5}$$

$$4B + 3D \rightleftharpoons 2E + C \tag{13.6}$$

$$E \rightleftharpoons A + B + D \tag{13.7}$$

$$2A + C \rightleftharpoons 2B + D \tag{13.8}$$

$$2E \rightleftharpoons 4A + C + D \tag{13.9}$$

The divariant assemblages for a three-component system are expressed chemographically using an equilateral triangle that has one component at each apex. The chemography for the three-component system in question is shown in Figure 13.4. (For rules governing the construction of this figure, see Section 2.3.)

Figure 13.5 presents a topology showing how these reactions could be related. The four-phase assemblages make up the five univariant reactions in this system. Chemographic projections are used to determine where three-phase or two-phase assemblages are stable. The chemographic representation of a divariant assemblage in a three-component system is a triangle that has as its apices the three stable phases. Figure 13.6 provides an illustration of how a topology of a three-phase system is interpreted using chemographic projections.

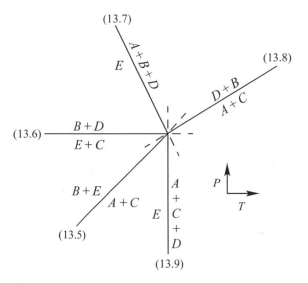

Figure 13.5 Topology for the hypothetical three-component system x–y–z.

Consider the field outlined by Equations 13.5 and 13.6 in Figure 13.6. From Equation 13.6 ($4B + 3D \rightleftharpoons 2E + C$) one can see that the tie line E–C is stable, and from Equation 13.5 ($3A + C = B + E$) it is evident that the tie line B–E is stable. Having established two tie lines, how does one fill in the rest of the triangle? Note that if Equation 13.6 operated on a bulk assemblage with abundant D and small amounts of B, B would be consumed by this reaction, leaving some D remaining along with the products $E + C$. This implies that in the divariant field between Equations 13.5 and 13.6 there must be a tie line from E to D. Similar reasoning concerning Equation 13.5 implies that there must be a tie line from E to A. Thus, the

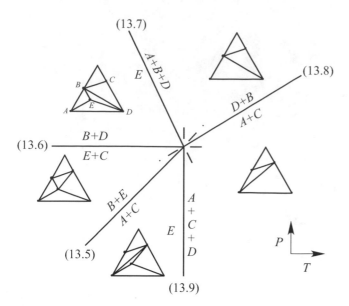

Figure 13.6 Topology for the hypothetical three-component system x–y–z, with divariant assemblages identified using chemographic projections.

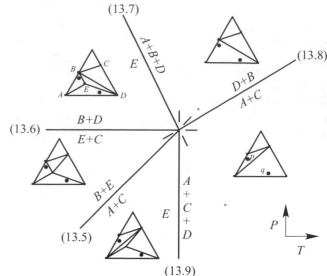

Figure 13.7 Same topology as Figure 13.6, showing the assemblages found in two hypothetical bulk compositions, p and q.

divariant field between Equations 13.5 and 13.6 contains the following assemblages: B–C–E, C–E–D, A–B–E, and A–E–D.

Once one has determined the divariant assemblages present in one field, the relations in other fields are determined by performing the operation implied by each univariant reaction. For example, in crossing Equation 13.6, the tie line from E to C is broken and a tie line from B to D forms instead. This leads to the chemography shown in the divariant field between Equations 13.6 and 13.7. Note that all the other tie lines shown in the field between Equations 13.5 and 13.6 are unchanged. What happens when one goes from the field bounded by Equations 13.6 and 13.7 to that bounded by Equations 13.7 and 13.8? Equation 13.7 says that E reacts out to form A + B + D. Chemographically, that means E disappears within the triangle A–B–D. Since E has disappeared, the tie lines from E to A, B, and D disappear also (Figure 13.6).

A careful study of Figure 13.6 reveals two types of reactions. One type has two or more phases on each side of a reaction. The chemographic representation of this reaction is a quadrilateral in which two tie lines have an interior intersection (e.g. Equation 13.6 in Figure 13.6). This type of reaction is known as a **tie-line flip reaction**. When a reaction such as this occurs, all the phases remain intrinsically stable; it is merely the stable *assemblages* that change. The second type of reaction occurs when only one phase occurs on either side of the reaction.

The chemographic representation of this type of reaction is a triangle with an interior phase (e.g. Equation 13.7 in Figure 13.6). A reaction such as this is called a **terminal reaction**. In crossing a terminal reaction, a phase disappears in all bulk compositions.

Rarely are minerals consumed by terminal reactions. Most of the time, metamorphic reactions involve tie-line flip reactions. As an example, consider the stability of chlorite. Chlorite is a common mineral in low-grade metamorphism of most protoliths; it disappears with increasing metamorphic grade. However, chlorite is not restricted to low metamorphic temperatures. Chlorite disappears because, with increasing temperatures, tie-line flip reactions restrict its occurrence to increasingly limited bulk compositions. The terminal reaction for chlorite (chlorite ⇌ olivine + orthopyroxene + spinel + H_2O) occurs at temperatures around 700 °C, which means that in ultramafic rocks chlorite may be stable up to very high temperatures.

In addition to indicating the stable phases in a given divariant field, chemographic projections also show the assemblage present in a rock of a given bulk composition. To understand this concept, consider Figure 13.7, which shows the same topology as that given in Figure 13.6 along with the chemographic diagrams for the compositions of two hypothetical rocks, p and q. Rock p has a composition very close to mineral B, whereas rock q has a composition that lies close to the A–D tie line. In the divariant field bounded by Equations 13.7 and 13.8, both bulk compositions will have the same assemblage: A–B–D. Rock p will have a lot of B and only a small amount of A and D, whereas rock q will have mostly D and A and only minor amounts of B. As Equation 13.8 (D + B = A + C) proceeds, each assemblage behaves differently. Mineral D will be depleted from rock p, which is dominated by mineral B, whereas mineral B will be depleted first from rock q. These relations are shown in the chemographic projection in the field bounded by Equations 13.8 and 13.9, where composition p lies in the triangle A–C–B, whereas composition q lies in the field A–C–D. The reader is encouraged to determine the assemblages for these bulk compositions in the other fields in Figure 13.7 as an exercise.

13.3.3 Chemographic Projections in Systems with Four and More Components

In a system with four components, the phases can be expressed using a three-dimensional tetrahedron, but in systems of five or more components it is not possible to display relations chemographically because to do so it would require more than three dimensions. There are two ways to handle this problem. One way is to explore the system in n-dimensional vector space, using matrix algebra. In other words, the problem is addressed mathematically rather than graphically. This approach, while useful, will not be addressed in this text. The other way is to use projections. The extra dimensions of a complex system can be eliminated by assuming that one or more phases are always present in an assemblage. By projecting from the common phase, it is possible to reduce a multidimensional chemography to a triangular chemographic projection. Such diagrams are called pseudoternary projections and are used precisely like the chemographic diagrams described for a three-component system. The difference is that these projections only depict reactions that contain the projecting phases. Crossing a reaction that does not contain all the projecting phases does not change the relations shown on the projection. Pseudoternary projections have been used previously in this text to discuss basalt melting and the chemistry of mid-ocean ridge basalt (MORB), where relations were projected from plagioclase onto the olivine–clinopyroxene–quartz triangle (see Figures 6.5 and 7.8).

Figure 13.8 illustrates a chemographic projection from a ternary system to a binary one. Figure 13.8A shows chemographic relations in the system MgO–SiO₂–H₂O. If one assumes that H₂O is always present during metamorphism of a serpentinite, we can show the same information on a

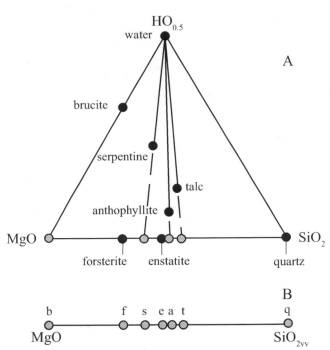

Figure 13.8 (A) Chemography for the system MgO–SiO₂–H₂O showing projection from H₂O to the MgO–SiO₂ plane. Dark points plot mineral compositions, gray points are projection points. (B) Pseudobinary projection from H₂O for the same system.

linear chemography projected from H₂O (Figure 13.8B). The coordinates used in the ternary projection are obtained by taking the ratio of each of the three cations (Mg^{2+}, Si^{4+}, H^+) to the total number of cations ($Mg^{2+} + Si^{4+} + H^+$). Similarly, the coordinates used in the projection from H₂O are obtained by taking the ratio of each of two cations (Mg^{2+} and Si^{4+}) to the sum of these two ions (Table 13.1). As long as H₂O is involved in a reaction, the binary and ternary chemographies show the same information (Figure 13.8A), although all reactions will appear as terminal reactions on the binary system. Indeed, they are terminal if H₂O is always in excess.

Ternary projections have strict rules in application to topologies and phase diagrams. If the plot of an assemblage on a chemographic diagram produces crossing tie lines or has a phase lying within a triangle outlined by three other phases, then it is clear there is a reaction relationship among the phases. There is, however, another type of projection that doesn't have such a strict application, but which is nevertheless very handy in understanding mineral relations in chemically complex rocks. This is the *ACF* projection of Eskola (1920). This is

Table 13.1 *Coordinates for phases in the system* *MgO–SiO$_2$–H$_2$O*

Phase	Formula	MgO:SiO$_2$: H$_2$O	MgO:SiO$_2$
Anthophyllite	Mg$_7$Si$_8$O$_{22}$(OH)$_2$	0.41:0.47:0.12	0.47:0.53
Brucite	Mg(OH)$_2$	0.33:0.0:0.67	1.0:0
Enstatite	MgSiO$_3$	0.5:0.5:0	0.5:0.5
Forsterite	Mg$_2$SiO$_4$	0.67:0.33:0	0.667:0.33
Quartz	SiO$_2$	0:1:0	0:1.0
Serpentine	Mg$_3$Si$_2$O$_5$(OH)$_4$	0.33:0.222:0.44	0.6:0.4
Talc	Mg$_3$Si$_4$O$_{10}$(OH)$_2$	0.33:0.44:0.22	0.43:0.57
Vapor	H$_2$O	0:0:1.0	

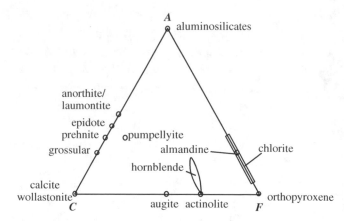

Figure 13.9 ACF diagram plotting common metamorphic minerals.

a projection from quartz, plagioclase, orthoclase, and fluid onto a plane defined by $A = (Al_2O_3 + Fe_2O_3 - Na_2O - K_2O)$, $C = CaO$, and $F = (FeO + MgO)$ (Figure 13.9). Since this projection is from feldspars, the A coordinate involves subtraction of Na$_2$O and K$_2$O to account for aluminum that would be bound in the feldspars. Eskola (1920) used these *ACF* diagrams to define metamorphic facies, painting a rough picture of the reactions involved in metamorphism. However, because of the large number of other components in natural rocks that are not considered in *ACF* projections, they cannot be used strictly to define mineral reactions.

Summary

- Chemographic projections assist in visualizing what phases are stable in divariant fields of a phase diagram.

- The chemographic projection of a two-component system is a line. A reaction is indicated by the appearance or disappearance of a phase between two other phases on this line.

- In a three-component system, the chemographic projection is a triangle.

- Two types of reactions are recognized in three-component (or higher) systems:

  ○ In a tie-line flip reaction, all phases remain stable; the reaction is simply manifested by a change in assemblage.

  ○ In a terminal reaction, one phase reacts away.

- In a four-component or higher system, relations can be shown by projection from one or more phases, assumed present in all reactions.

Questions and Problems

Problem 13.1. Show that for typical metamorphic rocks, the phase rule can be expressed as $F = C$. You may wish to refer to Chapter 2, where the phase rule is introduced. Under what circumstances might $F > C$?

Problem 13.2. The following questions refer to the figure below.

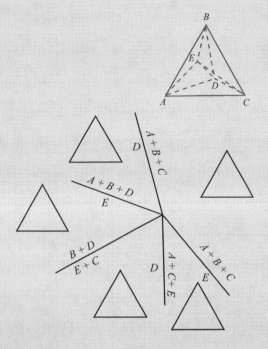

a. Fill in the triangles showing the assemblages in each divariant field.

b. Label which reactions are tie-line flip reactions and which are terminal reactions.

Problem 13.3. Answer the following questions, referring to the figure below.

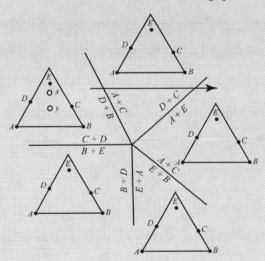

a. Indicate the stable assemblages present in each of the divariant fields above by filling in the appropriate tie lines on the chemographic triangles.

b. For each reaction note whether it is a tie-line flip reaction or a terminal reaction.

c. Which of the two bulk compositions, x or y, is more sensitive to changes in temperature along the path indicated by the arrow?

Problem 13.4. The following questions refer to the diagram below.

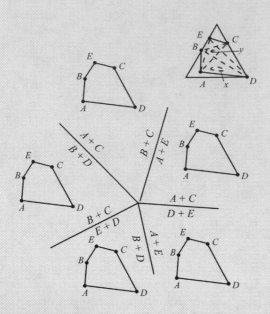

a. Fill in the chemographic triangles showing the stable assemblages in each field.

b. Given rocks (x and y) with two distinct bulk compositions (see plot above), note what assemblages are found in these rocks in each divariant field. Inset shows the relation between the composition of y and the D–B and A–C tie lines.

c. Which bulk composition is the most sensitive monitor of metamorphic conditions? Explain your answer.

Problem 13.5. Below is a topology for a portion of the system Al_2O_3–SiO_2–H_2O, which is projected onto the triangle Al_2SiO_5–SiO_2–H_2O. The phases involved are A = andalusite (Al_2SiO_5), K = kaolinite ($Al_2Si_2O_5(OH)_4$), P = pyrophyllite ($Al_2Si_4O_{10}(OH)_2$), Q = quartz, W = water.

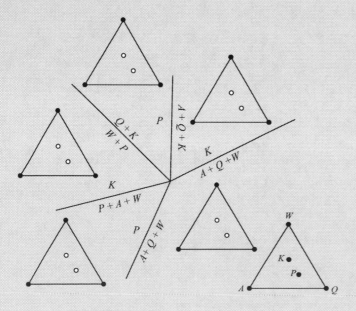

a. Indicate the stable assemblages present in each of the divariant fields by filling in the appropriate tie lines on the chemographic triangles.

b. For each reaction note whether it is a tie-line flip reaction or a terminal reaction.

Further Reading

Ernst, W. G., 1976, *Petrologic Phase Equilibria*. W. H. Freeman, San Francisco, CA.

Thompson, J. B., Jr., 1982. Composition space: An algebraic and geometric approach. *Reviews in Mineralogy*, 10, 1–31.

Thompson, J. B., Jr., 1982. Reaction space: An algebraic and geometric approach. *Reviews in Mineralogy*, 10, 33–52.

Chapter 14

Metamorphic Facies and the Metamorphism of Mafic Rocks

14.1 Introduction

In the late nineteenth and early twentieth centuries, petrologists found it challenging to characterize the conditions of metamorphism because they lacked the thermodynamic data that today allow researchers to calculate the pressures and temperatures at which assemblages formed. Metamorphic facies, which were formulated by Eskola (1915), remain useful, even though their pressure and temperature (P–T) conditions were initially very poorly defined. Indeed, metamorphic facies were simply defined as "a series of mineral assemblages that are found repeatedly associated in time and space" (Eskola, 1915). The definitions naturally carried an implication about the P–T conditions of metamorphism, but at the time petrologists had no way to specify these conditions. For example, they recognized that, with increasing metamorphic grade, actinolite turns into hornblende, and that at even higher grades hornblende breaks down to orthopyroxene and augite, but researchers had no way to assign absolute temperatures and pressures to these assemblages.

This chapter details the mineral changes that occur in mafic rocks with increasing temperature and pressure and how these produce the assemblages that define the various facies of metamorphism. Chapter 18 describes how petrologists use thermodynamics to characterize the pressure and temperature limits to metamorphism. That chapter also discusses the pressure and temperature conditions that characterize the facies discussed in this chapter.

Table 14.1 *Metamorphic facies as defined by mineral assemblages in mafic rocks (phases in parentheses may be present under certain conditions)*

Metamorphic facies	Ferromagnesian minerals	Calcium–aluminum silicates
Zeolite	Chlorite	Zeolites, (prehnite), (epidote)
Prehnite–pumpellyite	Chlorite	Epidote, pumpellyite, (prehnite)
Lawsonite–albite–chlorite	Chlorite	Lawsonite
Greenschist	Actinolite	Epidote, (plagioclase)
Blueschist	Sodic amphibole, chlorite, almandine	Epidote, lawsonite
Amphibolite	Hornblende, (garnet)	Plagioclase, (epidote)
Granulite	Orthopyroxene, augite, (garnet)	Plagioclase
Eclogite	Omphacite, almandine (in garnet)	Grossular (in garnet)

14.2 Definition of Metamorphic Facies

ACF diagrams, as noted in the previous chapter, can graphically depict the assemblages in metamorphic rocks because they can accommodate a wide range in bulk compositions, including pelitic, mafic, and calc-silicate rocks (Figure 14.1). Eskola (1915) defined metamorphic facies based on the assemblages found in mafic rocks, which have bulk compositions that lie in the middle of the *ACF* triangle (Figure 14.1; Table 14.1). Because of its bulk composition, a given assemblage in mafic rocks defines the tie lines within the *ACF* triangle and thus constrains assemblages in protoliths of other bulk compositions. The protolith of metamorphosed mafic rocks (also called **metabasites**) is basalt or gabbro. These rocks consist mainly of plagioclase (usually around $An_{60\ 80}$), and mafic minerals (pyroxene and olivine). A good way to remember the mineralogy that characterizes the individual metamorphic facies is to consider what minerals substitute for the primary minerals from a gabbro in each metamorphic facies. To examine the mineral assemblages that define these facies this chapter first discusses the mineral changes that occur in mafic rocks during prograde metamorphism.

14.3 Facies of Regional Metamorphism

In the original definition, Eskola distinguished between facies formed in areas of regional metamorphism (Figure 14.2) and facies formed at low pressures due to metamorphism by shallow igneous intrusions, that is, facies of contact metamorphism. Because the assemblages found in mafic rocks are essentially the same regardless of

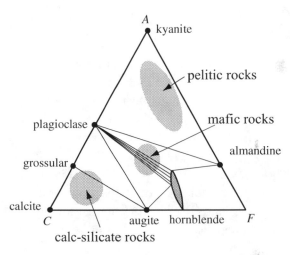

Figure 14.1 *ACF diagram for mineral stabilities at amphibolite facies, showing the approximate fields where pelitic, mafic, and calc-silicate rocks occur.*

pressure and because the boundary between regional and contact metamorphism in *P–T* space is very poorly defined, this text focuses on the facies of regional metamorphism. Contact metamorphism is discussed briefly in Section 14.4.

14.3.1 Greenschist Facies

In most metamorphic belts, the lowest-grade metamorphism encountered is greenschist facies. This name reflects the fact that, under these metamorphic conditions, mafic rocks contain minerals that tend to give them a greenish color in hand sample. Chief among these minerals is actinolite, although epidote and chlorite are also commonly present (Figure 14.3). Under greenschist facies, plagioclase is usually nearly pure albite, although at

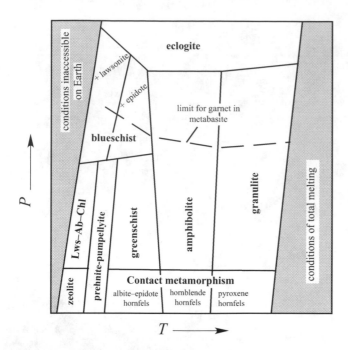

Figure 14.2 Relative *P–T* stabilities for the metamorphic facies.

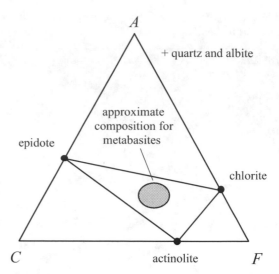

Figure 14.3 ACF diagram showing the stable assemblages in mafic rocks in greenschist facies.

low pressures a somewhat more calcic plagioclase may be present (Starkey and Frost, 1990). Because epidote and chlorite are distinctly silica-poor (both have 35 percent SiO_2 or less), the formation of these minerals tends to release silica. Consequently, quartz is common in greenschists and their non-foliated counterparts, greenstones, even though the igneous protolith almost certainly did not contain quartz. Common accessory minerals include titanite, muscovite (or phengite), and biotite. At low pressures, greenschist rocks may contain oligoclase or even andesine coexisting with epidote, whereas high-pressure, low-temperature greenschists may contain pumpellyite. Pumpellyite-bearing greenschists are sometimes considered as belonging to a separate facies, the pumpellyite–actinolite facies.

14.3.2 Blueschist Facies

With increasing pressure, actinolite disappears as the major amphibole and a sodic amphibole appears in its stead. The sodic amphibole turns the rock blue, hence the term *blueschist*. With increasing pressure, the first blue amphibole to appear contains nearly as much ferric iron as alumina and has the informal name crossite. The high ferric iron content gives it a deep purple color in thin section. With increasing pressure, the ferric iron content decreases and alumina increases, with the result that the amphibole becomes glaucophane and its color becomes pale lavender in thin section.

Blueschist facies was originally defined as the glaucophane–lawsonite schist facies, and some authors (e.g. Turner, 1968) prefer to restrict the facies to those assemblages that contain high-pressure phases such as lawsonite, and to relegate crossite-bearing assemblages to greenschist facies. A broader interpretation is advocated by Evans and Brown (1987) whereby blueschist facies is defined as conditions where metabasites contain any blue amphibole. Evans (1990) identifies two subfacies of blueschist facies, depending on the Ca-Al silicate present. Lawsonite blueschists occur at relatively low temperatures and relatively high pressures, whereas epidote blueschists occur at higher temperatures and lower pressures (see Figure 14.2). At the highest pressures of blueschist facies, a jadeite pyroxene may form from albite; this marks the beginning of the transition to eclogite facies. Other minerals potentially present in mafic rocks in blueschist facies include pumpellyite, almandine garnet, rutile, aragonite, and stilpnomelane. Because the amphibole in blueschist facies has a high sodium content the mineral assemblages for blueschist facies are not easily expressed on an ACF projection, where sodium is not a participating component.

14.3.3 Amphibolite Facies

Subjecting weakly metamorphosed metabasites to increasing temperature allows more aluminum to substitute in the tetrahedral site of the amphiboles, causing both actinolite and glaucophane to change to hornblende. The appearance of hornblende in a mafic rock defines the amphibolite facies. Because the substitution of aluminum in hornblende is also accompanied by ferric iron, the

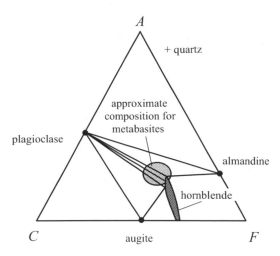

Figure 14.4 *ACF* projection showing mineral relationships in mafic rocks in amphibolite facies.

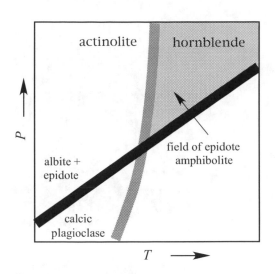

Figure 14.5 Schematic diagram comparing the *P–T* stabilities of anorthite-forming and hornblende-forming reactions.

color of the amphibole in hand sample changes from green to black. In thin section, hornblende has a brownish color in plane-polarized light (PPL). In addition to changes in amphibole composition, increasing temperature stabilizes calcium in the plagioclase and causes the disappearance of epidote. Because hornblende has such a variable composition, amphibolites are classic high-variance rocks (meaning they have many degrees of freedom), and the diagnostic assemblage is simply hornblende + plagioclase (Figure 14.4). Those amphibolites that are relatively more calcic than typical basalt may contain diopside, while those that are relatively less calcic may contain chlorite, garnet, or, more rarely, cummingtonite.

Although amphibolites are renowned for their high variance, some mineralogic characteristics indicate the metamorphic conditions within amphibolite facies at which the rocks formed. One feature is the pleochroism of the hornblende. Hornblende pleochroism apparently is dependent on the amount of Fe^{3+} and TiO_2 it contains. These minor components usually increase with increasing temperature. Consequently, hornblende from high-temperature amphibolite facies has an olive-brown to brown pleochroism, while hornblende from low-temperature amphibolite facies will have a green pleochroism with a hint of olive brown. Hornblende from high-pressure, low-temperature amphibolites (i.e. those from grades slightly higher than that of blueschist) will have a distinct bluish tone to its pleochroism. Two other indicators of metamorphic conditions in amphibolite facies are the presence of epidote or garnet. Although the reactions governing the transition from actinolite to hornblende and the reaction of albite + epidote to calcic

plagioclase are not well defined, petrologists recognize they are not parallel in *P–T* space (Figure 14.5). As a consequence, in low-pressure amphibolites, epidote disappears from the rock before hornblende forms, while in high-pressure rocks, epidote can survive into amphibolite facies. Such rocks are known as epidote amphibolites and indicate relatively high pressure.

At high pressures, the anorthite component of plagioclase ($CaAl_2Si_2O_8$) reacts to grossular ($Ca_3Al_2Si_3O_{12}$) in garnet. Thus, at high pressures, generally above 5–6 kbar, a calcic almandine garnet appears in amphibolites (Figure 14.6). Garnet amphibolites are good indicators of relatively high pressure of metamorphism. In many samples, such as that shown in Figure 14.6, the garnet is rimmed by plagioclase. This texture indicates that the rock underwent decompression late in its metamorphic history with the result that the grossular component of the garnet reacted to plagioclase.

14.3.4 Very Low-Temperature Metamorphism

Before mafic igneous rocks can form equilibrium mineral assemblages in low-grade metamorphism, they must undergo hydration. Many weakly metamorphosed mafic rocks are marked by the survival of the primary igneous minerals. Indeed, augite in particular seems resistant to low-temperature hydration. Zeolite and prehnite–pumpellyite facies were originally defined from mineral assemblages in sedimentary protoliths, where hydration of the original igneous mineralogy occurs more readily than in igneous rocks (Coombs, 1954, 1960). Many of the subsequent studies of very low-temperature

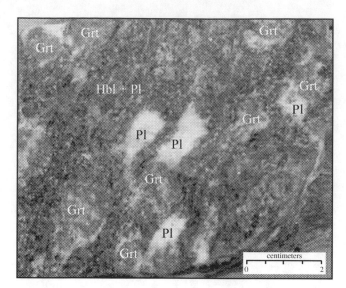

Figure 14.6 Photograph of a garnet amphibolite with plagioclase rims (Pl) around the garnet (Grt). The plagioclase probably formed from calcium released from the garnet during decompression. From near Mica Dam, British Columbia, Canada.

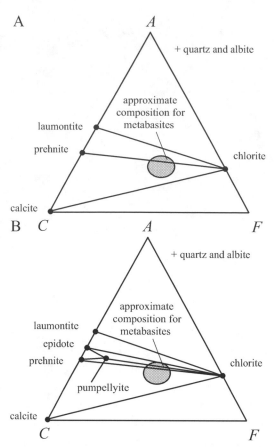

Figure 14.7 *ACF* diagrams showing mineral relations in weakly metamorphosed mafic rocks. (A) Zeolite facies; (B) prehnite–pumpellyite facies.

metamorphic changes in mafic rocks have been made in the amygdaloidal portions of basaltic flows.

Zeolite Facies. Zeolite facies is defined by the presence of zeolites, of which there are a bewildering variety. Zeolites are tectosilicates that are analogous to feldspars, but because of their open lattice they accommodate structural water. Near the upper temperature limit of zeolite facies, the most common zeolite is laumontite. In some areas of high heat flow (i.e. relatively low pressure and high temperature), wairakite, a mineral with the same composition as laumontite but with less H_2O, may also occur. Other minerals in mafic rocks metamorphosed to zeolite facies include calcite, albite, quartz, chlorite, epidote, prehnite, and pumpellyite (Figure 14.7A).

Prehnite-Pumpellyite Facies. The reactions that eliminate laumontite from mafic rocks are those that produce the assemblage chlorite–epidote–quartz (Frost, 1980) (Figure 14.7B). Assemblages containing prehnite or pumpellyite are also likely to be present (Figure 14.7B). Actinolite is distinctive by its absence. It is important to note that in some bulk compositions laumontite and wairakite may survive into prehnite–pumpellyite facies. If they do so, however, they will not occur with prehnite or pumpellyite. The high-pressure limits of prehnite–pumpellyite facies are poorly defined. At high pressures, prehnite–pumpellyite facies is transitional into blueschist facies, since pumpellyite is found in some blueschists. The transition from prehnite–pumpellyite facies to greenschist

facies is marked by the appearance of actinolite and the disappearance of prehnite and pumpellyite. At low pressures, pumpellyite disappears at lower temperatures than does prehnite, resulting in a narrow temperature range where prehnite and actinolite coexist. At high pressures, there is a stability field for pumpellyite and actinolite.

Lawsonite–Albite–Chlorite Facies. A separate facies distinguished by the assemblage lawsonite–albite–chlorite may exist, a possible high-pressure equivalent to zeolite facies (Coombs et al., 1976). This facies lies at temperatures below those of prehnite–pumpellyite facies and at pressures below those of blueschist facies. Lawsonite–albite–chlorite facies is indicative of very low temperatures and higher pressures of metamorphism and is found in only a few locations, such as on Crete, where subduction of a cold slab is postulated to have occurred (Theye et al., 1992).

14.3.5 Granulite Facies

Granulite facies is marked by the breakdown of hornblende to orthopyroxene and augite (Figure 14.8).

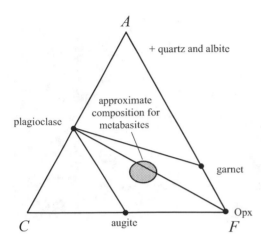

Figure 14.8 *ACF* diagram showing mineral relations in mafic rocks in granulite facies.

Because hornblende can assume a range of compositions, this reaction takes place over a range of temperatures, with the result that most granulites contain both hornblende and its breakdown products, pyroxenes. Mafic granulites are mineralogically very simple. At relatively low pressures, the assemblage is orthopyroxene–clinopyroxene–plagioclase ± hornblende. In some rocks olivine may also occur. Because a mafic granulite has the same mineral assemblage as a gabbro, the distinction between the two is purely textural. A gabbro has an igneous texture, which is commonly manifested as tabular plagioclase with tabular or interstitial pyroxene (see Figure 1.8A, C). In contrast, in a mafic granulite both the pyroxenes and the plagioclase form equant grains (see Figure 12.3A). In fact, the term granulite used to be a textural term, describing the granular texture of the rock; it has been superseded by the term granofels.

At higher pressures (above about 6 kbar), garnet is typically present as well. At even higher pressures, orthopyroxene reacts out by the reaction:

$$\text{orthopyroxene} + \text{plagioclase} \rightleftharpoons \text{augite} + \text{garnet} \qquad (14.1)$$

resulting in the high-pressure assemblage garnet–clinopyroxene–plagioclase. The plagioclase in this assemblage gets progressively more sodium-rich with increasing pressure as the anorthite component reacts to grossular in garnet. With increasing pressure, the albite component of plagioclase reacts to jadeite by the reaction:

$$\underset{\text{in plagioclase}}{\text{NaAlSi}_3\text{O}_8} \rightleftharpoons \underset{\text{in proxene}}{\text{NaAlSi}_2\text{O}_6} + \underset{\text{quartz}}{\text{SiO}_2} \qquad (14.2)$$

The jadeite component in this reaction dissolves into the clinopyroxene. As Equation 14.2 progresses, plagioclase abundance decreases; the rock becomes an eclogite when plagioclase has been consumed.

14.3.6 Eclogite Facies

As noted above, with increases in pressure, the anorthite component of plagioclase dissolves into garnet and the albite component dissolves into pyroxene. Thus, at very high pressures, rocks of basaltic composition are made up of dominantly garnet and omphacite (sodium-rich augite), both of which show extensive solid solution. A rock with this assemblage is known as an eclogite. Other common phases likely to occur in eclogite include kyanite, phengite, and rutile. As shown in Figure 14.2 eclogite facies lies on the high-pressure side of blueschist, amphibolite, and granulite facies.

Eclogites occur in two high-pressure environments. One is at low temperatures in subduction zones associated with blueschists. In some fossil subduction zones, such as in the Alps, where Europe was partially subducted under Africa, the transition from blueschist facies to eclogite facies can be observed in the field (see Map 19.2). The other environment for eclogite lies at high temperatures, where they are associated with granulites. Such eclogites are found in continental collisions, at the base of crustal cross-sections, continental-scale thrust faults that expose deep crust, and as xenoliths sampled from the lower crust by alkali basalts.

14.4 Facies of Contact Metamorphism

Contact metamorphism can be considered a separate type of metamorphism, which occupies the low-pressure fields in Figure 14.1. Contact metamorphic facies are designated by the same mineral assemblages as facies of regional metamorphism, but the contact metamorphic environment is recognized on the basis of textural features and field relations rather than by mineral assemblages. The contact metamorphic facies shown in Figure 14.2 include albite–epidote hornfels, essentially similar to greenschist; hornblende hornfels, equivalent to a low-pressure amphibolite facies; and pyroxene hornfels, which has the assemblages of low-pressure granulite facies.

14.5 Textural Changes during Metamorphism

Weakly metamorphosed mafic rocks commonly preserve relics of the primary igneous texture, including tabular plagioclase and blocky crystals of iron–titanium oxides

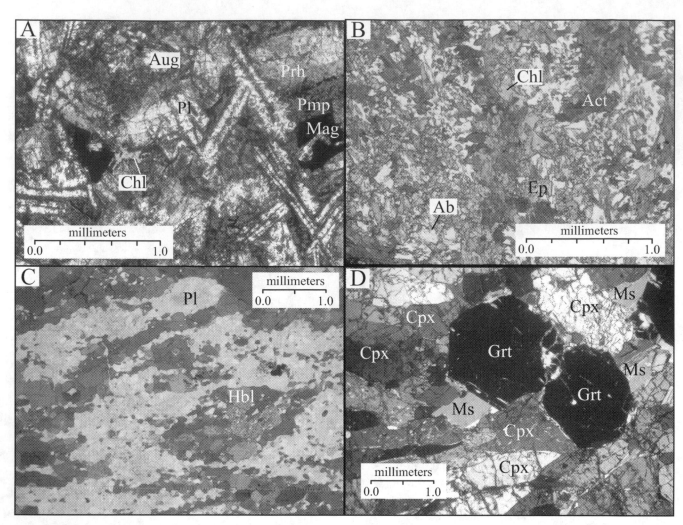

Figure 14.9 Photomicrographs showing textures typical of metabasic rocks. (A) Metabasalt in prehnite–pumpellyite facies in PPL. Note that the basaltic texture is still evident. The tabular plagioclase (Pl) is albite with clusters of pumpellyite on the margin. Primary augite (Aug) and Ti-magnetite (Mag) are still present locally. Other metamorphic minerals include prehnite (Prh), pumpellyite (Pmp), and chlorite (Chl). From the Karmutsen basalts, Vancouver Island, Canada. (B) Greenschist containing actinolite (Act), epidote (Ep), chlorite (Chl), and albite (Ab) in PPL. Note that the albite (white in this figure) is arranged in clusters, which might be relict from primary plagioclase. From the Ligurian Alps, Italy. (C) Amphibolite containing hornblende (Hbl) and plagioclase (Pl) in PPL. The rock has a relict diabasic texture similar to the matrix in Figure 12.2A. In thin section, the plagioclase maintains a suggestion of tabular shape, but it has recrystallized into a matrix of small grains. From the Laramie Mountains, southeastern Wyoming. (D) Eclogite containing garnet (Grt), omphacite (Cpx), muscovite (Ms) in XPL. Sesia zone, Italian Alps.

(Figure 14.9A). The extent to which the igneous textures survive metamorphism depends on the deformation history of a rock, but these relict textures may be present in rocks metamorphosed to moderate grades, which allows a geologist to characterize a mafic rock as a metabasalt, metadiabase, or a metagabbro. Figure 14.9B shows a metabasalt from greenschist facies that retains a subtle hint of tabular feldspar. The tabular shape is partially obscured because the calcium component of the feldspar has reacted to epidote and is only recognized by considering the shape of the colorless areas in the figure (which are mostly small grains of albite). Figure 14.9C shows a thin section of a metadiabase similar to that shown in Figure 12.2A, which shows white plagioclase grains in hand sample. In thin section, it is clear these plagioclase grains have recrystallized into a mosaic of fine grains. In the eclogite shown in Figure 14.9D there is no trace of the original igneous texture. However, a relict gabbroic texture may be found in some eclogites, as displayed by the large-scale distribution of garnet, clinopyroxene, and rutile.

Table 14.2 *Changes in mineralogy in mafic rocks at very low temperature*

Original Minerals	Zeolite Facies	Prehnite-Pumpellyite Facies	Greenschist
Orthopyroxene Augite	Chlorite	Pumpellyite	Actinolite
Olivine	Chlorite	Chlorite	Chlorite
Anorthite	Laumontite	Prehnite Epidote	Epidote
Albite	Albite Analcime at v. low T	Albite	Albite
Other	±Quartz ± Calcite	±Quartz ± Calcite	±Quartz

Table 14.3 *Changes in mineralogy in mafic rocks with increasing temperature*

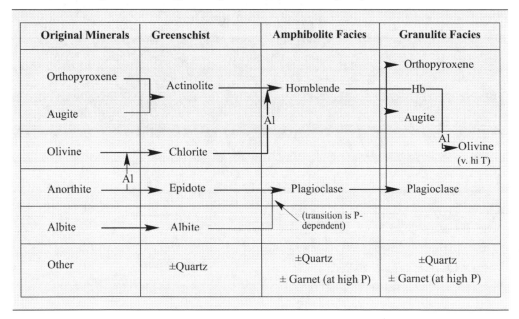

Original Minerals	Greenschist	Amphibolite Facies	Granulite Facies
Orthopyroxene Augite	Actinolite	Hornblende	Orthopyroxene Hb Augite Olivine (v. hi T)
Olivine	Chlorite		
Anorthite	Epidote	Plagioclase	Plagioclase
Albite	Albite	(transition is P-dependent)	
Other	±Quartz	±Quartz ± Garnet (at high P)	±Quartz ± Garnet (at high P)

14.6 Mafic Mineral Assemblages at Increasing Temperature and Pressure

Rather than memorizing a series of assemblages, a preferable way to understand metamorphism of mafic rocks is to follow the mineralogical changes with increasing temperature and pressure. The unmetamorphosed mafic protolith consists of plagioclase, pyroxenes, and perhaps olivine. These minerals contain chemical components that are redistributed during metamorphism – an albite component (sodium, aluminum), an anorthite component (calcium, aluminum), and a ferromagnesian component (calcium, magnesium, iron). Because metamorphism covers a wide range of pressure and temperature conditions, they are summarized in three tables. Table 14.2 covers the mineralogical changes that occur from zeolite to greenschist facies, Table 14.3 covers the changes that

Table 14.4 *Changes in mineralogy in mafic rocks with increasing pressure at low temperature*

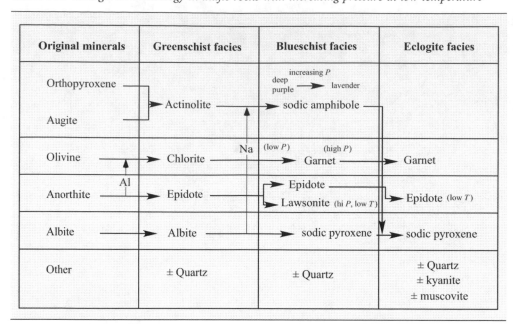

occur with increasing temperature from greenschist facies to granulite facies, and Table 14.4 covers the changes that occur with increasing pressure from greenschist to blueschist facies.

14.6.1 Relations at Very Low Temperatures

At low temperatures and pressures, the stable plagioclase is albite (although note that at the very lowest temperatures, the albite component may be present as a zeolite, analcime) (Table 14.2). In zeolite facies, the calcium–aluminum component of plagioclase is present as laumontite, in prehnite–pumpellyite facies it occurs as prehnite, and in greenschist facies it occurs in epidote. Some of the calcium released by the breakdown of anorthite may occur in calcite, which is common in zeolite and prehnite–pumpellyite facies, but less so in greenschist facies. The major host to iron and magnesium in zeolite facies is chlorite; the aluminum in chlorite probably is derived from the breakdown of anorthite, which has more aluminum than other calcium aluminum silicates. In prehnite–pumpellyite facies, the iron and magnesium may also be hosted in pumpellyite (which also contains considerable calcium and aluminum). In greenschist facies, pumpellyite is replaced by actinolite, producing the common greenschist assemblage albite–epidote–chlorite–actinolite. Because some of the minerals in low-grade metabasites are silica-poor (especially chlorite), quartz may be present in all these assemblages.

14.6.2 Relations at Low Pressure with Increasing Temperature

At low pressures, increasing temperature causes aluminum to dissolve into the actinolite, making hornblende. The appearance of hornblende marks the entry into amphibolite facies. At approximately the same temperature, epidote begins to break down, releasing CaO that dissolves into the plagioclase (Table 14.3). The epidote-out reaction is more strongly pressure-dependent than the hornblende-forming reaction, with the result that epidote may be present in high-pressure amphibolites. Garnet also may be present in high-pressure amphibolites. Hornblende breaks down by the reaction hornblende $\rightleftharpoons$ orthopyroxene + clinopyroxene + plagioclase, so the appearance of pyroxenes in amphibolites marks the upper limit of amphibolite facies. The presence of augite in a mafic rock doesn't necessarily indicate amphibole breakdown because clinopyroxene may be present in relatively calcic amphibolites even at relative low temperature. For this reason, the beginning of granulite facies is marked by the appearance of orthopyroxene in metabasites. Hornblende changes its composition as it breaks down, becoming increasingly enriched in refractory elements such as titanium, ferric iron, and fluorine. This change in composition allows hornblende to persist to very high temperature in granulite facies. As in amphibolites, high-pressure granulites may contain garnet.

14.6.3 Relations at Low Temperature with Increasing Pressure

At low temperatures, increasing pressure causes actinolite to accommodate more sodium (Table 14.4). The appearance of sodium-bearing amphiboles with a bluish color (and pleochroism) marks the low-pressure limits of blueschist facies. Such low-pressure blueschists have a similar mineralogy to greenschist (sodic amphibole–albite–epidote–chlorite), except for the appearance of sodic amphibole rather than actinolite. Increasing pressure causes garnet to form from chlorite. At still higher pressures and somewhat lower temperatures, lawsonite forms from epidote. Finally, at still higher pressures, albite reacts to the jadeite component of pyroxene. This reaction produces the assemblage for high-pressure blueschists of glaucophane–garnet–clinopyroxene–epidote. With increasing temperature or pressure, the amphibole breaks down, producing the assemblage garnet–clinopyroxene–epidote, distinctive of low-temperature eclogite facies. High-temperature eclogite facies lacks epidote. As discussed in Section 14.3.5, high-temperature eclogites form from granulites as increasing pressure drives plagioclase components (anorthite and albite) into garnet and pyroxene.

Summary

- A useful way to understand metamorphic facies in mafic rocks is to consider the ferromagnesian minerals present or absent.

- **Amphibole-bearing facies**. Three metamorphic facies contain amphibole as a critical phase.

  - In greenschist (or the albite–epidote hornfels) facies the amphibole is actinolite

  - In amphibolite (or hornblende hornfels) facies the amphibole is hornblende

  - In blueschist facies the amphibole is sodic amphibole

- **Amphibole-free facies**. Five metamorphic facies do not contain amphibole as a critical phase, although some relict amphibole may be present.

  - In granulite (or pyroxene hornfels) facies, hornblende is in the act of breaking down to orthopyroxene and clinopyroxene

  - In eclogite facies, the stable minerals are omphacite (sodium-rich augite) and garnet. Minor amounts of sodic amphibole may be present at low temperatures

  - In prehnite–pumpellyite facies, assemblages including chlorite and pumpellyite appear instead of actinolite

  - In zeolite facies, assemblages including chlorite and laumontite ($\pm$ prehnite, laumontite, or calcite) appear instead of actinolite

  - In albite–chlorite–lawsonite facies, albite–lawsonite and chlorite are present instead of actinolite

Questions and Problems

Problem 14.1. What minerals give the green color to rocks in greenschist facies? What minerals are responsible for the blue that gives the name to blueschist facies?

Problem 14.2. From the assemblages below, give the metamorphic conditions as precisely as the assemblage allows. (For example, the assemblage clinopyroxene–garnet–epidote would be in low-temperature eclogite facies.) Explain how you reached your conclusion.

a. glaucophane–garnet–albite–lawsonite

b. albite–epidote–actinolite–chlorite

c. hornblende–plagioclase–garnet

d. hornblende–orthopyroxene–clinopyroxene–plagioclase

e. clinopyroxene–garnet–orthopyroxene–plagioclase

Problem 14.3. Below are lists of minerals that occur in a series of mafic rocks. For each list determine if the assemblage could represent an equilibrium assemblage and, if so, identify the facies to which it belongs. If it is not an equilibrium assemblage, determine what minerals are out of equilibrium and to what facies the rest of the minerals belong.

a. hornblende–plagioclase–garnet–prehnite

b. orthopyroxene–augite–hornblende–plagioclase

c. augite–plagioclase–garnet–chlorite

d. omphacite–garnet–glaucophane

e. prehnite–chlorite–epidote–quartz

Further Reading

Bucher, K. and Frey, M., 2002, *Petrogenesis of Metamorphic Rocks*. Springer, Heidelberg, Chapter 9: Metamorphism of mafic rocks.

Cooper, A. F., 1972, Progressive metamorphism of metabasic rocks from the Haast Schist group of southern New Zealand. *Journal of Petrology*, 13, 457–92.

Evans, B. W., 1990, Phase relations of epidote-blueschists. *Lithos*, 25, 3–23.

Philpotts, A. R. and Ague, J. J., 2009, *Principles of Igneous and Metamorphic Petrology*. Cambridge University Press, Cambridge, Chapter 16: Metamorphism and metamorphic facies.

Winter, J. D., 2010, *Principles of Igneous and Metamorphic Petrology*, 2nd edn. Prentice Hall, New York, NY, Chapter 25: Metamorphic facies and metamorphosed mafic rocks.

Yardley, B. W. D., 1989, *An Introduction to Metamorphic Petrology*, Longman, London, Chapter 4: Metamorphism of basic igneous rocks.

Chapter 15

Metamorphism of Peridotitic Rocks

15.1 Introduction

Most peridotitic rocks originally formed as portions of the upper mantle and oceanic crust. During interaction with seawater on the sea floor or during tectonic events occurring at subduction zones, metaperidotite, originally composed of olivine and pyroxenes, may be wholly or partially converted to serpentine $[(Mg,Fe)_3Si_2O_5(OH)_4]$. Prograde metamorphism initiates a series of mineralogical changes in these serpentinites that make them effective markers of metamorphic conditions. Thus, although metamorphosed serpentinites and peridotites make up only a small portion of most metamorphic belts, their significance is greater than their abundance might suggest. Because their very existence in a sequence indicates that those rocks incorporated ultramafic material, either from the mantle or from the cumulate horizon of a layered mafic intrusion, the presence of metaperidotite provides clues about the geologic history of the rock sequence within which it lies. In addition, metaperidotite mineral assemblages are very useful in estimating the temperature of metamorphism. Finally, because peridotites are chemically simple rocks, they are a good protolith to illustrate how mineral assemblages can estimate metamorphic conditions.

15.2 The Process of Serpentinization

The processes by which mantle peridotites become serpentinites must be discussed before describing the metamorphism of serpentinite. The description of oceanic magmatism in Chapter 7 noted how faulting commonly exposes peridotites on the sea floor. Those peridotites exposed on or within a few kilometers of the sea floor react with seawater to form serpentinite. Serpentinization plays a major role in modifying the rheology of the oceanic crust, defines an important geochemical process on the sea floor, and produces the most reducing, alkaline fluids of any natural environment on Earth.

When, at low temperatures, water gains access to peridotite, the rock undergoes serpentinization. The modal mineralogy of the peridotite controls the products of low-temperature metamorphism. If the original rock consisted of more than 58 volume percent olivine, as is typical of most peridotites, the products of hydration are serpentine and brucite. If, on the other hand, the protolith was a pyroxenite with less than 58 volume percent olivine, the hydrated rock consists of serpentine and talc. Although peridotites and serpentinites have a relatively simple chemical composition the process of serpentinization is complex. Part of this complexity comes from the fact that, given a source of silica, brucite [$Mg,Fe(OH)_2$] will react to make serpentine (Frost and Beard, 2007) by Equation 15.1:

$$3Mg(OH)_2 + 2SiO_{2(aq)} + 2H_2O \rightarrow Mg_3Si_2O_5(OH)_4$$

$$(15.1)$$

The silica needed for this reaction can come from many sources, including alteration of orthopyroxene within the peridotite or alteration of plagioclase in associated gabbro. Another source of silica is from hydration of diopside in peridotite to make serpentine (Equation 15.2):

$$3CaMgSi_2O_6 + 5H_2O \rightarrow Mg_3Si_2O_5(OH)_4 + 4SiO_2 + 3Ca(OH)_2 \quad (15.2)$$

This reaction produces abundant silica that is available for the conversion of brucite to serpentine, and it also produces abundant $Ca(OH)_2$. Calcium hydroxide is a strong base, which explains why hydration of peridotite produces alkaline fluids (which may reach pH of 12). When these calcium-bearing fluids encounter a rock that is not ultramafic, the calcium reacts with this rock to form a **rodingite**. No matter whether the protolith to rodingite was basalt or gabbro, pelitic rocks, or granite,

the minerals formed are the same: grossular, epidote, prehnite, and other calcium-aluminum silicates. Although in some ways rodingites are petrologic oddities, they provide important clues that help petrologists reconstruct the tectonic history of an area. In high-grade metamorphic terrains, they may provide the only evidence that the metaperidotites with which they are associated were once serpentinites.

Serpentinization almost always involves magnetite formation. Since almost all the iron in olivine is ferrous, magnetite formation requires significant oxidation. Seawater likely provides the necessary oxygen, leaving free hydrogen in the fluid. The exact process by which magnetite is formed is still a matter of considerable debate, but the significant consequence is that the fluids released from serpentinites create some of the most reducing conditions on Earth. Early in Earth's history, the reducing conditions in serpentinizing environments may have been conducive to the synthesis of reduced organic compounds that combined to produce the earliest lifeforms (Russell et al., 2010).

Serpentinization causes a 40–50 percent increase in the volume of the rock, which may affect ocean-floor bathymetry in areas underlain by serpentinites. Because this large volume change becomes a volume *decrease* when serpentinite breaks down during metamorphism in subduction zones, Dobson et al. (2002) propose that some deep-focus earthquakes could be caused by dehydration of serpentinite in subduction zones (Box 15.1).

15.3 Prograde Metamorphism of Serpentinite: Reactions in the System $CaO–MgO–SiO_2–H_2O$

Serpentine is a mineral group that includes chrysotile, lizardite, and antigorite. Serpentine group minerals consist of interlayered tetrahedral [$(Si_2O_5)^{-2}$] and octahedral [$Mg_3(OH)_6$] sheets. Charge balance is maintained because two of the oxygen atoms in the tetrahedral sheet substitute for two OH^- ions in the octahedral sheet. However, the two sheets are not the same size; the octahedral sheet is somewhat larger than the tetrahedral sheet. This misfit can be accommodated in three ways. One way is for the misfit to spread throughout the structure, producing the mineral lizardite. Another way is for the misfit to accommodate by curvature of the sheet, making a fiber, the mineral form of chrysotile. The third

Table 15.1 *Reactions occurring during metamorphism of peridotitic rocks*[1]

Reaction	Equation number
$17Ctl \rightleftharpoons 3Brc + Atg^2$	15.3
$Atg + 20Brc \rightleftharpoons 34Fo + 51H_2O$	15.4
$Atg + 8Di \rightleftharpoons 18Fo + 4Tr + 27H_2O$	15.5
$Atg \rightleftharpoons 18Fo + 4Tlc + 27H_2O$	15.6
$4Fo + 9Tlc \rightleftharpoons 5Ath + 4H_2O$	15.7
$2Fo + 2Ath \rightleftharpoons 9En + 2H_2O$	15.8
$2Fo + 2Tlc \rightleftharpoons 5En + H_2O$	15.9
$2En + Tlc \rightleftharpoons Ath$	15.10
$Fo + Tr \rightleftharpoons 3En + 2Di + H_2O$	15.11
$Tc + Mgs = 4Fo + 5CO_2 + H_2O$	15.12
$3Mgs + 2Qz + H_2O \rightleftharpoons Tlc + 3CO_2$	15.13
$Atg + 20Mgs \rightleftharpoons 34Fo + 31H_2O + 20CO_2$	15.14
$17Tlc + 45Mgs + 45H_2O \rightleftharpoons 2Atg + 45CO_2$	15.15
$Brc + CO_2 \rightleftharpoons Mgs + H_2O$	15.16

[1] Mineral abbreviations used in this text are defined in Table A.1.
[2] Capitani and Mellini (2004) give the composition of antigorite as $Mg_{2.82}Si_2O_5(OH)_{3.64}$. Multiplying by 17 yields the composition $Mg_{48}Si_{34}O_{85}(OH)_{62}$, which was used in balancing reactions on this table.

way is for the structure to modulate such that some areas have the octahedral sheet pointing up and others have it pointing down. When the sheets flip, a certain amount of the octahedral sheet is omitted to make it better fit to the shape of the tetrahedral sheet, making the mineral antigorite, which is not, strictly speaking, a polymorph of serpentine because it is slightly more silica-rich (Capitani and Mellini, 2004).

Antigorite is the high-temperature phase, as indicated by the fact that it is slightly less hydrous than lizardite or chrysotile. The formation reaction is chrysotile (or lizardite) $\rightleftharpoons$ brucite + antigorite (Equation 15.3 in Table 15.1). One of the first metamorphic effects recognizable in serpentinites is the replacement of lizardite by antigorite. Although it is difficult to distinguish serpentine polymorphs in thin section, antigorite has a clearly bladed appearance (though it is very fine-grained), whereas lizardite does not (compare Figures 15.1A and 15.1B).

Most of the mineralogical changes encountered during metamorphism of peridotitic rocks can be modeled using the system CaO–MgO–SiO₂–H₂O. Although Al_2O_3 and

FeO are present, they do not affect the basic petrogenetic relationships. The mineral reactions encountered in the system CaO–MgO–SiO₂–H₂O at low and medium temperatures of metamorphism are listed in Table 15.1, and the occurrence of these reactions in pressure–temperature (P–T) space is shown in Figure 15.2. All phase diagrams in this chapter, including Figure 15.2, were calculated from the thermodynamic database of Berman (1988). The mineralogical changes encountered in the system CaO–MgO–SiO₂–H₂O are controlled by dehydration reactions that have steep slopes on a P–T diagram. As a result, metaperidotites are very sensitive indicators of changes in metamorphic temperature but are generally poor indicators of metamorphic pressure. Notice that in the pure magnesium system, anthophyllite in metaperidotites is restricted to pressures below 7 kbar (Equation 15.8, Figure 15.2). Furthermore, the reaction of forsterite + talc to enstatite + H₂O (Equation 15.9) has a slightly negative slope. At very high pressures (around 15 kbar), Equation 15.9 intersects antigorite $\rightleftharpoons$ forsterite + talc + H₂O (Equation 15.6), truncating the stability field for talc in metaperidotites. The reaction of antigorite directly to forsterite + enstatite may occur deep in subduction zones, but it is only recognized at one location in the world (Padrón-Navarta et al., 2011). The discussion in this chapter therefore focuses only on relations at crustal pressures shown in Figure 15.2.

The chemography of peridotitic rocks can be represented by projecting from H₂O, assuming that water is always present, allowing the reactions to be projected onto the triangle CaO–MgO–SiO₂ (Figure 15.3). Since all reactions of interest occur within the triangle defined by brucite (or periclase)–diopside–quartz, only that part of the system defined by the triangle brucite–diopside–quartz is shown for the assemblages **a** through **h** in Figure 15.2. Primary peridotites have a relatively restricted composition, consisting mostly of olivine with lesser amounts (10–20 percent) of enstatite and diopside (see shaded area in the triangle CaO–MgO–SiO₂). The limited range of composition means there will be only one divariant assemblage stable in metaperidotites at any given temperature. These assemblages (**a** through **h**) are shown in Figure 15.3 as gray fields.

At low temperatures, the stable phases in a serpentinite include chrysotile (or lizardite), brucite, and diopside (assemblage **a**, Figure 15.3). The brucite may be cryptic, meaning it may not be visible in thin section and can only be distinguished by X-ray diffraction. The diopside may

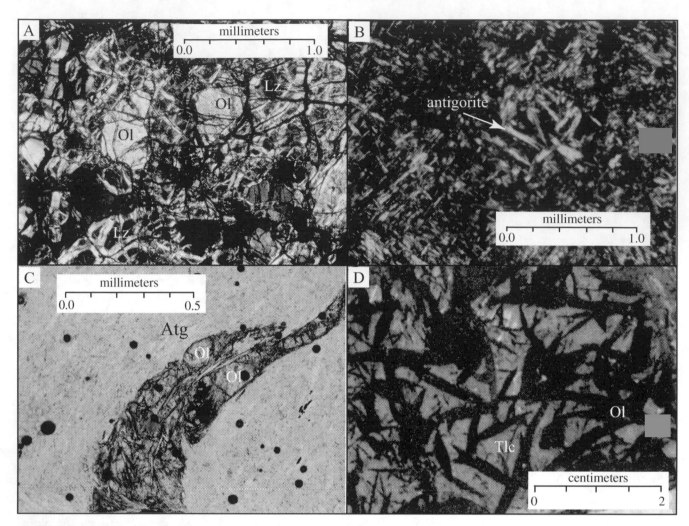

Figure 15.1 Photographs and photomicrographs of textures in metaperidotites. (A) Photomicrograph in crossed polarized light (XPL) showing lizardite (Lz) pseudomorphs after olivine (Ol). Notice onion-skin texture of the lizardite pseudomorphs. Paddy-Go-Easy pass, Central Cascades, Washington. (B) Photomicrograph in XPL showing antigorite serpentinite. Notice how the antigorite is acicular. Paddy-Go-Easy pass, Central Cascades, Washington. (C) Photomicrograph in XPL of olivine (Ol) in equilibrium with antigorite (Atg) from the Zermatt–Saas area, Swiss Alps. Black circles are bubbles in the mounting epoxy. (D) Photograph of elongate olivine crystals, showing characteristic "jack-straw" texture, in olivine–talc (Tlc) rocks, Central Cascades, Washington.

be relict after the primary diopside in the rock, although stable metamorphic diopside may be present as well. As noted earlier, the first change observed during prograde metamorphism is the change from chrysotile or lizardite to antigorite (Equation 15.3), assemblage **b** on Figure 15.3). Brucite may appear at this point, because antigorite is slightly more siliceous than chrysotile or lizardite. Although this reaction is shown as a line in Figure 15.2, its location in P–T space is poorly constrained. Furthermore, the reaction is apparently very sluggish, since detailed X-ray studies show that chrysotile persists in some rocks to olivine-bearing grades.

Olivine appears in metaperidotites by the model reaction antigorite + brucite $\rightleftharpoons$ forsterite + H_2O (Equation 15.4), leading to the assemblage antigorite–forsterite–diopside (assemblage **c**, Figure 15.3). Texturally, the olivine produced by this reaction is distinct from relict olivine. Generally, it has rational boundaries with antigorite (i.e. the boundary is parallel to 001 plane in antigorite) (Figure 15.1C). If olivine grains impinge each other during growth they will form typical 120° grain boundaries without any crystallographic evidence of serpentine. Such a texture is exceedingly unlikely in rocks containing serpentine and relict olivine.

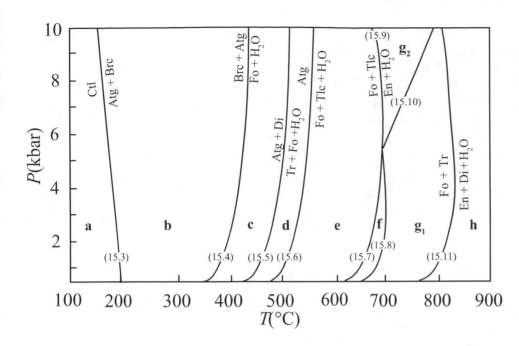

Figure 15.2 *P–T* diagram showing mineral stabilities in the system CaO–MgO–SiO₂–H₂O. Numbers refer to equations listed in Table 15.1.

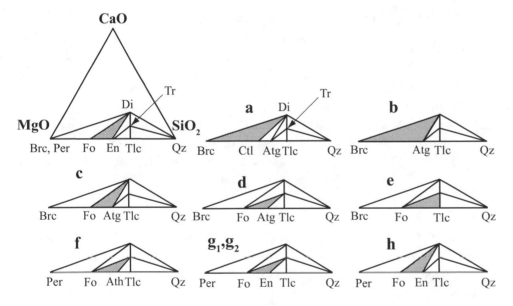

Figure 15.3 Chemographic projections showing assemblages in the divariant fields in Figure 15.2. Shaded area in the CaO–MgO–SiO₂ triangle shows the compositional field for peridotites. Shaded area in projections **a** through **h** indicate the assemblages present in divariant fields shown in Figure 15.2. Ath = anthophyllite, Atg = antigorite, Brc = brucite, Ctl = chrysotile, En = enstatite, Di = diopside, Per = periclase, Qz = quartz, Tlc = talc, Tr = tremolite.

The next change encountered with increasing metamorphic grade involves the disappearance of diopside by the reaction diopside + antigorite ⇌ forsterite + tremolite + H₂O (Equation 15.5), forming the assemblage antigorite–forsterite–tremolite (assemblage **d**, Figure 15.3). Note that this reaction not only produces tremolite, but also increases the amount of forsterite relative to antigorite. Antigorite disappears at a slightly higher metamorphic grade by the reaction antigorite ⇌ forsterite + talc + H₂O (Equation 15.6). In metaperidotites, this reaction produces the assemblage forsterite–talc–tremolite (assemblage **e**, Figure 15.3). In some metamorphic rocks, olivine forming by this reaction will be distinctively elongate

(Figure 15.1D). Some workers have mistaken Equation 15.6 for the reaction that marks the first appearance of olivine. However, in a typical metaperidotite, olivine appears by Equation 15.4 at temperatures more than 100 °C lower than those marked by Equation 15.6. Indeed, given a typical protolith (80 percent olivine, 10 percent enstatite, and 10 percent diopside), at temperatures just below those of Equation 15.6, the rock will have 68 percent olivine, 16 percent tremolite, and only 16 percent antigorite. Thus, the dehydration of serpentine to peridotite is not a single step that releases large amounts of H₂O at a given temperature, but rather a process that takes place over a temperature span of 100 °C.

In the pure system, the assemblage forsterite–talc–tremolite is limited at high temperatures by the reaction forsterite + talc $\rightleftharpoons$ anthophyllite + H_2O (Equation 15.7) or, at higher pressures by the reaction forsterite + talc $\rightleftharpoons$ enstatite + H_2O (Equation 15.9). Equation 15.7 produces the assemblage forsterite–anthophyllite–tremolite (**f**, Figure 15.3). Because most metaperidotites in this temperature range contain tremolite, it is important to distinguish anthophyllite from tremolite in thin section. The easiest way to accomplish this is to search for amphiboles with the highest birefringence. In grains with this orientation, the petrographer is looking nearly down the Y index and, therefore, the b crystallographic axis. In such an orientation, anthophyllite will have parallel extinction, whereas tremolite will have extinction of about $15°$.

At temperatures slightly higher than those of the reaction given in Equation 15.7, anthophyllite reacts out by Equation 15.8, anthophyllite + forsterite = enstatite + H_2O, producing the assemblage forsterite–enstatite–tremolite (assemblages **g1** and **g2**, Figure 15.3). This is the same assemblage that would be produced by Equation 15.9 at higher pressures. Finally, at somewhat higher temperatures, tremolite + forsterite reacts to form enstatite + diopside (Equation 15.11). This produces the ultramafic assemblage forsterite–enstatite–diopside (assemblage **h**, Figure 15.3), which is the same as that of the mantle protolith for metaperidotites.

15.4 Role of Minor Components

15.4.1 Iron

The preceding section described reactions in metaperidotites assuming they were composed solely of CaO, MgO, SiO_2, and H_2O. Olivine in the average peridotite contains $X_{Fe} \sim 0.1$, so it is important to consider the effect that small amounts of iron have on the reactions shown in Figure 15.2. This question can be addressed using two simple chemical principles. The first is that chemical species, such as iron and magnesium, are seldom distributed evenly between phases. Some minerals have structural sites that are more suitable for magnesium, which is a slightly smaller ion, than for iron. Consequently, there is a distribution of iron and magnesium between coexisting silicates in natural systems, with the result that X_{Mg} [Mg/(Mg + Fe)] is different for different minerals in a rock. Figure 15.4 shows a plot of X_{Mg} for various silicates in metaperidotites compared to X_{Mg} of coexisting olivine. From this diagram it is evident that anthophyllite is always more iron-rich than olivine, while enstatite is only

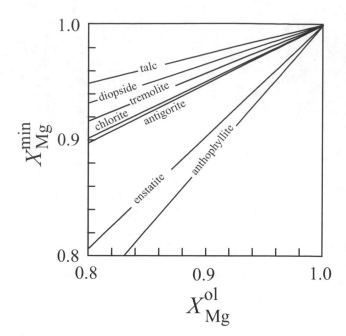

Figure 15.4 Diagram showing the relation between X_{Mg} of olivine and X_{Mg} of other silicate minerals (min). Data from Trommsdorff and Evans (1972).

slightly more magnesian. Talc, diopside, tremolite, chlorite, and antigorite are considerably more magnesian than coexisting olivine, with talc the most magnesian mineral found in metaperidotites.

The second chemical principle, Le Chatelier's principle, explains the significance of this iron distribution. Le Chatelier's principle states that a chemical equilibrium responds to any disturbance by trying to undo the effects of that disturbance. To see how this operates, consider the reaction talc $\rightleftharpoons$ enstatite + quartz + H_2O. Figure 15.4 shows that enstatite accepts iron more readily than does talc. If a small amount of iron is added to this system, Le Chatelier's principle indicates that the reaction will adjust itself to counter the effects of this addition. Because enstatite can accept iron more readily than can talc, enstatite is stabilized, driving the reaction talc $\rightleftharpoons$ enstatite + quartz + H_2O to lower temperatures.

The addition of FeO to the system CaO–MgO–SiO_2–H_2O has another effect, as indicated by the phase rule. Because FeO dissolves into the silicate phases without producing a new phase, the addition of FeO also adds an additional degree of freedom in all reactions. This means that a reaction, which is univariant in the pure system MgO–SiO_2–H_2O, is divariant in the system containing FeO. This increase in variance means the reaction between talc, enstatite, and quartz can only be used to

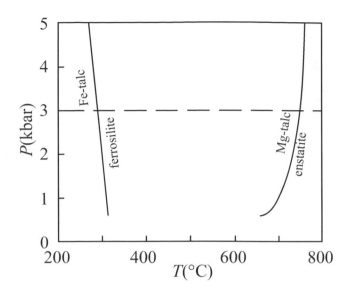

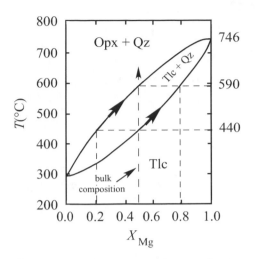

Figure 15.5 *P–T* diagram comparing the breakdown of Mg-talc with Fe-talc. The dashed line is the pressure at which the isobaric section in Figure 15.6 is constructed.

Figure 15.6 Isobaric T–X_{Mg} diagram for the reaction talc $\rightleftharpoons$ orthopyroxene + quartz + H_2O at 3 kbar.

estimate temperature when pressure is known and X_{Mg} can be determined.

To see how this works, consider again the reaction talc $\rightleftharpoons$ orthopyroxene + quartz + H_2O. Figure 15.5 shows the *P–T* conditions at which both pure magnesium and pure iron reactions occur (this example is for illustrative purposes; in reality, the pure iron reaction is metastable; iron amphiboles would form instead of orthopyroxene and iron-rich talc). As indicated previously, because of the strong tendency for iron to favor orthopyroxene, the iron end member of the reaction lies at a lower temperature than does the magnesium end member. In a natural system containing both iron and magnesium, talc must break down at some temperature between the extremes shown in Figure 15.5. A helpful way to visualize the effect of composition on talc breakdown is to display relations on a T–X_{Mg} diagram that is constructed at fixed pressure (i.e. isobaric conditions) (Figure 15.6). At the pressure of 3 kbar, iron-end member talc (Fe-talc) breaks down at 292 °C and magnesium-end member talc (Mg-talc) breaks down at 746 °C. Between these two temperatures, a loop represents the reaction in the Fe–Mg-bearing system. The two curves on this projection show the compositions of talc and orthopyroxene solid solutions in equilibrium with quartz at any given temperature. Because orthopyroxene is always more iron-rich than coexisting talc, it is clear that the high-iron curve represents the composition of orthopyroxene, while that at higher magnesium represents the talc composition.

Figure 15.6 can be used to understand how reactions take place in natural, divariant, or even multivariant systems. Consider metamorphism of a hypothetical assemblage consisting of talc having $X_{Mg} = 0.5$ (with or without quartz). When the temperature reaches 440 °C, talc begins to break down and produces a small amount of orthopyroxene with composition $X_{Mg} = 0.2$. As temperature increases, talc will continue to break down and both talc and orthopyroxene become richer in magnesium. The reaction ceases at 590 °C, when the final amounts of talc are consumed. At this temperature, orthopyroxene has the composition $X_{Mg} = 0.5$ while the last talc to disappear has the composition $X_{Mg} = 0.8$.

In this example, the assemblage orthopyroxene–quartz–talc would be stable over a temperature range of 150 °C, and hence it would not be a sensitive indicator of metamorphic grade. The addition of iron affects the *P–T* stability fields of mineral assemblages in metaperidotites, but it is not an extreme effect because ultramafic rocks are very magnesian. The composition of olivine from nearly all metaperidotites falls in the range of $X_{Mg} = 0.85$–0.95 (although the range within each individual metaperidotite body is much smaller). The effects of this range in olivine composition on the phase relations of the ultramafic system are shown in Figure 15.7. Except for the stability field for the assemblage olivine–anthophyllite (assemblage **f**), the effects are minor. The largest change has been the expansion of the field for the assemblage olivine + anthophyllite at the expense of olivine + talc and, to a lesser extent, olivine + orthopyroxene. This

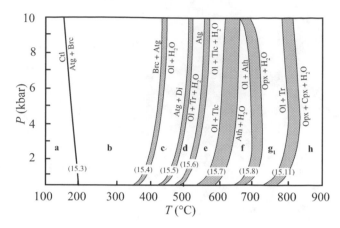

Figure 15.7 *P–T* diagram showing the divariancy of reactions in natural metaperidotites due to the presence of iron.

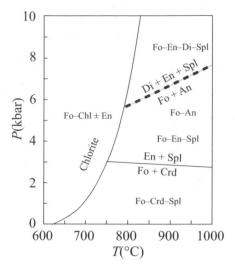

Figure 15.8 *P–T* diagram showing important reactions in high-temperature peridotites. An = anorthite, Chl = chlorite, Crd = cordierite, Spl = spinel, other abbreviations as in Table 15.1. Data from Kushiro and Yoder (1966) and Chernosky (1974).

reflects the fact that iron is far more compatible with anthophyllite than with talc or enstatite.

15.4.2 Aluminum

In discussing mineral assemblages in low and medium metamorphic grades, the role of aluminum has not been considered because the aluminous phase, chlorite, is stable throughout this temperature range. At higher temperatures, however, chlorite breaks down to make cordierite or spinel (Figure 15.8). Figure 15.8 predicts that cordierite is the stable aluminous phase in metaperidotites at low pressures. Cordierite, however, is very rare in metaperidotites (Arai, 1975) because of the relatively high chromium content of metaperidotites. Spinel is very receptive to chromium, and therefore Le Chatelier's principle predicts that in natural systems, the reaction enstatite + spinel ⇌ forsterite + cordierite will lie at very low pressures and that a chromium-bearing spinel will be stable in metaperidotites at almost all pressures.

As a result of the wide stability range of spinel, in most metaperidotites the reaction that eliminates chlorite from the rock is chlorite ⇌ forsterite + enstatite + spinel + H_2O. This occurs at temperatures slightly above those for the appearance of enstatite in the aluminum-free system. When chlorite breaks down, the calcic amphibole in the metaperidotite becomes hornblende rather than tremolite. Because metaperidotites are markedly poor in elements such as ferric iron and titanium that produce the pleochroism in hornblende from mafic rocks, the hornblende in most metaperidotites is colorless.

Figure 15.8 also predicts that plagioclase (shown as An in the figure), not spinel, is stable in metaperidotites at pressures below about 6 kbar. However, like cordierite, plagioclase is destabilized by the presence of chromium in the spinel. Consequently, plagioclase is found in metaperidotites only in low-pressure environments. There are two environments where plagioclase peridotites form. One is in low-pressure contact metamorphic environments, where the plagioclase has formed from complex reactions involving spinel, hornblende, or chlorite (see Frost, 1976). The other environment is in suboceanic mantle that has been infiltrated by basaltic melt at relatively shallow crustal depths (i.e. at low pressures), a process known as "refertilization." Plagioclase peridotites may form from depleted peridotite by incorporation of up to 12 percent of MORB-like melt (Müntener et al., 2010). Such "refertilized" peridotites have been recognized in the mantle sections of ophiolites from around the world (Müntener et al., 2010).

At mantle pressures (ca. 15–20 kbar), spinel plus orthopyroxene react to make garnet (Figure 15.9). The garnet produced by this reaction is pyropic but contains a substantial grossular component. These reactions separate the upper mantle into low-pressure and high-pressure regions. At low pressures, the peridotites of the mantle contain spinel, whereas at high pressure garnet is the aluminous phase instead (Figure 15.9). Although garnet

Table 15.2 *Relationship between metamorphic facies and mineral assemblages in metaperidotites*

Facies	Assemblage
Zeolite	Lizardite/chrysotile
Prehnite–pumpellyite	Serpentinites
"Low" greenschist	Antigorite–brucite–diopside
Blueschist	Antigorite–brucite–diopside
"High" greenschist	Antigorite–olivine–diopside
"Low" eclogite	Antigorite–olivine–diopside
Amphibolite	Tremolite-bearing metaperidotite
Hornblende hornfels	Tremolite-bearing metaperidotite
Pyroxene hornfels	Plagioclase lherzolite
Granulite	Spinel lherzolite
"High" eclogite	Garnet lherzolite

Source: After Evans (1977).

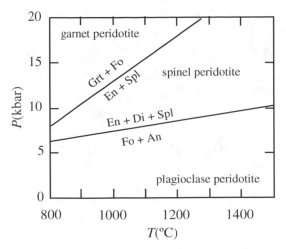

Figure 15.9 Phase diagram showing the stability fields of garnet, spinel, and plagioclase in high-temperature peridotites, based on experimental results of Kushiro and Yoder (1966) and Danckwerth and Newton (1978).

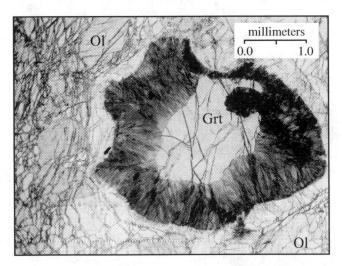

Figure 15.10 Photomicrograph in plane-polarized light (PPL) of garnet in a garnet peridotite. Note the garnet is rimmed by a dark halo of spinel and orthopyroxene produced by decompression. From Alpe Arami, Swiss Alps.

velocity than do spinel peridotites. Second, garnet has a greater affinity for rare-earth elements (REEs), which means magmas derived from the garnet peridotite field will have a different REE pattern from those derived from the spinel field.

15.5 Metaperidotites and Metamorphic Facies

It is evident from Figures 15.2 and 15.8 that mineral assemblages in metaperidotites are more sensitive indicators of temperature than are the metamorphic facies. However, since the metamorphic facies classification is so widely used in the literature, it is important to relate the assemblages found in metaperidotites with the metamorphic facies. This is done in Table 15.2.

The transition from chrysotile to antigorite is approximately equivalent to the boundary between prehnite–pumpellyite and greenschist or blueschist facies. The assemblage antigorite–diopside (with brucite or olivine) is found in greenschist, blueschist, and even low-temperature eclogite facies, making most serpentinites a poor indicator of metamorphic pressure. The appearance of tremolite in metaperidotites marks the transition to amphibolite facies. The transition to granulite facies is more difficult to determine. Evans (1977) places it at the reaction tremolite + forsterite ⇌ enstatite + diopside + H_2O. These indicators of amphibolite and granulite facies in metaperidotites are similar to those in mafic rocks. In metaperidotites, the

peridotites are considered a major component in the deep mantle, this assemblage is not commonly found in the crust. In most garnet peridotites exposed in the crust, the garnet is haloed by fine-grained spinel and orthopyroxene, which formed as the hot mantle was exhumed to lower pressures (Figure 15.10). The division of the mantle into spinel and garnet peridotites is geologically important for two key reasons. First, because garnet is denser than spinel, garnet peridotites have a higher seismic presence of tremolite marks amphibolite facies, whereas the assemblage orthopyroxene–clinopyroxene–olivine + spinel marks granulite facies. Finally, one must note the correspondence of high-temperature eclogite facies with the garnet lherzolite assemblage.

15.6 Role of CO$_2$ in Metamorphism of Peridotites

Up to now it has been assumed that the fluid phase present during metamorphism of serpentinites is pure H$_2$O. In the absence of any evidence to the contrary, this is a reasonable assumption. Dehydration of serpentinite produces 12 weight percent H$_2$O. This large amount of H$_2$O would overwhelm any other fluid species originally present in the rock and drive the fluid toward H$_2$O saturation. However, many metaperidotites, particularly serpentinites, contain carbonate, clear evidence that at least some CO$_2$ must have been present in the fluid. Metaperidotites emplaced into metasedimentary sequences commonly show evidence of CO$_2$ infiltration from the country rock. Alternatively, the parent rock may have been a carbonate-bearing serpentinite from the sea floor, in which case the CO$_2$ was produced by reactions with primary carbonate material. In either instance, it is important to consider both H$_2$O and CO$_2$ as volatile species. Carbon dioxide has only limited solubility in water under surface conditions, as any soda or beer drinker knows. This is because of the great difference between the properties of water, a liquid, and gaseous CO$_2$.

Under increasing temperatures, water behaves more and more like a gas. As shown on the phase diagram for the system H$_2$O in Figure 15.11, at low pressures and temperatures, H$_2$O can exist as two distinct phases, liquid and vapor. Increasing temperature or decreasing pressure can cause a liquid to boil (see paths A–B and A–C, Figure 15.11). Boiling involves an abrupt increase in the volume of H$_2$O with the appearance of a vapor phase and marks the phase boundary between vapor and liquid. As temperature and pressure increase, the volume difference between the vapor phase and the liquid phase gets progressively smaller until no distinction can be made between the two phases. This point is known as the **critical point** and it marks the termination of the phase boundary between liquid and vapor. At temperatures above those of the critical point, both pressure and temperature can change to a great extent

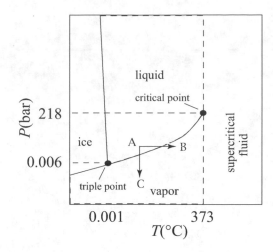

Figure 15.11 $P–T$ diagram for the system H$_2$O showing the relation between liquid, vapor, and supercritical fluid. Axes are not to scale.

without observing a discontinuity in the properties of H$_2$O. Under such conditions, H$_2$O is said to exist as a **supercritical fluid**. The critical point for H$_2$O is at 373 °C and 218 bar. Thus, over most of the metamorphic temperature range, H$_2$O will exist as a supercritical fluid, the density of which approaches that of liquid water at surface conditions. Thus, when metamorphic petrologists speak of "water," they generally mean an H$_2$O-rich supercritical fluid.

A $T–X_{CO2}$ diagram (Figure 15.12) shows that at 2 kbar and temperatures below 270 °C there is limited solubility of CO$_2$ in H$_2$O. For example, at 200 °C the aqueous fluid can dissolve only 8 percent CO$_2$ while the vapor coexisting with this fluid contains 86 percent CO$_2$ and 14 percent H$_2$O (see dashed lines in Figure 15.12). As temperature increases, CO$_2$ becomes increasingly soluble in H$_2$O until at temperatures above 280 °C there is complete solubility between H$_2$O and CO$_2$. Up to about 5 kbar, the solvus is only weakly dependent on pressure, meaning that in the pure system, CO$_2$ and H$_2$O are completely miscible in medium- and high-grade metamorphic environments. Of course, the fluid in metamorphic rocks is seldom pure H$_2$O. Metamorphic fluids probably contain variable amounts of soluble species, the most abundant of which is NaCl. This salt is highly soluble in aqueous fluids, but it is nearly insoluble in CO$_2$. Consequently, the presence of NaCl in fluids will expand the H$_2$O–CO$_2$ solvus to higher temperatures. If the brines are saline enough, the H$_2$O–CO$_2$ solvus may extend up to temperatures where granitic rocks melt. In the discussions that follow, this point will be ignored and relations will be discussed

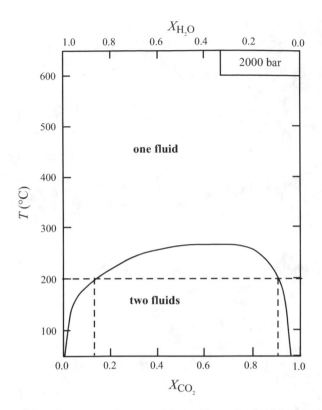

Figure 15.12 *T–X* diagram showing the mutual solubilities of H$_2$O and CO$_2$. Drawn from data of Tödheide and Frank (1963).

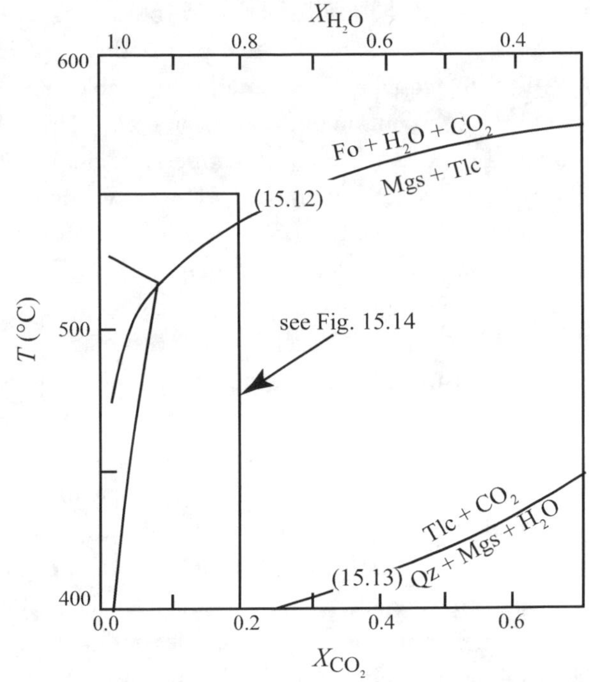

Figure 15.13 *T–X* diagram showing mineral relations in the system MgO–SiO$_2$–H$_2$O–CO$_2$ at 3 kbar. Reactions show the upper stability of carbonate in metaperidotite. Numbers refer to reactions listed in Table 15.1.

as if the fluids involved in metamorphosed peridotites were pure H$_2$O + CO$_2$.

The addition of CO$_2$ to the ultramafic system adds another degree of freedom to most reactions, for as with iron, CO$_2$ does not produce an additional phase but dissolves into one of the phases already present. To take this extra degree of freedom into account, phase relations are shown on an isobaric *T–X*$_{CO2}$ diagram. For the sake of simplicity, relations in the system MgO–SiO$_2$–H$_2$O–CO$_2$ will be discussed (Figure 15.13); the phase diagram is far more complex in the calcium-bearing system (Trommsdorff and Evans, 1977), but the general relations are the same.

The important feature about phase relations in the system MgO–SiO$_2$–H$_2$O–CO$_2$ is that the X_{CO2} of the fluid with which forsterite is stable is very dependent on temperature. At a temperature near 600 °C, forsterite may coexist with a fluid containing more than 80 mole percent CO$_2$, whereas at 500 °C forsterite reacts to form magnesite if the fluid has more than 2 percent CO$_2$. Low-temperature relations are shown in detail in Figure 15.14. A very important observation to draw from this figure is that antigorite and brucite are stable only in fluids that contain small amounts of CO$_2$. The fact that small

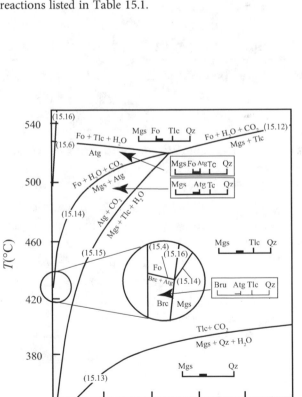

Figure 15.14 *T–X* diagram showing mineral relations in the system MgO–SiO$_2$–H$_2$O–CO$_2$ at 3 kbar and low temperatures.

amounts of CO_2 in the fluid stabilize carbonates means that the assumption made in the first part of this chapter about P_{H2O} being equal to P_{total} is quite reasonable in any rock containing serpentine but no carbonate.

Figure 15.14 shows that, at low temperature, influx of fluid containing even a small amount of CO_2 into a serpentinite will produce carbonate-bearing assemblages. At low temperatures (ca. 400 °C), increasing X_{CO2} will produce the following sequence of stable assemblages in metaperidotite: antigorite–brucite, antigorite–magnesite, and talc–magnesite. At somewhat higher temperatures (480 °C and 3000 bar), the stable assemblages will be antigorite–forsterite, antigorite–magnesite, and talc–magnesite. At very high X_{CO2}, the assemblage magnesite–quartz may be encountered (see Figure 15.13). Zoning due to infiltration of CO_2-bearing fluid is commonly found in metaperidotites, with talc–magnesite along the outer margin of the metaperidotitic body, where X_{CO2} was likely to have been relatively high, and antigorite–magnesite or antigorite–forsterite found in the core, where CO_2-bearing fluids did not penetrate (Figure 15.14).

15.7 Metasomatism of Peridotites

Metaperidotite bodies that display evidence of CO_2 metasomatism commonly show margins subjected to metasomatism of other chemical species. Because metaperidotite bodies that are exposed on the continents are chemically very different from the rocks that surround them, huge chemical gradients are developed at the margins. This results in intense metasomatism, involving the movement of chemical species into and out of ultramafic bodies. Although such zoning patterns are highly variable, they have many things in common. Figure 15.15 shows schematically two examples of metasomatized metaperidotite that are described in the literature and Figure 15.16 shows an example of a metasomatic reaction zone in the field.

Figure 15.15A shows a reaction sequence formed at greenschist grade. The inner zone of this body has the assemblage serpentine–talc with serpentine making up more than 91 percent of the assemblage. This rock could have undergone slight SiO_2 metasomatism or it might represent a pyroxene-rich protolith. Outward from this is a zone of talc + magnesite, precisely what one would predict to form from CO_2 metasomatism (Figure 15.14). Further outward is a zone of

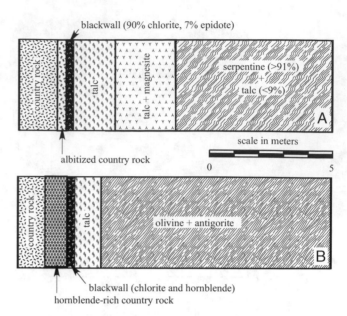

Figure 15.15 Diagrams showing metasomatic zones found at the margin of metaperidotite bodies. (A) greenschist facies (Chidester, 1962); (B) amphibolite facies (Sandiford and Powell, 1986).

monomineralic talc (such a rock is also known as steatite or soapstone). To make such a rock from a metaperidotite requires addition of SiO_2. This suggests the steatite marks an area of SiO_2 metasomatism. Talc commonly forms during low-grade metamorphism of peridotites in response to SiO_2 and CO_2 metasomatism. Although the presence of talc is commonly thought to indicate low-temperature metamorphism of metaperidotite, this is a misconception, because in isochemically metamorphosed peridotitic rocks talc appears only in middle amphibolite facies.

Figure 15.15B shows a reaction sequence formed at amphibolite grade. Many details differ between this reaction sequence and that shown in Figure 15.15A, but two aspects are similar. One is the presence of soapstone, marking SiO_2 metasomatism, and the other is the presence of an external blackwall zone. Blackwall zones are common around metaperidotite at most metamorphic grades. Depending on metamorphic grade and the nature of the country rock, a blackwall may contain chlorite and actinolite, or hornblende, or phlogopite. The high variance of blackwall rocks makes them of limited value in determining the metamorphic grade of an ultramafic body. However, in many terrains, the original peridotitic rocks have undergone such a high degree of digestion that only the blackwall mineral assemblages remain. In such a situation, the occurrence of blackwall is important

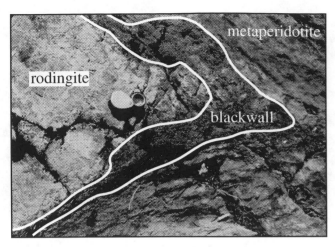

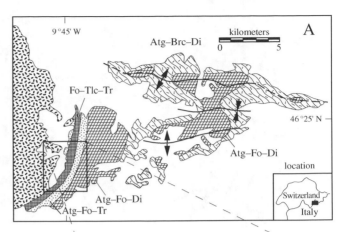

Figure 15.16 Metasomatic blackwall zone between metagabbro (now rodingite) and metaperidotite. Paddy-Go-Easy pass, central Cascade Range, Washington.

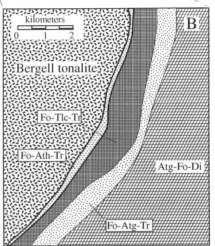

Map 15.1 Metamorphic geology of the Val Malenco area of Italy and Switzerland. (A) The regional and contact metamorphic isograds. (B) Detailed map of the contact aureole. Modified after Trommsdorff and Evans (1972) and Peretti (1988).

because it alerts a petrologist to the fact that at some point in its history the blackwall was a metaperidotite.

Figure 15.16 provides a good example of the persistence of metasomatic zones adjacent to peridotites to high metamorphic grades. This figure shows a block of metagabbro that reacted with metaperidotite during serpentinization. Calcium moving out of the peridotite during serpentinization (as described in Section 15.2) altered the gabbro to a pale green rodingite, while Al_2O_3 moving out of the metagabbro altered the surrounding serpentinite to chlorite-rich blackwall. Long after these metasomatic reactions took place, the rocks underwent metamorphism that converted the serpentinite back to peridotite. However, the presence of the gabbro that has been altered to rodingite and a surrounding blackwall demonstrates the existence of a previous serpentinization event.

15.8 Examples of Metaperidotites in the Field

Because metaperidotites are so sensitive to changes in temperature, mineral assemblages in metamorphosed serpentinites make ideal monitors for the thermal structure of areas that have undergone regional or contact metamorphism. Two classic areas are described in this section.

15.8.1 Malenco Serpentinite

Some of the best examples of metaperidotite come from the Malenco serpentinite in the Swiss and Italian Alps (Map

15.1). The Malenco serpentinite is a slab of the Tethyan ocean floor that originally stood between Africa and Europe and that was thrust onto the European continent during the Alpine orogeny. The thrusting was associated with upper greenschist-grade metamorphism. This regional metamorphism produced two assemblages exposed in Val Malenco: antigorite–brucite–diopside and antigorite–forsterite–diopside. The isograd that separates these assemblages, antigorite + brucite ⇌ forsterite + H_2O, has an irregular distribution across the serpentinite body (Map 15.1A). The distribution of the isograds reflects the late-stage folding of the Malenco serpentinite, with the higher-temperature assemblages occupying cores of anticlines and the lower-grade assemblages lying in synclines.

During the waning stages of the Alpine orogeny, the western margin of the Malenco serpentinite was intruded by the Bergell tonalite. This intrusion produced a contact aureole that overprinted the regional isograds in the

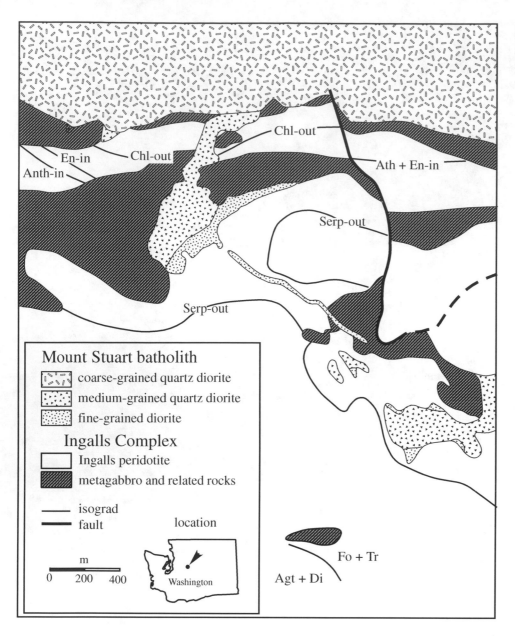

Map 15.2 Geologic map of the contact aureole at Paddy-Go-Easy pass, Central Cascades, Washington. After Frost (1975).

Malenco serpentinite. The aureole measured from the contact to the tremolite-in isograd (Equation 15.5 in Figure 15.7) measures about 2 km wide. If the westernmost boundary of the antigorite–brucite–diopside zone is considered part of the contact metamorphism, then the aureole is 5 km wide. The highest-grade assemblage in the Bergell aureole is forsterite–anthophyllite–tremolite, which is indicative of temperatures of around 650 °C (assemblage **f** in Figure 15.7).

15.8.2 Ingalls Peridotite

The contact aureole produced by the intrusion of the Mount Stuart batholith into the Ingalls peridotite serves as a good comparison to the aureole around the Bergell

tonalite. The Ingalls peridotite lies in the Central Cascades in Washington State and is part of the complex series ophiolites emplaced onto the western portion of North America during the late Mesozoic. The Ingalls peridotite was extensively serpentinized before or during emplacement, though in many places the primary mantle assemblage (olivine–orthopyroxene–augite–chromite) persists. It has been intruded and metamorphosed by the Cretaceous Mount Stuart batholith; the aureole is best exposed where it crosses the Wenatchee Mountains at Paddy-Go-Easy pass (Frost, 1975) (Map 15.2). Here, the highest metamorphic assemblage encountered is forsterite–enstatite–tremolite–spinel, which, at approximately 700 °C, is much hotter than in the Bergell aureole

(Map 15.1). The quartz diorite of the Mount Stuart batholith had similar magmatic temperature to the Bergell tonalite, so the difference in metamorphic grade in the two aureoles does not reflect differences in the temperatures of the intruding magma.

One clue to the cause of the higher temperatures recorded in the aureole at Paddy-Go-Easy pass lies in the complex shapes of the isograds. Unlike the aureole around the Bergell tonalite (cf. Map 15.1), where the isograds are relatively planar, the isograds at Paddy-Go-Easy pass form irregular patterns that extend for kilometers away from the contact of the main batholith. The irregular isograd pattern reflects the complex intrusive history of the Mount Stuart batholith in this area. The first intrusive event was the emplacement of isolated plutons of quartz diorite as well as dikes that dehydrated the serpentinite to the assemblage forsterite–talc–tremolite and that caused the irregular shape to the serpentine-out isograd. As a result of these early intrusions, the main stage of Mount Stuart batholith was emplaced into a country rock that was already hot and partially dehydrated. This meant that in the high-grade portions of the aureole the contact metamorphism produced the reactions forsterite + anthophyllite $\rightleftharpoons$ enstatite + H_2O and chlorite $\rightleftharpoons$ forsterite + enstatite + spinel + H_2O. Neither reaction is present in the Bergell aureole.

BOX 15.1 | PETROLOGY AND GEOPHYSICS

Although they are generally studied separately, petrology and geophysics are actually sister sciences. Geophysics uses information including seismic velocity, density, magnetic susceptibility, and electrical potential to infer the structure and composition of the crust and mantle. Interpretation of the geophysical data requires petrologic information that links geophysical properties to particular rock types and structures. Similarly, petrology can tell geologists what minerals, rocks, and structures are present on the surface of Earth, but geologists must rely on geophysical techniques to project these materials to depth with any degree of accuracy. Nowhere is this more evident than in the study of the oceanic mantle.

During flow within the mantle, olivine crystals assume a preferred orientation. Depending on physical conditions, several types of olivine orientation are possible, but it is common for the b-axis of olivine to lie normal to the foliation plane while the a- and c-axes lie within the plane. Olivine is an orthorhombic mineral, and like light waves, the velocity of seismic waves within olivine vary with crystallographic orientation. The maximum P-wave velocity lies in the plane of the a-axis, whereas the minimum P-wave velocity lies parallel to the b-axis. Therefore, if the mantle has an olivine fabric in which the b-axis lies normal to the foliation, then the seismic waves will travel faster in the plane of the foliation than across it. This property is called **seismic anisotropy** and is a common property of the mantle. By studying the olivine fabrics in peridotites exposed in ophiolites, petrologists and geophysicists can model the seismic anisotropy that those rocks would have had in the mantle (Salisbury and Christensen, 1985). By examining the rocks directly, petrologists also can determine what processes may produce changes in mantle anisotropy (Vauchez and Garrido, 2001; Michibayashi et al., 2006).

Serpentinite has a distinctly lower density, lower seismic velocity, and higher magnetic susceptibility than the parent mantle peridotite. This means that zones of serpentinized mantle should be readily imaged by geophysical surveys (Schroeder et al., 2002). Because of the low seismic velocity of serpentine relative to fresh olivine, Kamimura et al. (2002) interpreted low velocity in the mantle wedge above the Iszu-Bonin subduction zone, which is in the western Pacific, south of Japan, to be caused by serpentinization due to dewatering of the subducting slab. In addition, Jung (2011) argues that the trench-parallel seismic anisotropy seen in the mantle wedge above many trenches is produced by serpentine with a strong preferred orientation.

Summary

- Serpentinization of peridotites, which occurs mostly on the sea floor, produces fluids that are highly reducing and very alkaline.

- Prograde metamorphism of serpentinites is marked by the following assemblages (Figure 15.3):

  - For assemblages with serpentine: brucite + diopside → olivine + diopside → olivine + tremolite

  - For assemblages with olivine: antigorite + diopside → antigorite + tremolite → talc + tremolite → anthophyllite + tremolite → enstatite + tremolite → enstatite + diopside

- At low temperatures, the presence of even minor amounts of CO_2 in the fluid produces carbonate: magnesite, calcite, or dolomite (Figures 15.13 and 15.14).

- At granulite grade, the aluminous phase in lherzolites is a monitor of pressure: plagioclase = low pressure, spinel = moderate pressure (6–12 kbar), garnet > 12 kbar.

Questions and Problems

Problem 15.1. Is serpentinization a metamorphic process? Explain why or why not.

Problem 15.2. Number the following ultramafic assemblages in order of increasing grade:

a. antigorite–brucite–diopside–chlorite

b. antigorite–forsterite–tremolite–chlorite

c. forsterite–enstatite–tremolite–spinel

d. forsterite–talc–tremolite–chlorite

Problem 15.3. Number the following assemblages in order of increasing pressure:

a. forsterite–enstatite–diopside–garnet

b. forsterite–enstatite–diopside–anorthite

c. forsterite–enstatite–diopside–spinel

Problem 15.4. Below is a list of minerals in a series of rocks. For each rock, determine if the association could represent a stable assemblage. If it cannot be a stable assemblage, what mineral in the list is out of equilibrium?

a. forsterite–enstatite–diopside–lizardite

b. forsterite–antigorite–talc–chlorite

c. antigorite–forsterite–magnesite–talc

d. forsterite–orthopyroxene–clinopyroxene–chlorite–garnet

Problem 15.5. Why do many geologists believe life probably began in sea-floor vents, derived from serpentinizing fluid?

Problem 15.6. To what metamorphic facies do the following assemblages belong? (Note there may not be a unique answer to some questions.)

a. antigorite–brucite–diopside–chlorite

b. forsterite–orthopyroxene–clinopyroxene–garnet

c. forsterite–anthophyllite–tremolite–chlorite

d. forsterite–orthopyroxene–clinopyroxene–spinel

Problem 15.7. Why is monomineralic talc more commonly associated with metasomatism rather than regional metamorphism of ultramafic rocks?

Further Reading

Evans, B. W., 1977, Metamorphism of alpine peridotite and serpentinite. *Annual Reviews of Earth and Planetary Sciences*, 5, 397–447.

Evans, B. W., Hattori, K., and Barronet, A., 2013, Serpentinite: What, why, where? *Elements*, 9, 99–106.

Guillot, S. and Hattori, K., 2013, Serpentinites: Essential roles in geodynamics, arc volcanism, sustainable development, and the origin of life. *Elements*, 9, 95–8.

Chapter 16

Metamorphism of Pelitic Rocks

16.1 Introduction

Perhaps the most sensitive protolith to changes in temperature and pressure during metamorphism is pelitic schist. In addition, pelitic rocks (i.e. metamorphosed shales) are widespread in most orogenic belts. Therefore, many studies have used pelitic schists to map out geographic variations in metamorphic conditions and the pressure and temperature history of orogenic belts. However, pelitic rocks are a chemically complex system composed of at least 10 components: $Na_2O-K_2O-CaO-FeO-Fe_2O_3-MgO-Al_2O_3-TiO_2-SiO_2-H_2O$. Because of this chemical complexity, pelitic rocks can crystallize into a bewildering array of mineral assemblages, making the study of them appear daunting. For more than a century, petrologists recognized that different minerals appeared in pelitic rocks during prograde metamorphism. A major breakthrough in metamorphic petrology occurred when J. B. Thompson showed how to project the complex chemical system of pelitic rocks onto two-dimensional chemographic diagrams (Thompson, 1957). These diagrams allowed petrologists to visualize how the mineral assemblages in pelitic rocks relate to reactions within the pelitic chemical system. As soon as these projections appeared, petrologists began to use these diagrams to characterize continuous and discontinuous reactions that occur during metamorphism and to show how they can be used to estimate temperature and pressure of metamorphism. This chapter endeavors to build on the introduction to chemographic projections presented in Chapter 13 and to present the array of assemblages found in pelitic rocks in a systematic way so that the student understands how these assemblages can be used to monitor changes in temperature and pressure within a mountain belt.

This chapter first discusses the difference between continuous and discontinuous reactions and how they are represented in chemographic diagrams. It then discusses the derivation of chemographic diagrams for pelitic systems and how to use them to characterize metamorphic assemblages in pelitic rocks. Finally, the chapter discusses how mineral assemblages in pelitic rocks change as a function of temperature and pressure and how this variation can be used to understand the tectonic environment where the assemblages formed.

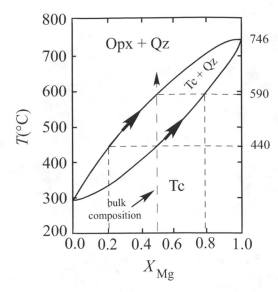

Figure 16.1 Isobaric T–X diagram showing the breakdown of talc to orthopyroxene + quartz (reproduction of Figure 15.6).

16.2 Chemographic Projections for Pelitic Systems

16.2.1 Chemographic Projections for Continuous Reactions

The description of reactions in metaperidotites presented in Chapter 15 made the simplifying assumption that variations in iron and magnesium could be neglected because metaperidotites are invariably very magnesian (X_{Mg} [Mg/(Mg + Fe)] = 0.9). This assumption does not hold for pelitic rocks because they are characterized by a wide range in X_{Mg}. To construct chemographic projections for pelitic rocks, it is necessary to understand how to show continuous reactions (i.e. reactions involving changes in X_{Mg} as well as changes in pressure and temperature) graphically.

As noted in the previous chapter, the presence of additional components introduces additional degrees of freedom to any reaction. For example, Section 15.4.1 observed that in ultramafic rocks the addition of iron to the magnesium end-member system causes a reaction such as talc $\rightleftharpoons$ orthopyroxene + quartz + H_2O to occur over a range of temperatures (Figure 15.6; reproduced here as Figure 16.1). Such a reaction is known as a **continuous reaction.** We can show this reaction chemographically on a projection from H_2O onto the FeO–MgO–SiO$_2$ plane of the FeO–MgO–SiO$_2$–H_2O tetrahedron, resulting in an MgSiO$_3$–FeSiO$_3$–SiO$_2$ diagram (Figure 16.2). On this chemographic projection the minerals present do not have a fixed composition. Rather, they occur as lines with a fixed SiO$_2$/(FeO + MgO + SiO$_2$) ratio. For example, the

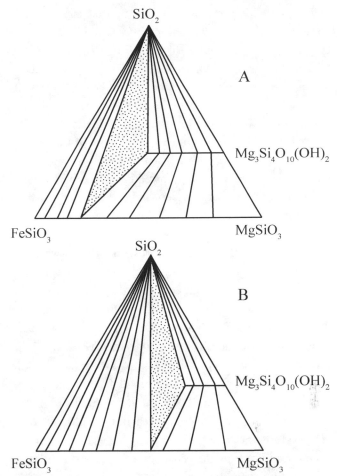

Figure 16.2 Isobaric, isothermal FeSiO$_3$–MgSiO$_3$–SiO$_2$ chemographic projections displaying the continuous nature of the reaction talc $\rightleftharpoons$ orthopyroxene + quartz + H_2O at (A) 440 °C and (B) 590 °C. Stippled fields show composition range where talc, orthopyroxene, and quartz coexist with an aqueous fluid.

magnesium end-member composition of talc, Mg$_3$Si$_4$O$_{10}$(OH)$_2$, lies on a point between the SiO$_2$ and MgSiO$_3$ apices of the triangle. The addition of iron moves the composition into the triangle, defining a horizontal line representing the range of talc compositions. Tie lines in this figure show compositions of talc and orthopyroxene that coexist (and also tie lines of talc or orthopyroxene that coexist with quartz). In actuality, there are an infinite number of tie lines possible, but for the sake of clarity only a few are shown in Figure 16.2. Figure 16.2A shows the relations at 440 °C, just at the point where, as described in the previous chapter, talc with a composition X_{Mg} = 0.5 begins to react to orthopyroxene. As shown in Figure 16.1, at this temperature, talc with X_{Mg} less than 0.5 no longer is stable, so the line indicating talc composition terminates at $X_{Mg} = 0.5$. Figure 16.1 also tells us that talc with this composition coexists with an orthopyroxene with $X_{Mg} = 0.2$; in the chemography this is the line connecting talc ($X_{Mg} = 0.5$) with orthopyroxene ($X_{Mg} = 0.2$). Since both talc and orthopyroxene in this assemblage coexist with quartz, the assemblage is represented by the stippled triangles on the chemographic diagram in Figure 16.2. This triangle is a chemographic representation of the assemblage that occurs within the reaction loop on Figure 16.1, namely talc, orthopyroxene, and quartz.

On the kind of diagram depicted in Figure 16.2, where one or more phases exhibits a solid solution, there are two types of fields. The two-phase field is marked by tie lines between two phases and represents a high-variance assemblage (i.e. the Fe/Mg ratio of the silicates may vary over a wide range without changing the assemblage). The second is a three-phase field such as the one marked by the stippled field in Figure 16.2A. The additional phase means this field displays one fewer degree of freedom. This decrease in variance is indicated by the fact that the X_{Mg} of each phase in a three-phase field is fixed. The three-phase field in a chemographic triangle is thus a representation of a continuous reaction. With increasing temperature, the three-phase field changes location in the chemographic triangle. For example, with increasing temperature, the three-phase triangle in Figure 16.2A migrates toward higher magnesium values, as indicated in Figure 16.1. This temperature dependence is seen in Figure 16.2B, which depicts phase relationships at 590 °C, the temperature at which the last trace of the talc in the original assemblage has been consumed.

Because most metamorphic reactions involve devolatilization, the three-phase triangles may propagate either toward the iron-rich side of diagram or the magnesium-rich side, depending on whether iron or magnesium is incorporated preferentially in the hydrous phase. Figure 16.3 shows the two possible chemographies for continuous reactions. Figure 16.3A shows a situation where the three-phase field has migrated toward the short legs of the triangle (i.e. toward phase A). Phase A_1 is incorporated within the new triangle A_2–B–C, meaning that the continuous reaction is $A_1 \rightarrow A_2 + B + C + H_2O$. In Figure 16.3B, the three-phase field has migrated toward the long leg of the triangle (i.e. away from phase A). This implies the reaction has an opposite sense, namely $B + C + A_1 \rightarrow A_2 + H_2O$.

In addition to continuous reactions, chemographic projections with solid solution can define tie-line flip and terminal reactions. These reactions are distinct from

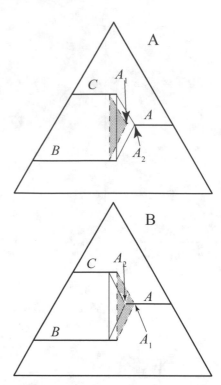

Figure 16.3 Diagram showing the behavior of chemographic projections in continuous reactions. (A) The behavior of the three-phase field in a system where the reaction is $A_1 \rightarrow A_2 + B + C + H_2O$. (B) The same behavior for the reaction $B + C + A_1 \rightarrow A_2 + H_2O$. Dashed tie lines = T_1, solid lines = T_2, where $T_1 < T_2$.

the continuous reactions and are referred to as **discontinuous reactions** because they occur at a fixed temperature and pressure. A discontinuous reaction occurs when two continuous reactions intersect. An example in Figure 16.4 shows four hypothetical solid solutions: A, B, C, and D. At temperatures below T_1 the two-phase field B–D is stable and we have the three-phase fields A–B–D and B–C–D. If, with increasing temperature, the three-phase field A–B–D migrates to the right and the three-phase field B–C–D migrates toward the left, the two-phase field B–D will shrink. If this happens, then at some temperature, taken as T_1 in Figure 16.4A, the two-phase field shrinks to a single tie line. At this temperature the compositions of A, B, C, and D are fixed and the discontinuous tie-line flip reaction $B + D \rightarrow A + C$ will occur. Increasing temperature causes the two-phase field $A + C$ to expand while the three-phase field A–C–D migrates left and A–B–C migrates right.

A similar sequence produces discontinuous terminal reactions, as indicated in Figure 16.4B. This shows four hypothetical solid solutions with mineral D lying in the

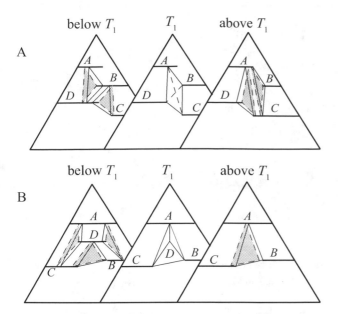

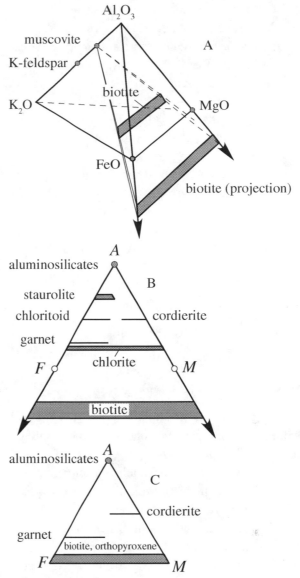

Figure 16.4 Diagram showing the relation between continuous reactions and discontinuous reactions. Stippled areas show the location of the three-phase fields at temperatures a small increment above the temperature given in each diagram. (A) shows a tie-line flip reaction, and (B) shows a terminal reaction.

triangle defined by phases A, B, and C. At temperatures below T_1, the stable assemblages are A–C–D, A–B–D, and B–C–D. With increasing temperature, the three-phase field of A–C–D migrates toward right while the three-phase field A–B–D migrates left. The result is that the compositional range for D shrinks. At T_1, the compositional field for D has shrunk to a point, the compositions of A, B, C, and D are fixed, and the terminal reaction $D \rightarrow A + B + C$ (+ H_2O) takes place. With further increases in temperature above T_1, this system will have one continuous reaction involving the three-phase field A–B–C.

16.2.2 *AFM* Projections for Pelitic Rocks

As observed earlier, the metamorphic reactions in pelitic rocks are complex because of the number of components in the pelitic system. This complexity is manifested by the many continuous reactions in pelitic rocks. Representing pelitic mineral assemblages and continuous reactions on chemographic projections is possible by making the following simplifying assumptions (Thompson, 1957):

1. The variations in Na_2O and CaO are fixed by the presence of plagioclase in the rock.

2. Variations of TiO_2 and Fe_2O_3 are controlled by the presence of magnetite and ilmenite in the rock.

3. Variation in SiO_2 is controlled by the presence of quartz in the rock.

4. $P_{H2O} = P_{total}$

In this way, the 10-component system Na_2O–K_2O–CaO–FeO–MgO–Al_2O_3–Fe_2O_3–TiO_2–SiO_2–H_2O can be reduced to K_2O–FeO–MgO–Al_2O_3 and can be displayed in three dimensions as a tetrahedron (Figure 16.5).

Figure 16.5 Construction of the *AFM* projection for metapelitic compositions. (A) The K_2O–FeO–MgO–Al_2O_3 tetrahedron showing the projection from muscovite to the FeO–MgO–Al_2O_3 plane. (B) *AFM* projection showing where common minerals plot when projected from muscovite. (C) *AFM* projection showing where common minerals plot when projected from orthoclase.

Because a three-dimensional figure is difficult to portray graphically, this chemography is commonly projected into two dimensions. Most minerals in pelitic rocks are potassium-free (i.e. garnet, staurolite, aluminosilicates, chlorite, cordierite) and can be expressed in terms of the system $FeO-MgO-Al_2O_3-SiO_2-H_2O$. In a projection from quartz and H_2O these minerals will lie on the $Al_2O_3-FeO-MgO$ (AFM) plane. Thus, the best projection would be from a potassium-rich phase onto the $Al_2O_3-FeO-MgO$ plane. Such a chemography is achieved by an *AFM* diagram, which projects from muscovite or from potassium feldspar (Thompson, 1957). Because at all except the highest metamorphic grades, muscovite, rather than orthoclase, is the stable potassium phase in pelitic schists, petrologists use a muscovite projection for most metamorphic conditions. On an *AFM* projection from muscovite, one plots the following:

$$A = (Al - 3K)/(Al - 3K + Fe + Mg)$$

$$F = Fe/(Al - 3K + Fe + Mg)$$

$$M = Mg/(Al - 3K + Fe + Mg)$$

In this calculation, subtracting three times the moles of potassium from the number of moles of aluminum accounts for the projection from muscovite [$KAl_3Si_3O_{10}(OH)_2$], which has three moles of aluminum for each mole of potassium. For a projection from orthoclase ($KAlSi_3O_8$), $A = (Al - K)/(Al - K + Fe + Mg)$. By inspection of these expressions, it is evident that those minerals that do not contain potassium will plot on the Fe–Mg–Al face of the K–Al–Fe–Mg tetrahedron and their position will be unaffected by the choice of projection, either from muscovite or K-feldspar. Potassium-bearing phases, such as biotite, will be projected onto the Fe–Mg–Al plane (Figure 16.5A).

Because the *AFM* projection includes a plane involving iron and magnesium, the *AFM* projection will have properties similar to the projection described for the system $FeO-MgO-SiO_2$; the diagrams will depict both continuous and discontinuous reactions. *AFM* projections can be used to visualize mineral reactions if the distribution of iron and magnesium between common pelitic phases is known. Mineral analyses from many rocks show that the tendency to accept iron relative to magnesium is: hercynite (spinel) > garnet > chloritoid > staurolite > orthopyroxene > biotite > chlorite > cordierite. Figures 16.5 B and 16.5C show where the typical pelitic minerals plot on *AFM* projections from muscovite

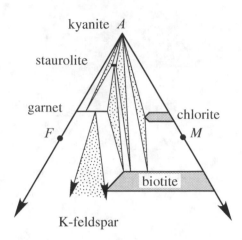

Figure 16.6 A typical *AFM* projection from muscovite for middle amphibolite facies (kyanite zone). Stippled areas show the three-phase fields.

and orthoclase, respectively. Figure 16.6 shows the tie lines between coexisting minerals for a projection from muscovite for conditions in middle amphibolite facies.

16.3 Progressive Metamorphism of Pelitic Rocks: Barrovian Metamorphism

16.3.1 The Protolith: The Mineralogy of Shale

The protolith for pelitic schists is typically shale. Prior to metamorphism, shales consist of kaolinite, montmorillonite, carbonate, usually ankerite [$Ca(Fe,Mg)(CO_3)_2$], and very fine-grained detrital quartz and feldspar. Shales range in color from red to black, recording the presence or absence of graphite and the oxidation state of the iron. Red shales are rich in hematite, whereas black shales are rich in organic matter. Organic matter, which contains sulfur as well as carbon, reduces any iron in the rock so that iron is present as pyrite, rather than hematite.

16.3.2 Low-Grade Metamorphism of Pelitic Rocks

The reactions by which these protolith minerals react to form the mineral assemblages diagnostic of greenschist facies are complex and not well known, largely because weakly metamorphosed pelitic rocks are fine-grained. Many of the low-grade minerals, such as kaolinite, illite, pyrophyllite, and muscovite look the same in thin section; usually they are grouped under the general term "sericite." Consequently, mineral assemblages in low-grade pelitic rocks can be inferred only by X-ray diffraction techniques or with a scanning electron microscope.

From studies of weakly metamorphosed pelitic rocks we can infer that prograde metamorphism of these rocks involves the following reactions:

1. *Disappearance of montmorillonite*: Montmorillonite (a complex clay with sodium or calcium in the interlayer site) disappears early in diagenesis, with chlorite and illite forming at its expense. Montmorillonite is usually gone by middle zeolite facies.

2. *Formation of chlorite*: Chlorite forms at grades around those of zeolite facies. The formation reactions are not well known. One reaction may involve the breakdown of montmorillonite, another reaction is probably:

$$\text{kaolinite} + \text{ankerite} + \text{quartz} + H_2O$$
$$\rightarrow \text{chlorite} + \text{calcite} + CO_2 \qquad (16.1)$$

3. *Formation of illite*: Illite can be thought of as closely related to muscovite. Muscovite has the composition $KAl_2Si_3AlO_{10}(OH)_2$. As the formula indicates, muscovite contains tetrahedral aluminum substituting for silica. The charge imbalance produced by this substitution is accommodated by the substitution of potassium between the layers of the phyllosilicate. At low temperatures, the substitution of aluminum for silica is not energetically favored, so the formula for the potassic phyllosilicate becomes $K_{(1-x)}Al_2Si_{(3+x)}Al_{(1-x)}O_{10}(OH_2)$; this mineral is illite. Illite probably forms by reactions between detrital orthoclase and kaolinite and appears in lowest zeolite facies. With increasing grade, illite becomes increasingly enriched in aluminum and potassium. In the process, illite approaches muscovite in composition and by greenschist facies muscovite appears in its stead.

As a result of these three reactions, by the beginning of greenschist facies the mineral assemblage in pelitic rocks has become muscovite, chlorite, quartz, and detrital feldspar (mostly albite and orthoclase). Other phases may be pyrophyllite [$Al_2Si_4O_{10}(OH)_2$], the sodic mica paragonite, the calcic mica margarite, carbonates, graphite, iron–titanium oxides, and pyrite or pyrrhotite.

16.3.3 Barrovian Metamorphism of Pelitic Schists

One of the first studies of the metamorphism of metapelitic rocks was undertaken in the southern Highlands of Scotland, where George Barrow introduced the concept of metamorphic zones (Barrow, 1893). Barrow studied the Dalradian Supergroup, an Eocambrian–Cambrian sequence of sediments with minor volcanic rocks. During the Devonian Caledonian orogeny this group of rocks was metamorphosed to grades ranging from greenschist to upper amphibolite facies. The southeastern margin of this metamorphic belt is marked by the Highland Boundary fault, which juxtaposes weakly metamorphosed Dalradian metasedimentary rocks against unmetamorphosed Devonian sedimentary rocks (Map 16.1).

The Dalradian rocks just north of the fault are slates that preserve much of their sedimentary structure. Within a short distance from the Highland Boundary fault the rocks increase in grade to become garnet-bearing and change from phyllite to schist. Transecting

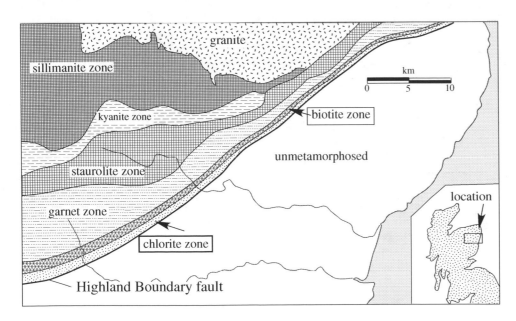

Map 16.1 Map showing the distribution of metamorphic zones in southeastern Scotland as defined by assemblages in pelitic rocks. Modified from Barrow (1912).

farther northwest the metamorphic grade increases from garnet, through staurolite and kyanite to become sillimanite-bearing; a sequence now known as Barrovian metamorphism. There is an area in the eastern portion of the region where the kyanite-bearing rocks disappear, and the staurolite-bearing rocks abut directly against those with sillimanite. A decrease in pressure to the northeast, which would eliminate kyanite as a participating phase, could cause the observed variation in mineralogy. Another possibility is that it simply reflects how rocks in this region lacked the proper composition to form kyanite, crystallizing staurolite instead.

Barrow (1893) recognized three metamorphic zones in the Dalradian Supergroup, and Tilley (1925) later identified three more low-grade zones (Map 16.1). These zones were defined in terms of the first appearance of key minerals. With increasing grade, the zones are:

chlorite

biotite

garnet

staurolite

kyanite

sillimanite

Until the appearance of the *AFM* projection (Thompson, 1957), these zones were the basis for describing and comparing the metamorphism of pelitic schist. It is, therefore, reasonable to start the discussion of pelitic schists by showing how the Barrovian metamorphic zones are expressed in terms of *AFM* projections.

Biotite Zone

As described earlier, metamorphism of pelitic rocks to lower greenschist grade results in the common assemblage muscovite–chlorite–quartz. The reaction between the chlorite and the biotite zone is shown in *AFM* diagrams on Figure 16.7. In relatively aluminous rocks, the common assemblage also contains pyrophyllite. Because of the optical similarity of muscovite and pyrophyllite it is usually impossible to distinguish between the two without detailed X-ray analysis. In rocks that are less aluminous, the common assemblage contains potassium feldspar. Although the stable potassium feldspar at this grade is microcline, in many rocks the K-feldspar present is detrital orthoclase.

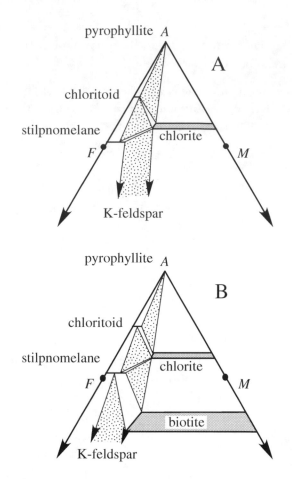

Figure 16.7 *AFM* projections showing transition from the chlorite zone (A) to the biotite zone (B).

Increasing grade leads to the appearance of biotite by the model reaction:

$$\text{chlorite} + \text{potassium feldspar} \rightarrow \text{biotite} + \text{quartz} + H_2O \tag{16.2}$$

This reaction cannot easily be balanced because chlorite in weakly metamorphosed pelitic rocks has a highly variable composition. The aluminous assemblage with pyrophyllite is not changed by this reaction. The less aluminous assemblage is, however, replaced by the assemblage muscovite–quartz–chlorite–biotite, the dominant assemblage in the biotite zone because most pelitic rocks contain biotite. Equation 16.2 is important because it marks the disappearance of K-feldspar from pelitic rocks. As shown later in this chapter, only at the very highest limits of amphibolite facies does K-feldspar appear again.

Very iron-rich rocks at biotite grade may contain stilpnomelane, an iron–magnesium phyllosilicate that looks like biotite but is more siliceous, less potassic, and more hydrous. Because of its iron- and water-rich

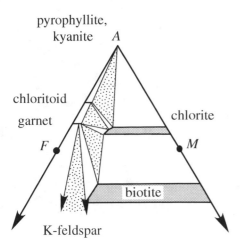

Figure 16.8 *AFM* projection for the garnet zone.

composition stilpnomelane is restricted to iron-rich bulk compositions and low temperatures of metamorphism.

Garnet Zone

The first appearance of garnet in pelitic rocks is not well understood. Because garnet and chlorite have roughly the same Mg/Al ratio, the reaction can be written:

$$\text{chlorite} + \text{quartz} \rightarrow \text{garnet} + H_2O \qquad (16.3)$$

The first garnet to appear in pelitic rocks, however, is rich in manganese as well as iron. Thus, the garnet-in reaction must also involve a manganese-bearing phase, typically ilmenite. As metamorphic grade increases above the garnet-in isograd, the manganese content of garnet decreases. This decrease is often seen in individual garnets, which are commonly zoned, and have cores rich in manganese and rims rich in iron. The variable manganese content of garnet cannot be shown on an *AFM* projection because MnO is a minor component that is not included in this projection.

The *AFM* projection for rocks in the garnet zone is shown in Figure 16.8. Garnet first appears on the iron-rich side of the projection. The most common assemblage in this zone is muscovite-quartz–garnet–chlorite–biotite. Rocks with high iron and low aluminium contents may have the assemblage muscovite–quartz–garnet–biotite–K-feldspar. Rocks with this assemblage are not truly pelitic because pelitic rocks necessarily contain aluminous minerals other than (or in addition to) garnet. Rocks that do not contain these aluminous minerals are called *semi-pelitic*.

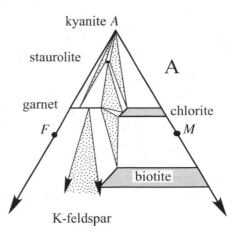

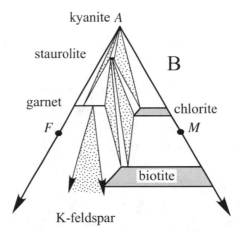

Figure 16.9 *AFM* projections showing the transition from the garnet to the staurolite zones. (A) Upper garnet zone, and (B) staurolite zone.

In highly aluminous rocks, the assemblage is muscovite–quartz–chlorite–pyrophyllite (± chloritoid). Pyrophyllite dehydrates to kyanite + quartz in this temperature range. As noted earlier, because biotite is usually found in pelitic rocks, pyrophyllite-bearing assemblages are not common. However, it is important to note that in aluminous quartzites, which may have originally been deposited as quartz + kaolinite, kyanite may appear in greenschist facies.

Staurolite Zone

The staurolite zone is marked by the first appearance of staurolite in biotite-bearing pelitic rocks. In aluminous compositions, the reaction that leads to the first appearance of staurolite is likely something like (Figure 16.9A):

$$\text{kyanite} + \text{garnet} + \text{chlorite} \rightarrow \text{staurolite} + H_2O \qquad (16.4)$$

Staurolite is likely to appear in typical pelitic rocks, in which biotite is common, by the reaction (Figure 16.9B):

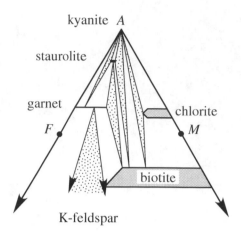

Figure 16.10 *AFM* projection for the kyanite zone.

garnet + chlorite + muscovite
$$\rightarrow \text{staurolite} + \text{biotite} \pm \text{quartz} + H_2O \qquad (16.5)$$

In modeling this reaction (or any other reaction based upon *AFM* chemographies) note that, apart from muscovite, biotite is the only K-bearing phase in pelitic rocks. Thus, to balance any reaction, muscovite, which is a projection point, must occur on the opposite side of a reaction from biotite. It is not often easy to determine merely by inspection on which side of the reaction quartz occurs. However, since pelitic rocks typically have excess quartz, it doesn't matter on which side of the reaction quartz occurs.

Kyanite Zone

The first appearance of kyanite in biotite-bearing pelitic rocks results from the reaction:

staurolite + muscovite + chlorite
$$\rightarrow \text{kyanite} + \text{biotite} \pm \text{quartz} + H_2O \qquad (16.6)$$

The *AFM* projection resulting after this reaction has occurred is shown in Figure 16.10. This tie-line flip reaction represents the disappearance of chlorite from many pelitic rocks. With increasing metamorphic grade, the stability field of chlorite shrinks to progressively higher magnesium compositions, first by discontinuous reactions, and then by the continuous reaction (Equation 16.7). In a few rare rocks, magnesium-rich chlorite may coexist with sillimanite, making the continuous reaction:

chlorite + muscovite $\rightarrow$ kyanite (or sillimanite)
$$+ \text{biotite} + H_2O \qquad (16.7)$$

In most terrains, therefore, chlorite disappears from pelitic rocks when the Mg/Fe ratio required to stabilize

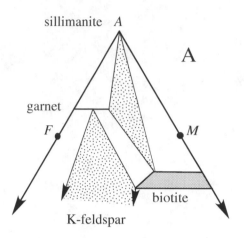

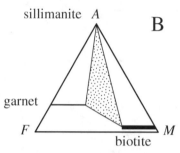

Figure 16.11 *AFM* projections for the sillimanite zone: (A) with muscovite, (B) with K-feldspar.

chlorite becomes higher than the Mg/Fe ratio found in the rocks themselves.

Sillimanite Zone

The transition from the kyanite zone to the sillimanite zone in the type area for Barrovian metamorphism is marked by two mineralogic changes. Not only is there a change from kyanite to sillimanite as the major aluminosilicate, there is also the disappearance of staurolite by the reaction:

$$\text{staurolite} + \text{muscovite} \rightarrow \text{sillimanite} + \text{garnet} +$$
$$\text{biotite} \pm \text{quartz} + H_2O \qquad (16.8)$$

The *AFM* projection resulting from this reaction is shown in Figure 16.11A, and it illustrates that assemblages in high-grade metapelites are comparatively simple; the dominant assemblage is muscovite–quartz–biotite–sillimanite–garnet. It is interesting that the two minerals often considered diagnostic of pelitic rocks, garnet and aluminosilicates, don't coexist until the higher grades of amphibolite facies.

Although Barrow did not recognize a grade of metamorphism higher than sillimanite, later workers discovered that at higher grades of metamorphism muscovite disappears by the reaction:

$$muscovite + quartz \rightarrow sillimanite + K\text{-}feldspar + H_2O$$
$$(16.9)$$

In a Barrovian sequence, this reaction is often called the *second sillimanite isograd*. This isograd marks the reappearance of K-feldspar in pelitic rocks and the disappearance of muscovite. The assemblages in rocks crystallizing at metamorphic temperatures above those of Equation 16.9 must be shown on K-feldspar projections, rather than muscovite projections (compare Figure 16.11B with Figure 16.11A).

At pressures below about 4 kbar, the second sillimanite reaction proceeds without melting, but above 4 kbar, the potassium feldspar produced by the reaction resides in a granitic melt (Figure 16.12A). As noted in Section 11.3, this kind of a reaction is called dehydration melting and crossing these reactions through decompression can produce significant amounts of crustal melt. In some areas, it is possible to use the mineral assemblage in the melted zones to identify the reaction involved in the melting. For example, the rock in Figure 16.12B contains large garnets surrounded by K-feldspar halos. The K-feldspar halos likely represent crystallized granitic melt and indicate the reaction that formed this texture as: biotite + sillimanite + quartz + plagioclase → garnet + melt.

16.4 Pressure–Temperature Conditions for Metamorphic Assemblages in Metapelitic Rocks

For years after Barrow's work in Scotland, Barrovian metamorphism was considered typical for pelitic rocks. Developments after World War II showed that, rather than being common, Barrovian metamorphism indicated unusually high pressures. A prograde metamorphic sequence occurring at a lower pressure would have the same sequence of zones as in Barrovian metamorphism, except that andalusite may be present rather than kyanite. At even lower pressures, cordierite is found in metapelitic rocks. Rocks with the assemblage cordierite and staurolite are restricted to pressures of 3–4 kbar. At yet lower pressures, staurolite destabilizes and prograde reactions involve garnet, cordierite, andalusite, and sillimanite.

16.4.1 Metapelitic Assemblages and Metamorphic Facies

Zeolite and Prehnite–Pumpellyite: As noted previously, mineral reactions in low-grade pelitic rocks are not well defined. In general, pelitic rocks metamorphosed at subgreenschist conditions contain abundant clays, mostly illite.

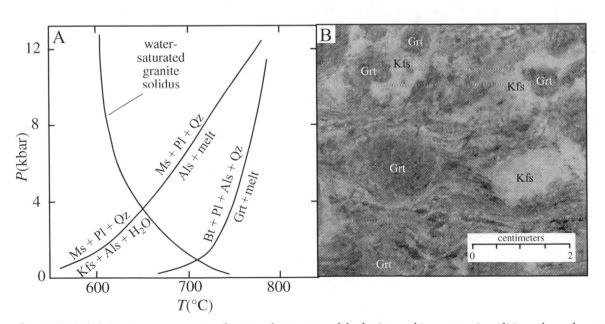

Figure 16.12 (A) Pressure–temperature diagram showing two dehydration melting curves in pelitic rocks and a possible path followed during decompression melting (reproduction of Figure 11.1). (B) photomicrograph of pelitic gneiss that underwent melting via the reaction biotite + plagioclase + sillimanite + quartz = garnet + melt. From central Massachusetts.

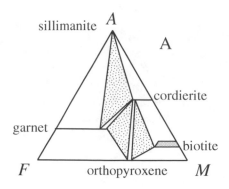

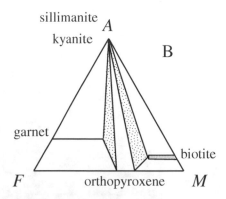

Figure 16.13 *AFM* projections for pelitic rocks in granulite facies: (A) relatively low pressure, (B) relatively high pressure.

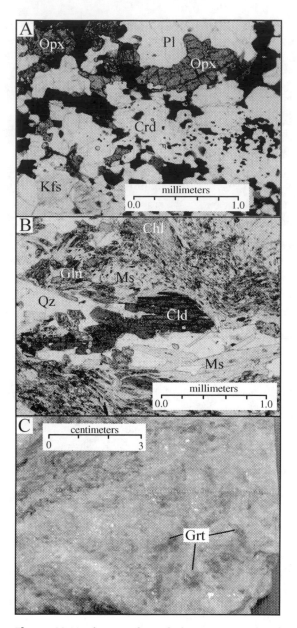

Figure 16.14 Photographs and photomicrographs of pelitic rocks in granulite, blueschist, and eclogite facies. (A) Photomicrograph in plane-polarized light (PPL) of a pelitic rock in granulite (pyroxene hornfels) facies containing cordierite (Crd), K-feldspar (Kfs), orthopyroxene (Opx), and Plagioclase (Pl). Contact aureole of the Laramie anorthosite complex, southeast Wyoming. (B) Photomicrograph in PPL of pelitic rock in blueschist facies containing chlorite (Chl), chloritoid (Cld), glaucophane, (Gln), muscovite (Ms), and quartz (Qz). Note the total lack of biotite. From the Bering Peninsula, Alaska. (C) Photograph of whiteschist (eclogite facies) containing pale pink garnet (Grt) in a matrix of muscovite, kyanite, and quartz. From Casa Parigia, Italian Alps.

Greenschist: Greenschist facies corresponds to the chlorite and biotite zones of pelitic rocks.

Amphibolite: Most of the reactions described in the Barrovian sequence occur in amphibolite facies. Indeed, it is because pelitic rocks are so reactive in amphibolite facies that they are especially useful in subdividing the pressure and temperature conditions of amphibolite facies.

Granulite: Granulite facies can be divided into two assemblages (Figure 16.13). At pressures below about 10 kbar, the distinctive granulite assemblage is: K-feldspar–garnet–cordierite±biotite±orthopyroxene. At around 10 kbar, the garnet–cordierite tie line breaks, forming tie lines between orthopyroxene and sillimanite instead. At 12–14 kbar, the diagnostic assemblage becomes orthopyroxene–kyanite. A good example of a granulite assemblage (although it comes from a contact aureole) is the assemblage cordierite–orthopyroxene–biotite–K-feldspar–plagioclase, shown in Figure 16.14A.

Blueschist: The distinctive feature about blueschist metamorphism of pelitic rocks is the lack of biotite. Garnet is present and some iron-rich rocks

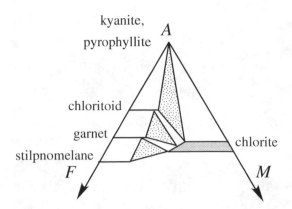

Figure 16.15 *AFM* projection for common pelitic assemblages in blueschist facies. Note that sodic amphibole may be present but cannot be shown in this projection.

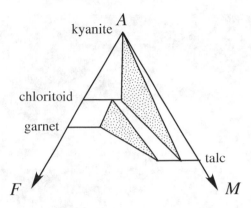

Figure 16.16 *AFM* projections for mineral assemblages in whiteschist (i.e. eclogite-facies pelitic rocks).

may have stilpnomelane. In addition, chloritoid, which has a rather limited stability at low pressures, is much more common in blueschist facies (Figures 16.14B and 16.15). Sodic amphibole may also be present, but it is not shown on an *AFM* projection because sodium, which is usually incorporated in albite, is a projection point.

Eclogite: Pelitic rocks metamorphosed in low-temperature eclogite facies are similar to those in blueschist facies, with chlorite, phengite, and quartz being dominant. Until the 1980s, the existence of pelitic rocks in high-temperature eclogite facies was unknown. Discoveries of unusual pelitic rocks known as *whiteschists* (Chopin, 1981) has led to the recognition that pelitic rocks have been metamorphosed under conditions consistent with high-temperature eclogite facies. The distinctive feature of whiteschists is the instability of both biotite and chlorite. Instead, the assemblage kyanite and talc are stable (Figure 16.16). In addition, there is magnesium-rich garnet, which in some localities contains inclusions of coesite, the high-pressure polymorph of silica, indicative of pressures in excess of 24 kbar.

16.4.2 Pressure Information from Metapelitic Rocks

Barrovian metamorphic zones provide a template for understanding how assemblages in pelitic rocks change with respect to temperature and pressure. With decreasing pressure, andalusite will take the place of kyanite. The presence of andalusite indicates that the pressure of metamorphism was below that of the aluminosilicate triple point (550 °C and 4.5 kbar, see Figure 2.1). At low pressures, cordierite may occur in place of garnet or staurolite. In metapelitic

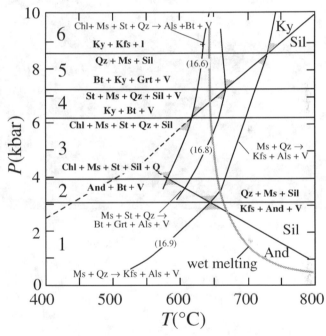

Figure 16.17 Pressure–temperature diagram showing the location of bathograds and bathozones. Shaded areas represent where the phase assemblages on either side of the bathograd are stable. And = andalusite, Bt = biotite, Chl = chlorite, Grt = garnet, Ky = kyanite, Kfs = K-feldspar, l = liquid (granitic melt), Ms = muscovite, Qz = quartz, Sil = sillimanite, St = staurolite, V = vapor. After Pattison (2001).

rocks metamorphosed in amphibolite facies, cordierite is generally restricted to pressures of 4 kbar or below (Holdaway and Lee, 1977). Cordierite stability is a function of the Fe/Mg ratio of the rock; cordierite is stable to higher pressures in the more magnesian rocks. The maximum pressure

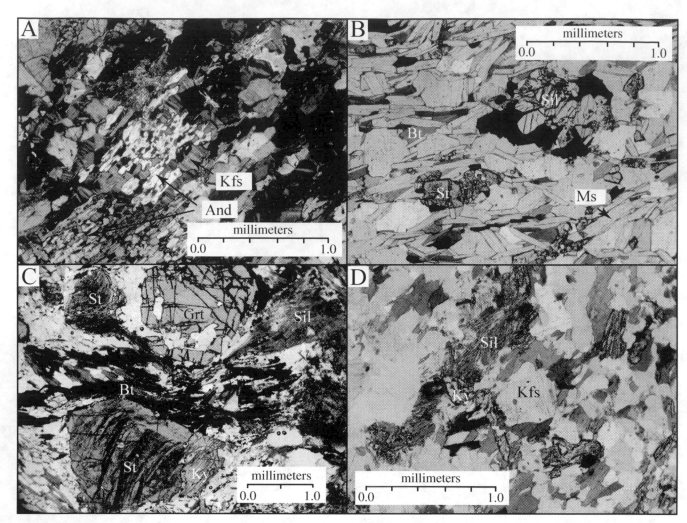

Figure 16.18 Photomicrographs of pelitic rocks from a range of bathozones. (A) Photomicrograph in cross-polarized light (XPL) showing coexisting andalusite (And) and K-feldspar (Kfs), indicative of bathozone 1 and pressures of 3 kbar or below. Contact aureole of the Laramie anorthosite complex, southeastern Wyoming. (B) Photomicrograph in PPL of the assemblage biotite (Bt)–muscovite (Ms)–staurolite (St)–sillimanite (Sil), indicating bathozones 3 or 4, pressures of 4–7 kbar. Grenville province of southern Ontario, Canada. (C) Photomicrograph in PPL of pelitic schist with the assemblage biotite (Bt)–garnet (Grt)–kyanite (Ky)–staurolite (St) with late sillimanite (Sil). This assemblage is indicative of bathozone 5 or pressures of 7–8.5 kbar. Central Laramie Mountains, Wyoming. (D) Photomicrograph in PPL of the assemblage Kyanite (Ky)–K-feldspar (Kfs) with late sillimanite indicative of bathozone 6. This assemblage is diagnostic of pressures in excess of 8.5 kbar. From the Grenville of southern Ontario, Canada.

of cordierite also increases with increasing temperature; in granulite-facies pelitic rock its stability may extend to temperatures of 900 °C and 10 kbar (Carrington, 1995).

Pressures associated with particular mineral assemblages in pelitic rocks can be limited more precisely through the use of **bathograds** (Carmichael, 1978; Pattison, 2001). Ordinary isograds are dependent on both pressure and temperature and, hence, it is not possible to know in a prograde terrain whether the increasing metamorphism is a function merely of increasing temperature

or of increases in both temperature and pressure. A bathograd, on the other hand, consists of two intersecting isograds and, therefore, depends only on pressure. The bathograd net as modified by Pattison (2001) is constructed by combining the aluminosilicate-in reaction (Equation 16.6), the staurolite-out reaction (Equation 16.8), the muscovite-out reaction (Equation 16.9), and the aluminosilicate phase diagram (Figure 16.17). The reactions shown by Equations 16.6, 16.8, and 16.9 intersect the kyanite–sillimanite and the andalusite–sillimanite reactions. As a result, there are five intersections defined as bathograds, and these bathograds define six bathozones, which are regions in pressure–temperare space delimited by the reactions between the bathograds.

Bathozone 1: At the lowest pressure (below about 3 kbar), quartz + muscovite react to form K-feldspar while andalusite is still stable. Thus, the assemblage andalusite–K-feldspar (Figure 16.18A) is diagnostic of the low-pressure bathozone. Cordierite is common in such regimes, while staurolite is very rare. Bathograd 1 is marked by the intersection of the muscovite-out reaction (Equation 16.9) with the andalusite–sillimanite phase boundary. This bathograd is defined by the reaction:

K-feldspar + andalusite + vapor
  → muscovite + sillimanite + quartz

At higher pressures (about 3 kbar), muscovite + quartz react to K-feldspar + sillimanite rather than to andalusite. The assemblage muscovite–quartz–sillimanite–K-feldspar is not particularly diagnostic of pressure because it can occur in bathozones 2, 3, 4, or 5.

Bathozone 2: Bathozone 2 is defined as a terrain in which andalusite and biotite are stable at low pressure and sillimanite and muscovite are stable at higher pressure. At low pressures, this bathozone is bounded by bathograd 1, whereas at high pressures it is bounded by the reaction of chlorite + staurolite to muscovite + biotite + sillimanite. Cordierite may be present in rocks from this zone, but it will be restricted to bulk compositions that have a relatively high Mg/(Fe + Mg) ratio.

Bathozone 3: Bathozone 3 is marked by the breakdown of the assemblage muscovite + staurolite + chlorite to sillimanite + biotite (i.e. the sillimanite-bearing version of the reaction shown in Equation 16.6). Andalusite is not stable with biotite in this bathozone.

Bathozone 4: In bathozone 4, staurolite + chlorite reacts to kyanite + biotite so that there is a stability field for kyanite + biotite at grades below those where sillimanite appears. At the higher temperatures indicative of bathozone 4, staurolite + muscovite react to garnet + biotite + sillimanite. In other words, the reaction shown in Equation 16.6 occurs in the presence of kyanite, whereas that in Equation 16.8 occurs in the presence of sillimanite. The assemblage biotite–muscovite–staurolite–sillimanite (Figure 16.18B) is stable in this bathozone and in bathozone 3. In bathozone 3, the assemblage lies upgrade of assemblages containing muscovite–chlorite–staurolite, whereas in bathozone 4 it lies upgrade of a zone where kyanite is stable.

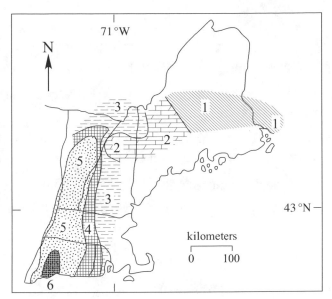

Map 16.2 Bathozones in New England (after Carmichael, 1978, reprinted by permission of the *American Journal of Science*). Numbers refer to Figure 16.17.

Bathozone 5: In bathozone 5, staurolite + muscovite react to garnet–kyanite–biotite. Figure 16.18C shows the assemblage garnet–staurolite–kyanite–biotite. This is the relationship found in Barrovian metamorphism.

Bathozone 6: Bathozone 6 is marked by the coexistence of kyanite and K-feldspar (Figure 16.18D) and occurs at pressures above those where the muscovite breakdown reaction (Equation 16.9) crosses the kyanite–sillimanite boundary. In H_2O-saturated systems, upon which the bathograd classification is predicated, this reaction accompanies partial melting, so bathozone 6 is defined by kyanite-bearing migmatites and is indicative of pressures greater than 9 kbar.

The validity of the bathograd classification is debatable because variations in water pressure or changes in bulk composition might allow chlorite, muscovite, or staurolite to survive to slightly higher temperatures in one rock than they do in another. Such variations in the staurolite-out and the muscovite-out reactions necessarily add a degree of uncertainty to the pressures ascribed to the various bathograds. However, they do not decrease the utility of the bathograd or bathozone concept when it is applied on a regional scale. For example, a bathozone map of the northern Appalachians (Carmichael, 1978) (Map 16.2) shows a distinct, north–south-trending high-pressure area defined by bathozones 5 and 6. The high-pressure zone probably marks the area of crustal thickening produced by the Taconic orogeny.

Summary

· ·

- In complex systems like pelitic rocks both *continuous* and *discontinuous* reactions take place.

- *Continuous reactions* take place over a range of temperatures as the compositions of the participating minerals change.

- *Discontinuous reactions* occur when the compositions of the minerals are fixed by the appearance of another mineral. There are two types of discontinuous reactions, as discussed in Chapter 13:

  ◦ Tie-line flip reactions

  ◦ Terminal reactions

- A chemical description of the system of pelitic rocks requires at least 10 components. To portray this system on the *AFM* chemographic projections requires a number of simplifying assumptions.

  ◦ Plagioclase is present, fixing the activities of Na_2O and CaO

  ◦ Quartz is present, fixing the activity of SiO_2

  ◦ Iron–titanium oxides are present, fixing the activity of TiO_2

- Low-temperature metamorphism of shale involves:

  ◦ The formation of illite; with increasing temperature illite becomes muscovite

  ◦ The formation of chlorite

- Progressive metamorphism of shale at high pressures is marked by seven Barrovian zones: chlorite, biotite, garnet, staurolite, kyanite, sillimanite, and sillimanite–K-feldspar (or second sillimanite). The chlorite and biotite zones occur in greenschist facies. The garnet, staurolite, kyanite, sillimanite, and sillimanite–K-feldspar zones are found in amphibolite facies.

- At low pressures, andalusite appears in place of kyanite and cordierite appears in place of staurolite.

- At the upper temperature limits of amphibolite facies, muscovite breaks down to sillimanite + K-feldspar and pelitic rocks begin to melt.

- Bathozones are one way to monitor the variation of pressure in pelitic terrains:

  ◦ Bathozone 1 = coexisting andalusite + K-feldspar

  ◦ Bathozone 2 = staurolite coexisting with andalusite

  ◦ Bathozone 3 = staurolite + muscovite + chlorite breakdown to sillimanite + biotite

  ◦ Bathozone 4 = staurolite appears in the presence of kyanite but breaks down in the presence of sillimanite

  ◦ Bathozone 5 = staurolite breaks down in the presence of kyanite

  ◦ Bathozone 6 = muscovite breaks down in the presence of kyanite (i.e. kyanite + K-feldspar are stable.)

- In granulite facies, orthopyroxene is stable.

- Blueschist facies is marked by the presence of garnet and sodic amphibole and the absence of biotite.

- Eclogite facies is marked by whiteschists, where talc, quartz and garnet are stable.

Questions and Problems

Problem 16.1. The following questions refer to Figure 16.9, reproduced below:

1. For the following assemblages (each has both muscovite and quartz) note whether it is stable in Figure A, Figure B, both Figure A and B, or neither Figure A or B:

   a. garnet–staurolite–biotite_____

   b. staurolite–kyanite–biotite_____

   c. garnet–biotite–chlorite_____

   d. kyanite–chlorite–staurolite_____

   e. biotite–kyanite–chlorite_____

2. What is the reaction that separates Figure A and Figure B?

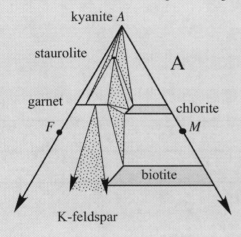

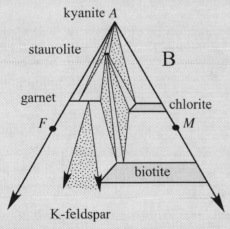

Problem 16.2. Buffo, a budding metamorphic petrologist, has asked you to help him with a mapping project. He has collected a number of samples of pelitic schists and has done the petrology on them but the isograd map he has produced (figure below), on which isograds have the ticks on the high-grade side), is downright confusing. He hasn't a clue as to what to do next. Please help him out! Below is a list of assemblages he found and a map showing where the samples were collected.

1. quartz–muscovite–biotite–garnet

2. quartz–muscovite–biotite–staurolite–chlorite

3. quartz–muscovite–biotite–staurolite–kyanite

4. quartz–muscovite–biotite–staurolite–chlorite

5. quartz–muscovite–biotite–kyanite–chlorite

6. quartz–muscovite–biotite–garnet

    a. Plot each assemblage below on an *AFM* projection.

    b. Determine the reactions that relate the various assemblages (hint: there is only one!)

    c. Plot the location of the reaction on the map (shown below) as an isograd.

    d. Explain why Buffo had things so very wrong.

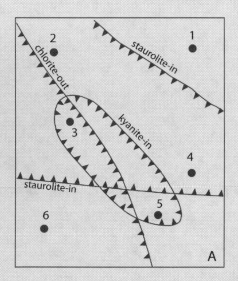 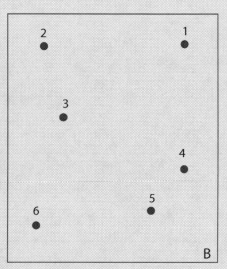

Problem 16.3. Plot the following assemblages on *AFM* diagrams (assume all have muscovite and quartz):

a. biotite–staurolite–kyanite

b. biotite–kyanite–chlorite

c. biotite–cordierite–garnet

d. biotite–andalusite–cordierite

Problem 16.4. Below is a list of minerals found in a series of rocks. For each rock determine the bathozone and relative pressures at which the assemblage formed. Assume all rocks contain quartz and plagioclase.

a. feldspar–biotite–kyanite–garnet

b. biotite–muscovite–andalusite–staurolite

c. biotite–muscovite–kyanite–staurolite–garnet

d. biotite–muscovite–staurolite–sillimanite–garnet

e. biotite–muscovite–andalusite–K-feldspar–garnet

Problem 16.5. Determine to what metamorphic facies the mineral assemblages listed below belong. Constrain the pressure–temperature conditions further, if possible:

a. quartz–plagioclase–microcline–chlorite–muscovite

b. quartz–muscovite–talc–garnet–kyanite

c. quartz–plagioclase–biotite–muscovite–staurolite–kyanite

d. quartz–plagioclase–K-feldspar–sillimanite–garnet

e. quartz–plagioclase–K-feldspar–cordierite–garnet

Problem 16.6 While mapping you find the following mineral assemblages, which are keyed to the map below:

A. K-feldspar–quartz–biotite–sillimanite–cordierite

B. muscovite–quartz–biotite–andalusite–cordierite

C. muscovite–quartz–biotite–andalusite

D. muscovite–quartz–biotite–andalusite–cordierite

E. K-feldspar–quartz–biotite–andalusite–cordierite

F. K-feldspar–quartz–bioitite–sillimanite–cordierite

G. K-feldspar–quartz–biotite–andalusite

   a. What are the reactions that separate these assemblages?

   b. Draw the reactions as isograds in the map below

   c. What is the pressure of metamorphism?

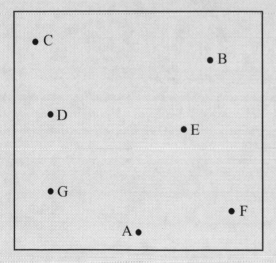

Further Reading

Bucher, K. and Frey, M., 2002, *Petrogenesis of Metamorphic Rocks*. Springer, Heidelberg, Chapter 7: Metamorphism of pelitic rocks (metapelites).

Philpotts, A. R. and Ague, J. J., 2009, *Principles of Igneous and Metamorphic Petrology*, 2nd edn. Cambridge University Press, Cambridge, Chapter 18: Graphical analysis of metamorphic mineral assemblages.

Winter, J. D., 2010, *Principles of Igneous and Metamorphic Petrology*, 2nd edn. Prentice Hall, New York, NY, Chapter 28: Metamorphism of pelitic sediments.

Yardley, B. W. D., 1989, *An Introduction to Metamorphic Petrology*, Longman Scientific and Technical, London, Chapter 3: Metamorphism of pelitic rocks.

Chapter 17

Metamorphism of Calcareous Rocks and the Role of Fluids in Metamorphism

17.1 Introduction

Chapters 14 through 16 described how the mineral assemblages in mafic, ultramafic, and pelitic rocks record the temperature and pressure of metamorphism. In addition to indicating temperature and pressure, mineral assemblages in metamorphic rocks can monitor fluid composition. Previous chapters assumed that the metamorphic fluid phase consisted entirely of H_2O. This approximation was valid because most metamorphic reactions in mafic, ultramafic, and pelitic rocks involve dehydration of hydrous silicates; carbonates are uncommon in these rock types. Chapter 15 did mention the role of fluid composition in altering carbonate-bearing serpentinites because small amounts of CO_2 in the fluid are sufficient to produce carbonate in serpentinite. As long as carbonate is absent in serpentinites, assuming that the major fluid species is H_2O provides a reasonable simplification. In discussing the metamorphism of carbonates, however, CO_2 is a fluid phase of critical importance because reactions in these rocks produce (or consume) both H_2O and CO_2.

This chapter discusses the mineral assemblages that form during progressive metamorphism of a dolomitic limestone that originally contained minor quartz and how the assemblages found in these rocks reflect changes in temperature and in fluid composition. It then shows how mineral assemblages can be used to monitor changes in the composition of other fluid species, including oxygen, sulfur, and pH.

17.2 Metamorphism of Impure Dolomitic Marble

Before considering the role of fluids, consider first the mineralogic changes that accompany increasing temperature in a carbonate rock that originally consisted of calcite, dolomite, and quartz (Figure 17.1). Because the protolith has the assemblage calcite-dolomite-quartz, the bulk composition for these rocks occupies only half of the CaO–MgO–SiO$_2$ triangle (Figure 17.1A). There are a few bedded magnesite deposits in the world that may have sedimentary origin, but they are very unusual. Most natural sediments do not have bulk compositions with an Mg/(Mg+Ca) ratio greater than 0.5; in other words, magnesium in the protolith is usually tied up in dolomite. In Figure 17.1, all areas whose assemblages are inaccessible because of this compositional restriction are shaded.

In mafic, ultramafic, and pelitic protoliths, increasing metamorphic grade progressively dehydrates the silicates. In metamorphosed carbonate rocks, however, increasing metamorphism produces reactions between the silicates and the carbonates that release mostly CO$_2$. These reactions drive calcium and magnesium from the carbonate minerals into the silicates. As a result, progressive metamorphism of impure dolomitic marbles produces calcium–magnesium silicates with decreasing amounts of silica. As metamorphic grade increases, these minerals are: talc, tremolite, diopside, forsterite and periclase. At very high temperatures, calcite + quartz react to make a series of calcium silicates. The first of these is wollastonite. At temperatures around 1000 °C, wollastonite is replaced by a sequence of calcium silicates that show a trend of decreasing silica content with increasing metamorphic grade, but because this sequence is uncommon it won't be considered further.

In some metamorphic terrains, talc is the first silicate to form during metamorphism of impure dolomitic limestones (Tilley, 1948), and with increasing metamorphic grade talc reacts with calcite to form tremolite. In many terrains, talc does not occur at all and tremolite forms directly from dolomite + quartz. In amphibolite facies, tremolite may dehydrate to diopside (Figure 17.1B). Diopside is the highest-grade silicate found in many metamorphosed carbonates, but in silica-poor marbles, forsterite forms by a reaction between dolomite and tremolite or diopside (Figure 17.1C).

In upper amphibolite and granulite facies conditions, where no hydrous minerals are present in metamorphosed carbonates, three assemblages are possible depending on the relative amount of silica (Figure 17.1C). In relatively siliceous rocks the assemblage is calcite–quartz–diopside. Although these rocks may be considered relatively siliceous, the protolith did not necessarily have high quartz content; it merely had a high quartz/dolomite ratio. Because all the magnesium in the rock is contributed by dolomite, if the abundance of dolomite was low then quartz remains after dolomite is depleted. Rocks with slightly higher dolomite/quartz ratio in the protolith have the assemblage calcite–diopside–forsterite, whereas those that have the highest dolomite/quartz ratio have the assemblage calcite–dolomite–forsterite. Figures 17.1B and C illustrate an important concept in metamorphosed siliceous dolomites: if a magnesium-bearing silicate is stable, be it tremolite, diopside, or forsterite, then quartz and dolomite are incompatible. Thus, in a carbonate rock that contains magnesium silicates and quartz, it is quite likely that all carbonate is calcite, and dolomite is not present.

At extremely high temperatures, usually in contact metamorphic environments, two additional minerals appear, periclase and wollastonite. They both indicate very high temperatures and usually low pressures but

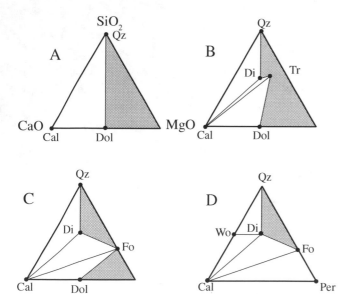

Figure 17.1 Ternary diagrams showing the minerals that form in impure dolomitic limestones during metamorphism. Gray areas are compositionally inaccessible in sedimentary rocks. (A) Composition range in the protolith. (B) Mineral assemblage in low amphibolite facies. (C) Mineral assemblages in upper amphibolite facies. (D) Mineral assemblages in high-grade contact aureoles.

these two oddities are mutually exclusive (Figure 17.1D). Wollastonite forms in quartz-excess rocks by the reaction:

$$CaCO_3 + SiO_2 \rightarrow CaSiO_3 + CO_2 \qquad (17.1)$$

whereas periclase forms in dolomite-excess rocks by the reaction:

$$CaMg(CO_3)_2 \rightarrow MgO + CaCO_3 + CO_2 \qquad (17.2)$$

It is exceedingly common for the periclase, formed by the reaction above (Equation 17.2), to hydrate to brucite during cooling.

17.2.1 Stability of Metamorphic Assemblages in T–X Space

This section describes the sequence of minerals found during metamorphism of dolomitic limestones that contained only quartz as a major non-carbonate mineral. Metamorphosed dolomitic limestones that originally contained other detritus, such as kaolinite, plagioclase, K-feldspar, or iron–titanium oxides, may also contain spinel, scapolite, grossular, phlogopite, or titanite.

Although the general variability of mineral assemblages in impure dolomitic limestones during prograde metamorphism was an early observation in metamorphic petrology (Bowen, 1940), detailed mineral relations weren't understood until isobaric T–X_{CO_2} diagrams were developed, several decades later (Greenwood, 1967; Skippen, 1974). Inspection of Table 17.1 shows that some reactions in metamorphosed siliceous dolomites consume H_2O and release CO_2, some release CO_2 and do not involve H_2O, and others release various proportions of both CO_2 and H_2O. These reactions cannot be shown on a pressure–temperature (P–T) diagram without somehow fixing the fluid composition. One approach is to hold pressure constant and show relations on an isobaric T–X_{CO_2} diagram (Figure 17.2). Because the reactions in Table 17.1 evolve (or consume) variable amounts of CO_2 and H_2O they produce a complex pattern in Figure 17.2.

Figure 17.2 illustrates how individual mineral assemblages in impure metamorphosed dolomites may be stable over a wide temperature range, depending on the composition of the fluid. The figure also indicates that if the CO_2 content of the fluid is high enough, tremolite, which is a hydrous mineral, is never stable and prograde metamorphism produces first diopside and then forsterite. If the fluid is H_2O-rich then all reactions occur over a

Table 17.1 *Mineral reactions in impure dolomitic marbles*[1]

Reaction	Equation number
$CaCO_3 + SiO_2 \rightleftharpoons CaSiO_3 + CO_2$	17.1
$CaMg(CO_3)_2 \rightleftharpoons MgO + CaCO_3 + CO_2$	17.2
$5Dol + 8Qz \rightleftharpoons H_2O = Tr + 3Cal + 7CO_2$	17.3
$Dol + 2Qz \rightleftharpoons Di + 2CO_2$	17.4
$3Cal + 2Qz + Tr \rightleftharpoons 5Di + H_2O + 3CO_2$	17.5
$3Cal + Tr \rightleftharpoons Dol + 4Di + H_2O + CO_2$	17.6
$3Dol + Di \rightleftharpoons 4Cal + 2Fo + CO_2$	17.7
$11Dol + Tr \rightleftharpoons 13Cal + 8Fo + H_2O + 9CO_2$	17.8
$5Cal + Tr \rightleftharpoons 11Di + 2Fo + 3H_2O + 5CO_2$	17.9
$3Dol + Kfs + H_2O \rightleftharpoons Phl + 3Cal + 3CO_2$	17.10
$5Phl + 6Cal + 24Qz \rightleftharpoons 3Tr + 5Kfs + 6CO_2 + 2H_2O$	17.11
$5Tr + 6Cal + Kfs \rightleftharpoons Phl + 12Di + 6CO_2 + 2H_2O$	17.12

[1] Mineral abbreviations used in this text are defined in Table A.1.

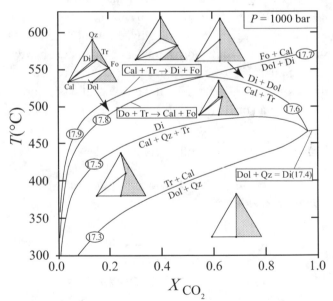

Figure 17.2 T–X diagram showing mineral stabilities during metamorphism of impure dolomitic marbles. Numbers indicate reactions in Table 17.1.

very small temperature range. Not only did the introduction of T–X_{CO_2} diagrams allow petrologists to understand the controls on mineral assemblages during metamorphism of impure carbonates, they also provided a tool to

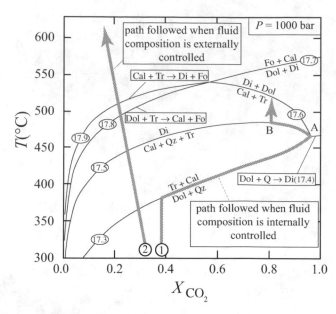

Figure 17.3 *T–X* diagram showing the paths followed during (1) internal control of fluid composition (buffering) and (2) external control of fluid composition (infiltration). Numbers indicate reactions in Table 17.1.

monitor how fluid composition in metamorphosed carbonates changed during metamorphism.

Two end members are recognized for the behavior of the fluid during metamorphism. In one end member the fluid is internally controlled or **buffered**. In this instance, fluid is entirely generated by the devolatization reactions, producing a small volume of fluid relative to the volume of rock. In this situation, the fluid composition follows the paths dictated by the mineral reactions and is thought of as a "rock-dominated" fluid (path 1 in Figure 17.3). For example, consider a rock containing dolomite–quartz–tremolite–calcite. As temperature increases, the fluid follows a reaction path along the reaction dolomite + quartz → tremolite + calcite, to a point where the reaction intersects another reaction, dolomite + quartz → diopside (Figure 17.3) (point A in Figure 17.3). At this point either dolomite or quartz will be depleted from the rock, thereby initiating a new reaction governed by the minerals that remain in the rock. Figure 17.3 assumes that dolomite is depleted at invariant point A and the fluid follows the reaction shown in Equation 17.5. Once the fluid attains the composition that marks the maximum stability of a reaction (point B in Figure 17.3), one of the phases becomes depleted from the reaction. The fluid composition remains unaffected by changes in temperature unless another reaction is reached.

The other end member for metamorphic fluid behavior involves external control of fluid composition, or **infiltration**. In this instance, there is a large volume of fluid relative to the volume of rock and the fluid composition is controlled by an external reservoir. The fluid path then is the mixture of the composition of the fluid in the rock and the composition of the fluid in the reservoir (path 2, Figure 17.3). Because of the proportionately large volume of externally derived, typically hydrous, fluid, the X_{CO_2} will change little during interaction with the fluids derived by reactions in the rock. As a result, most fluid paths cross the reaction curves at a high angle and the individual reaction assemblages occupy only a small temperature range. For example, in Figure 17.3, rocks undergoing metamorphism with an internally controlled fluid composition (path 1) will have the stable assemblage dolomite–quartz–calcite–tremolite over a temperature range of 75 °C. In contrast, rocks undergoing metamorphism during infiltration will have the same assemblage over a much more limited temperature range as they cross the reaction boundary.

17.2.2 Examples of How Mineral Assemblages Can Monitor Fluid Flow in Aureoles

Two examples illustrate how the mineral assemblages can be used to monitor fluid flow in metamorphosed carbonates. The first is the contact aureole around the Alta stock (Map 17.1). The Alta stock is one of a series of Tertiary plutons that intruded Paleozoic sedimentary rocks in the Wasatch Mountains, east of Salt Lake City, Utah. The country rock largely consists of impure dolomitic marbles, making this locale an ideal area to study the effect of contact metamorphism on carbonaceous rocks. Moore and Kerrick (1976) recognize three isograds in the area – tremolite-in, forsterite-in, and periclase-in. The lower-grade isograds parallel the intrusive contact with the stock, suggesting that they record a simple thermal gradient. In contrast, the periclase zone forms a long arm extending more than a kilometer north of the intrusive contact. The distribution of the periclase zone suggests either that this arm of the zone outlines an area of extremely high heat flow or a place where the roof of the Alta stock lies just below the present erosional surface.

Low-variance assemblages (i.e. assemblages with many phases) are rare in the aureole, suggesting the fluid composition was externally controlled. The occurrence of periclase further indicates this fluid relation, because this

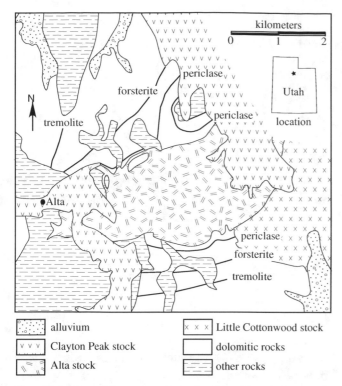

Map 17.1 Map of the contact aureole of the Alta stock in Utah, showing the tremolite-, forsterite-, and periclase-in isograds. Modified after Moore and Kerrick (1976), with permission of the *American Journal of Science*.

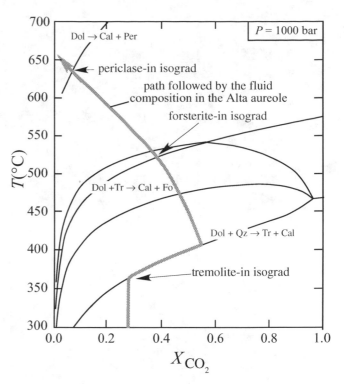

Figure 17.4 Isobaric *T–X* diagram, showing the fluid path followed in the contact aureole of the Alta stock. Modified after Moore and Kerrick (1976), with permission of the *American Journal of Science*.

mineral is restricted to very water-rich conditions (Figure 17.4). From the mineral assemblages, Moore and Kerrick (1976) inferred that the distal portion of the aureole fluids were internally buffered and the fluids became enriched in CO_2 as a reaction akin to Equation 17.3 proceeded (Figure 17.4). In contrast, the occurrence of periclase near the contact of the stock indicates that the fluid was very water-rich proximal to intrusion. The water-rich fluid most likely sourced from the magma of the Alta stock, because extensive skarns (massive calc-silicate rocks with grossular, epidote, and pyroxene) developed along the contact between the dolomitic marbles and the stock. Between the distal portion of the stock, where the assemblage dolomite–quartz–tremolite–calcite was stable, and in the areas near the granite contact, the fluid was externally controlled by water infiltrating out of the granite (Figure 17.4). This is indicated by the forsterite-in isograd, which appears as a single line, instead of by a band as would happen if the fluid had been buffered.

The map of the contact aureole around the Marysville stock in Montana (Rice, 1977) (Map 17.2) looks similar to the aureole around the Alta stock. In detail, however,

evidence for very different processes occurring in the two aureoles is readily apparent. The Marysville stock is a small Cretaceous pluton emplaced into the Proterozoic Helena dolomite. Unlike the highly irregular periclase-in isograd in the Alta aureole, the isograds around the Marysville stock are parallel to the contact. In addition, although the isograds around the Alta stock are easily designated by the appearance of single minerals, the isograds in the Marysville stock are much more complex and must be designated by the appearance or disappearance of mineral assemblages. This complexity indicates that the fluid was internally buffered in the Marysville stock. The Helena dolomite contains detrital K-feldspar alongside quartz, so the phase diagram summarizing the mineral reactions in the Marysville aureole (Figure 17.5) is slightly more complex than the phase diagram for calcite–dolomite–quartz (Figure 17.2), but the buffering process is the same in both depictions. During metamorphism in the aureole of the Marysville stock, the fluid in the distal portion of the aureole was buffered by the reaction dolomite + K-feldspar → phlogopite + calcite (Equation 17.10) until it hit invariant point I, where tremolite appears with K-feldspar. The fluid was buffered

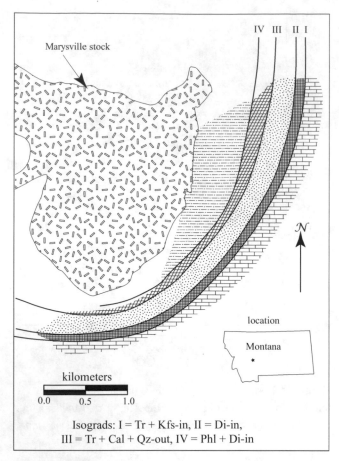

Map 17.2 Geologic map of the contact aureole around the Marysville stock, Montana, USA. Modified after Rice (1977), with permission of the *American Journal of Science*.

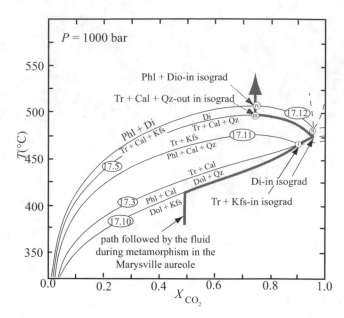

Figure 17.5 Isobaric *T*–*X* diagram, showing the path followed by the fluid during metamorphism in the Marysville stock. Modified after Rice (1977), with permission of the *American Journal of Science*.

along the reaction dolomite + quartz → tremolite + calcite until it reached invariant point II, where diopside appeared. Subsequently, the fluid was buffered along reaction shown in Equation 17.6 until the assemblage tremolite–calcite–quartz was depleted from the rock (isograd III). The fluid composition at this isograd was $X_{CO_2} = 0.75$, because both reactions shown in Equations 17.5 and 17.12 have a maximum at $X_{CO_2} = 0.75$. Therefore, no change in the fluid occurred during the reaction shown in Equation 17.12, which produced the assemblage phlogopite + diopside (isograd IV).

17.3 Buffering of Other Fluid Components

Water and CO_2 are not the only fluids that may be buffered or monitored by mineral assemblages in metamorphic rocks. If a reaction involving a fluid species can be defined by the minerals seen in a rock, then those mineral assemblages reflect the fluid composition. An example is given by the reaction zone that exists around the sulfide ore bodies in Ducktown, Tennessee (Nesbitt and Kelly, 1980). The ore bodies, consisting mainly of pyrrhotite and pyrite with minor chalcopyrite, sphalerite, and magnetite, are hosted in a metagreywacke schist with the assemblage quartz–plagioclase–biotite–muscovite–ilmenite–pyrrhotite–graphite. Metamorphism is inferred to have occurred at 6 kbar and 550 °C (Nesbitt and Kelly, 1980). Traversing a distance of 100 m from the ore body to the country rock, the following changes are observed. First, magnetite disappears, then rutile appears, then pyrite disappears, then ilmenite replaces rutile, and finally graphite appears (Figure 17.6). The appearance of rutile reflects the change in bulk composition between the ore and the host schist; the other changes are governed by the mineral reactions listed in Table 17.2.

The key to understanding what caused this sequence of assemblages is the realization that, in the schist, the reaction of graphite to CO_2 (Equation D) consumes oxygen and makes the rock and the fluid associated with it highly reducing. In contrast, the fluid in the ore body will be rich in S_2 and, compared to the schist, relatively enriched in O_2. At peak metamorphism when the rocks were immersed in fluid, the O_2 and S_2 would have diffused outward from the ore body into the schist. The outward migration of oxygen is reflected by the disappearance of magnetite in the ore body (Equation A), which releases

Table 17.2 *Reactions around the Ducktown ore body*

Reaction	Equation number
$Fe_3O_4 + 3FeS_2 \rightleftharpoons 6FeS + 2O_2$	A
$2FeS_2 \rightleftharpoons 2FeS + S_2$	B
$2TiO_2 + 2FeS + O_2 \rightleftharpoons 2FeTiO_3 + S_2$	C
$C + O_2 \rightleftharpoons CO_2$	D

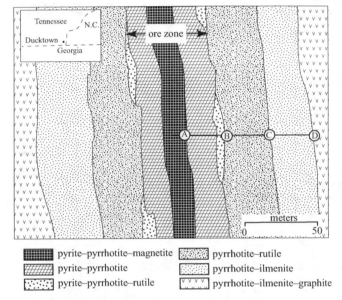

Legend:
- ▓ pyrite–pyrrhotite–magnetite
- ▒ pyrrhotite–rutile
- ▨ pyrite–pyrrhotite
- ░ pyrrhotite–ilmenite
- ⁙ pyrite–pyrrhotite–rutile
- ᵛ pyrrhotite–ilmenite–graphite

Figure 17.6 Sketch map of the mineral assemblages surrounding the ore body at Ducktown, Tennessee. Arrow connects assemblages found in a traverse from the ore zone into the country rock schist.

oxygen. The disappearance of pyrite reflects the reaction shown in Equation B, producing S_2, which migrated outward into the schist. The reaction of rutile and pyrrhotite to ilmenite (Equation C) consumes O_2 released via Equation A and produces S_2. Any remaining O_2 released via Equation A is consumed by the breakdown of graphite (Equation D).

17.4 Buffering of pH

Understanding the reactions between a rock and an infiltrating fluid is integral to metamorphic petrology; it is also critical to the field of economic geology. Many ore deposits form via reactions between a hydrothermal fluid and the country rock. Economic geologists study these alteration reactions around ore deposits as a way to understand the composition of the fluid that deposited the ore, including its pH. Rocks have the ability to buffer

the pH of the fluid moving through them. Good examples of buffering reactions are:

$$3KAlSi_3O_8 + H_2O + 2HCl \rightarrow KAl_3Si_3O_{10}(OH)_2$$
$$\underset{\text{K-spar}}{} \quad \underset{\text{fluid}}{} \quad \underset{\text{muscovite}}{}$$
$$+ \underset{\text{quartz}}{6SiO_2} + \underset{\text{fluid}}{2KCl} \qquad (17.13)$$

$$2KAl_3Si_3O_{10}(OH)_2 + 2HCl + 4H_2O \rightarrow 3Al_2Si_2O_5(OH)_4$$
$$\underset{\text{muscovite}}{} \qquad \underset{\text{fluid}}{} \qquad \underset{\text{kaolinite}}{}$$
$$+ \underset{\text{fluid}}{2KCl} \qquad (17.14)$$

These reactions are balanced using HCl and KCl rather than H^+ and K^+ because at metamorphic temperatures HCl and KCl are highly associated (rather than disassociated, as they are at low temperature). The reactions show that as acidic fluids move through a granitic rock, muscovite forms from K-feldspar as potassium is leached from the rock. This type of alteration is called sericitic alteration. If environments are acidic enough, even more potassium will be leached from the rock, forming kaolinite. At higher temperatures, pyrophyllite [$Al_2Si_4O_{10}(OH)_2$] will form instead of kaolinite. Both minerals record argillic alteration.

Figure 17.7 shows how acidic alteration affected a rhyolite. Figure 17.7A is a flow-banded rhyolite from the Yerington porphyry copper district, Nevada. Figure 17.7B shows the same unit that has been exposed to infiltrating acidic fluids. The gray region contains sericite and hence underwent sericitic alteration. The irregular white veins contain pyrophyllite and record the fluid pathways producing argillic alteration.

Neutralization reactions like this are important in economic geology because metals, sulfur, and chlorine behave incompatibly during crystallization of granitic melts. The fluid liberated late in the crystallization of granitic melts will be acidic (or at very least chlorine-rich) and sulfidic. Because metal ions are moderately soluble when they complex with chlorine, these fluids will also be laden with metals. When these fluids interact with surrounding rocks the reactions neutralize the fluid, altering the original mineralogy from K-feldspar to muscovite or clay, and simultaneously depositing metal sulfides. Hydrothermal alteration similar to that shown in Figure 17.7 may cover areas of many square kilometers (Box 17.1). Geologists searching for ore deposits carefully map areas of hydrothermal alteration, because this alteration is commonly the clue that an ore deposit is hidden at depth.

Figure 17.7 (A) Photo of flow-banded porphyritic rhyolite from Yerington District, Nevada. (B) Sample of the same unit that has undergone intense alteration from acidic fluids. Gray areas contain sericite and hence record sericitic alteration. Irregular white veins contain pyrophyllite and record argillic alteration from late fluids. Width of samples is about 5 cm. Photos courtesy of Simone Runyon.

BOX 17.1 | HYDROTHERMAL ORE DEPOSITS AND HYDROTHERMAL ALTERATION

Many ore deposits are formed when metals are deposited from hot aqueous fluids, commonly referred to as *hydrothermal fluids*. There are many classes of hydrothermal ore deposits. Some, like porphyry copper deposits (see Box 8.2), form from fluids derived directly from igneous intrusions. In others, such as those associated with black smokers (Box 7.2), igneous intrusions provide the heat for the fluids, but the fluids are derived mostly from seawater or groundwater. Still other deposits derive from the circulation of groundwater: the heat involved is not magmatic, it is geothermal. The formation of all of these deposits is associated with wall-rock alteration, referred to as *hydrothermal alteration*. Economic geologists study hydrothermal alteration because it provides important information about the formation of the ore, including the composition of the fluid that deposited the ore, the reactions that could have caused the deposition of the ore, and the temperature at which the deposition occurred.

The most common type of hydrothermal alteration is the kind found around porphyry copper deposits (Box 8.2) and epithermal gold deposits (Box 11.1). These types of alteration indicate the movement of acidic fluids through granitic rocks. With increasing acidity, the alteration assemblages include:

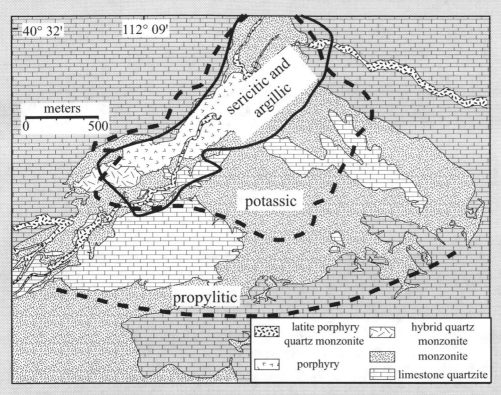

Box Figure 17.1 Geologic map of the area around the Bingham ore body in Utah, showing the alteration around the ore body. Modified from Parry et al. (2002).

Propylitic: alteration to chlorite, epidote, calcite, and other minerals that are indicative of greenschist facies in relatively neutral fluids;

Potassic: alteration to K-feldspar and biotite, indicative of neutral fluids or relatively acidic fluids at high temperature;

Sericitic: muscovite-rich alteration, indicative of acidic conditions; and

Argillic: clay-rich alteration, indicative of highly acidic conditions.

These altered rocks often form on a regional scale, as indicated by the alteration around the Bingham ore body in Utah (Box Figure 17.1), and thus provide helpful clues in the exploration of ore bodies.

Fluids that are not particularly acidic, or fluids interacting with rocks other than granitic rocks, can alter in ways other than the types listed above. Chief among these are:

Silicic: Because silica is highly soluble at temperatures above 300 °C, silicic alteration can simply indicate extensive movement of fluid through the rock with attendant deposition of quartz.

Chloritic: Chloritic alteration occurs mostly in mafic rocks. Leaching of alkalis by acidic fluids will leave behind iron, magnesium, and aluminum, which will manifest as chlorite.

Carbonate: This kind of alteration contains large amounts of calcite or dolomite, which indicates that the fluid contained CO_2. Figures 15.13 and 17.2 show that, at low temperature, it is not necessary for the fluid to have contained large amounts of CO_2 to form carbonate.

Summary

- Metamorphosed calcareous rocks give petrologists information on the composition of the ambient fluid, as well as the temperature of metamorphism.

- There are two end members for the behavior of fluid during metamorphism:

  - When fluids are *buffered*, or *internally controlled*, their compositions are controlled by the mineral reactions in the rocks. Such rocks usually have a comparatively large number of minerals present.

  - When fluids are *infiltrated*, or *externally controlled,* the mineral assemblages are governed by the composition of the fluid. Such rocks usually have a relatively small number of minerals present.

- A rock was buffering the fluid composition if one can write a reaction among the phases present using fluid species as the only unknown. In some areas, these reactions can monitor the flow of fluids.

- Rocks buffer pH, which is an important process in the formation of ore deposits.

Questions and Problems

Problem 17.1. A metamorphic aureole in impure dolomitic marble records the prograde assemblages listed below. Was the fluid internally or externally controlled during this metamorphism? Show the likely path on Figure 17.2 (reproduced below).

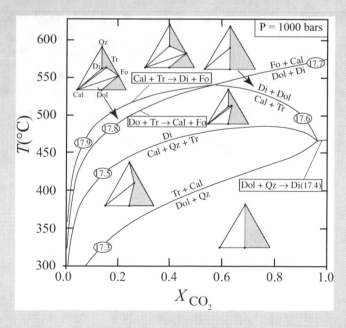

a. calcite–dolomite–tremolite–quartz

b. calcite–dolomite–diopside–tremolite–quartz

c. calcite–tremolite–diopside–quartz

d. calcite–diopside–quartz

Problem 17.2. A metamorphic aureole in impure dolomitic marble records the prograde assemblages listed below. Was the fluid internally or externally controlled during this metamorphism? Show the likely path on Figure 17.2 (reproduced again below).

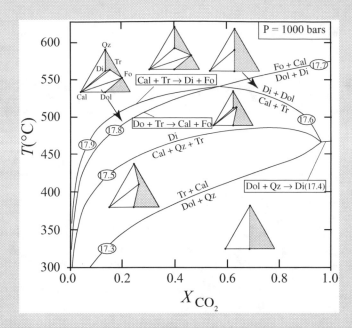

a. calcite–dolomite–tremolite–quartz

b. calcite–tremolite–quartz

c. calcite–diopside–quartz

Further Reading

Bucher, K. and Frey, M., 2002, *Petrogenesis of Metamorphic Rocks.* Springer, Heidelberg, Chapter 6: Metamorphism of dolomites and limestone and Chapter 8: Metamorphism of marls.

Rose, A. W. and Burt, D. M., 1979, Hydrothermal alteration. In *Geochemistry of Hydrothermal Ore Deposits*, 2nd edn, ed. H. L. Barnes. John Wiley & Sons, New York, 173–235.

Winter, J. D., 2010, *Principles of Igneous and Metamorphic Petrology*, 2nd edn. Prentice Hall, New York, NY, Chapter 29: Metamorphism of calcareous and ultramafic rocks.

Yardley, B. W. D., 1989, *An Introduction to Metamorphic Petrology*, Longman, London, Chapter 5: Metamorphism of marbles and calc-silicate rocks.

Chapter 18

Thermobarometry and the Conditions of Metamorphism

18.1 Introduction

Metamorphic facies and metamorphic assemblages provide a qualitative estimate of the conditions of metamorphism. This information can be obtained by application of thermodynamic databases. These data were obtained from experimental petrology, which provides pressure and temperature information on the mineral reactions which produce the assemblages that are characteristic of the metamorphic facies. Over the past 50 years, thermodynamic data for end members and for solid solutions of common metamorphic minerals have been extracted from these experimental studies. The resulting database can be used for calculation of the pressures and temperatures of mineral assemblages from most rocks. These data have been used in two ways. *Thermobarometry* calculates the pressure and temperature at which a mineral assemblage equilibrated using mineral chemistry. **Metamorphic assemblage diagrams** use whole-rock geochemical analyses to calculate the temperature and pressure ranges over which various mineral assemblages would form in a rock of that composition.

Many of the details of these calculations are beyond the scope of this text. This chapter introduces how these techniques provide quantitative information about metamorphic conditions. It then describes how the results of these calculations provide quantitative constraints on the conditions of the metamorphic facies discussed in Chapter 14. The chapter ends with a discussion of the limits of metamorphism.

18.2 Review of Thermodynamics

18.2.1 Free Energy

As was introduced in Chapter 1, the key variable in thermo-dynamics is free energy (G), which is the energy produced by a chemical reaction that is available to do work. As noted in Chapter 1, the free energy for a reaction is defined as:

$$\Delta G = \Delta H - T\Delta S \qquad (18.1)$$

where Δ refers to the difference in the thermodynamic function between the products and reactants, H refers to enthalpy, and S represents entropy. As noted in Chapter 1, *enthalpy* is the heat released (or consumed) by a reaction due to the breaking of bonds and the formation of new bonds. *Entropy* is the heat tied up in the bonds of the various phases involved in the reaction. Equation 18.1 says the amount of energy released by a reaction is equal to the amount of energy released (or consumed) when the bonds of the reactant minerals are reorganized to form the bonds of the product minerals minus the amount of heat consumed (or produced) by changing the configur-ation of the elements within the mineral structures.

If the ΔG in Equation 18.1 is less than zero, then the free energy of the reactants is greater than that of the products and the reaction can proceed spontaneously, evolving heat in the process. If ΔG is greater than zero, the reaction cannot proceed unless further heat is added to the system. If ΔG equals zero, the products and reactants of the reac-tion are in equilibrium. This does not mean that nothing is happening in the reaction; it merely means that the rate of the reaction producing the products from the reactants is the same as the rate of the reaction producing the reactants from the products.

18.2.2 Effect of Changes in Pressure and Temperature on ΔG

How free energy changes with respect to pressure and temperature is given by Equation 18.2:

$$dG = -SdT + VdP \qquad (18.2)$$

where V refers to volume.

If we are dealing with a reaction, as we usually are in metamorphic petrology, Equation 18.2 becomes:

$$d\Delta G = -\Delta SdT + \Delta VdP \qquad (18.3)$$

At equilibrium, Equation 18.3 becomes:

$$0 = -\Delta SdT + \Delta VdP \qquad (18.4)$$

Rearranging gives:

$$\Delta VdP = \Delta SdT \qquad (18.5)$$

and

$$dP/dT = \Delta S/\Delta V \qquad (18.6)$$

Equation 18.6 is called the Clausius–Clapeyron equation and it indicates that the slope of a reaction in a P–T diagram is dependent on the ratio of $\Delta S/\Delta V$ of the reaction.

The Clausius–Clapeyron equation explains why dehy-dration reactions on a P–T diagram generally have an increasing slope with increasing pressure. Water released by dehydration reactions has a large volume at low pres-sures (see Figure 3.4), leading to a large denominator in Equation 18.6 and a correspondingly low positive slope in P–T space. As pressure increases, the ΔV also decreases, and the slope, as governed by Equation 18.6, becomes progressively steeper. At very high pressures, some dehy-dration reactions, such as the reaction shown in Figure 18.1, have a negative slope.

18.2.3 Equilibrium Constant

The expression for the equilibrium constant, also known as the law of mass action, is the key step in determining pressure and temperature of metamorphism from min-eral analyses. Consider an equilibrium:

$$nA + mB \rightleftharpoons pC + qD \qquad (18.7)$$

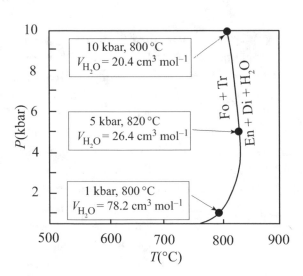

Figure 18.1 P–T diagram showing how the decrease in the molar volume of water affects the slope of the dehydration reaction forsterite + tremolite $\rightleftharpoons$ enstatite + diopside + H_2O. Water volume data from Burnham et al. (1969).

where A, B, C, and D are phases and n, m, p, and q are stoichiometric coefficients of those phases in the reaction. The equilibrium expression for this reaction is:

$$K = \frac{\alpha_A^n \alpha_B^m}{\alpha_C^p \alpha_D^q} \tag{18.8}$$

where K is the equilibrium constant and α_i is the **activity** of phase i, a point discussed in the next section (18.2.4). K is related to the free energy of Equation 18.7 by the following expression:

$$\ln K = \frac{-\Delta G}{RT} \tag{18.9}$$

where R = the gas constant. Substituting Equation 18.1 into Equation 18.9:

$$\ln K = \frac{-\Delta H}{RT} + \frac{\Delta S}{R} \tag{18.10}$$

On a plot of ln K versus $1/T$, Equation 18.10 yields a straight line with a slope of $-\Delta H/R$ and an intercept of $\frac{\Delta S}{R}$.

Assuming that temperature is fixed, and substituting Equation 18.3 into Equation 18.9, yields:

$$\ln K = \frac{-\Delta V}{RT} \tag{18.11}$$

Finally, combining Equations 18.10 and 18.11 gives the basic equation for thermobarometry:

$$\ln K = \frac{-\Delta H}{RT} + \frac{\Delta S}{R} - \frac{\Delta V}{RT} \tag{18.12}$$

18.2.4 Activity–Composition Relations

The key to using mineral analyses to calculate thermometry or barometry lies in evaluating the activity of a component in the mineral. The simplest definition of activity is that it is the "effective concentration" of a component in a solution. If a solution is ideal, the activity is the same as the mole fraction (X_i). As noted in Section 2.3, X_i is defined as the number of moles of component i divided by the total number of moles of components in solution. Calculating mole fraction is somewhat more complex when dealing with a solid solution, such as the olivine from New Caledonia given in Table 18.1. Table 18.1 contains two parts. The upper portion gives the analysis for olivine in terms of weight percent oxide. The lower portion of the table gives the cation proportion of the olivine that is normalized to four oxygens, which is the number of oxygens in the olivine formula.

Table 18.1 *Analysis of olivine from New Caledonia*

SiO_2	40.75
Al_2O_3	0.00
FeO	8.70
MnO	0.11
NiO	0.37
MgO	50.22
CaO	0.02
Total	100.17
Cation proportions on a basis of four oxygens	
Si	0.994
Al	0.000
Fe	0.177
Mn	0.002
Ni	0.007
Mg	1.825
Ca	0.000
Total	3.006
X_{Mg}	0.908
Mg/(Mg+Fe)	0.912
X_{Fe}	0.088
Fe/(Fe+Mg)	0.088

In calculating the mole fraction of forsterite in this olivine, one can neglect silica because magnesium and iron (and minor amounts of nickel and manganese) substitute for each other only on the octahedral sites in olivine. Silica fills the tetrahedral site in olivine regardless of how much iron or magnesium is present in the octahedral sites. Mole fraction of magneium (X_{Mg}) is then calculated as the cation ratio Mg/(Mg + Fe + Mn + Ni). Because, in the vast majority of silicates, iron and magnesium are the major cations in the octahedral site many people equate X_{Mg} with Mg/(Mg+Fe). For the olivine listed in Table 18.1, these two ratios are nearly equivalent. Because there are two sites for magnesium in olivine, the activity of Mg_2SiO_4 in olivine (assuming an ideal solution) is given by X_{Mg}^2, which means that (assuming ideality) the activity of Mg_2SiO_4 in this olivine is 0.82.

Unfortunately, solution of individual ions into mineral structures is rarely ideal. Some energy is required or evolved when various ions substitute for each other in a site. To convert mole fraction to activity, one uses the **activity coefficient** (γ). Thus, in real solutions activity becomes:

$$\alpha_i = X_i \gamma_i \qquad (18.13)$$

Activity coefficients must be determined empirically for each mineral. Most modern thermobarometry programs have these determinations built into the calculation; the inputs for such programs usually are the mole fractions of various components.

18.3 Thermobarometers

Equation 18.2 implies that a chemical reaction that has a relatively low ΔV and a high ΔS will be highly temperature-dependent and therefore makes a good thermometer, whereas a reaction with a relatively low ΔS and high ΔV will be pressure-dependent and, accordingly, will make a good barometer. As the field of thermobarometry developed in the 1970s and decades following, a series of reactions were calibrated as thermometers and barometers.

18.3.1 Geothermometry

Ion-Exchange Thermometry
The main type of reaction for geothermometry is an **ion-exchange reaction**. This reaction type involves the exchange of two ions (usually Fe^{2+} and Mg^{2+}) between two phases. A classic example is the distribution of Fe^{2+} and Mg^{2+} between garnet and biotite. This reaction can be written as:

$$\underset{\text{in garnet}}{Fe_3Al_2Si_3O_{12}} + \underset{\text{in biotite}}{KMg_3AlSi_3O_{10}(OH)_2}$$
$$\rightleftharpoons \underset{\text{in garnet}}{Mg_3Al_2Si_3O_{12}} + \underset{\text{in biotite}}{KFe_3AlSi_3O_{10}(OH)_2} \qquad (18.14)$$

Because garnet and biotite appear on both sides of the reaction, the ΔV of this reaction will be small, making a good thermometer. Garnet strongly favors iron over magnesium (Figure 18.2), such that at 450 °C biotite with even a moderate amount of iron will coexist with a very iron-rich garnet. As the temperature increases, garnet can accommodate more magnesium and the tie lines between garnet and biotite increasingly steepen. Consider the bulk composition of a rock that contains equal amounts of garnet and biotite and a bulk composition of $X_{Fe} = 0.70$, which is represented by the gray point in Figure 18.2. At 450 °C this rock will contain a garnet with $X_{Fe} = 0.90$

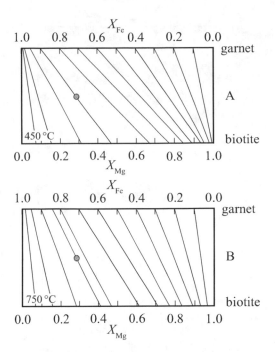

Figure 18.2 Diagram showing how the iron and magnesium contents of biotite and garnet vary as a function of temperature. (A) Relations at 450 °C, (B) relations at 750 °C. Gray circle gives the hypothetical bulk composition of a rock. See text for details. Calculations from the data of Ferry and Spear (1978).

coexisting with a biotite with $X_{Fe} = 0.52$. At 750 °C, the same rock will contain garnet with $X_{Fe} = 0.84$ and biotite with $X_{Fe} = 0.59$.

The tie lines in Figure 18.2 were calculated using the garnet–biotite thermometer of Ferry and Spear (1978). More recent formulations of the thermometer have incorporated various solution models and are more complex than that derived by Ferry and Spear (1978). However, Figure 18.2 shows graphically how the garnet–biotite tie lines change orientation with increasing temperature. By measuring the composition of coexisting biotite and garnet and knowing how the orientations of the tie lines change with temperature, the temperature at which these minerals last equilibrated can be determined.

Solvus Thermometry
A special type of ion-exchange thermometer involves the equilibrium between two phases on either side of a solvus. For example, calcite and dolomite are compositionally distinct at low temperatures but, with increasing temperature, dolomite can accommodate increasing amounts of calcium and more magnesium can be accommodated in the calcite structure until, at temperatures around 1100 °C, there is but a single carbonate with the composition

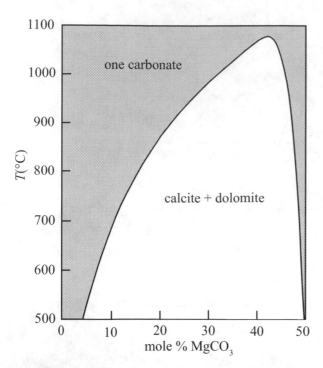

Figure 18.3 Calcite–dolomite solvus, after Goldsmith and Heard (1961).

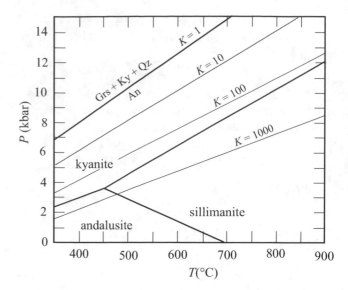

Figure 18.4 Location of the equilibrium anorthite $\rightleftharpoons$ grossular + kyanite + quartz. Lines show the displacement of this equilibrium in response to changes in the compositions of plagioclase and garnet. After Koziol and Newton (1988).

(Ca,Mg)CO$_3$ (Figure 18.3). Figure 18.3 shows that the magnesium content in calcite that coexists with dolomite is a sensitive thermometer. From the composition of calcite coexisting with dolomite, this solvus can be used to determine the temperature at which the two minerals last equilibrated. Theoretically, the calcium content of dolomite may also serve as a thermometer; however, because the limb of the solvus is so steep on the dolomite side, the calcium content of dolomite is not as sensitive to temperature changes as is the magnesium content of calcite.

18.3.2 Geobarometry

Reactions that are good barometers are those that involve large changes in volume and do not involve volatiles because, as shown in Figure 18.1, dehydration reactions have steep slopes in P–T space. A good example of a barometer is the reaction:

$$\underset{\text{in plagioclase}}{3CaAl_2Si_2O_8} \rightleftharpoons \underset{\text{in garnet}}{Ca_3Al_2Si_3O_{12}} + \underset{\text{kyanite}}{2Al_2SiO_5} + \underset{\text{quartz}}{SiO_2}$$

(18.15)

The equilibrium constant for this reaction (assuming quartz and kyanite are pure phases) is:

$$K = \frac{\alpha_{Gtr}^{gar}}{\left(\alpha_{An}^{Pl}\right)^3}$$

(18.16)

Assuming that $\alpha_{An}^{Pl} = X_{Ca}^{Pl}$ and, because there are three atoms of calcium in a formula unit of grossular, $\alpha_{Grs}^{Grt} = \left(X_{Ca}^{Grt}\right)^3$ (in other words assuming the plagioclase and garnet solutions are ideal), Equation 18.16 becomes:

$$K = \frac{\left(X_{Ca}^{Gtr}\right)^3}{\left(X_{Ca}^{Pl}\right)^3}$$

(18.17)

The reaction shown in Equation 18.15 has many properties that make it an ideal geobarometer. First, it has a relatively flat slope in P–T space (Figure 18.4). Second, garnet and plagioclase have a wide range of solid solution, which means the assemblage garnet–plagioclase–kyanite (or sillimanite or andalusite) is widespread. At low pressures, where K is large, garnet will have very little calcium and plagioclase will be relatively calcic. With increasing pressure, calcium will be transferred from plagioclase to garnet until at high pressures, the garnet will be relatively calcic and the plagioclase will be sodic. The barometer determined by this reaction is called the GASP barometer (for garnet–aluminosilicate–silica–plagioclase). Because the reaction of the anorthite component of plagioclase to form the grossular component into garnet, with increasing pressure, is a common petrologic process, there are many barometers in addition to GASP that rely on the distribution of calcium between plagioclase and garnet.

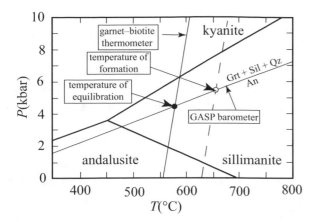

Figure 18.5 Thermobarometry from an assemblage sillimanite–garnet–biotite–plagioclase–quartz. Filled circle gives conditions of final equilibrium. Open circle gives possible conditions of formation followed by iron–magnesium exchange between garnet and biotite.

Because barometers involve reactions that transfer a component from one phase to another, they are called **mass-transfer reactions**. As a mass-transfer reaction (such as in Equation 18.15) proceeds under conditions of increasing pressure, there is volumetrically less plagioclase in the rock and more garnet. This differs substantially from ion-exchange reactions, such as the reaction shown in Equation 18.14. As this reaction proceeds with increasing temperature the amount of garnet and biotite remains the same, but iron in garnet is exchanged for magnesium in biotite.

18.3.3 Thermobarometry

Because thermometers have a steep slope and barometers have a gentle slope, the intersection of a thermometer and a barometer should provide a unique determination of the temperature and pressure at which an assemblage equilibrated. Figure 18.5 shows the results from a hypothetical assemblage of sillimanite-garnet-biotite-plagioclase-quartz. Analyses of garnet and plagioclase from this sample provide the location of the GASP barometer whereas analyses of biotite and garnet provide a temperature from the garnet biotite thermometer. The intersection of these two curves (filled circle at 550 °C and 4.5 kbar in Figure 18.5) is the point where the assemblage last equilibrated.

The major problem with thermobarometry is that, particularly at temperatures of upper amphibolite and granulite facies, mineral compositions in a rock tend to reset on cooling. Thermometers that involve simple ion

Table 18.2 *Equilibria in the assemblage orthopyroxene–clinopyroxene–garnet–plagioclase–quartz*[1]

Reaction	Equation number
Qz + Di + Alm $\rightleftharpoons$ An + En + Prp	18.1
3Qz + Grs + 2Alm $\rightleftharpoons$ 3An + 6Fs	18.2
3Qz + 3Di + 4Alm $\rightleftharpoons$ 3An + 12Fs + Prp	18.3
Alm + 3Di $\rightleftharpoons$ Grs + 3Fs + 3En	18.4
Qz + Grs + 3En $\rightleftharpoons$ An + 2Di	18.5
Alm + 3En $\rightleftharpoons$ Prp + 3Fs	18.6
Qz + Prp + Di $\rightleftharpoons$ An + En	18.7
2Alm + 3Di $\rightleftharpoons$ Prp + Grs + 6Fs	18.8
3Qz + Prp + 2Grs $\rightleftharpoons$ 3An + Di	18.9
Prp + 3Di $\rightleftharpoons$ En + Grs	18.10
3Qz + 2Prp + Grs = 3An + 6En	18.11

[1] Mineral abbreviations used in this text are defined in Table A.1.

exchange tend to reset readily and may freeze in at temperatures significantly below those at which the assemblage formed. Barometers, in contrast, involve dissolution of components from plagioclase, for example, and reprecipitation of those components in garnet. This is a much more complex process, and these types of equilibria tend to freeze in at much higher temperatures than ion-exchange equilibria. Thus, it is impossible to tell, without additional information, whether the garnet–sillimanite–biotite–plagioclase–quartz shown in Figure 18.5 actually formed at 550 °C and 4.5 kbar or whether it formed at a higher temperature and pressure (such as 650 °C and 5.5 kbar shown by the open circle in Figure 18.5) and the ion-exchange reactions continued to reset as the rock cooled (Frost and Chacko, 1989).

One way to deal with this uncertainty is to determine the equilibrium conditions for a rock using all the possible equilibria in an assemblage (Berman, 1991). For example, given the assemblage orthopyroxene–clinopyroxene–garnet–plagioclase–quartz, and using iron and magnesium end members of these solid solutions, Berman (1991) recognized 11 equilibria (Table 18.2). Note that some reactions (Equations 18.6, 18.8, and 18.4) involve simple exchange between iron and magnesium end members and are thermometers, and others (Equations 18.2, 18.5, 18.9, 18.11) involve equilibria between grossular and anorthite and are barometers. Other equilibria involve a mixture of ion-exchange and mass-transfer reactions. To show the

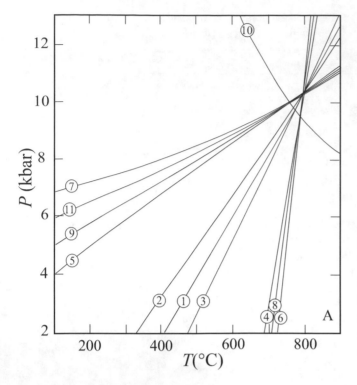

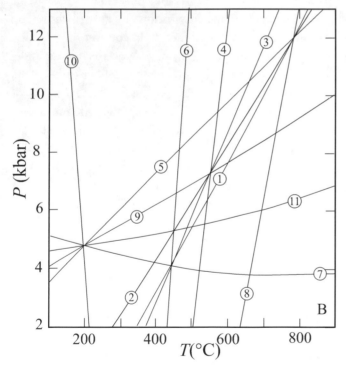

Figure 18.6 *P–T* diagrams showing multi-equilibria thermometry of a granulite with the assemblage orthopyroxene–clinopyroxene–garnet–plagioclase–quartz. (A) Calculations using the compositions of texturally early phases. (B) Calculations using compositions of a mixture of texturally early and young phases. From Berman (1991). Numbers refer to reactions given in Table 18.2.

utility of his multi-equilibria approach, Berman calculated the *P–T* location of these 11 equilibria using the compositions of the texturally earliest assemblage in the rock (Figure 18.6A). All the equilibria intersect in a small area around 800 °C and 10 kbar. When he used the composition of early phases mixed with the compositions of later-formed phases in the rock, which were clearly not in equilibrium with the earlier minerals, he obtained no common intersection (Figure 18.6B).

One can use Berman's (1991) multi-equilibria process to obtain an estimate of the original equilibrium conditions for an assemblage by back-calculating to take the retrograde iron–magnesium exchange into account. Pattison et al. (2003) did this by taking a series of granulites and adjusting the iron–magnesium contents of the phases until the equilibria among the minerals intersected at a single point. By doing this, they found that granulites formed at temperatures around 800 °C, rather than the 700 °C commonly recorded in the reset thermometers.

18.3.4 Metamorphic Assemblage Diagrams (Pseudosections)

The development of thermodynamic databases and sophisticated mineral solution models has led to the common use of metamorphic assemblage diagrams. These were originally called **pseudosections** by Powell et al. (1998) but were renamed *metamorphic assemblage diagrams* by Spear and Pattison (2017). This text uses the term metamorphic assemblage diagrams because that name accurately describes what they are: a graphical representation of the mineral assemblages found in a rock of given composition over a range of temperature and pressures.

Metamorphic assemblage diagrams differ from conventional thermobarometry in that they are based upon the whole-rock geochemical composition of a rock rather than upon the compositions of the minerals that the rock contains. Figure 18.7A shows a metamorphic assemblage diagram for the average Archean shale of Condie (1993). The assemblages in this rock that are found in the fields in Figure 18.7A are listed in Table 18.3. Like many metamorphic assemblage diagrams, this figure is very complex because, over the pressure and temperature range of metamorphism, many assemblages may be stable in the rock. Petrologists commonly simplify the metamorphic assemblage diagram by highlighting the assemblages present in the rock or the inferred *P–T* path of the

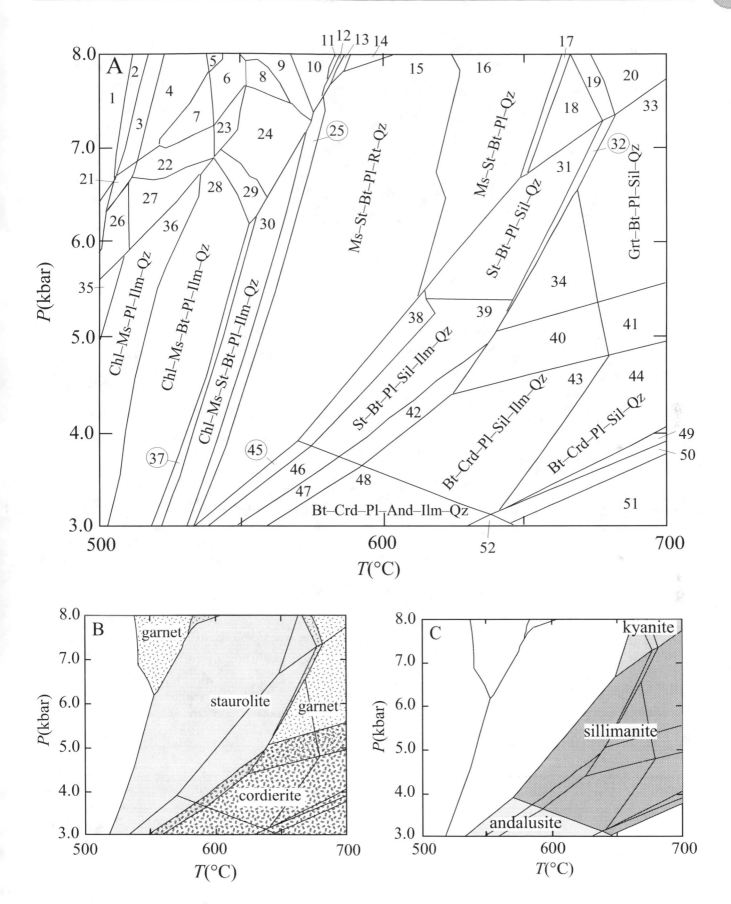

Figure 18.7 (A) Metamorphic assemblage diagram (or pseudosection) for the average Archean shale of Condie (1993). Numbers in each field refer to assemblages listed in Table 18.3. (B) The *P–T* stability field for the iron–magnesium–aluminum silicates in Figure 18.7A. (C) The *P–T* stability of the Al$_2$SiO$_5$ minerals in A and B.

Table 18.3[1] *List of mineral assemblages found during metamorphism of the average Archean shale. Numbers refer to fields shown on Figure 18.7.*

1	Chl–Pg–Ms–Czo–Ttn–Rt–Qz
2	Chl–Pg–Ms–Bt–Czo–Rt–Qz
3	Chl–Pg–Ms–Bt–Czo–Rt–Ilm–Qz
4	Chl–Pg–Ms–Bt–Czo–Ilm–Qz
5	Chl–Grt–Pg–Ms–Bt–Czo–Ilm–Qz
6	Chl–Grt–Ms–Bt–Czo–Ilm–Qz
7	Chl–Ms–Bt–Czo–Ilm–Qz
8	Chl–Grt–Ms–Bt–Ilm–Qz
9	Chl–Grt–Ms–Bt–Rt–Ilm–Qz
10	Chl–Grt–Ms–Bt–Rt–Qz
11	Chl–Grt–Ms–St–Bt–Rt–Qz
12	Chl–Grt–Pg–Ms–St–Bt–Rt–Qz
13	Grt–Pg–Ms–St–Bt–Pl–Rt–Qz
14	Grt–Ms–St–Bt–Pl–Rt–Qz
15	Ms–St–Bt–Pl–Rt–Qz
16	Ms–St–Bt–Pl–Qz
17	Ms–St–Bt–Pl–Ky–Qz
18	St–Bt–Pl–Ky–Qz
19	Grt–St–Bt–Pl–Ky–Qz
20	Grt–Bio–Pl–Ky–Qz
21	Chl–Ms–Bt–Pl–Czo–Rt–Ilm–Qz
22	Chl–Ms–Bt–Pl–Czo–Ilm–Qz
23	Chl–Grt–Ms–Bt–Pl–Czo–Ilm–Qz
24	Chl–Grt–Ms–Bt–Pl–Ilm–Qz
25	Chl–Ms–St–Bt–Pl–Rt–Qz
26	Chl–Ms–Pl–Czo–Rt–Ilm–Qz
27	Chl–Ms–Pl–Czo–Ilm–Qz
28	Chl–Ms–Bt–Pl–Ilm–Qz
29	Chl–Grt–Ms–Bt–Pl–Ilm–Qz
30	Chl–Ms–St–Bt–Pl–Ilm–Qz
31	St–Bt–Pl–Sil–Qz
32	Grt–St–Bt–Pl–Sil–Qz
33	Grt–Bt–Pl–Sil–Qz
34	Grt–Bt–Pl–Sil–Ilm–Qz
35	Chl–Ms–Pl–Rt–Ilm–Qz
36	Chl–Ms–Pl–Ilm–Qz
37	Chl–Ms–St–Bt–Pl–Ilm–Qz
38	St–Bt–Pl–Sil–Rt–Qz
39	St–Bt–Pl–Sil–Ilm–Qz
40	Grt–Bt–Crd–Pl–Sil–Ilm–Qz
41	Grt–Bt–Crd–Pl–Sil–Qz
42	St–Bt–Crd–Pl–Sil–Ilm–Qz
43	Bt–Crd–Pl–Sil–Ilm–Qz
44	Bt–Crd–Pl–Sil–Qz
45	St–Bt–Pl–And–Rt–Qz
46	St–Bt–Pl–And–Ilm–Qz
47	St–Bt–Crd–Pl–And–Ilm–Qz
48	Bt–Crd–Pl–And–Ilm–Qz
49	Bt–Crd–Pl–Sil–Kfs–Qz
50	Bt–Crd–Pl–Sil–Kfs–Ilm–Qz
51	Bt–Crd–Pl–Kfs–Ilm–Qz
52	Bt–Crd–Pl–And–Kfs–Ilm–Qz

[1] Mineral abbreviations used in this text are defined in Table A.1.

rock. A simplified version of Figure 18.7A, showing the stability fields for iron–magnesium–aluminum silicates, is presented in Figures 18.7B and C. Figures 18.7B and C indicate that at high pressures the stable iron–magnesium–aluminum silicates in this rock follow the Barrovian sequence of garnet–staurolite–kyanite–sillimanite (see Section 16.3.3). Andalusite and cordierite occur only at lower pressures and, as noted in Chapter 16, the upper pressure limit to the cordierite stability field increases with increasing temperature.

Metamorphic assemblage diagrams have several advantages over ion-exchange thermobarometry. Most importantly, they allow one to recover the range of metamorphic conditions over which a rock formed even if mineral compositions have been reset during cooling or if retrogression has obliterated much of the original mineral assemblage. Furthermore, if the rock studied contains relicts of the earlier assemblages, possibly preserved as inclusions in porphyroblasts, part of the *P–T* path followed by the rock during metamorphism can be reconstructed.

Metamorphic assemblage diagrams also have some important limitations. Most importantly, the key assumption in the use of these diagrams is that the composition of

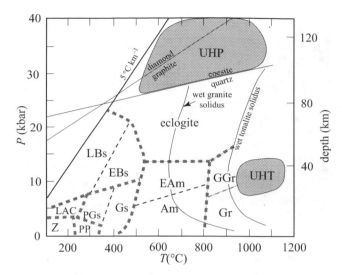

Figure 18.8 *P–T* conditions of metamorphism, showing the conditions for ultra-high-temperature (UHT) and ultra-high-pressure (UHP) metamorphism and the approximate conditions for the metamorphic facies. Am = amphibolite, EAm = epidote–amphibolite, EBs = epidote–blueschist, LBs = lawsonite–blueschist, Gr = granulite, GGr = garnet–granulite, Gs = greenschist, PGs = pumpellyite–greenschist, PP = prehnite–pumpellyite, LAC – lawsonite–albite–chlorite, Z = zeolite. Data from Ringwood (1972), Moody et al. (1983), Liou et al. (1987), Evans (1990), Green and Harley (1998), and Liou et al. (2004).

the rock has not changed during metamorphism (i.e. that the metamorphism has been isochemical). Usually this is a reasonable assumption, at least at low to medium grades where the only components that are typically lost during metamorphism are H_2O and CO_2. However, this assumption cannot be applied to very high-grade rocks that have either melted or interacted with a melt, and consequently would have followed a prograde or retrograde path that was likely to be associated with changes in composition. In rock that had obviously undergone melting, the metamorphic assemblage diagrams could only be used to estimate the maximum temperature of formation; any evidence of the metamorphic path would be lost.

The metamorphic assemblage diagrams also assume that equilibrium is attained at all times. Spear and Pattison (2017) point out that there is considerable evidence that the equilibrium conditions of a dehydration reaction in a metamorphic rock may be considerably overstepped before the reaction takes place. In such situations the *P–T* conditions implied by a metamorphic assemblage diagram would be erroneous.

18.4 Conditions of Metamorphism

The thermodynamic data extracted from experimental petrology have allowed petrologists to determine the temperatures and pressures where the various facies boundaries lie (Figure 18.8). The boundaries in Figure 18.8 are shown in dashed lines because the locations of the reactions that determine the facies boundaries are very strongly dependent on bulk rock composition (Green and Ringwood, 1972; Evans, 1990). Despite this inherent uncertainty, Figure 18.8 gives a good indication of where the various metamorphic facies lie in *P–T* space.

18.4.1 *P–T* Conditions for the Metamorphic Facies

In Figure 18.8, the low-temperature limit for metamorphism is set at the geothermal gradient of 5 °C km^{-1} since only rare examples of cold subduction have assemblages indicating metamorphism under geothermal gradients lower than this (Ravana et al., 2010). At low pressures and temperatures, metamorphism grades into diagenesis. Zeolites, particularly laumontite, are common diagenetic minerals in sandstones (Noh and Boles, 1993). In some locations, diagenesis extends to temperatures around 200 °C, well into the field of zeolite facies (Helmold and Van de Kamp, 1984). The boundary between diagenesis and metamorphism is poorly defined. As noted in Section 12.2, a reasonable distinction between diagenesis and metamorphism is that diagenesis involves reactions in the pores of a sandstone, whereas metamorphic reactions also involve the detrital grains. Clearly, the detrital grains in a sandstone will begin to react at lower temperatures if the grains are reactive, such as plagioclase or pyroxene, than if the grains are inert, like quartz.

The low-temperature limit for zeolite facies is difficult to constrain because zeolite stability is strongly dependent on the fluid composition. Zeolites may not appear at all if the fluid is too rich in CO_2. Even the presence of moderate amounts of CO_2 in the fluids causes kaolinite + calcite to form instead of laumontite (Frost, 1980). In many low-grade terrains, metasedimentary rocks contain the assemblage clay–calcite–quartz; zeolites are never encountered. Some zeolite assemblages, especially those in which analcime has replaced albite, have formed at temperatures as low as 125 °C (Miron et al., 2012). The high-temperature limits of zeolite facies are marked by the disappearance of zeolites from basaltic rocks. Frost

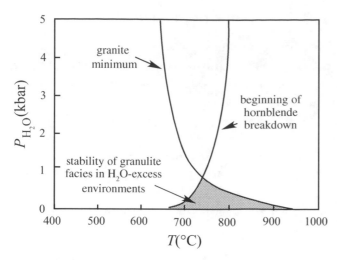

Figure 18.9 Comparison of the $P-T$ conditions for granite minimum melt (data from Luth et al., 1964) and the beginning of hornblende breakdown (data from Spear, 1981).

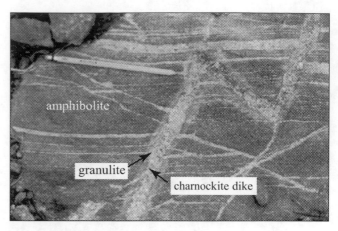

Figure 18.10 Photograph of charnockite dike in amphibolite in the Wind River Range, Wyoming. Intrusion of anhydrous charnockitic magma dehydrated the adjacent amphibolite, producing a granulite facies selvage along the margins of the dike. On a larger scale, this process may be responsible for the formation of granulite terrains.

(1980) argues these conditions coincide with the formation of the epidote–chlorite–quartz tie line (see Figure 14.7B) at temperatures around 250 °C.

Prehnite–pumpellyite facies is stable from around 250 °C to around 300–350 °C, where actinolite appears, marking the transition to greenschist facies. The high-temperature limit to greenschist facies is around 400–450 °C. The high-pressure limit of greenschist facies, where it is bounded by blueschist facies, is around 6 kbar. The high-temperature limit of blueschist facies, where, like greenschist facies, it is bounded by amphibolite facies, is somewhat higher than the high-temperature limit of greenschist facies, perhaps 450–500 °C. The high-pressure limit of blueschist facies, where it transitions to eclogite facies, is strongly temperature- and composition-dependent and ranges from 10 kbar to 20 kbar.

Amphibolite facies covers the whole range of conditions from ca. 400–500 °C to 800 °C. The cores of many metamorphic terrains are marked by areas of abundant granite, usually intermixed with various types of metamorphic rocks. These field relations indicate the high-grade areas were flooded with granitic melt. In most of these areas, hornblende has remained stable in mafic rocks and is present in many of the surrounding granites. Thus, in most environments, amphibolite facies extends into the realm of igneous crystallization. This is consistent with experimental studies (Figure 18.9), which show that when $P_{H_2O} = P_{total}$ hornblende begins to break down to pyroxenes at temperatures around 800 °C. Thermometry indicates that granulites may equilibrate at 700 °C, although Pattison et al. (2003) argue this temperature is due to resetting of

the thermometers and that the low-temperature limit to granulite metamorphism is around the 800 °C that is indicated by experimental studies. The transition from amphibolite (or epidote amphibolite) facies to eclogite facies at high pressure is rarely exposed. Figure 18.9 shows the transition occurs somewhere around 12–14 kbar, although this limit is poorly constrained.

Figure 18.9 shows that when $P_{H_2O} = P_{total}$ pyroxenes are stable with a granitic melt only at very low pressures. From this, it is evident that granulite facies requires unusually anhydrous conditions for formation. The cause of these anhydrous conditions was the subject of intense debate throughout the 1980s. Some authors maintain that granulites display evidence of pervasive CO_2 streaming from the mantle (Collerson and Fryer, 1978); others believe the dehydration is a natural consequence of crustal melting (Brown and Fyfe, 1970); while others argue that granulite metamorphism is associated with the passage of dry magmas through the lower crust (Frost and Frost, 1987; Figure 18.10). The theory of CO_2 streaming is partially constructed on the conclusion from thermobarometry that granulites form at relatively low temperatures (ca. 700 °C). If this temperature estimate is true, then something external, such as the addition of CO_2, is needed to suppress the temperature of hornblende dehydration. The observation by Pattison et al. (2003) that the lower temperature limits of granulite facies is closer to 800 °C obviates the need for anhydrous fluids to suppress amphibole stability, meaning the

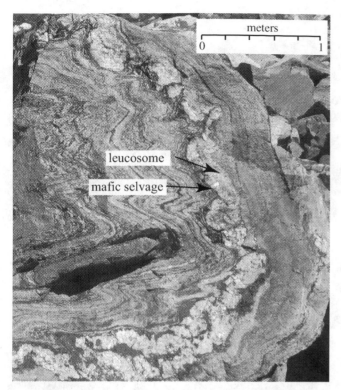

Figure 18.11 Photograph of migmatitic paragneiss, Teton Range, Wyoming. The existence of a mafic selvage around the leucosome indicates that at least some of the granitic material in the leucosome was extracted from the host rock.

theories of Brown and Fyfe (1970) and Frost and Frost (1987) are more likely to be true. This suggests that granulite-grade mineral assemblages formed under unusually hot conditions in the lower and middle crust, likely in response to igneous activity.

The lower pressure limits to eclogite facies are shown in Figure 18.8 to lie between 15 and 20 kbar, although there is a great uncertainty in this. Because mafic rocks have a wide range in composition, the lower pressure limit of eclogite facies will occur at different pressure for different mafic rocks. Depending on the bulk composition of the rock, the lower limit of eclogite facies lies between 12 kbar and 20 kbar (De Paoli et al., 2012).

18.4.2 Upper Temperature Limits to Metamorphism and Migmatites

The upper temperature limit of metamorphism lies at the wet tonalite solidus, where basaltic rocks melt to make tonalite magma. Most metamorphic terrains lie at temperatures below this, but a few record temperatures of metamorphism in excess of 1000 °C. This extreme condition has been called **ultra-high-temperature (UHT) metamorphism** and indicates heating of the lower crust

by either basaltic melts or diapiric emplacement of hot asthenosphere (Harley, 1998).

Even at temperatures well below those of high- and ultra-high-temperature metamorphism, rocks will begin to melt. In fact, most rocks in UHT terrains are refractory rocks from which granitic melt has been extracted. These rocks are called *restites*. Most melting takes place under conditions of upper amphibolite facies where rocks contain water. Subjection of high-temperature metasedimentary rocks, mostly biotite-rich psammitic and pelitic schists, to decompression can produce considerable magma through dehydration melting (see Figure 11.1). If the granitic melt is segregated into large bodies it could produce leucogranites, such as the Himalayan granite discussed in Chapter 11. If the melt is not efficiently segregated from the restite, melting produces a **migmatite**. A migmatite is a "mixed rock," which contains granite, or granitic gneiss, interlayered with a more mafic gneiss (Figure 18.11). The granitic layers are called **leucosome** (for light layer) and the more mafic layers are called **melanosome** (for dark layer) (Mehnert, 1968). The leucosome usually is granitic, since this rock melts at the lowest temperature, but it may range in composition from granite to tonalite. The melanosome may range from pelitic, wherein it will contain biotite and aluminum-rich silicates such as garnet, cordierite, or sillimanite, to psammitic, wherein it will contain mainly biotite as the ferromagnesian silicate, although hornblende and garnet may be present.

A key question when interpreting a migmatite is whether the rock represents a closed system, in which the leucosome and melanosome separated from the protolith through melting or metamorphic segregation, or the leucosome represents granitic material injected from an external source. In most migmatite localities this is difficult to determine because no sample of the unmelted protolith survives. In some localities a melting reaction that relates the leucosome (i.e. the melt) with the melanosome (i.e. the restite) is apparent (Milord et al., 2001), but in most migmatites the reaction relationship is unclear. In many occurrences of migmatite it is difficult to tell whether the leucosome was extracted from the melanosome, even through the use of chemical analyses. One helpful feature is the presence of a **mafic selvage**, a concentration of mafic minerals (mostly biotite and hornblende) on the contact between the melanosome and the leucosome, such as is shown in Figure 18.11. The presence of a mafic selvage indicates that at least some portion of the leucosome was derived from the melanosome, leaving a selvage of mafic minerals behind.

BOX 18.1 | THE DYNAMIC EARTH

Since the plate tectonic revolution in the 1970s geologists have known that Earth is a highly dynamic planet. All students of introductory geology learn that crust is continually consumed at subduction zones and replenished at mid-ocean ridges. More recently, metamorphic petrology has shown just how dynamic the process of subduction and continental collision is. The first hint was the discovery of coesite in whiteschists from the western Alps (Chopin, 1984). Coesite is a high-pressure polymorph of quartz, which is only stable at pressures of more than 25 kbar (Figure 18.8). Coesite is now known in many high-pressure terrains from the Caledonian orogeny in Norway (Janak et al., 2012), in various orogenic belts in China (Han et al., 2011; Liu et al., 2011), and in the Himalaya (Mukherjee and Sachan, 2009). The presence of coesite-bearing eclogites means portions of mountain belts must have been exhumed from depths of nearly 100 km. Similar high pressures are reported from eclogites on the Zermatt–Saas ophiolite (Bucher and Grapes, 2009), but these are trivial compared to the possible 60-kbar pressures estimated for garnet peridotite inclusions in the Alps (Nimis and Trommsdorff, 2001). Together, these petrologic observations indicate that crustal material can be carried to depth of 100 km or more and can be exhumed rapidly enough that the mineral record of these conditions is preserved.

18.4.3 Upper Pressure Limit of Metamorphism

The upper pressure limit on metamorphism is mostly determined by the preservation of **ultra-high-pressure (UHP)** rocks. UHP rocks almost certainly represent fragments of continental crust that were subducted to mantle depths during continental collisions and were tectonically exhumed because of the inherent buoyancy of continental crust (Liou et al., 2004). Fragments of rocks subjected to pressures within the coesite stability field are present in many orogenic belts. Most of these UHP rocks were metamorphosed at pressures of 20–40 kbar, but some rocks preserve evidence of pressures of around 60 kbar (Nimis and Trommsdorff, 2001). The presence of rocks evincing UHP metamorphism is a dramatic indication of the dynamic nature of Earth's crust. Without the presence of rocks containing relict coesite there would be little indication that crustal rocks could be subducted to depths of 120 km or more and then return to the surface to tell their stories (see Box 18.1).

Summary

· ·

- *Thermobarometry* involves the use of thermodynamics to determine the temperatures and pressures at which mineral assemblages last equilibrated.

- *Geothermometers* are reactions that involve very small changes in volume over a range of temperatures. Good thermometers include

  ◦ Ion-exchange thermometers, in which ions such as Fe^{2+} and Mg^{2+} are exchanged between two minerals

  ◦ Solvus thermometers

- *Geobarometers* are reactions that involve large volume changes over a range of pressures.

- Combining a thermometer and a barometer should give the temperature and pressure at which an assemblage crystallized, unless the ion-exchange thermometer re-equilibrated on cooling.

- *Metamorphic assemblage diagrams*, also called pseudosections, show the calculated mineral assemblages found in a rock of a fixed composition over a range of temperatures and pressures.

- Calculations using thermodynamic data, determined from experimental petrology (thermobarometry and metamorphic assemblage diagrams), indicate that the temperature conditions of metamorphism ranges from around 200 °C to more than 900 °C. They also show that whereas most metamorphism occurs at pressures less than 8 kbar, some high-pressure rocks may have formed at pressures of 25 kbar or more.

Questions and Problems

Problem 18.1. How does a modern approach to determining metamorphic conditions compare to Eskola's metamorphic facies approach? What are the advantages and disadvantages of each?

Problem 18.2. A garnet from the Teton Range, Wyoming, has the following composition: 3.78% MgO, 34.95% FeO, 1.70% MnO, and 1.01% CaO.

a. What are the mole fractions of pyrope (Pyp), almandine (Alm), spessartine, and grossular in this garnet.

b. How valid is it to assume that for this garnet $X_{Fe} = X_{Fe}/(X_{Fe} + X_{Mg})$?

(Hint: The gram formula weights are: MgO = 40.3314, FeO = 71.8494, MnO = 70.9394, and CaO = 56.0794)

Problem 18.3. What kinds of reactions make a good barometer or thermometer?

Problem 18.4. What is the difference between mass-transfer and ion-exchange reactions?

Problem 18.5. Referring to Figure 18.4, determine whether ΔG for the reaction An $\rightleftharpoons$ Grs + Ky + Qz is positive or negative at 600 °C and 10 kbar.

Problem 18.6. Consider a rock with the composition of the average Archean shale and the assemblage quartz–biotite–garnet–sillimanite–ilmenite–cordierite–staurolite. The staurolite occurs only as inclusions in garnet and the cordierite occurs only as rims on the sillimanite and garnet.

a. Using Figure 18.7, determine what likely P–T path this rock followed.

b. What assumptions are required to produce this answer?

Further Reading

Berman, R. G., 1991, Thermobarometry using multi-equilibrium calculations: A new technique, with petrological applications. *Canadian Mineralogist*, 29, 833–55.

Clark, C., Fitzsimmons, I. C. W., Healy, D., and Harley, S. L., 2011, How does the continental crust get really hot? *Elements*, 7, 235–40.

Gilotti, J. A., 2013, The realm of ultrahigh-pressure metamorphism. *Elements*, 9, 255–60.

Powell, R. and Holland, T. J. B., 2008. On thermobarometry. *Journal of Metamorphic Geology*, 26, 155–79.

Sawyer, E. W., Cesare, B., and Brown, M., 2011, When the continental crust melts. *Elements*, 7, 229–34.

Spear, F. S. and Pattison, D. R. M., 2017, The implications of overstepping for metamorphic assemblage diagrams (MADs). *Chemical Geology*, 457, 38–46.

Chapter 19

Regional Occurrence and Tectonic Significance of Metamorphosed Rocks

19.1 Introduction

Over the past four decades, metamorphic petrology has become a significant tool for unraveling the thermal history of mountain belts and other tectonic environments. The conditions of metamorphism of a given sequence of rocks provide important clues about the pressures and temperatures at which these rocks formed, which in turn provide insights into the tectonic processes responsible for metamorphism. There are five major tectonic environments for regional metamorphism: (1) within continent–continent collisions; (2) along convergent plate margins; (3) within rifting terrains; (4) on the sea floor; and (5) in deep basins where burial metamorphism occurs. In addition, we recognize one environment where the tectonic causes are variable or debatable: (6) metamorphism in Precambrian shields. This chapter summarizes metamorphism in these environments.

19.2 Metamorphism in Continental Collisions

Regional metamorphism was first studied by George Barrow (1893) in northeastern Scotland (see Map 16.1). It was here that the Barrovian sequence of minerals that appear in pelitic rocks during prograde metamorphism was first identified. Petrologists now recognize that Barrovian metamorphism is diagnostic of continental collisions (Figure 19.1). The heat required for the metamorphism is largely geothermal heat that was produced by burial of rocks to depths of 20 km or more, although some heat may be conducted by the emplacement of leucogranites. The high pressure of the Barrovian sequence indicates a significant amount of the crust had to be eroded off to expose them, but a more precise understanding of their tectonic evolution didn't emerge until the 1980s.

Thermal modeling shows that during continent–continent collisions rocks should follow a clockwise pressure–temperature–time (P–T–t) path (England and Thompson, 1984; Thompson and England, 1984) (Figure 19.2). Such a path can be divided into three stages, which are marked on Figure 19.2. First, continental overthrusting increases pressure without much increase in temperature, producing blueschist- and possibly eclogite-facies metamorphism. The second stage involves nearly isobaric heating as the deep rocks slowly

adjust to the insulating effect of the overlying thrust pile. This high-temperature, high-pressure (Barrovian) metamorphism overprints the blueschist and eclogite metamorphism. The third stage involves relatively rapid uplift of the rocks as a result of erosion or tectonic exhumation. Some heating may continue early in this stage, but as the rocks get closer to the surface the cooling rate increases until any fluids in the rocks are consumed by retrogressive reactions. Once the fluids have been consumed, metamorphic reactions will cease. The uplift during this stage may be recorded in some rocks as decompressive reactions (such as sillimanite after kyanite and cordierite after garnet) or as minor retrogression. However, most of the cooling path may not be recorded in the rocks at all. The time dimension of the P–T–t path can be monitored through the use of petrochronology (see Chapter 5) of metamorphic minerals such as zircon, monazite, titanite, rutile, or apatite (Kohn, 2016).

The concept of the clockwise P–T–t path introduces a critical question: how should petrologists interpret the P–T conditions recorded in Barrovian metamorphism? Historically, geologists have interpreted that the conditions recorded in rocks across an orogenic belt, as in Scotland (Map 16.1), indicated the geothermal gradient

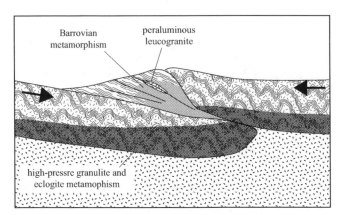

Figure 19.1 Sketch of metamorphic conditions encountered in continental collisions. The metasedimentary rocks that lie between the two plates will undergo prograde metamorphism, involving high pressures and high temperatures (Barrovian metamorphism). Deeply eroded exposures of the upper plate may contain high-pressure granulite and even eclogite-facies assemblages.

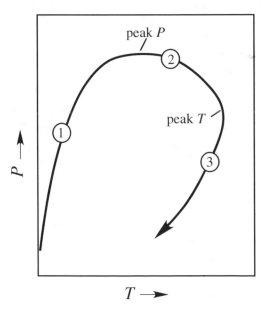

Figure 19.2 P–T diagram showing the clockwise path followed in a metamorphic terrain during continental overthrusting. Three stages in the path are labeled: 1 = rapid burial by thrusting, 2 = heating, 3 = uplift due to interaction between erosion or tectonic exhumation and isostasy (after England and Thompson, 1984).

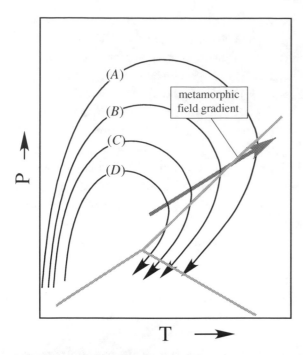

Figure 19.3 A comparison between the assemblages recorded in metamorphic terrain (the metamorphic field gradient) and the P–T path followed by the rocks during metamorphism. A, B, C, D = localities in a metamorphic belt in order of decreasing grade, for example A = sillimanite, B = kyanite, C = staurolite, and D = garnet zone. Heavy gray line is the metamorphic field gradient, light gray lines show the aluminosilicate phase diagram.

that was present at the time of metamorphism. However, if, as shown in Figure 19.2, rocks in a continental collision follow a clockwise P–T–t path, the maximum pressure is likely reached before the maximum temperature. As noted earlier, metamorphic reactions likely cease once the rocks start cooling after reaching maximum temperature because, during cooling, water is depleted from the rock and is no longer available to flux metamorphic reactions. Thus, the P–T conditions recorded in a rock are probably those of the maximum temperature, even though relicts of earlier, higher-pressure assemblages might survive or minor retrograde and decompressive effects might be present.

In a metamorphic belt that records a series of metamorphic assemblages, each assemblage would have followed a different P–T–t path, as shown in Figure 19.3 for locations A–D. A gradient defined by the highest temperature in each path (i.e. the temperature where metamorphic reactions would have ceased) defines a P–T path that none of the rocks in the terrain actually followed. The curve connecting the maximum

temperature conditions recorded across a terrain is now known as the **metamorphic field gradient**. As shown in Figure 19.3, the metamorphic field gradient may outline the maximum temperature recorded at each locality across a terrain, but it does not reflect a fossil geothermal gradient.

19.2.1 Examples of Continental Collisions

Barrovian metamorphism is recognized in most collisional orogens. As noted above, the type locality is the Caledonide belt of Scotland and Ireland (Map 16.1) and its continuation north into Norway and south into New England (which has since been displaced 4000 km to the west; Map 16.2). In the Caledonides most of the early high-pressure assemblages have been obliterated by the later high-temperature metamorphism. Evidence for relict blueschist metamorphism survives in only a few areas of low-grade metamorphism in Vermont (Laird and Albee, 1981) and in Ireland (Gray and Yardley, 1979).

High-pressure granulite metamorphism or even eclogite facies may be found on the hanging wall of continental collisions in deep orogens, such as the 1.1-Ga Grenville belt in Quebec (Rivers et al., 2002) and the Silurian–Devonian Scandian orogeny in Norway (Gee et al., 2012).

One prominent example of a continental collision is in the Alps, where the early high-pressure metamorphism has been extensively preserved (Map 19.1). The Alps formed when Europe subducted beneath the Apulian plate, a microplate, consisting mostly of Italy, which was trapped between the European and African plates. A series of ophiolites represent the remnants of the Tethyan sea floor that lay between the two continents. Structural relations for the construction of the Alps are beautifully exposed on the Matterhorn, where the upper portions of the mountain consist of continental crust of the Apulian plate, whereas the base of the mountain contains rocks of the Zermatt–Saas ophiolite (Figure 19.4). The Zermatt–Saas ophiolite contains metagabbros that locally record eclogite-facies metamorphism at 600 °C and record pressures up to 25 kbar (Bucher and Grapes, 2009), indicating that the rocks had been subducted to depths of more than 90 km!

It has long been known that the Alps have undergone two distinct metamorphic events, an earlier Eoalpine event and a later Alpine metamorphic event. The

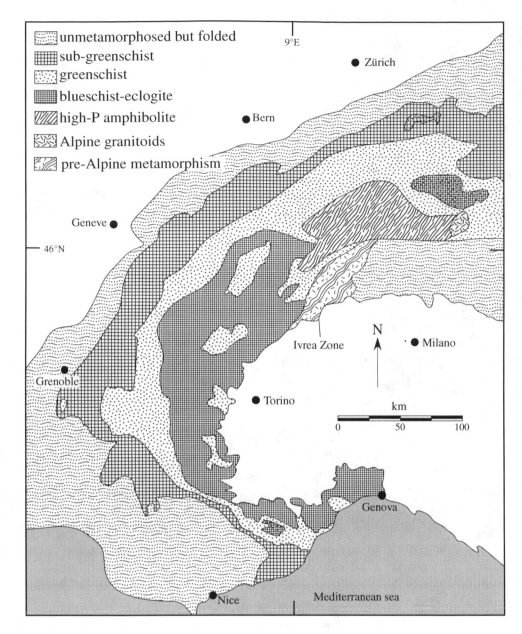

unmetamorphosed but folded
sub-greenschist
greenschist
blueschist-eclogite
high-P amphibolite
Alpine granitoids
pre-Alpine metamorphism

9°E

● Zürich

● Bern

Geneve ●

— 46°N

N

Ivrea Zone

· ● Milano

Grenoble

● Torino

km

0   50   100

Genova

● Nice   Mediterranean sea

Map 19.1 Map of rocks recording the Alpine metamorphism in the western Alps. Modified after Frey et al. (1974).

recognition that collisional orogenies follow a clockwise P–T–t path helps geologists to understand the relation between these events. The Eoalpine event was caused by the overthrusting of Apulian plate on the European plate (condition 1 in Figure 19.2). This event is recorded by an increase in metamorphic grade from zeolite facies to high-pressure blueschist or eclogite facies along an eastward line, from Grenoble toward Turino (Map 19.1). The Alpine metamorphic event is a later Barrovian metamorphic event, which occurs in a rather restricted area (Map 19.1), produced by tectonic emplacement of hot, high-pressure rocks to shallower levels, late in the Alpine orogeny (Berger et al., 2011).

19.3 Metamorphism along Convergent Plate Margins

Metamorphism occurs in four environments in convergent margins: (1) in the subduction zone, (2) in island arcs, (3) in accretionary wedges, and (4) in continental arcs.

19.3.1 Subduction-Zone Metamorphism

The geothermal gradient is suppressed over subduction zones due to the emplacement of cold oceanic crust into the mantle. For this reason, the rocks in subduction zones experience high-pressure–low-temperature

Matterhorn 4476 m

Dent Blanche nappe
(continental crust of
the Apulian plate)

Combin zone
(Mesozoic accretionary prism)

Zermatt–Saas ophilite nappe
(oceanic lithosphere of
the Tethys Ocean)

Figure 19.4 Structure of the Alpine orogeny as exemplified by exposures on the Matterhorn.

metamorphism, diagnostic of blueschist facies. As a slab of crust is subducted, the rocks, which may have originally been metamorphosed at zeolite or prehnite–pumpellyite facies, undergo large increases in pressure and moderate increases in temperature. This results in the formation of blueschist and, eventually, eclogite facies with increasing depth. These transitions are shown on Figure 19.5 as gradational because the exact depth at which these transitions occur is a function of the temperature of the down-going slab as well as the bulk composition of the rocks in the slab. If the ocean floor was old enough to allow hydration to penetrate through the crust, the underlying mantle may be partially serpentinized. During subduction, these serpentinites will undergo high-pressure dehydration, such as those from the Betic Cordillera in Spain described by Padrón-Navarta et al. (2011).

If subduction continued, the high-pressure rocks would never be exposed. However, if the plate orientation changed and subduction in the area stopped, the forces pulling the slab down would disappear and buoyant forces would allow portions of the slab to rise and be tectonically emplaced into the upper crust. In areas where subduction-zone metamorphism is exposed, such as in

the Sanbagawa belt of Japan (Map 19.2) the metamorphic grade increases from zeolite or prehnite–pumpellyite facies on the side nearest the ocean toward blueschist, and, locally, even eclogite facies on the side nearest the continent. Similar relations are shown in the Franciscan belt of California and the Shuksan belt in the North Cascades.

19.3.2 Metamorphism in Island Arcs

One might expect that metamorphism is not common in island arcs because most island arcs consist of unmetamorphosed volcanic rocks, but that is because only the tops of most arcs are exposed. Some arcs have been tectonically tilted so that deeper levels that have undergone amphibolite metamorphism are exposed (for example the island of Tobago, Section 8.2.4). A few arcs, such as the Kohistan arc in Pakistan and the Talkeetna arc in Alaska have been tectonically thrust onto continents; the lower portions of these arcs record granulite-facies metamorphism (Miller and Snoke, 2009).

19.3.3 Metamorphism in Accretionary Prisms

Accretionary prisms, the sedimentary rocks that have accumulated adjacent to the subduction zone (Figure 19.5), have usually been subjected to low grades of metamorphism, ranging from zeolite to prehnite–pumpellyite facies, although higher grades are seen locally. In fact, zeolite and prehnite–pumpellyite facies were defined in the accretionary prism in Otago, New Zealand (Coombs, 1954). One of the most well-known examples of an accretionary prism is the Franciscan Formation that lies along the western margin of California. The Franciscan is an impure metamorphosed sandstone to mudstone that contains tectonic blocks of chert, limestone, serpentinite, blueschist, and eclogite. This tectonic mixture of rocks types is known as a **mélange** and is postulated to have formed when exotic blocks from the hanging wall of the trench were tectonically incorporated into the downgoing accretionary prism. The matrix of the Franciscan Formation increases in metamorphic grade from zeolite facies to blueschist facies on moving eastward from the coast to the eastern margin of the formation (Coleman and Lanphere, 1971).

19.3.4 Metamorphism in Continental Arcs

As with oceanic arcs, continental arcs are mostly made up of volcanic rocks. Like oceanic arcs, continental arcs are

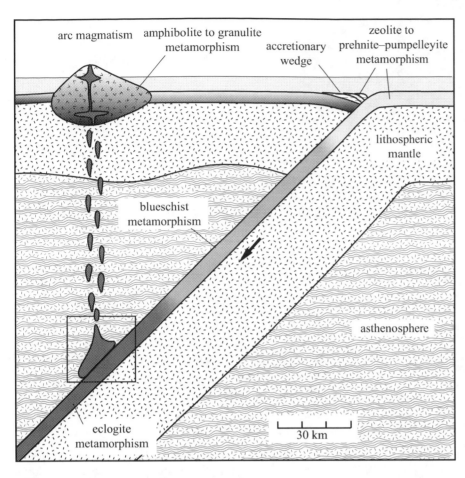

Figure 19.5 Metamorphic environments in island arcs. Sedimentary rocks in the accretionary wedge usually record zeolite to prehnite–pumpellyite-facies metamorphism. Rocks of the oceanic crust contain assemblages ranging from zeolite through prehnite–pumpellyite and, if subducted, blueschist facies to eclogite facies. Rocks in the deeper portions of the island arc may experience conditions equivalent to amphibolite to granulite facies, but they may not appear as metamorphic rocks unless they have been deformed.

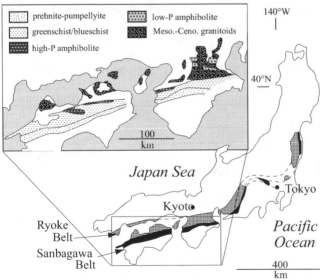

Map 19.2 Metamorphic belts from the island of Honshu, Japan. Inset shows the increase in metamorphic grade in the blueschist belt toward the continent. Modified after Miyashiro (1961) and Hashimoto et al. (1970).

characterized by areas of high heat flow due to the emplacement of large amounts of magma. The movement of this heat into the crust becomes a driving force for metamorphism (Fig. 19.6). Not uncommonly, the plutons that are part of the continental arc have roof pendants and contact aureoles that record low-pressure amphibolite metamorphism (Figure 19.6, area 1). With increasing depth, this gives way to high-pressure amphibolite facies and, if the exposure is deep enough, to granulite facies. As noted in Section 18.4.1, granulite facies is indicative of both high temperature and low water activity. Granulites associated with continental arcs may form in several ways. One process is by melting of the country rock, which extracts H_2O with granitic melt, leaving behind a dehydrated restite (Figure 19.6, area 2). Another way results from the fact that silicate melts are hygroscopic (see Chapter 3). This means that wherever possible they absorb H_2O from the surrounding rocks, dehydrating them to produce granulites. In addition, silicate melts absorb CO_2 at high pressures but will degas it with decreasing pressure. Thus, as a silicate melt moves through the crust, particularly at deep crustal levels, it will dehydrate the surrounding crust by emitting CO_2 and absorbing H_2O (Figure 19.6, area 3). The water will be carried by the migrating melt to shallow crustal levels where it will be exsolved when the melt crystallizes (Frost and Frost, 1987). Deeply exposed igneous rocks in arcs

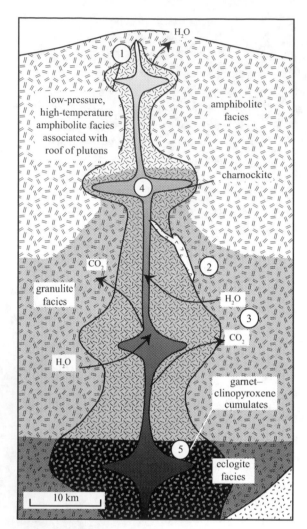

Figure 19.6 Metamorphic environments in continental arcs. (1) Low-pressure amphibolite-facies metamorphism in contact aureoles around plutons. (2) Granulite facies associated with deep melting of crustal rocks. (3) Granulite metamorphism due to extraction of H_2O and emission of CO_2 from melt. (4) Crystallization of pyroxene-bearing granitoid (i.e. charnockite), which if deformed becomes felsic granulite. (5) Deposition of garnet–clinopyroxene cumulates, which if deformed become eclogite. Modified from Frost and Frost (1987).

may contain pyroxenes instead of (or in addition to) biotite and hornblende and are known as *charnockite*. The orthopyroxene in charnockite indicates that the rock, though igneous, crystallized under the hot and dry conditions of granulite facies. If a charnockite recrystallizes and loses its igneous texture, it will become a felsic granulite (Figure 19.6, area 4). In the root of the arc, the magmas crystallize garnet–clinopyroxene cumulates, assemblages indicative of eclogite facies (Figure 19.6, area 5) (Saleeby et al., 2003).

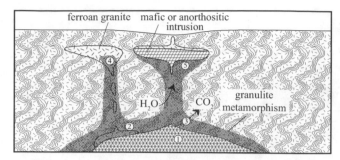

Figure 19.7 Sketch showing the metamorphic environments in continental rifts: (1) recrystallization of underplated gabbro; (2) partial melting of lower crust; (3) dehydration of lower crust by CO_2 exsolved from the melt; (4) formation of (and later recrystallization of) charnockite; and (5) contact metamorphism at the base of mafic intrusions. Modified from Frost and Frost (1987).

The higher-grade metamorphism is rarely preserved in arcs except where their roots have been exposed by later tectonism. One example is the Sierra Nevada batholith, which was formed in the Cretaceous Period. The southern portion subsequently tilted to expose deeper levels of the arc. In the north, the plutons of the batholith intrude coeval volcanic rocks and pressure is calculated to have been less than 2 kbar. In the south, the rocks contain granulite facies assemblages and pressure was 8–10 kbar (Saleeby et al., 2007). Much of the dense, eclogitic, cumulate rocks of the Sierra Nevada batholith, such as that pictured in Figure 19.6, area 5, was removed by delamination in the Miocene to Pliocene epochs, but their existence is recorded in garnet pyroxenite xenoliths, brought up by Miocene volcanic rocks (Saleeby et al., 2003).

19.4 Metamorphism in Rifting Terrains

One of the major revolutions in structural geology in the 1970s and 1980s was the recognition that extension plays a major role in the development of many tectonic belts. This observation was accompanied by the realization that high-grade metamorphism can be associated with rifting environments. In areas undergoing active extension, the thinning of the crust, the emplacement of hot asthenosphere at the base of the crust, and the intrusion of magmas into the crust produce high-grade metamorphic environments (Figure 19.7). Because the crust has been thinned in these environments, low-pressure (i.e. pressures of 5–6 kbar) granulites can form in such a manner (Wickham and Oxburgh, 1985).

Five mechanisms of granulite metamorphism can be recognized in rifting environments, many of which formed by processes that are similar to those that form granulites in arcs (Figure 19.7).

1. The gabbroic rocks that have crystallized at the base of the crust contain the assemblage clinopyroxene–orthopyroxene–plagioclase ($\pm$ olivine). If they undergo recrystallization that destroys the igneous texture, they become mafic granulites without ever having a hydrated protolith.

2. Melting of the crust around mafic rocks dehydrates the country rock as H_2O is extracted into the melt. Movement of these melts to shallower depth leaves behind a dehydrated source region.

3. Injection of CO_2 from the melt into the country rock and absorption of H_2O lowers water activity and produces granulites.

4. The lower portions of granitic plutons are pyroxene-bearing (i.e. a charnockite). If these rocks are recrystallized and lose their igneous textures, they will become felsic granulites.

5. Rocks at the base of mafic intrusions are likely to undergo granulite- or pyroxene–hornfels-facies metamorphism. In shallow intrusions, the metamorphic aureole is likely to be only on the scale of a kilometer or so and this is true contact metamorphism. However, in deep intrusions, such as around the Rogaland Anorthosite Province in Norway, the granulite aureole is on the order of 19 km wide (Westphal et al., 2003) and is more properly considered a type of regional metamorphism than contact metamorphism.

Rifting necessarily precedes the opening of any ocean, and hence is an early stage of the mountain-building cycle. When the ocean basin closes again, the high-grade metamorphic rocks produced by the earlier rifting event may be emplaced into a metamorphic and structural belt, formed through convergent tectonics. The Ivrea zone in the Alps (see Map 19.1 for location) preserves a record of high-grade metamorphism that formed during rifting. The metamorphic assemblages in the Ivera zone formed in response to the intrusion of mafic magmas at the base of the crust during the Permian opening of the Tethyan ocean (the ocean that separated Africa from Europe during the Mesozoic). The Ivrea zone was exposed in the Tertiary Period as the African continent collided with Europe to make the Alps. Similarly, the Proterozoic granulites of Norway were tectonically emplaced onto the Scandinavian shield during the Caledonian orogeny (Bingen et al., 2001).

19.5 Sea-Floor Metamorphism

The basalts emplaced at the mid-ocean ridges during sea-floor spreading are clearly out of equilibrium with the ocean water that surrounds them. Through time, the basalts hydrate in reactions with this water. These reactions are the driving force for sea-floor metamorphism (Figure 19.8A). Gabbros that are newly emplaced into the crust beneath the sea floor are hot. If they are deformed by extensional faults while still hot, they will recrystallize as mafic granulites (Figure 19.8B). As the crust moves away from the spreading center and cools, ocean water will percolate into the rocks, hydrating them, and producing sea-floor metamorphism that ranges in grade from zeolite facies to amphibolite facies. These metamorphic zones are postulated to move downward through the ocean crust as the crust ages and cools (Figure 19.8C).

Unfortunately, the sea floor is covered by several kilometers of water, limiting the view of the processes that occur there. Indeed, until the 1970s the concept of sea-floor metamorphism was essentially unknown. However, aided by a large number of Ocean Drilling Program (ODP) and Integrated ODP (IODP) drill holes and the study of ophiolites, which, as discussed in Chapter 6 are considered fragments of sea-floor thrust on to the continental crust, petrologists over the past few decades have outlined a general model of the processes governing sea-floor metamorphism.

For a few million years after emplacement of fresh basaltic magma, the new ocean crust lies above the thermal anomaly that is marked by the spreading center. This hot area acts as a driving force for the circulation of hot seawater through fractures in the ocean crust. This circulation has a significant effect on the rocks on the sea floor. Not only do the rocks alter by hydration reactions, they also undergo significant metasomatism. Magnesium and sodium are added and transition metals (iron, manganese, copper, and zinc) are removed. These transition metals may be reprecipitated on the sea floor as "black smokers," sulfide-rich hydrothermal deposits (McCaig et al., 2007) (see Box 7.2).

The effect of seawater circulation through the ocean crust may be recorded in ophiolites. Where complete

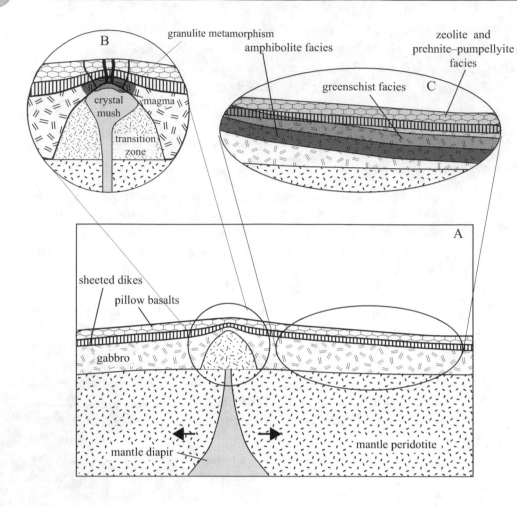

Figure 19.8 (A) Environments of sea-floor metamorphism. (B) Environment near the rift showing where granulite-facies metamorphism may form during slow rifting. (C) The metamorphic facies tend to migrate to deeper portions of the crust as the crust migrates away from the rift.

sections of ophiolites are exposed, a general gradient in metamorphism is observed through the section. The upper portion of the ophiolites, occupied by pillow basalts, is highly altered. The degree of alteration varies from ophiolite to ophiolite; the metamorphic grade typically is prehnite–pumpellyite; more rarely, zeolite facies is found. The influence of circulating fluid is indicated by the common occurrence of veins. Not uncommonly, these rocks record a range of metamorphic conditions; for example, host rocks containing prehnite–pumpellyite-facies assemblages may be cut by zeolite-facies veins. This relation documents the cooling of the circulating fluids as the crust migrates away from the mid-ocean ridge heat source. Further down in the ophiolite section, generally in the sheeted dike horizon, the metamorphic grade increases to greenschist facies. In the gabbroic section, the metamorphism is seen as late hornblende growth, indicative of amphibolite facies, although commonly the lower portions of the gabbros are unaltered.

The same sequence of metamorphic assemblages has been documented in cores from holes drilled into fast-spreading oceanic crust. The deep IODP hole 1256D, for example, contains clay as the major alteration mineral in the upper portion of the hole (Teagle et al., 2011). Somewhere between 1000 and 1200 m below the sea floor, the clay gives way first to greenschist assemblages and then to amphibolite facies a few hundred meters lower (Figure 19.9). The alteration occurs mostly in veins rather than pervasively, as is typical in ophiolites. The crust penetrated by hole 1256D is only 15 million years old. As the crust ages, one would expect that the alteration would become more pervasive.

Two other deep drill holes in the ocean crust (735B and 1309D) were drilled into slow-spreading crust, where deep oceanic crust was tectonically exhumed during spreading (Dick et al., 2000; Blackman et al., 2011). Hole 735B comes from crust that is 12 million years old; the crust in hole 1309D is only about 2 million years old. These drill holes exhibit a different style of metamorphism than the drill hole in fast-spreading oceanic crust. The alteration pattern in these holes is very complex (Figure 19.9). In both of them, the assemblages occur as

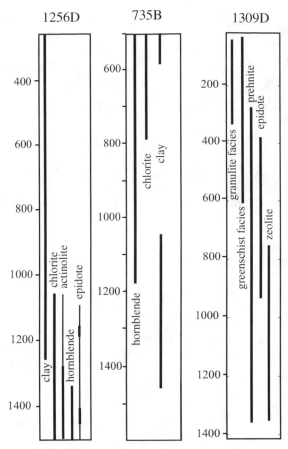

Figure 19.9 Metamorphic alteration in IODP holes 1256D, 745B, 1309D. Locations for these holes are given in Map 6.2. Sources of data as in Figure 7.7.

isolated veins in otherwise relatively fresh rock. In both holes, high-grade assemblages, such as granulite in 1309D, occur in localized shear zones high in the hole, and low-grade assemblages, such as clay and zeolites, are more prevalent toward the bottom. This inversion reflects the fact that the different metamorphic assemblages occurred at different times. The formation of the oceanic core complexes into which both hole 735B and 1309D were drilled involved deformation while the crust was still hot. The granulite-facies assemblages seen locally in the upper portion of 1309D formed during these extensional events. The rocks in the lower part of the core were also hot at this time, but they escaped the deformation that was localized in the upper part of the core and hence did not develop metamorphic textures, indicative of granulite metamorphism. After tectonic exhumation, the upper portion of the crust cooled rapidly. The low-grade metamorphism documented in the lower portions of holes 735B and 1309D are assemblages produced by reactions that were taking place at the time the holes were drilled.

Most ophiolites appear to be fragments of young crust thrust onto the continents within a few million years after they formed. No ophiolites have been identified that are composed of oceanic crust that was more than 100 million years old at the time it was obducted. It is likely that the circulation of seawater through the ocean crust continues throughout the life of that crust, with the water penetrating progressively deeper as the crust cools and moves away from the spreading center (Figure 19.8C). If this is true, metamorphic zones should migrate downward through the crust with time, with lower amphibolite-facies rocks (i.e. serpentinites) possibly forming the upper mantle in the oldest regions of the sea floor. Such a correlation of increased degree of metamorphism with age of oceanic crust is suggested by the magnetic anomalies in Cretaceous oceanic crust, which are best explained by having a contribution from magnetite in serpentinized (or partially serpentinized) mantle (Dyment et al., 1997).

19.6 Burial Metamorphism

Metasedimentary rocks or basalt flows that have been buried deeply in a basin may undergo metamorphism due to the thermal gradient within the basin. This type of metamorphism commonly is associated with very little deformation and is called burial metamorphism. For sedimentary rocks, burial metamorphism is an up-grade extension of diagenesis. For example, shales collected from depths of 2–5 km in drill holes adjacent to the Gulf of Mexico show the beginning of metamorphism with the occurrence of chlorite and illite (Land et al., 1997). Likewise, sandstone from Lower Cretaceous rocks that were buried to depths of 6–9 km in the Great Valley Sequence in California contain abundant laumontite, indicative of zeolite facies (Dickinson et al., 1969). Basalts in the North Shore Volcanic Group in Minnesota record assemblages ranging from zeolite facies to greenschist facies, all due to burial metamorphism (Schmidt, 1993).

19.7 Metamorphism in Archean Terrains

Precambrian shields are large terrains composed mostly of metamorphic rocks that form the cores of all continents. In many areas, the Archean shield rocks are surrounded by Proterozoic orogenic belts. The tectonic and metamorphic style of the Proterozoic Eon is generally similar to that of the Phanerozoic, but that of the Archean Eon is markedly different. A good example is in the

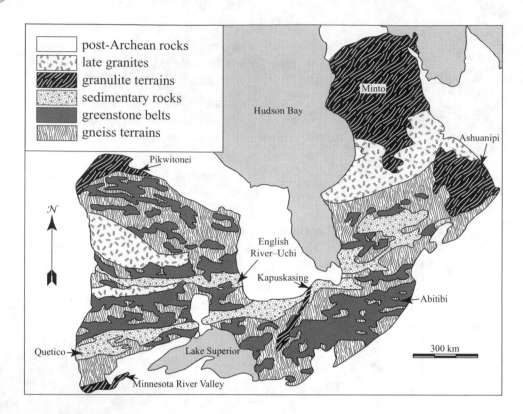

Map 19.3 Distribution of terrains within the Superior province, with the granulite terrains labeled. After Thurston (2002).

Legend:
- post-Archean rocks
- late granites
- granulite terrains
- sedimentary rocks
- greenstone belts
- gneiss terrains

Superior province, the largest Archean craton in the world (Map 19.3). The Superior province consists of four types of metamorphic terrains: (1) granulite terrains, (2) greenstone belts, (3) gneiss terrains, and (4) metasedimentary belts. As noted in Sections 19.3.4 and 19.4, the granulite terrains represent either deep crustal rocks that have been tectonically exhumed or terrains of hot and dry magmatism. The rest of the province contains individual blocks of greenschist-facies metabasalts and related rocks (greenstone belts), interspersed within quartzo-feldspathic gneiss. The greenstone belts are separated by east–west trending belts of supracrustal rocks (Map 19.3).

The Superior province contains five major granulite terrains (Map 19.3). The Pikwitonei and Kapuskasing are crustal cross-sections that were formed at 1.8 Ga during continental collision between the Superior craton and cratons to the north and east. Metamorphic pressures of the Pikwitonei increase from 3 kbar to 9 kbar on moving northwestward across the block (Mezger et al., 1990). Similarly, the granulites in the Kapuskasing block record 9-kbar pressures, which is distinct from the 4–6-kbar pressures in amphibolite-facies rocks to the west of the block (Percival and McGrath, 1986). In contrast, the Minto and Ashuanipi blocks both show pressures of 6–6.5 kbar and contain a mixture of granulite-facies metamorphic rocks and charnockitic plutons (Percival,

1991; Percival et al., 1992), indicating that the granulites in these terrains formed during intrusion of hot, dry magmas. Metamorphism in the Minnesota River is also relatively low pressure (4.5–7.5 kbar), suggesting granulites here too were produced by magmatic processes (Moecher et al., 1986).

It is possible that the low-grade Archean metamorphic terrains formed by plate tectonic processes similar to those occurring in Earth today. However, because the nature of Archean tectonism is still a matter of considerable debate, this chapter considers the metamorphism of Archean terrains to be a separate topic.

The elongate metasedimentary belts, which run for thousands of kilometers across the province, appear to represent sedimentary basins that formed between the various greenstone–gneiss blocks. The metasedimentary rocks are immature lithic-rich sandstones, which appear to have come from a volcanic source region (Percival, 1989). They appear similar to sedimentary rocks found in modern accretionary prisms, which are composed of marine sediment, scraped off the downgoing oceanic plate at convergent margins. However, the metamorphic conditions and the pattern of metamorphism of the metasedimentary belts in the Superior province do not match the conditions found in modern accretionary prisms. Both the Quetico and the English River–Uchi belts in the Superior province were metamorphosed at low pressures (Thurston and Breaks, 1978; Percival, 1989). Pressure is generally 3–5 kbar, although pressure in the Quetico belt gets as high as 6 kbar near the Kapuskasing uplift. Andalusite and cordierite are common throughout; kyanite is only found locally. Temperature increases from the margins of the belts, where it may be as low as 450 °C, toward the middle, where it is more than 700 °C, and locally it reaches granulite conditions (Thurston and Breaks, 1978; Percival, 1989). Percival (1989) suggests that the Quetico belt was an accretionary prism trapped in a dextral transpressional boundary between two continental blocks. and that thickening of the sedimentary package in the core of the belt produced melting. Movement of these melts to a shallow level brought the heat necessary for the high-grade metamorphism.

The continental blocks on either side of the metasedimentary belts consist of irregularly shaped greenstone belts within gneiss terrains. These two rock types are present in most exposed Archean cratons and there is considerable debate as to their origin.

19.7.1 Greenstone Belts

Greenstone belts are sequences of relatively weakly metamorphosed supracrustal rocks. As the name implies, greenstone belts generally contain green-colored amphiboles, typically hosted in greenschist- to amphibolite-facies mafic volcanic rocks. In addition, most greenstone belts also contain significant amounts of immature sediments, mostly greywacke (i.e. lithic arenite), with minor amounts of iron formation. Within any given greenstone belt, the proportion of volcanic rocks relative to the sedimentary rocks tends to decrease with stratigraphic height. In addition, at higher stratigraphic levels, the volcanic rocks tend to become more felsic and the sedimentary rocks incorporate more detritus from continental sources. Greenstone belts, such as the Abitibi in Quebec and Ontario (Map 19.3), are economically important because they host large deposits of base and precious metals.

Greenstone belts generally occur as isolated and irregularly shaped synformal basins, infolded between granitic batholiths or domes of gneiss. Classic examples of greenstone belts occur in the Pilbara craton of northwestern Australia where, because of the lack of vegetation in the area, the structure is evident in satellite images (Figure 19.10). The metamorphic grade in greenstone

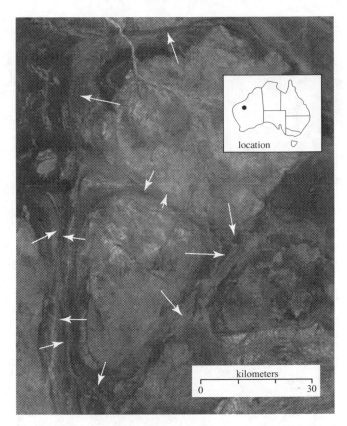

Figure 19.10 Google Earth image of gneiss terrains (light) and greenstone terrains (dark) in the Pilbara craton of northwestern Australia. Arrows indicate the dip direction of the foliations in the gneiss and in the greenstone. Structural trends after Van Kranendonk et al. (2002).

belts, for example in the Abitibi (Jolly, 1978) and in the greenstone belts in the Yilgarn craton of Australia (Archibald et al., 1978), tends to increase on moving outward from the core of the synforms (Map 19.4). The lowest grade recorded is often greenschist facies, although prehnite–pumpellyite facies is found in some greenstone belts. The margins of the belts are usually in amphibolite facies. Pressure is moderate to low, commonly around 3–4 kbar; certainly, no blueschists have been found in the Precambrian Era.

19.7.2 Gneiss Terrains

Gneiss terrains commonly surround greenstone belts. These terrains consist dominantly of quartzo-feldspathic gneisses, the composition of which ranges from tonalite to granite. Fragments of metasedimentary rocks are commonly interfolded in these gneisses. These metasedimentary rocks usually have a mature composition,

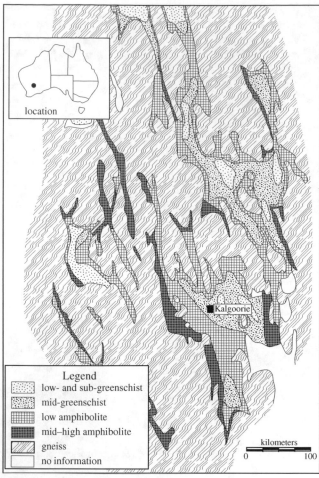

Map 19.4 Distribution of metamorphic facies in the Eastern Goldfields subprovince of the Yilgarn Archean province. After Binns et al. (1976).

19.7.3 Tectonic Interpretation of Archean Metamorphic Belts

There are some similarities between the greenstone belt–gneiss belt metamorphic suite and Phanerozoic paired metamorphic belts. Both greenstone belts and blueschist terrains are dominated by mafic volcanic rocks of sea-floor affinity. However, greenstone belts lack any evidence of high-pressure metamorphism. Furthermore, the metamorphic grade increases in both directions outward from the core of the synforms that characterize greenstone belts, rather than being unidirectional in fossil subduction zones (compare Map 19.1 to Map 19.4). In addition, many greenstone belts form roughly circular synformal bodies, as opposed to the markedly linear nature of blueschist terrains, and tend to grade toward volcanic and sedimentary rocks of a continental affinity at their tops. For this reason, skeptics maintain that, if greenstone–gneiss terrains represent paired metamorphic belts, then plate tectonics operated on different scales and resulted in different thermal regimes in the Archean Eon than it does today.

Smithies et al. (2003) postulate that the Archean greenstone–gneiss terrains formed by melting of over-thickened oceanic plateaus in a manner unlike any tectonic process found on Earth today. In their theory, plate tectonics did not operate on the early Earth and ocean-floor basalts were erupted from relatively fixed vents, producing thick accumulations of basalt. Because the Archean Earth had a higher heat flow than modern Earth, the base of these basalt plateaus eventually became hot enough to melt, producing a tonalitic magma. Magma intruded the basalt pile and left remnants as deep synforms between the tonalitic plutons. This theory of formation explains the shapes of the greenstone belts and the pattern of metamorphic assemblages they show. If the tonalitic magmatism produced both plutons and extrusive rocks, it would also explain why the greenstone belts include more felsic lavas in the higher stratigraphic levels.

with metapelitic rocks, quartzites, marbles, and iron formations usually represented. Metamorphic grade is typically amphibolite and, unlike the greenstone belts, where some changes in metamorphic grade are recognized on a regional scale, many gneiss belts have a uniform metamorphic grade over many thousands of square miles.

Summary

Regional metamorphism occurs in five tectonic environments:

- In continent–continent collisions

- Along convergent plate margins

- In rifting terrains

- On the sea floor

- In deep basins where burial metamorphism occurs

Continental collisions are characterized by a clockwise *P–T–t* path that involves:

- Low-temperature, high-pressure metamorphism during continental collision

- Increasing temperature at high pressure as the thermal structure of the crust adjusts, producing Barrovian metamorphism

- Decreasing pressure at high temperature as the metamorphic belt is unroofed by erosion or tectonic processes

- Granulite and eclogite facies may occur on the hanging wall of deeply eroded continental collision belts

Convergent margins may contain five environments of metamorphism:

- *Subduction zones*, where one may find metamorphism, ranging from prehnite–pumpellyite, through blueschist, rarely to eclogite facies

- *Arcs*, both oceanic and continental

  - Shallow levels of arcs will contain low-pressure, high-temperature metamorphism around plutons

  - Arcs that have been tectonically tilted can display amphibolite, granulite, and, in continental arcs, eclogite facies

- *Accretionary prisms*, which may display zeolite to prehnite–pumpellyite facies and, locally, even blueschist facies

Continental rifts where granulite-facies metamorphism form by several mechanisms:

- Underplated gabbros if deformed will become granulites

- The base of the crust may undergo granulite-facies metamorphism due to melting or heat and fluids from the underplating magma

- Crystallization and later deformation of pyroxene-bearing felsic rocks

- Metamorphism from rifting must be exposed by later tectonic processes

- *Sea-floor metamorphism* occurs as the fresh oceanic crust hydrates in reactions with the ocean floor; this hydration continues as long as the ocean crust is in contact with seawater and may reach into the mantle in old crust

- *Archean terrains* contain two types of metamorphic belts not common in Phanerozoic environments

- *Greenstone terrains*: Greenstone belts are tightly folded synformal structures that usually expose weakly metamorphosed rocks (prehnite–pumpellyite or zeolite facies) in the cores of the folds and amphibolite facies on the margins

- *Gneiss terrains*: Gneiss terrains surround the greenstone belts and contain orthogneiss with enclaves of supracrustal rocks in amphibolite facies

Questions and Problems

Problem 19.1. Consider rocks with the following assemblages. Minerals are listed in decreasing abundance. What was the likely protolith for each rock? What kinds of conclusions can be drawn about the tectonic environments in which the rocks formed?

a. quartz–plagioclase–biotite–muscovite–andalusite

b. plagioclase–orthopyroxene–clinopyroxene–quartz–garnet

c. quartz–muscovite–chlorite–garnet–glaucophane

d. quartz–plagioclase–biotite–muscovite–kyanite–garnet

e. albite–chlorite–prehnite–pumpellyite–quartz

Problem 19.2. In what ways are greenstone belts similar to blueschist terrains and in what way are they different? What evidence suggests it might be a mistake to equate the two?

Problem 19.3. What are the requirements for the formation of a granulite terrain? In what geologic environments might granulite terrains form?

Problem 19.4. In what way is sea-floor metamorphism different from other types of metamorphism? What features would you look for in a metamorphic terrain to determine if the metamorphism formed on the sea floor?

Problem 19.5. What characteristics would you look for in a suite of metamorphic rocks to determine if they underwent burial metamorphism?

Further Reading

Card, K. D., 1990, A review of the Superior province on the Canadian Shield, a product of Archean accretion. *Precambrian Research*, 48, 99–156.

Kohn, M. J., 2014, Himalayan metamorphism and its tectonic implications. *Annual Review of Earth and Planetary Sciences*, 42, 381–419.

Miyashiro, A., 1975, *Metamorphism and Metamorphic Belts.* John Wiley & Sons, New York, NY, pp. 327–425.

Appendix: Review of Mineralogy

A.1 Introduction

An important skill necessary to the mastery of igneous and metamorphic petrology is the ability to recognize minerals in hand sample and thin section. Most mineralogy texts follow the system pioneered by James Dana in his 1848 book *The Manual of Mineralogy* whereby minerals are discussed in order of increasing chemical complexity. In this scheme, native elements are discussed first, then oxides, then carbonates, and finally silicates. This appendix reviews minerals in a different order, one that emphasizes the major rock-forming minerals found in igneous and metamorphic rocks. First, the leucocratic minerals that dominate rocks from gabbro through granite and most metamorphic rocks are described: quartz, feldspars, feldspathoids, and carbonates. Next, ferromagnesian minerals are reviewed, then the minerals common in metamorphic rocks, including the aluminum-rich minerals and calcium–aluminum silicates. Finally, the reader will find a description of minerals that occur in minor abundance in most rocks: opaque minerals and accessory minerals.

A.2 Leucocratic Rock-Forming Minerals

A.2.1 Quartz

Quartz (SiO_2) is the most common mineral in Earth's crust. As such it is common in many igneous and metamorphic rocks. It is usually found as anhedral grains that make up the framework of the rock. The hexagonal crystal shape that is diagnostic in drusy quartz that crystallizes on rock surfaces is rarely exhibited when quartz crystallizes in a rock, although euhedral quartz may be an early crystallizing phase in some siliceous volcanic rocks or in some quartz veins.

Hand-Sample and Optical Properties. Quartz in hand sample has a gray color, greasy luster, conchoidal fracture, and lack of cleavage. It is identified in thin section by its uniaxial (+) optic sign, its low birefringence, and its low relief. Other features that readily identify quartz in thin section include: (1) undulatory extinction, (2) lack of alteration, and (3) common occurrence of fluid inclusions.

(1) *Undulatory extinction*: At high temperatures (above ca. 350 °C), quartz is very easily deformed by crystal-plastic flow. Any quartz-bearing rock that is even slightly deformed above these temperatures will show the effects of this deformation by development of undulatory extinction. This means extinction sweeps across a grain as the microscope stage is rotated, instead of the whole grain going to extinction at the same time.

(2) *Lack of alteration*: Other low-relief, low-birefringence minerals that can be confused with quartz (feldspars, nepheline, or cordierite) alter easily at low temperatures. Even a small amount of alteration will make these crystals cloudy in plane-polarized light. In contrast, quartz is particularly resistant to alteration and remains clear.

(3) *Fluid inclusions*: Because silica is highly soluble at high temperatures, fractures in quartz (even on a micron scale) tend to be quickly healed by re-precipitation of silica. Often, during this healing process, micro-sized bubbles of the ambient fluid are left behind as fluid inclusions. In many rocks, trails of fluid inclusions cutting the quartz are common features. In some rocks the inclusions are so fine that, except at the highest magnification, they look like dust. Fluid inclusions are rare in feldspars or nepheline, so any low-relief mineral with fluid inclusions is likely quartz. Beware, however, apatite, which typically forms only tiny grains, may trap fluid inclusions, and can look like quartz. However, apatite has a much higher relief than quartz.

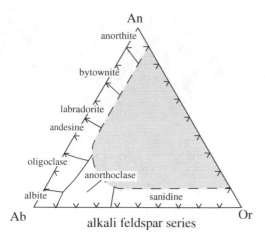

Figure A.1 The system Ab–An–Or, showing the names of the various feldspars. Boundary within the ternary diagram is dashed because the location is temperature-dependent. Modified from Dyar et al. (2008). Mineral abbreviations as on Table A.1.

A.2.2 Feldspars and Feldspathoids

The feldspars are the most common mineral group in the crust. Few indeed are the rocks that do not contain a feldspar; many common rocks contain two. The feldspar group consists of three end members – albite (Ab, NaAlSi$_3$O$_8$), anorthite, (An, CaAl$_2$Si$_2$O$_8$), and orthoclase (Or, KAlSi$_3$O$_8$) – arranged into two solid solutions, plagioclase (albite–anorthite) and alkali feldspars (albite–orthoclase) (Figure A.1). Only a limited amount of calcium can substitute for potassium so any rock with a bulk composition that plots in the shaded portion of Figure A.1 contains two feldspars, both a plagioclase and an alkali feldspar.

In hand sample, the feldspars are white, gray, pink, or black in color, distinguishable from quartz, which is glassy and gray. In distinguishing the feldspars from one another, it is helpful to look for twinning and exsolution. Plagioclase feldspar characteristically shows albite twins, which appear as striations parallel to the (010) plane. K-feldspar (sanidine, orthoclase, or microcline) shows only the Carlsbad twin, which forms a single twin, usually in the middle of a feldspar lath. Microcline in slowly cooled rocks may contain fine, usually irregularly shaped, exsolution blebs of plagioclase. These exsolved alkali feldspars are called *perthite*. When the plagioclase dominates in an exsolved feldspar, it is called *antiperthite*.

Table A.1 *Abbreviations for minerals*

Mineral	Abbreviation	Mineral	Abbreviation
Actinolite	Act	Lawsonite	Lws
Albite	Ab	Leucite	Lct
Almandine	Alm	Lizardite	Lz
Andalusite	And	Magnesite	Mgs
Andradite	Adr	Magnetite	Mag
Anorthite	An	Muscovite	Ms
Anthophyllite	Ath	Nepheline	Nph
Antigorite	Atg	Olivine	Ol
Augite	Aug	Orthoclase	Or
Biotite	Bt	Orthopyroxene	Opx
Brucite	Brc	Paragonite	Opx
Calcite	Cal	Periclase	Per
Chrysotile	Ctl	Phlogopite	Phl
Chlorite	Chl	Pigeonite	Pgt
Chloritoid	Cld	Plagioclase	Pl
Clinozoisite	Czo	Prehnite	Prh
Cordierite	Crd	Pumpellyite	Pmp
Diopside	Di	Pyrite	Py
Dolomite	Dol	Pyrope	Prp
Enstatite	En	Pyroxene	Px
Epidote	Ep	Pyrrhotite	Po
Fayalite	Fa	Quartz	Qz
Ferrosilite	Fs	Rutile	Rt
Forsterite	Fo	Sillimanite	Sil
Garnet	Grt	Spessertine	Sps
Glaucophane	Gln	Sphalerite	Sp
Grossular	Grs	Spinel	Spl
Grunerite	Gru	Staurolite	St
Hedenbergite	Hd	Talc	Tlc
Hornblende	Hbl	Titanite	Ttn
Hypersthene	Hyp	Tremolite	Tr
Ilmenite	Ilm	Water vapor	V
K-feldspar	Kfs	Wollastonite	Wo
Kyanite	Ky		

Plagioclase Series

The plagioclase series is one of the most important groups of igneous minerals; almost all igneous rocks contain plagioclase of some composition. Plagioclase is a solid solution between albite Ab and An. (These and other mineral abbreviations are listed in Table A.1.). The solution series has been broken into compositional ranges, each of which has been assigned a mineral name (Figure A.1). Apart from labradorite, which locally shows iridescence, most of these names are nearly obsolete because the microprobe allows petrologists to determine the exact composition of plagioclase. Thus, instead of using mineral names, modern petrologists designate plagioclase composition by the mole fraction (i.e. mole percent) of An or Ab. For example, a plagioclase with the composition $(Na_{0.4}Ca_{0.6}Al_{1.6}Si_{2.4}O_8)$ can be expressed either as An_{60} or Ab_{40}.

Calcic plagioclase tends to be found in mafic rocks, such as basalt and gabbro, whereas sodic plagioclase is more common in granodiorite and granite. In igneous rocks it is common for plagioclase to be tabular, elongate along the c-axis, and to show euhedral zoning (see Figure 1.8C). If the core of the plagioclase is more calcic than the rim, then the zoning is called *normal zoning* (because this is the pattern one would expect to form during cooling of an igneous melt). If the core is more sodic than the rim, the zoning is called *reverse zoning*. Igneous plagioclase may show alternating normal and reverse zoning; such zoning is called *oscillatory zoning*.

The diagnostic feature of plagioclase, both in hand sample and thin section, is its twinning (Figure A.2). The most common twin in plagioclase is the albite twin; this has the (010) plane for its twin plane. In plagioclase the albite twin is *polysynthetic*, which means that usually multiple albite twins are present in a given plagioclase crystal. Plagioclase may also show Carlsbad and pericline twins. Carlsbad twins have the same twin plane as albite twins but a feldspar crystal can have only one Carlsbad twin. Pericline twins are twinned along a plane that is oblique to c and a but parallel to b. When viewed down the a-axis, pericline twins are parallel to the 001 cleavage plane. In plagioclase, pericline twins are not polysynthetic.

Plagioclase in metamorphic rocks, unlike igneous plagioclase, is generally equant; it is never tabular. Tabular plagioclase forms when it grows unconstrained in a melt. Thus the existence of tabular plagioclase in a

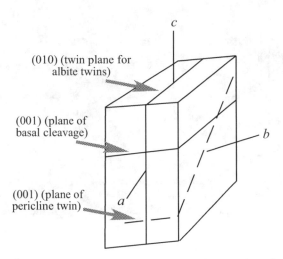

Figure A.2 Crystallographic sketch of plagioclase showing the orientation of albite twins (010), the basal cleavage (001), and pericline twins (which form on a rhombic section).

metamorphic rock is a clear indication that the rock had an igneous protolith. Metamorphic plagioclase may be untwinned and it is rarely zoned.

Optical Properties. Plagioclase of all compositions has similar optical properties. All are triclinic, have a low birefringence (usually first-order gray), and have diagnostic tabular twinning on (010) (albite twinning). Compositions from albite to anorthite are accompanied by changes in relief and changes of optic orientation:

(1) *Changes in relief*: Albite and oligoclase have a low negative relief, andesine and labradorite have a low positive relief, and bytownite and anorthite have a moderate positive relief.

(2) *Changes of optic orientation*: The optic orientation of plagioclase changes as a function of composition, which means plagioclase grains with different compositions will have different extinction angles. This is important because it provides a technique for estimating plagioclase composition in thin section called the a-normal method.

The a-Normal Method: This way to estimate plagioclase composition under the microscope requires that you find a plagioclase crystal oriented normal to the a-axis. In these crystals, the extinction angle will help identify the plagioclase composition. The easiest approach to finding an appropriately oriented crystal is to look for twin planes that are razor sharp and for crystals in which the twins on either side of the twin plane have the same birefringence

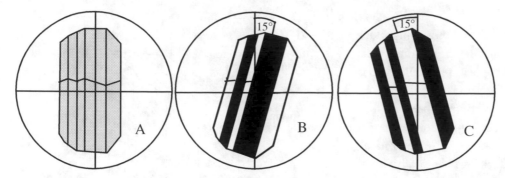

Figure A.3 Figure showing what a plagioclase grain would look like when oriented normal to *a*. Both (010) cleavage and the albite twin plane are razor sharp and both sets of twins show the same birefringence (A). The twins will go to extinction with the same amount of rotation clockwise (B) and counter-clockwise (C). Using the chart shown on Figure A.4, the composition of this plagioclase can be identified as An_{34}.

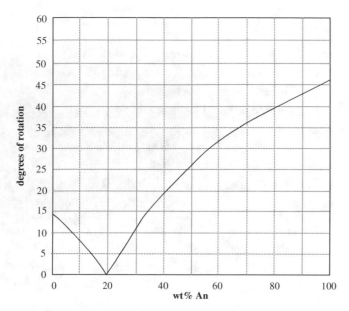

Figure A.4 Graph showing the relation between plagioclase composition and rotation angle to extinction when looking down the *a*-axis. From Nesse (1991).

(Figure A.3A). However, crystals with these characteristics can be oriented at any angle normal to the (010) plane. To be certain of a crystal orientation normal to the *a*-axis, the petrographer must identify a section perpendicular to (001). Fortunately, plagioclase has two features that lie along (001), the (001) cleavage and the pericline twin (narrow twins that lie at high angles to the albite twins). A section in which the albite twin *and* the (001) cleavage or the pericline twin plane both are razor sharp is a section oriented normal to *a*.

Another property of the section normal to (010) is that the alternating twins will show symmetrical extinction; one set will go to extinction with a certain rotation clockwise and the other set will go to extinction with the same rotation counter-clockwise (Figures A.3B and C). To determine the anorthite content of plagioclase in a thin section, the petrographer finds a grain that shows the proper orientation and then measures the angle of extinction clockwise and counter-clockwise for the twin lamellae. They should agree to within 5°. If the results are not exactly the same (but within 5°), then average them and, using Figure A.4, read off the anorthite composition. There is one slight complication with Figure A.4 – for rotations of 0–20° there are two possible compositions. For example, the plagioclase with an extinction angle of 10° could have the composition of either An_8 or An_{28}. The correct value can be determined from the relief of the plagioclase against quartz or epoxy. Remember that plagioclase with $An_{<20°}$ has a low negative relief, while plagioclase with $An_{>20°}$ has a positive relief. Thus, if the plagioclase in question has an extinction angle of 10° and a negative relief it is An_8; if it has a positive relief it is An_{28}.

Because extinction angle changes with composition, one can easily determine whether plagioclase in a rock is zoned. Unzoned plagioclase will have uniform extinction, such as the plagioclase shown in Figure A.3. In zoned plagioclase, extinction will sweep across the grain, and in oscillatory-zoned plagioclase the zoning will be seen as light and dark bands running parallel to the margins of the grain (see Figure 1.8C).

Alkali Feldspars

Alkali feldspars are those that are solid solutions of $(Na, K)AlSi_3O_8$ and $KAlSi_3O_8$ with minor amounts of $CaAl_2Si_2O_8$ (Figure A.5). In igneous rocks the phase relations among the alkali feldspars depend on the water content and cooling rate of the magma. Under conditions of low water pressures, felsic magmas solidify at temperatures higher than the top of the solvus between Ab and Or

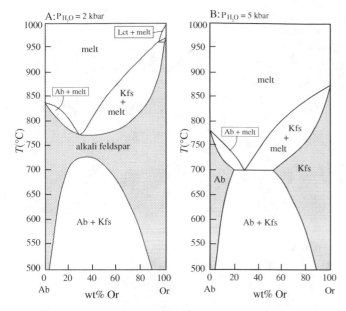

Figure A.5 (A) Phase diagram for the system albite–orthoclase at low (2 kbar) water pressure (Bowen and Tuttle, 1950). At these conditions the liquidus and solidus lie above the solvus and a single alkali feldspar crystallizes. At high temperature and very K-feldspar-rich compositions, leucite is present. The small leucite field is eliminated at >2.5 kbar. (B) Phase diagram for the system albite–orthoclase at high (5 kbar) water pressure (Morse, 1970). At these conditions the temperature of the liquidus is depressed to intersect the solvus, and two alkali feldspars crystallize.

(Figure A.5A). In these *hypersolvus* granites only a single feldspar crystallizes. Igneous melts that contain abundant water, fluorine, or boron may remain molten until temperatures below those of the alkali feldspar solvus (Figure A.5B). In such *subsolvus* granites, two alkali feldspars – an albite-rich feldspar and an orthoclase-rich feldspar – crystallize directly from the melt. In slowly cooled rocks, the feldspars re-equilibrate along the solvus, exsolving albite from K-feldspar to produce perthite.

K-feldspar. There are three polymorphs of K-feldspar ($KAlSi_3O_8$): *sanidine*, *orthoclase*, and *microcline*. The differences between them are determined by the extent to which aluminum and silicon are ordered on the tetrahedral site. Sanidine, the disordered alkali feldspar, is found in rapidly cooled volcanic rocks. Orthoclase may occur in some volcanic rocks or in relatively quickly cooled plutonic rocks. The ordered feldspar, microcline, occurs in slowly cooled plutonic rocks.

Optical Properties. All K-feldspars have a low negative relief (relief is always lower than coexisting albite) and a low birefringence, with maximum colors of first-order gray.

Microcline is completely ordered and is triclinic. Like the triclinic plagioclase feldspars, it shows abundant twinning. Carlsbad, albite, and pericline twins are all present, although only the albite and pericline twins occur as multiple twins. Unlike plagioclase, where albite twins are dominant and pericline twins are rare, the albite and pericline twins in microcline are present in equal abundance. Furthermore, they are repeated on a fine scale, giving microcline a pattern of cross-hatched or "tartan" twinning under crossed polarized light. Because microcline occurs in slowly cooled rocks, it is usually nearly pure $KAlSi_3O_8$. The sodium expelled from the structure commonly forms inclusions with the microcline. Microcline containing exsolution blobs of albite is perthitic. This exsolution is commonly seen in thin section, and in some rocks it is coarse enough to be recognized in hand sample.

Orthoclase and sanidine are progressively more disordered and are monoclinic. Because they are monoclinic, they cannot show albite and pericline twins; only Carlsbad twins may be present. Both sanidine and orthoclase are biaxial (–); the 2V is a function of degree of ordering. 2V ranges from 40° to 70° for orthoclase; K-feldspar with a 2V less than 40° is called sanidine. Orthoclase may show exsolution of albite, but sanidine never does.

Sanidine, being a high-temperature, disordered form of K-feldspar, is found mainly in volcanic rocks and usually contains considerable amounts of dissolved $NaAlSi_3O_8$. It is recognized by its lack of twinning and low 2V.

Anorthoclase. Anorthoclase, $(Na,K)AlSi_3O_8$, is a high-temperature alkali feldspar that, like sanidine, occurs only in volcanic rocks. Indeed, one might consider it a sodic variant of sanidine (Figure A.1). Most samples of anorthoclase also contain more than 5 mole percent $CaAl_2Si_2O_8$. Like plagioclase, anorthoclase is triclinic, but unlike plagioclase, the pericline twins are equally as abundant as albite twins, much as they are in microcline. However, unlike microcline, the twins in anorthoclase are repeated on an extremely fine scale, giving the anorthoclase a felted appearance in thin section. It is also easy to distinguish anorthoclase from microcline because anorthoclase only occurs in volcanic rocks and microcline is found only in plutonic rocks.

Feldspathoids

Feldspathoids are minerals that are characteristic of igneous rocks which are so poor in silica that quartz is not present and feldspars are destabilized. These rocks will contain nepheline ($NaAlSiO_4$) or sodalite ($Na_8Al_6Si_6O_{24}Cl_2$) in place of or in addition to albite ($NaAlSi_3O_8$) and perhaps leucite ($KAlSi_2O_6$) in place of K-feldspar. They are very rarely found in metamorphic rocks.

Nepheline. Nepheline is the most common feldspathoid. It substitutes for albite in silica-undersaturated, sodic rocks. In hand sample it looks similar to quartz, being greasy gray with no cleavage. With a hardness of 5.5 to 6, it is softer than quartz. In thin section it looks like quartz and untwinned feldspars, with a low birefringence and a low negative relief. Unlike feldspars it is uniaxial. Nepheline can be distinguished from quartz by the fact that it is uniaxial (–) whereas quartz is uniaxial (+). Nepheline, like feldspars, is easily altered. One common alteration product is *cancrinite*, a complex hydrated tectosilicate, similar in many respects to zeolites. Cancrinite is pink in hand sample. In thin section it is colorless with a moderate first-order birefringence. Alteration of nepheline to cancrinite is a diagnostic feature in hand sample and thin section.

Leucite. Leucite, the potassic feldspathoid, is not nearly as abundant as nepheline. Leucite is a common mineral in potassic, alkaline magmas. At high temperatures, leucite is cubic and crystallizes into diagnostic hexoctahedral phenocrysts. On cooling, the structure inverts to tetrahedral. This is accommodated by the formation of twins in the original hexoctahedra. Leucite has such a low birefringence that it looks isotropic; sometimes the twins can only be seen by inserting the first-order plate. This is because the eyes are much more sensitive to minor changes in blue than minor changes in black.

Sodalite. Sodalite is chemically equivalent to 6 nepheline + 2 NaCl, which indicates that it is likely to be found in alkaline intrusions that are also rich in chlorine. In a few localities, sodalite is deep blue in hand sample. Because those localities have supplied geology labs for years, many people assume this is a diagnostic feature of sodalite. However, sodalite can assume a wide variety of colors from bluish gray to green to yellow. In some rocks, sodalite has a distinct cubic shape; in others, it is anhedral. In thin section it is identified by its isotropic nature and a moderate negative relief. The only mineral with similar properties is fluorite, but fluorite is characterized by octahedral cleavage.

A.2.3 Carbonates

In addition to all the silicates discussed previously, the other important framework-forming minerals found in

Table A.2 *Carbonate minerals commonly found in metamorphic rocks*

Mineral	Composition
Calcite	$CaCO_3$
Dolomite–ankerite	$CaMg(CO_3)_2$–$CaFe(CO_3)_2$
Magnesite	$Mg(CO_3)$
Siderite	$Fe(CO_3)$
Aragonite	$CaCO_3$

metamorphic (and much more rarely in igneous) rocks are the carbonates. Carbonates form the framework of metamorphic rocks derived from some metasedimentary hosts, but there are also igneous rocks, known as carbonatites, that also are composed mostly of carbonates. There is a large number of carbonates, but only those listed in Table A.2 are commonly rock-forming minerals. Of these, the first four have similar crystal structure and optical properties. They form in the hexagonal system and have distinct rhombohedral cleavage. All are uniaxial (–) and have an extreme birefringence and thus are pearly white under crossed polarized light. Calcite, dolomite, and magnesite have one index of refraction, ε, that is lower than 1.53. Therefore, on rotation, the relief of those grains that have the *c*-axis near the plane of the thin section goes from low negative to high positive. On rotation of the stage, therefore, the relief changes dramatically. This phenomenon is known as "twinkling."

There is no easy way to distinguish among these carbonate minerals in thin section. This is a problem because calcite and dolomite commonly occur together in metamorphosed limestones and dolomites; calcite, dolomite, and magnesite may coexist in carbonatized metaperidotites; and calcite, ankerite, and siderite may occur together in iron formations. Numerous staining techniques are used to distinguish the carbonates.

Here are a few ways by which calcite, dolomite, magnesite, and siderite may be distinguished.

- Dolomite, since it may contain small amounts of iron, can be somewhat turbid or stained by iron oxide in thin section, whereas calcite will not have such stains.

- Dolomite tends to be more idioblastic than the other carbonates.

- Siderite can be distinguished from calcite by its relief, which is higher than quartz in all orientations.

Another carbonate, *aragonite*, occurs in shells, fresh deep-water sediments, and in high-pressure, low-temperature metamorphosed rocks. Superficially it looks like calcite. It has a similar extreme birefringence and it twinkles on rotation of the stage. However, aragonite is orthorhombic and therefore it lacks the rhombohedral cleavage of calcite. Furthermore, aragonite gives a biaxial figure rather than a uniaxial one.

A.3 Ferromagnesian Minerals

A.3.1 Olivine

Olivine is made up of two major end members: forsterite (Fo, Mg_2SiO_4) and fayalite (Fa, Fe_2SiO_4). Magnesium-rich olivine melts at a higher temperature than iron-rich olivine, thus forsteritic olivine is usually found in the mantle or in igneous rocks directly extracted from the mantle, whereas iron-rich olivine is found in evolved rocks such as fayalite granites or fayalite rhyolites. Magnesium-rich olivine is also found in metamorphosed magnesian rocks, such as serpentinites or impure marbles, whereas fayalite may be found in metamorphosed iron-rich rocks such as iron formations. Over most of its composition range olivine is incompatible with quartz; pyroxenes form instead of olivine and quartz. However, because the iron-rich pyroxene ferrosilite is not stable at crustal pressures, fayalite can be present along with quartz in some granites and metamorphosed iron formations. Olivine is usually a nearly pure iron–magnesium solution, with only minor amounts of other elements. Forsteritic olivine may contain as much as 1 percent nickel oxide (NiO) and fayalite olivine may contain up to several percent manganese oxide (MnO). Unlike plagioclase, olivine rarely preserves zoning.

In hand sample, the color of olivine is a rough indication of its Fe/(Fe + Mg) ratio. Pure forsterite (such as forsterite in metamorphosed impure dolomitic marbles) may be white, and increasingly iron-rich olivine may range from green to honey yellow through brown to black. Regardless of its color, olivine has a distinct glassy luster and lacks cleavage. Magnesium-rich olivine weathers to a pale reddish brown; the rock name *dunite* alludes to this color. Iron-rich varieties weather to a rusty brown surface. Olivine is orthorhombic, but because olivine crystals are usually anhedral and show no cleavage, the crystallographic system can only be recognized where olivine occurs as euhedral phenocrysts in volcanic rocks.

In thin section, olivine is distinguished by its high relief, which increases with iron content. It is colorless (though very iron-rich olivine may be slightly yellow), has a high birefringence, and lacks cleavage. Magnesium-rich olivine is recognized by its positive optic sign and very high 2V, and fayalitic olivine by its negative optic sign and a 2V of around 50–60°. Olivine is distinguished from pyroxene by its higher relief, lack of cleavage, higher birefringence (third-order compared to second-order for pyroxene) and often anhedral habit.

Olivine is easily altered and this alteration is often helpful in identifying the presence of olivine (if it hasn't obliterated the olivine entirely). Magnesian olivine alters to serpentine, which may occupy grain boundaries or cracks. More iron-rich olivine (Fo_{80} or less) will form a red-brown isotropic material known as *iddingsite*. Iddingsite is a mixture of clay, chlorite, and goethite and, like serpentine, commonly forms on grain boundaries and cracks in olivine.

A.3.2 Pyroxenes

Pyroxenes are among the most common minerals in igneous rocks, second only to the feldspar-group minerals. They are particularly important in igneous petrology because the composition of pyroxenes in a rock can tell the petrologist a lot about the crystallization conditions of that rock. Pyroxenes also occur in metamorphic rocks; in most bulk compositions they are indicative of high temperatures.

Pyroxenes have the general formula XYZ_2O_6, where:

X = six–eight coordinated cations that occupy the M2 site, which include the larger cations like Ca^{2+} and Na^+, in addition to Fe^{2+}, Mg^{2+}, or Mn^{2+};

Y = octahedrally coordinated cations that occupy the M1 site, which include Fe^{2+}, Mg^{2+}, Ti^{2+}, Al^{3+}, and Mn^{2+}; and

Z = tetrahedrally coordinated cations in the T site such as silicon or aluminum.

The pyroxenes can be divided into two major series:

1. *Quadrilateral pyroxenes*: These are pyroxenes with the general formula $(Ca,Fe,Mg)(Fe,Mg)Si_2O_6$. The quadrilateral pyroxenes are so named because their composition can be projected onto a quadrilateral with the apices $CaMgSi_2O_6$, $CaFeSi_2O_6$, $Mg_2Si_2O_6$, and $Fe_2Si_2O_6$ (Figure A.6). The majority of naturally occurring pyroxenes occur in this family.

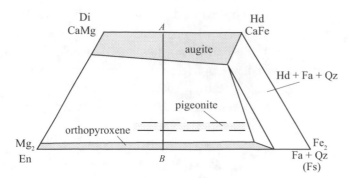

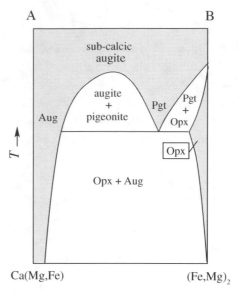

Figure A.6 Diagram showing the approximate composition ranges for the three pyroxene quadrilateral solution series. *A–B* shows the line of section displayed as a *T–X* diagram in Figure A.7. Mineral fields from Lindsley and Frost (1992).

2. *Sodic pyroxenes*: These are pyroxenes that contain sodium substituting for calcium and include relatively uncommon pyroxenes such as jadeite and aegirine.

Quadrilateral Pyroxenes

The quadrilateral pyroxenes have the following end members:

Diopside (Di)	$CaMgSi_2O_6$
Hedenbergite (Hd)	$CaFeSi_2O_6$
Enstatite (En)	$Mg_2Si_2O_6$
Ferrosilite (Fs)	$Fe_2Si_2O_6$

The quadrilateral pyroxenes can be divided into three solid-solution series (Figure A.6). Along the bottom of the pyroxene quadrilateral are the calcium-poor pyroxenes, which form a solid solution between enstatite and ferrosilite. Because all of these are orthorhombic, they are called orthopyroxenes (Opx). As noted earlier, iron-rich orthopyroxene is not stable at low pressures; fayalite is found instead.

Along the top of the quadrilateral are the calcium-rich pyroxenes, which form a solid solution between diopside and hedenbergite. Intermediate members are called augite or ferroaugite. The incorporation of the relatively large Ca^{2+} ion distorts the pyroxene lattice, hence calcium-rich pyroxenes have lower symmetry than calcium-poor pyroxenes. The calcium-rich pyroxenes are all monoclinic. In the middle of the quadrilateral is the calcium-poor clinopyroxene series, pigeonite.

Figure A.7 Schematic *T–X* diagram showing the phase relations between augite, pigeonite, and orthopyroxene along section *A–B* in Figure A.6.

To understand how the three solution series – augite (Aug), pigeonite (Pgt), and orthopyroxene (Opx) – are related, consider relations on a *T–X* slice across the quadrilateral (Figure A.7). Several features are key to understanding these minerals. One is that pigeonite is a high-temperature phase that, on cooling, inverts to a mixture of orthopyroxene and augite. The second is that augite and pigeonite have the same structure and if temperatures are high enough the two merge to form *sub-calcic augite*. The third is that the calcium content of pyroxene in the assemblages augite–pigeonite, pigeonite–orthopyroxene, and augite–orthopyroxene is a sensitive function of temperature. This makes quadrilateral pyroxenes important geothermometers.

The inversion of pigeonite to orthopyroxene + augite does not occur at a fixed temperature; rather, it is strongly dependent on iron content of the pyroxene. For the pure magnesium system (left side of Figure A.6), the inversion temperature is above the melting temperature of basaltic rocks, so pure magnesium pigeonite is not found in nature, whereas for iron-rich pyroxenes the inversion occurs around 850 °C (Figure A.8). At 800 °C and 3-kbar pigeonite is not stable at all and the assemblage orthopyroxene–augite occurs across the pyroxene quadrilateral, except for the most iron-rich compositions where olivine + augite + quartz (Qz) occur instead

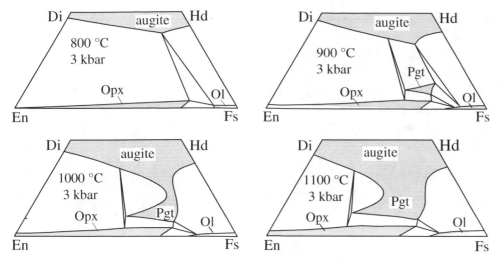

Figure A.8 Isothermal sections of the pyroxene quadrilateral at 800 °C, 900 °C, 1000 °C, and 1100 °C, showing how the pigeonite stability field expands with increasing temperature. Modified from Lindsley and Frost (1992).

(Figure A.6). Since metamorphic rocks rarely attain temperatures as high as 850 °C, this means pigeonite is mostly an igneous mineral; it is absent in all but the most iron-rich, high-temperature metamorphic rocks. At 900 °C and 3 kbar there is a small stability field for pigeonite in iron-rich compositions. By 1000 °C, the pigeonite stability field has expanded and there is a narrow composition range where pigeonite + augite react to form a single clinopyroxene, sub-calcic augite. The stability fields for both pigeonite and sub-calcic augite expand to more magnesian compositions with increasing temperature, as shown by the isothermal section at 1100°C in Figure A.8.

In hand specimen, pyroxenes may be distinguished from amphibole (almost exclusively hornblende in igneous rocks) on the basis of cleavage (90° as opposed to 120°) and on the basis of color. Whereas hornblende always appears black on the weathered surface, in magnesium-rich rocks such as gabbros, clinopyroxene is green and orthopyroxene is brown. In more iron-rich rock, both ortho- and clinopyroxenes weather brown in color. In thin section, pyroxenes are colorless with a high relief. They have a distinct (110) cleavage that, when looking down the c-axis, intersects at approximately 90° angles. Augite is monoclinic, (+) with a 2V of about 45°, and the angle between the c-axis and gamma is equal to about 45°. Pigeonite looks like augite, except it has a smaller 2V (0–36°). Both have a second-order birefringence. Orthopyroxene is distinguished from clinopyroxene by its lower birefringence (usually low first order) and parallel extinction.

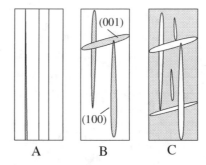

Figure A.9 Sketches of exsolution textures in pyroxenes. (A) Orthopyroxene with narrow lamellae of augite. (B) Orthopyroxene with wide lamellae of augite (inverted pigeonite). (C) Augite host with exsolution of orthopyroxene.

Pyroxenes in quickly cooled rocks, such as volcanic or hypabyssal rocks, may retain their high-temperature compositions, but pyroxenes in plutonic rocks will change composition as the rocks cool. Usually this is accommodated by exsolution. The exsolution textures produced in this process provide important keys to the composition of the original pyroxene (Figure A.9)

For example, an orthopyroxene host with fine lamellae of augite (sometimes the lamellae are so fine that they appear only as lines in thin section) indicates that at high temperature this mineral was orthopyroxene that had incorporated small amounts of CaO (Figure A.9A). If, however, an orthopyroxene contains abundant augite lamellae, then at high temperature this orthopyroxene had incorporated a lot of CaO. Exsolution usually takes place along specific crystallographic orientations. In pyroxene, the most common exsolution planes are roughly parallel to (100) or (001). Sometimes both orientations of lamellae can be seen. If, as in Figure A.9B, these orientations intersect at an angle *other* than 90°, then the exsolution took place when the mineral was a monoclinic pyroxene, rather than an orthopyroxene. In other words, this is evidence that the pyroxene formed as pigeonite, exsolved augite as it cooled, and then inverted to orthopyroxene. If the exsolved pyroxene consisted of an augite host with orthopyroxene exsolution (as in Figure A.9C), then the original pyroxene was augite with a certain amount of orthopyroxene dissolved in it. If the exsolution lamellae are particularly abundant, then the original pyroxene may have been a sub-calcic augite.

Sodic Pyroxenes

In addition to the quadrilateral pyroxenes, the other important pyroxene group is the sodic pyroxenes, which contain sodium substituting for calcium. Since sodium is univalent and calcium is divalent, this substitution must occur coupled with substitution of a trivalent ion for a divalent ion. The most common substitutions are:

$$\underset{\text{augite}}{Ca(FeMg)Si_2O_6} + NaAl \rightleftharpoons \underset{\text{jadeite}}{NaAlSi_2O_6} + Ca(Fe, Mg)$$

$$\underset{\text{augite}}{Ca(FeMg)Si_2O_6} + NaFe^{3+} \rightleftharpoons \underset{\text{aegirine}}{NaFe^{3+}Si_2O_6} + Ca(Fe, Mg)$$

The compositions of the non-quadrilateral pyroxenes can be shown on a triangular diagram (Figure A.10), in which the jadeite–aegirine solid solution is shown across the base of the triangle, and compositions with Ca^{2+}, Mg^{2+}, and Fe^{3+} in the M2 site lie closer to the apex.

Jadeite. Jadeitic pyroxenes are restricted to high pressures. With increasing pressure, the relatively open structure of albite becomes unstable relative to the denser pyroxene structure by the reaction:

$$\underset{\text{albite}}{NaAlSi_3O_8} \rightleftharpoons \underset{\text{jadeite}}{NaAlSi_2O_6} + \underset{\text{quartz}}{SiO_2}$$

Pyroxenes formed at low pressure and in equilibrium with plagioclase have a relatively low jadeite component, whereas those at increasingly higher pressures have progressively more of the jadeite component. Pyroxenes of intermediate composition between augite and jadeite are known as *omphacite*. Omphacitic pyroxenes have many properties similar to calcic pyroxenes. They have a slightly higher 2V and a pale green, non-pleochroic color. These two properties are diagnostic. Pure jadeite may be

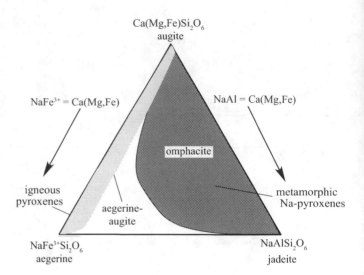

Figure A.10 Compositional range of sodic pyroxenes in igneous rocks and in metabasites. Compiled from data in Deer et al. (1978).

fibrous with a low anomalous birefringence. It may be confused with clinozoisite, which has a higher relief.

Aegirine (or Acmite). The aegirine solid solution typically occurs in igneous rocks that are enriched in sodium. If there is more sodium in a rock than can be accommodated in feldspars, the excess sodium becomes incorporated in pyroxene as the aegirine component. Compositions intermediate between augite and aegirine are known as *aegirine–augite*.

As ferric iron is added to augite the pleochroism increases from colorless in augite to deep green in aegirine. Increasing ferric iron also changes its 2V and optic sign; augite is (+) with 2V = 45°, aegirine–augite has a 2V around 90°, and aegirine is (−) with a 2V around 60–70°. Aegirine and aegirine–augite may occur with quartz in alkalic granites, but they are more commonly associated with nepheline.

A.3.3 Amphiboles

Amphiboles are double-chain silicates that contain OH as part of the structure. Because of this they are more stable at lower temperatures and consequently they are more common in metamorphic environments than in igneous ones. Amphiboles have the basic formula $AX_2Y_5Z_8O_{22}(OH)_2$, where:

A = the "vacant" site. It is so called because it is vacant in amphiboles such as tremolite, anthophyllite, and glaucophane. It is a large eight-fold site that can accommodate Na^+ or K^+.

X = a large six-fold site that can accommodate Ca^{2+}, Fe^{2+}, or Mg^{2+}. Na^+ is found in this site in high-pressure environments.

Y = a small six-fold site that accommodates Fe^{2+} and Mg^{2+}, and also ferric iron, titanium, and aluminum (such aluminum in octahedral coordination is known as Al^{VI}).

Z = tetrahedral site. This site is usually filled with silicon but, at high temperatures, variable amounts of aluminum (Al^{IV}) may also be present.

The amphiboles can be divided into three series:

(1) *Quadrilateral amphiboles*, which consist of amphiboles containing calcium, iron, and magnesium but with only minor amounts of aluminum substituting for silicon. They are analogous to the quadrilateral pyroxenes.

(2) *Sodic amphiboles*, in which sodium substitutes for calcium of actinolite. This series is analogous to the sodic pyroxenes.

(3) *Hornblende*, a calcic amphibole that contains significant amounts of aluminum in the tetrahedral site. There is no analogous mineral in the pyroxenes.

Quadrilateral Amphiboles

The quadrilateral amphiboles, like the quadrilateral pyroxenes, are composed of calium, iron, and magnesium. They have the following end members:

Tremolite (Tr)	$Ca_2Mg_5Si_8O_{22}(OH)_2$
Actinolite (Act)	$Ca_2Fe_5Si_8O_{22}(OH)_2$
Anthophyllite (Ath)	$Mg_7Si_8O_{22}(OH)_2$
Grunerite (Gru)	$Fe_7Si_8O_{22}(OH)_2$

These form three solid-solution series, the tremolite–actinolite series, the cummingtonite–grunerite series, and the orthoamphibole series (Figure A.11). There is only limited solid solution between the tremolite–actinolite series and the low-calcium amphiboles.

Tremolite–Actinolite Series. Minerals in this series occur exclusively in metamorphic rocks. Tremolite is found in weakly to moderately metamorphosed siliceous marbles and in metaperidotites up to very high grades of metamorphism. Actinolite is found in weakly metamorphosed mafic rocks and gives way to hornblende with increasing metamorphic grade. Both tremolite and

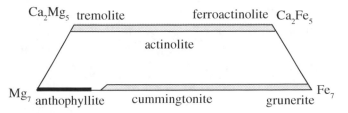

Figure A.11 Diagram showing the composition range of the quadrilateral amphiboles.

actinolite are recognized by their acicular habit. In hand sample, tremolite is colorless in marbles but may be dark colored (dark green to black) in metaperidotites because the rock itself has a dark color. Actinolite is a deep green color; this causes rocks containing abundant actinolite (such as greenschists) to be colored green. Both minerals are biaxial (–) with a 2V near 85° and a second-order birefringence. Both are monoclinic with a maximum extinction angle relative to the *c* crystallographic axis of 10–20°. In thin section, tremolite is colorless and distinguished from diopside (with which it may be associated) by its optically (–) character and its lower angle of extinction. Actinolite is pleochroic green and has a higher relief than chlorite but a lower relief than pumpellyite, two green minerals with which it might be confused.

Calcium-Free Amphiboles. Calcium-free amphiboles form two solid-solution series, an orthorhombic and a monoclinic series. The orthorhombic series is magnesian and consists of *anthophyllite* and an aluminous anthophyllite called *gedrite*. The aluminum-poor anthophyllite is most commonly found in metaperidotites while gedrite is found in medium-grade metamorphosed pelitic rocks and in metamorphosed, hydrothermally altered, mafic rocks. The monoclinic series extends from $Fe/(Fe + Mg)$ = 0.4 to $Fe/(Fe + Mg)$ = 1.0. The relatively magnesian end of this series is called *cummingtonite*, whereas the iron-rich end is called *grunerite*. Cummingtonite is the only quadrilateral amphibole that may occur in igneous rocks; it is found in some dacitic tuffs. Grunerite is found mostly in metamorphosed iron formations. Anthophyllite in metaperidotites is identified by its acicular habit, although in hand sample it is impossible to distinguish it from tremolite. In other rock types, a low-calcium amphibole can be identified by its acicular nature and honey-brown color. In thin section, anthophyllite and gedrite are colorless to pale brown with a high relief and a high birefringence. They are the only amphiboles that have parallel extinction. Both have a high 2V; anthophyllite is biaxial (–) and gedrite is biaxial (+).

In hand sample, cummingtonite has the same acicular habit and brownish color as gedrite. Grunerite in iron formations tends to be pale green and acicular. In thin section, minerals of the cummingtonite–grunerite series are colorless to pale green or brown in color. They have a high birefringence and an inclined extinction similar to that of tremolite. The diagnostic character of these amphiboles is the presence of polysynthetic twins on (100). The 2V of this series changes as a function of iron content, with the result that the relatively magnesian portion of the series (i.e. cummingtonite) is optically (+) while the iron-rich portion (grunerite) is optically (−).

Sodic Amphiboles

Sodic amphiboles have two types of substitution:

$$Ca_2(Fe, Mg)_5Si_8O_{22}(OH)_2 + 2NaAl \rightleftharpoons$$

tremolite−actinolite

$$Na_2(Fe, Mg)_3Al_2Si_8O_{22}(OH)_2 + Ca(Fe, Mg)_3$$

glaucophane

$$Ca_2(FeMg)_5Si_8O_{22}(OH)_2 + 2NaFe^{3+} \rightleftharpoons$$

tremolite−actinolite

$$Na_2(Fe, Mg)_3Fe^{3+}_2Si_8O_{22}(OH)_2 + Ca(Fe, Mg)_3$$

riebeckite

As was true of the sodium–aluminum-bearing pyroxenes, the sodium–aluminum amphibole, glaucophane, is restricted to metamorphic rocks. The sodic–ferric amphiboles are important minerals in sodic igneous rocks.

Glaucophane–Crossite Series. Sodic amphiboles are diagnostic of low-temperature and high-pressure metamorphism, where a complete solid-solution series from actinolite to glaucophane may be stable. Minerals intermediate in this series are called *crossite* (Figure A.12). They are found most commonly in mafic rocks but may also occur in metapelitic rocks and more rarely in metamorphosed impure carbonates. With increasing pressure, the stable amphibole in mafic rocks is first actinolite, then crossite, and finally glaucophane.

The distinctive blue color of sodic amphiboles makes them easy to identify, both in hand sample and in thin section. No other minerals have the amphibole habit and the blue–lavender–purple pleochroism of the sodic amphiboles. Because of its lower ferric iron content glaucophane generally has a less intense pleochroism than does crossite.

Riebeckite. Riebeckite is most easily identified in thin section by its intense blue or purple pleochroism. In some

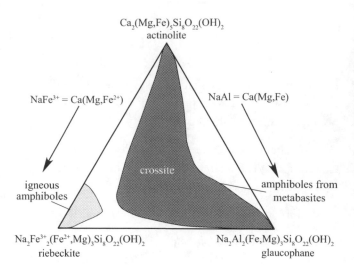

Figure A.12 Diagram showing the composition range of sodic amphiboles in igneous and metamorphosed mafic rocks. Fields defined by data from Brown (1974) and Deer et al. (1997).

rocks, the color is so intense that only on the thin edges of grains can one see that the grain is translucent and not opaque. Riebeckite is most common in alkaline rocks, where it is usually associated with aegirine. It is also found in some weakly metamorphosed iron formations. In highly sodic rocks, a substitution of $NaFe^{2+}$ for Fe^{3+} in riebeckite produces *arfvedsonite* – $NaNa_2(Fe,Mg)_4Fe^{3+}Si_8O_{22}(OH)_2$. There is a complete solid solution between riebeckite and arfvedsonite and they are optically very similar. Therefore, both are considered part of a riebeckite family.

Hornblende

Hornblende is an amphibole that has one to two atoms of aluminum substituting for silica; rarely does a hornblende have more than two atoms of aluminum in the tetrahedral site. Substitution of trivalent aluminum for tetravalent silicon produces a charge imbalance that can be satisfied through the substitution of Na^+ (or K^+) in the A site or aluminum for iron–magnesium. The two simplest exchange operations involve:

(1) *Si = NaAl or (NaAlSi$_{-1}$)*. In this substitution aluminum substitutes for silicon in the tetrahedral site while sodium moves into the vacant site. The resulting amphibole is called *edenite* with the formula $NaCa_2(Fe,Mg)_5Si_7AlO_{22}(OH)_2$.

(2) *Al$_2$ = (Fe,Mg)Si or (Al$_2$(Fe,Mg)$_{-1}$Si$_{-1}$)*. In this substitution the charge imbalance produced by substituting

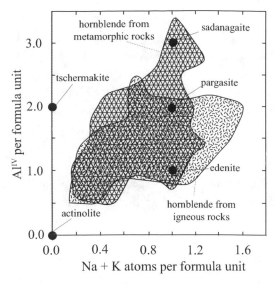

Figure A.13 Diagram showing the compositional range of hornblende from igneous and metamorphic rocks. Fields defined by data from Deer et al. (1997).

aluminum for silicon is accommodated by substituting aluminum for divalent iron or magnesium. The resulting mineral end member (*tschermakite*) has the formula $Ca_2(Fe,Mg)_4AlSi_7AlO_{22}(OH)_2$.

(3) A combination of these two substitutions leads to the *pargasite* end member, $NaCa_2(Fe,Mg)_4AlSi_6Al_2O_{22}(OH)_2$.

Tschermakite is a hypothetical end member; no hornblende is known that does not have Na^+ or K^+ in the A site. However, a plot of the number of aluminium and sodium + potassium atoms per formula unit of amphibole can accommodate the composition of almost all natural amphiboles (Figure A.13). The substitution of aluminum for silicon in the tetrahedral site is an important thermometer because it reflects increasing temperature of formation.

At high temperatures, additional substitutions may occur that accommodate the charge unbalance caused by aluminum substituting for silicon in the tetrahedral site of amphibole. Of the many possibilities, the most important optically are the substitution of titanium and ferric ions for octahedral aluminum. Both of these elements have a strong coloring effect, thus as titanium and ferric iron increase in a hornblende the color becomes more intense. This tends to make hornblende black rather than green in hand sample. In thin section, low-temperature hornblende tends to have a green pleochroism, whereas high-temperature hornblende tends to be intensely colored in brown or red. Sodic hornblende, wherein Na^+ has substituted for some of the Ca^{2+}, has been found in some alkalic rocks. It is identified as a hornblende with a touch of blue to its pleochroism.

A.3.4 Phyllosilicates

Like amphiboles, phyllosilicates contain considerable water. As a result, they are most stable at lower temperatures and are more common in metamorphic rocks than in igneous rocks. They can occur in metamorphic rocks of all compositions and are important indicators of metamorphic grade.

Serpentine

Although commonly referred to as a single mineral, serpentine is actually a complex mineral group involving the phases *chrysotile*, *lizardite*, and *antigorite*. Chrysotile and lizardite have compositions approaching the ideal formula of $Mg_3Si_2O_5(OH)_4$ although lizardite may contain a small amount of aluminum. Antigorite has a distinctly more siliceous composition, which can be modeled by the formula $Mg_{48}Si_{34}O_{85}(OH)_{62}$. The serpentine minerals are the hydration products of peridotite. Chrysotile and lizardite, the low-temperature forms of serpentine, are formed at temperatures ranging from those of surface conditions to those of low-grade metamorphism. Antigorite, being somewhat less hydrous than the other two serpentine minerals, is a mineral of medium-grade metamorphism.

Serpentine is pale green and waxy in hand sample. It is colorless to pale green in thin section, has a low birefringence (first-order gray) and is generally very fine-grained. Generally it is impossible to obtain optic figures on serpentine. The only mineral with which serpentine can be confused is chlorite. However, the chlorite that occurs in metaperidotites is length-fast while serpentine is length-slow. Although it is impossible to distinguish between the three serpentine types with certainty without the use of X-ray diffraction, one can often make a fairly good guess at the serpentine type from their microscopic textures. Antigorite, being the high-temperature form, generally has a fairly strongly developed micaceous habit, whereas chrysotile is markedly fibrous. Lizardite may assume a range of habits, from nearly isotropic aggregates to moderately coarse-grained sheets. Usually, however, it doesn't have the tabular, interlocking habit of antigorite.

Greenalite

Greenalite is an iron analog of serpentine that occurs in weakly metamorphosed iron formations. It forms green to brown, fine-grained aggregates that commonly preserve sedimentary textures. The grains are so small that they may look isotropic under crossed polarized light. The only mineral that looks similar to greenalite is glauconite, which has similar habit of occurrence but is distinguished by its high birefringence.

Talc

Talc is usually very close to the end-member composition $Mg_3Si_4O_{10}(OH)_2$. Small amounts of ferrous iron may be present, but most talc contains 90 percent or more of the magnesian end member. Talc is a common mineral in metamorphosed peridotites. It is found less commonly in metamorphosed impure dolomites and is also a diagnostic mineral for the rare high-pressure metapelitic rock known as whiteschist.

Talc is easily distinguished in hand sample because of its soapy feel. It is also possible to determine if talc is present in small concentrations in a rock by feeling the head of the hammer after hitting the rock. If the head feels soapy, talc is present. In thin section, talc is a colorless mica with a high (third-order) birefringence. It may be mistaken for muscovite or phlogopite, but it has a higher birefringence than either. Muscovite usually has a higher 2V and, except in the most magnesian environments, phlogopite has a pale pleochroism.

Minnesotaite

Minnesotaite is an iron analog of talc that forms in metamorphosed iron formations. It is optically similar to talc, although it is usually very fine-grained and may have a faint yellow pleochroism. Its association with iron formations is diagnostic. The only other fibrous mineral in iron formations is grunerite, which, unlike minnesotaite, does not have parallel extinction.

Chlorite

Although it does not have the same structure as serpentine, the simplest way to think of the composition of chlorite is as a mineral with a serpentine base $(Mg,Fe)_6Si_4O_{10}(OH)_8$ that has a certain amount of aluminum substituting for silicon and magnesium–iron. The amount of aluminum substitution is variable, so it is difficult to write a simple formula for this mineral, but most chlorites have compositions near that of clinochlore: $(Mg,Fe)_5Al(Si_3Al)O_{10}(OH)_8$. Chlorite is a widespread mineral in low-grade rocks, where it can be found in serpentinites, metamorphosed pelitic rocks, mafic rocks, and impure dolomitic limestones. The stability of chlorite shrinks with increasing metamorphic grade, but in rocks of appropriate composition, specifically metaperidotite and magnesian metapelitic schist, chlorite may survive to upper amphibolite facies.

Chlorite is green and micaceous in hand sample. Unlike serpentine, which is extremely fine-grained, individual grains of chlorite are commonly recognizable in hand sample. In thin section, it is recognized by its micaceous habit, colorless to green pleochroism, and a low birefringence. The birefringence is commonly anomalous, with blue and brown appearing in place of first-order gray. The only minerals that chlorite is likely to be confused with are serpentine and green biotite. The chlorite found in serpentinites is length-fast whereas coexisting serpentine is length-slow. Chlorite is distinguished from green biotite by its low and anomalous birefringence.

Micas

Micas are phyllosilicates that contain alkali ions, usually K^+, more rarely Na^+ and Ca^{2+}, between the silicate sheets. The extra charge from the alkalis is accommodated by substitution of aluminum for silicon in the tetrahedral site. The inter-layer alkali ions stabilize the phyllosilicate structure to higher temperature, with the result that some micas, including biotite and muscovite, are stable in igneous rocks.

Biotite. The biotite series is made up of the end members *phlogopite* – $KMg_3AlSi_3O_{10}(OH)_2$ – and *annite* – $KFe_3AlSi_3O_{10}(OH)_2$. In addition to an iron–magnesium substitution, there is substitution of aluminum and titanium for iron and magnesium in the octahedral sites. Phlogopite is optically distinct from the rest of the biotite series and, therefore, it is often categorized as a separate mineral from the more iron-rich biotites. This is not true for annite, which looks like any other iron-bearing biotite.

Biotite is identified by its micaceous habit, brown, red-brown, or, rarely, green pleochroism, and high birefringence. Green biotite might be confused with chlorite, from which it is distinguished by a high birefringence. Phlogopite has similar properties to those of biotite except that it is weakly colored in thin section. Very magnesian phlogopite may be totally colorless, but most samples of phlogopite have pale tan pleochroism. There is, of course, complete gradation between the optical properties of phlogopite and those of biotite, but the convention is that one vibration direction in phlogopite is colorless whereas in biotite it is pale brown or pale green. Phlogopite is distinguished from muscovite by its small 2V (near 0) and from talc by a lower birefringence.

Phlogopite, being a magnesian biotite, is restricted to magnesium-rich rocks, such as metaperidotite and impure dolomitic limestone. The rest of the biotite series is widespread in occurrence. Biotite is a dominant mineral in metapelitic schist but it is also found in metamorphic mafic rock (metabasite) and in quartzo-feldspathic gneiss. Biotite is stable over a wide range of metamorphic grades and in most rock types it can survive to the highest temperatures of metamorphism. It is also found in a wide variety of igneous rocks, where it may be stabilized by high titanium and fluorine contents.

Muscovite. Muscovite has the general formula $KAl_2(AlSi_3O_{10})(OH)_2$. High-temperature muscovite may have a composition close to this model formula, but muscovite from low-temperature metamorphic rocks commonly deviates significantly from this composition. In many metamorphic rocks, particularly those formed at high pressures or low temperatures, there are more than three silicon atoms per formula unit. This substitution of silicon for aluminum in the tetrahedral site is accommodated by substitution of iron and magnesium for aluminum in the octahedral site. Such an iron- and magnesium-bearing muscovite is known as *phengite*. Muscovite is common in metapelitic rocks and in other metamorphic rocks derived from sediments that contained significant amounts of clay. It occurs over a wide range of metamorphic grades but breaks down to K-feldspar at the higher grades by the reaction:

$$KAl_3Si_3O_{10}(OH)_2 \rightleftharpoons KAlSi_3O_8 + Al_2O_3 + H_2O$$
$$\text{muscovite} \qquad \text{K-feldspar} \quad \text{melt} \quad \text{fluid}$$

Muscovite is not as widespread in igneous rocks as biotite because, as indicated by the reaction above, it only forms in rocks that have more Al_2O_3 than is necessary to make feldspars. Most granitic rocks are not that aluminous and thus contain biotite but lack muscovite.

Muscovite is identified in hand sample by its lack of color. In thin section, it is colorless with a high birefringence and a 2V (−) of 30–45°. Phengite is a muscovite with a small 2V. Muscovite is distinguished from

phlogopite by a higher 2V and from talc by a lower birefringence.

Paragonite. Paragonite, $NaAl_2(AlSi_3O_{10})(OH)_2$, is a sodium mica that may occur in weakly metamorphosed pelitic rocks or in low-temperature, high-pressure metamorphosed basalts. Paragonite has the same optical properties as muscovite and can be rarely distinguished optically. It is usually safe to assume that the major white mica in a low-grade rock is muscovite, but X-ray methods are required to distinguish whether or not small amounts of paragonite are also present.

A.4 Aluminum-Excess Minerals

Aluminum-excess minerals are found in rocks where Al > Na + K + 2Ca (in moles). Another way of saying this is that there is more aluminum than required to make feldspars. Such a situation is characteristic of metapelitic rocks, where sodium and calcium have been removed by weathering but may also be found in other rock types subjected to hydrothermal alteration.

A.4.1 Aluminosilicates (Andalusite, Kyanite, and Sillimanite)

The aluminosilicates have the formula Al_2SiO_5 and are quite pure, with only minor substitution by Fe^{3+} and Mn^{3+}. The stability fields of these minerals are related by a phase diagram (Figure A.14), with andalusite the

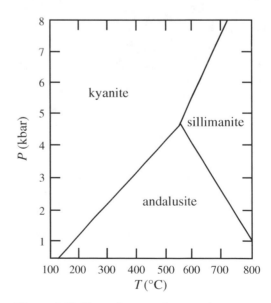

Figure A.14 Phase diagram showing the pressure–temperature stabilities of the aluminosilicate minerals. Data from Pattison (1992).

low-pressure polymorph, kyanite stable at high pressures, and sillimanite stable at high temperatures. The triple point, at which all three phases are stable, is at approximately 4.5 kbar and 550 °C (Pattison, 1992).

Andalusite

Andalusite typically forms tabular orthorhombic porphyroblasts that may be colorless, pale pink, or, rarely, pale green in hand sample. A common diagnostic feature is an accumulation of carbonaceous matter on the edges of the nearly square cross-section. This leads to a cross-section that looks like a cross; a mineral known as *chiastolite*. In thin section, the mineral has a low birefringence, a medium relief, and a high 2V (–). Andalusite is distinguished from sillimanite by a lower birefringence and a general tabular morphology. In andalusite, cleavage is either weak or missing. Rarely, andalusite shows a pale pink pleochroism that may be zoned.

Sillimanite

Sillimanite is typically seen in hand sample as fine, white, fibrous mats called fibrolite. Its optical properties include an orthorhombic crystal form, a second-order birefringence, a high relief, and a low 2V (+). In the lowest-grade metamorphic rocks in which it is stable, sillimanite forms fine-grained fibrous mats and individual crystals cannot be distinguished even in thin section. These fibrous mats are length-slow and are distinctive because nothing similar occurs in metapelitic schists. At higher grades, sillimanite may be coarser, but is usually more acicular than either andalusite or kyanite. It is distinguished from both by a higher birefringence, a distinctive cross-section when looking down the *c*-axis, and a low 2V.

Kyanite

In hand sample, kyanite is identified by its tabular shape. It is often white in color, but it commonly has touches of blue (hence its name); indeed gem-quality kyanite is deep blue. Kyanite is triclinic and is distinctive because it has two hardnesses. Parallel to the *c*-axis it has a hardness of five, and normal to the *c*-axis it is seven. Thus, a knife will scratch kyanite parallel to the *c*-axis but not normal to it. Optical properties include a high relief, a middle first-order birefringence, and a high 2V (–). Kyanite is distinctive because it has the highest relief of any of the aluminosilicates and is the only one with inclined extinction. Caution is warranted, however, because in some

orientations the extinction angle is only 2°. It also has a perfect (100) cleavage with a strong (001) parting. When present, the parting will distinguish it from andalusite and sillimanite.

A.4.2 Garnets

Garnets have the general formula $A_3B_2Si_3O_{12}$ where $A = 2+$ and $B = 3+$. There are two major series of rock-forming garnets:

1. *Aluminum garnets*, with various divalent ions in the A site. The major end members are:

A site	B site	
Mg^{2+}	Al^{3+}	Pyrope (Prp)
Fe^{2+}	Al^{3+}	Almandine (Alm)
Mn^{2+}	Al^{3+}	Spessartine (Sps)

This series is often referred to as *pyralspite* garnets.

2. *Calcic garnets*, with various trivalent ions in the B site. Major end members are:

A site	B site	
Ca^{2+}	Al^{3+}	Grossular (Grs)
Ca^{2+}	Fe^{3+}	Andradite (Adr)

This series is often referred to as *grandite* garnets.

In upper and middle crustal environments, the grandite and pyralspite garnets have two distinct types of occurrence. Pyralspites are found in metapelitic schists and, at higher pressures, in metabasites. Pyrope is not stable at low pressures. In olivine-bearing rocks, the assemblage enstatite + spinel occurs instead, and in quartz-bearing rocks, the assemblage cordierite + enstatite replaces pyrope. Grandites occur in metamorphosed and metasomatized marbles. The stability of garnet is strongly expanded by an increase in pressure to the extent that in lower crustal rocks and under mantle conditions there is a complete solid solution between grandites and pyralspites. Garnets from high-pressure environments will contain considerable amounts of calcium, magnesium, and iron.

Pyralspite garnets are recognized as equant red crystals that are commonly porphyritic and that may reach sizes on the order of centimeters. Grandite garnets are commonly xenoblastic and may range in color from colorless or pale yellow to black. In thin section, garnets are isotropic, with a very high relief ($n = 1.7$ to 1.9). Calcic garnets may have an anomalous low first-order birefringence. Pyralspite garnets are usually pale pink in transmitted light, while grandites can vary from colorless (for Al^{3+}-rich varieties) to deep orange (for Fe^{3+}-rich varieties).

A.4.3 Staurolite

Staurolite has the formula $(FeMg)_2Al_9Si_4O_{22}(OH)_2$, with iron usually more abundant than magnesium. The most important minor oxide in staurolite is ZnO, which may be present in abundances up to several weight percent. Staurolite is a relatively high-pressure mineral. It is more commonly found associated with kyanite than with andalusite. Staurolite is identified as tabular brown crystals that may form phenocrysts up to centimeters in size. In thin section, staurolite has a high relief and a low first-order birefringence. Staurolite is easily identified by its distinctive yellow pleochroism. It is usually poikiloblastic.

A.4.4 Cordierite

Cordierite is a cyclosilicate with the composition $(Mg,Fe)_2Al_4Si_5O_{18}$. Magnesium is generally more abundant than iron. Although the formula does not include water, cordierite almost invariably contains small amounts of H_2O and more rarely CO_2 and SO_2 in the broad channels defined by the rings of silica tetrahedra. Because of its open structure, cordierite is a large-volume mineral, favored by low pressure. It is a common mineral in contact metamorphic environments and is also found in low-pressure regional metamorphic terrains. Cordierite is found with andalusite and sillimanite but only very rarely with kyanite. It is characteristically a medium- to high-temperature mineral and also is found in some granites and rhyolites.

In low-grade occurrences cordierite may form anhedral porphyroblasts that produce a knotty appearance to the rock. Such rocks are commonly known as spotted schists or spotted hornfels. In higher-grade rocks, cordierite occurs as a matrix-forming mineral. Gem-quality cordierite is deep blue in color with no cleavage, a greasy luster, and a conchoidal fracture; it looks very much like quartz. A cordierite-rich rock is identified by its greasy luster and bluish color on a fractured surface. In thin section, cordierite looks very much like feldspar. It is biaxial (either ($+$) or ($-$)), has a low relief (either positive or negative) and a gray birefringence. It may be impossible to distinguish from orthoclase and untwinned plagioclase. Cordierite from high-grade rocks may be twinned. When these twins

are triangular-shaped, like slices of a pie, they are diagnostic. Cordierite is easily altered. Thus, the presence of a pale-yellow, non-pleochroic chlorite along grain boundaries or fractures is diagnostic of cordierite. Another distinguishing feature of cordierite is the presence of yellow pleochroic halos around uranium-bearing inclusions such as zircon or monazite. These are a reflection of oxidation of iron in the cordierite due to radiation damage from uranium decay. Since feldspars and quartz do not contain iron, zircon and monazite included in these minerals do not show pleochroic halos.

A.4.5 Chloritoid

Chloritoid is a tabular mineral with the formula $(Fe,Mg)_2Al_4Si_2O_{10}(OH)_4$. Iron is generally more abundant than magnesium. Chloritoid is common in low-grade metapelites. In iron-rich rocks it may be the first porphyroblastic phase to appear with prograde metamorphism. Its stability extends into medium grades of metamorphism; rarely, it is found with sillimanite. Similarly, its stability seems to expand with pressure, since it is found rather widely in high-pressure terrains and does not occur at all in contact metamorphic environments. In hand sample, chloritoid is recognized as a tabular green-to-gray porphyroblast with brittle fracture. In thin section, its optical properties are distinctive. It has a green-to-gray pleochroism and a low, frequently anomalous birefringence. It is commonly twinned normal to the c-axis. It is distinguished from chlorite by higher relief and twinning.

A.5 Calcium–Aluminum Silicates

Under many conditions of metamorphism the calcic plagioclase, anorthite, is not stable and another calcium–aluminum silicate occurs instead. For weakly metamorphosed mafic rocks the identity of the calcium–aluminum mineral is an important indicator of metamorphic grade.

A.5.1 Clinozoisite–Epidote

Zoisite and clinozoisite have the formula $Ca_2Al_3(SiO_4)_3(OH)$. Epidote is a variant of this mineral in which up to a third of the aluminum is replaced by ferric iron. Epidote is the major "anorthite substitute" mineral in metamorphic rocks. It occurs in a wide variety of protoliths at a wide range of metamorphic grades. Its stability increases with pressure to the extent that epidote is a stable igneous mineral at pressures of around 10 kbar.

The iron-poor varieties are found dominantly in meta-morphosed or metasomatized carbonate rocks.

Zoisite and clinozoisite are white to pink in hand sample, whereas epidote is a distinctive yellow–green. The color is reminiscent of that of pistachio nuts, hence the name pistachite is often given to epidote. In thin section, the minerals have a high relief ($n = 1.70$–1.78) and an anomalous birefringence (blue in place of first-order gray). Other properties are dependent on iron content. Clinozoisite is (+) with 2V = 14–$90°$. It is colorless with a low first-order birefringence. Epidote is (–) with 2V = 64–$89°$. It has a colorless to yellow pleochroism and up to a second-order birefringence. The change from optically (+) to optically (–) character marks the boundary between clinozoisite and epidote.

A.5.2 Prehnite

Prehnite has the formula $Ca_2Al_2Si_3O_{10}(OH)_2$, with up to half the aluminum replaced by ferric iron. Prehnite is diagnostic of very low-grade metamorphism. It is found most frequently in weakly metamorphosed basalts, commonly as an amygdule filling. It occurs more rarely as cement in sandstones.

In hand specimen, prehnite is white to pale green in color and, if massive, may show a botryoidal surface. When it is an amygdule filling, prehnite shows a distinctive radiating crystal shape. In thin section, it is colorless with a high relief and a high birefringence. It is optically (+) with a high 2V (although low-2V varieties are known). It is distinguished from epidote, with which it may occur, by a lower relief and a lack of pleochroism.

A.5.3 Pumpellyite

Pumpellyite has the formula $Ca_2(Mg,Fe)Al_2Si_3O_{11}(OH)_2 \cdot H_2O$. There is a wide range of substitution of ferrous iron for magnesium and ferric iron for aluminum. Like prehnite, pumpellyite is found in very low-grade metamorphic rocks. The stability of pumpellyite seems to be favored by increasing pressure, for unlike prehnite, pumpellyite is sometimes found coexisting with sodic amphiboles.

In hand sample, pumpellyite is recognized by its needle-like habit and dark green color. Its optical properties include high relief, biaxial (+) with a 2V ranging from $10°$ to $85°$, and an anomalous first- to second-order birefringence. In most occurrences pumpellyite has a deep-green pleochroism, but very magnesium-rich pumpellyite may be colorless to pale yellow and difficult to distinguish from epidote. It is distinguished from chlorite by a higher relief and a generally deeper pleochroism.

A.5.4 Lawsonite

Lawsonite has the formula $CaAl_2Si_2O_7(OH)_2 \cdot H_2O$ and in most rocks it shows little chemical variability. It is a high-pressure, low-temperature mineral that commonly occurs with sodic amphiboles. Lawsonite is orthorhombic, biaxial (+), with a high 2V and a moderate to high first-order birefringence. Lamellar twinning on (101) is common. Lawsonite is distinguished from zoisite and clinozoisite by its higher birefringence and the lack of anomalous birefringence.

A.5.5 Laumontite

Laumontite ($CaAl_2Si_4O_{12} \cdot 4H_2O$) is one of many zeolites. Zeolites form a complex mineral group that is stable at very low temperatures, in the range of diagenesis and the lowest temperatures of metamorphism. Because it is the most stable zeolite at relatively high temperatures, laumontite is the zeolite most likely to be encountered in metamorphic rocks. Laumontite is found both as amygdules and veins in weakly metamorphosed basalts (where it is studied by metamorphic petrologists) and also as cement in sandstones (where it is studied by sedimentary petrologists).

In thin section, laumontite is colorless with a low negative relief and a low first-order birefringence. It is biaxial (–), with a small to moderate 2V. It is found with other low-temperature calcium–aluminum silicates (prehnite, pumpellyite, and epidote). Its negative relief and occurrence are distinctive.

A.6 Oxide, Sulfide, and Other Nominally Opaque Phases

A.6.1 Iron–Titanium Oxides (Magnetite and Ilmenite)

Most igneous rocks contain one or both of the most common iron–titanium oxides, *magnetite* and *ilmenite*. Both of these minerals are complex solid solutions that have a wide range of stability. Magnetite is a spinel mineral with the formula $Fe^{2+}Fe^{3+}_2O_4$. It commonly is found to be nearly pure, but at high temperatures it forms the end member of a mineral series that shows complete solid solution with ulvöspinel (Fe_2TiO_4). Ilmenite has the

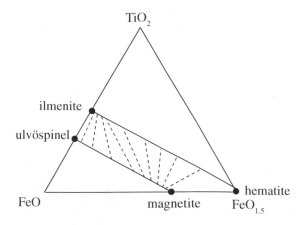

Figure A.15 Diagram showing the composition relations between the ilmenite and magnetite solid solutions at magmatic temperatures. Dashed tie lines show the approximate compositions of coexisting phases. Modified after Lindsley (1991).

formula FeTiO$_3$ and at high temperatures it has complete solid solution with hematite (Fe$_2$O$_3$). In most igneous and metamorphic rocks the ferrous–ferric iron ratio in these minerals monitors the fugacity (partial pressure) of oxygen in the rocks (Figure A.15).

Magnetite can be distinguished from ilmenite because it is magnetic. If the grains are large enough, ilmenite can be recognized by its perfect (001) cleavage. Since both magnetite and ilmenite are opaque phases, they are difficult to distinguish in thin section. Ilmenite may have a platy morphology and magnetite may form octahedra, but in many samples the grains are intergrown and show no distinct crystal form.

A.6.2 Other Spinel Minerals

Chromite [(Fe,Mg)Cr$_2$O$_4$] is common in ultramafic to basaltic rocks. In hand sample, it is identified as a black mineral, which is frequently octahedral and is non-magnetic. In thin section, it is isotropic, brown to nearly black in color. It is commonly the first mineral to crystallize in basalts and is often found as inclusions in olivine.

Green spinel [(Fe,Mg)Al$_2$O$_4$] is absent from quartz-bearing rocks because green spinel + quartz = cordierite (unless it is stabilized by considerable zinc). It is commonly found intergrown with magnetite in gabbros, in which the magnetite has exsolved out of, during cooling. It is also found as a metamorphic mineral in partially melted metapelitic rocks, in gabbroic rocks that have undergone high-pressure and high-temperature metamorphism, and in impure, dolomitic marbles (in these rocks the spinel may be pale green to colorless).

A.6.3 Iron Sulfides

Pyrite (FeS$_2$) and pyrrhotite [Fe$_{(1-x)}$S] may occur in minor amounts in many igneous and metamorphic rocks. Just as the iron–titanium oxides monitor oxygen fugacity in rocks, the iron sulfides monitor sulfur fugacity. With increasing temperature, or at low sulfur fugacities, pyrite breaks down to pyrrhotite by the reaction:

$$\underset{\text{pyrite}}{(1-x)\text{FeS}_2} \rightleftharpoons \underset{\text{pyrrhotite}}{\text{Fe}(1-x)\text{S}} + 1/2(1-x)\text{S}_2$$

This relationship indicates that, whereas pyrite may be stable in some relatively low-temperature rocks such as granites, pyrrhotite is the only sulfide found in high-temperature rocks such as basalt and gabbro.

Because they are opaque phases, pyrite and pyrrhotite cannot be distinguished from each other (or from oxides) in thin section. If an opaque phase in a thin section is large enough, sometimes it can be identified as a sulfide by examining the thin section with a hand lens and looking for light reflecting off the surface of the section.

A.6.4 Graphite

Graphite is an important constituent phase in many metamorphic rocks, where it forms by thermal decomposition of organic matter. In igneous rocks, where it is found only rarely, it forms by abiotic processes. A good way to determine whether a rock contains graphite is to hit the rock with a hammer and rub the hammerhead on your hand. Dark streaks on your hand will indicate the presence of graphite, even if only a small amount is present in the rock. Graphite can sometimes be distinguished in thin section as tabular flakes that have a "ragged" shape.

A.6.5 Rutile

Rutile (TiO$_2$) is a common oxide mineral in metapelitic rocks and metabasites. Rutile also is the major titanium-bearing mineral in eclogites. In addition, it is commonly found as a detrital mineral in sandstones. In hand sample, it is typically reddish brown with adamantine luster and forms acicular and prismatic crystals. In thin section, it is identified by its uniaxial (+) character and high birefringence (though this can be masked by the strong color of the mineral).

A.7 Accessory Minerals

Accessory minerals are minerals present in low abundance in most igneous rocks. The most important of these are zircon, titanite, apatite, and monazite. These minerals can be geochronometers because they contain minor amounts of uranium. They are rarely visible in hand specimen but are found in most igneous and metamorphic rocks.

A.7.1 Zircon

Zircon ($ZrSiO_4$) is an important mineral because it is one of the few that can accommodate zirconium. Zircon crystallizes out late in the evolution of many igneous rocks, and as it does, it incorporates uranium from the melt. As a result, zircon is one of the most important minerals for geochronology. Although it occurs in most rock types, it is rarely seen in hand sample because of its low abundance. In thin section, zircon is tetragonal with a high relief and a high birefringence. It is uniaxial (+), but rarely do you find grains large enough to give an interference figure. A diagnostic feature of zircon is that the decay of the uranium or thorium that it contains causes radiation damage to adjacent crystals. The radiation tends to oxidize the iron in surrounding ferromagnesian silicates. Thus, most zircon grains included in biotite or hornblende are surrounded by dark pleochroic halos, showing the extent of radiation damage.

A.7.2 Titanite (or Sphene)

Although ilmenite ($FeTiO_3$) is the major titanium-bearing mineral in most igneous rocks, titanite ($CaTiSiO_5$) is common in many plutonic rocks of intermediate composition (diorites, granodiorites, and some granites). It is also found in many metamorphic rocks. In hand sample, titanite is identified as small red to yellow grains with a distinct lozenge shape. In thin section, it is recognized by its high relief and extreme birefringence. It has a birefringence similar to that of calcite, but unlike calcite, the relief doesn't vary on rotation. Titanite can accommodate small amounts of uranium and is an important mineral for dating both igneous and metamorphic rocks.

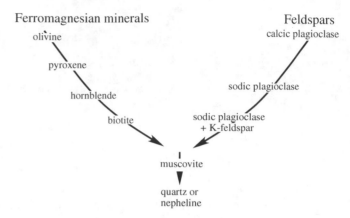

Figure A.16 Bowen's reaction series, showing the minerals that will crystallize during differentiation of a basaltic magma.

A.7.3 Apatite

Apatite [$Ca_5(PO_4)_3(F,OH,Cl)$] is one of the few minerals that can accommodate phosphorous. As a result, it is present in small quantities in almost every igneous rock. Apatite is uniaxial (−) and forms small rods that have a low first-order birefringence (usually gray) and a moderate relief. When looking down the c-axis, apatite forms small hexagonal grains that appear isotropic. Apatite can accommodate uranium but diffusion is fast enough in apatite that it has a relatively low closure temperature, which means that any age obtained from uranium–lead dating of apatite will give the time the rock cooled through that temperature, instead of the time the rock formed.

A.7.4 Monazite

Monazite ($CePO_4$) is a phosphate mineral that is rarer than apatite but is relatively common in metapelitic schists and in some granitic rocks. It is an important mineral for dating metamorphic rocks. Monazite is rarely seen in hand sample, but in thin section it forms small tabular grains that are distinctly yellow and have a pleochroic halo. Monazite is distinguished from zircon by its yellow color, but if the grains are fine enough, it may be difficult to tell whether a given pleochroic halo surrounds zircon or monazite.

Summary

The major minerals in igneous rocks are relatively few, and include quartz, feldspars, nepheline, olivine, pyroxenes, amphibole (mainly hornblende), and micas. Because igneous differentiation enriches the melt in H_2O, in sodium over calcium, and in iron over magnesium, there is a consistent change in mineralogy and mineral composition during differentiation. This change is summarized by Bowen's reaction series (Figure A.16). Although it is an oversimplification, Bowen's reaction series gives a first-order description of how the mineralogy of the crystallizing assemblage changes during the evolution of igneous magmas. The ferromagnesian side of the series consists of olivine, pyroxene, and, if the rock is hydrous, hornblende and biotite. The feldspar side of the series consists of plagioclase that initially is calcic and becomes progressively enriched in sodium. At some point in the differentiation history K-feldspar joins plagioclase in the crystallization history, and quartz is one of the last minerals to begin crystallizing.

Metamorphic minerals tend to be far more complex than igneous minerals. However, there are two simple rules for the stability of metamorphic minerals:

1. Dense minerals, such as garnet, are favored at high pressure.

2. The more hydrous a mineral is, the lower its likely temperature of stability.

Glossary

aa:	A Hawaiian term for basaltic lava flow with a blocky, rubbly surface. Compare with **pahoehoe.**
accretionary prism:	An accumulation of sedimentary material that has been scraped off the down-going oceanic plate and accreted onto the overriding plate at a subduction zone.
activity:	A thermodynamic term, abbreviated as a_i (where i = component), which describes the concentration of a component in a solution, assuming that the solution is ideal. See also **component.**
activity coefficient:	A coefficient, abbreviated as γ_i, which when multiplied with the mole fraction of a component in a solution yields the activity of that component. Mathematically, the activity of i is given as $a_i = \gamma_i X_i$. See also **mole fraction.**
alkalic:	See **alkali–lime index.**
alkali-calcic:	See **alkali–lime index.**
alkali–lime index:	A geochemical index introduced by Peacock (1931) that is defined by the silica content at which $Na_2O + K_2O$ intersect CaO for whole-rock analyses of a suite of igneous rocks. If the intersection lies at weight percent <51 percent SiO_2 the suite is **alkalic,** if it is between 51 percent and 56 percent it is **alkali-calcic,** if it is 56–61 percent it is **calc-alkalic,** and if it is over 61 percent the suite is **calcic.**
alkaline:	Alkaline rocks are defined as those that have higher sodium and potassium contents than can be accommodated in feldspar alone. See also **alkalinity index.**
alkali basalt:	A basalt that has normative nepheline. Alkali basalts are so named because they have higher content of alkalis Na_2O and K_2O than other basalt types.
alkalinity index (AI):	A geochemical index (Al – [Na + K]) established by Shand (1943) that indicates the relative abundances of aluminum and alkalis. If a rock has an AI > 1.0 it is **peralkaline.**
aluminum-saturation index (ASI):	The molecular ratio (Al/[Ca – 1.67P + Na + K]) (Shand, 1943; Zen, 1988). This index measures the abundances of aluminium in a rock relative to the amount of aluminum required to make feldspars. Rocks where ASI > 1.0 are **peraluminous,** those where ASI < 1.0 are **metaluminous.**
amygdule:	A spherical cluster of minerals filling a vesicle in metamorphosed basalt. See **vesicle.**
anatexis:	The process of partial melting of a rock.
anhedral:	A term describing the shape of a mineral in an igneous rock that is not bounded by crystallographic faces. Anhedral grains usually form late in the crystallization of a magma.
aphanitic:	An igneous texture in which the minerals are so fine-grained that they cannot be distinguished by the naked eye.
ash:	Fine pyroclastic material that is under 4 mm in diameter. Compare **lapilli, block** and **bomb.**
assemblage:	An association of metamorphic minerals that are assumed to have formed under equilibrium conditions.
assimilation:	The process whereby magma incorporates partial melts that have been generated from the wall rock.

autolith: An inclusion in an igneous rock that has been derived from a rock that is petrologically and geochemically related to the igneous rock enclosing it. Compare with **enclave** and **xenolith**.

bathograd: An imaginary line on a map that connects areas that record the same depth of metamorphism.

batholith: An igneous intrusion that outcrops over an area of more than 100 km² (40 mi²).

bathozone: An area on a map where the rocks contain a similar metamorphic assemblage and which is bounded by bathograds.

Benioff zone: A plane defined by an array of earthquakes that is a geophysical manifestation of a subduction zone.

bimodal volcanism: A term describing a suite of volcanic rocks that contains basaltic and felsic (rhyolitic to dacitic) volcanism with little or no intermediate lavas.

block: A rock measuring more than 32 mm in diameter that was ejected as a solid during a pyroclastic eruption. Compare with **bomb**, **lapilli**, and **ash**.

bomb: A rock measuring more than 32 mm in diameter that was ejected as a magma during a pyroclastic eruption. Compare with **block**, **lapilli**, and **ash**.

buffered: A term describing the condition in which the fluid composition in a metamorphic rock is controlled by the mineral reactions that were occurring during metamorphism. Compare with **infiltration**.

bulk-distribution coefficient: A function that measures the distribution of an element between a magma and the weighted average of the minerals that are crystallizing from it. See also **partition coefficient**.

calc-alkalic: See **alkali–lime index**.

calcic: See **alkali–lime index**.

charnockite: An orthopyroxene-bearing granite.

chemographic projection: A graphical means of showing the compositional relations among minerals in a metamorphic rock.

chilled margin: The zone along the margin of an igneous body that is markedly finer-grained than the core of the body because it crystallized relatively rapidly by chilling.

columnar jointing: Vertically oriented hexagonal columns in igneous rocks, most commonly basalt, that formed from contraction cracks associated with rapid cooling.

compatible element: An element in an igneous melt that is preferentially incorporated into the crystallizing minerals.

component: A chemical constituent of a system. See also **phase** and **phase rule**.

concordia: Adjective describing a curve on a diagram of $^{207}Pb/^{235}U$ versus $^{206}Pb/^{238}U$ along which the $^{207}Pb/^{235}U$ and $^{206}Pb/^{238}U$ chronometers give the same age.

congruent melting: A process by which the melting of a mineral produces a melt with the same composition as the mineral.

continuous reaction: A divariant reaction wherein the composition of the phases involved changes as the reaction proceeds. Compare with **discontinuous reaction**.

cotectic: A univariant line on a ternary phase diagram for igneous crystallization or melting where two solid phases coexist with a melt.

critical point: The end point on a univariant curve where two phases become identical. This can also refer to the point on a solvus where the solvus closes because the phases on either side of the solvus become compositionally the same.

crustal delamination: The process in which thickened, dense crust breaks off and sinks into the mantle.

crystal mush: A partially molten rock composed of an aggregate of melt and solid crystals in which the crystals form a continuous framework, throughout which melt is distributed.

cumulate: A textural term describing early formed crystals in an igneous rock.

decompression melting: Because many melting curves have a positive (dP/dT) slope, decompression of a rock near the melting temperature will cause the rock to cross the curve, producing melting.

dehydration melting: A melting reaction wherein the water necessary to flux the melt is generated by the dehydration of minerals involved in the melting.

diagenesis: The process of lithification by which a sediment becomes a sedimentary rock.

diffusion: The process by which a chemical species moves through a fluid, melt, or crystal as a result of a chemical gradient.

dihedral angle: The angle at which a mineral grain intersects two other grains in a metamorphic rock. In a monomineralic rock that formed under static conditions the dihedral angle should equal 120°.

dike: A tabular igneous body that cross-cuts bedding or foliation. Compare with **sill**.

dike swarm: A cluster of dikes with similar orientations and chemistry that are likely to have been intruded during a rifting event.

discontinuous reaction: A univariant reaction which occurs at fixed temperature and pressure. In reactions involving minerals with solid solution, the compositions of those phases are fixed. Compare with **continuous reaction**.

divariant: An adjective describing a reaction with two degrees of freedom, which is shown as a field on a phase diagram, rather than a line. See also **phase rule** and **univariant**.

enclave: An inclusion within an igneous rock that has an unspecified relation to the host. Compare with **autolith** and **xenolith**.

enthalpy: A thermodynamic term describing the heat energy released or consumed during a reaction or during a phase change. Compare with **entropy**.

entropy: A thermodynamic function related to the randomness of a system. In geologic reactions it is related to the strengths of the individual bonds within phases such that more weakly bonded phases have higher entropy. Compare with **enthalpy**.

equilibrium: A thermodynamic term describing a state where the rate at which the products are formed from the reactants is the same as the rate at which the reactants are formed from the products. Also, a calm state of mind.

equilibrium crystallization: A process by which the crystallizing minerals remain in contact and in thermodynamic equilibrium throughout the melting process. Compare with **fractional crystallization**.

equilibrium melting: Melting in which the melt and the residuum remain in communication and in thermodynamic equilibrium throughout the melting process. Compare with **fractional melting**.

euhedral: An igneous mineral that is bounded by crystal faces and which is assumed to have crystallized early from a magma. Compare with **anhedral**.

eutectic: An invariant point in an igneous phase diagram that represents the temperature at which the last remaining melt crystallizes.

feldspathoid silica-saturation index (FSSI): A geochemical index defined as the normative values of Q − (Lct + 2 Nph))/100. This index describes the degree to which a rock is oversaturated or undersaturated with respect to silica.

felsic:	A term describing a light-colored igneous rock. It is derived from the terms *fel*dspar (or *fel*dspathoid) + *si*lica. Compare with **mafic**.
fiamme:	Streaks of flattened pumice within a welded tuff.
filter pressing:	A type of igneous differentiation that is caused by separation of melt from crystals by pressure. Compare with **fractional crystallization**.
flow segregation:	A type of igneous differentiation wherein crystals are concentrated in zones of low shear and the melt is concentrated in high-shear zones.
foliation:	Any planar feature in a metamorphic rock. Compare with **gneissosity** and **schistosity**.
fractional crystallization:	A process in which crystallizing minerals are extracted from the melt as soon as they solidify and do not react further with the melt.
fractional melting:	A melting process in which partial melt is extracted from residual solid rock as soon as the melt forms.
free energy:	A term describing the amount of energy produced by a reaction that is available to do thermodynamic work.
geobarometry:	A term describing thermodynamic formulations that allow one to calculate the pressure at which an igneous and metamorphic mineral or mineral assemblage equilibrated.
geochronology:	The determination of age, typically by measurement of parent and daughter isotopes in minerals and rocks that allow one to calculate the time at which the system closed to isotopic re-equilibration.
geothermometry:	Thermodynamic and isotopic formulations that allow one to calculate the temperature at which an igneous and metamorphic mineral or mineral assemblage equilibrated.
glassy:	An adjective describing the texture in an igneous rock that has been rapidly chilled and where the melt has frozen to a glass.
gneissosity:	A planar feature in a metamorphic rock that is dominated by quartz and feldspar.
granoblastic:	An adjective describing a texture in a metamorphic rock in which the minerals show no preferred orientation.
heat capacity:	The quantity of heat required to raise the temperature of a system one degree Celsius at constant temperature and pressure.
heat of fusion:	The quantity of heat that must be added to a rock or mineral, already at the melting temperature, to produce one gram of melt.
hornfels:	A fine-grained contact-metamorphosed rock that breaks with conchoidal fractures.
hyaloclastite:	A volcanic deposit consisting of fine-grained fragments of glass.
hypabyssal:	An adjective describing a fine-grained igneous intrusion that formed at shallow depths or conditions under which such intrusions may occur. Compare with **plutonic**.
idioblastic:	A description of the shape of metamorphic minerals that are bounded by crystal faces. Compare with **euhedral**.
igneous petrogenesis:	The study of the origin and evolution of igneous rocks.
igneous petrography:	The description and classification of igneous rocks.
incompatible element:	An element that is not readily incorporated into crystallizing minerals and which is concentrated in the melt. Compare with **compatible element**.
incongruent melting:	The melting of a mineral that produces melt plus a mineral of different composition. Compare with **congruent melting**.
infiltration:	The condition in metamorphic rocks in which the composition of the fluid is fixed by fluid flowing in from an external source. Compare with **buffered**.

intrusion:	An igneous body emplaced into pre-existing rock.
invariant:	An adjective describing the situation wherein enough phases are present that temperature, pressure, and compositions of the phases are fixed.
ion-exchange reaction:	A reaction that involves the exchange of ions, such as Fe^{2+} and Mg^{2+}, between phases. As the reaction proceeds compositions of the minerals involved change, but the abundances do not. Compare with **mass transfer reaction**.
iron-enrichment index (Fe-index):	A geochemical index $[(FeO + 0.9\ Fe_2O_3)/(FeO + 0.9\ Fe_2O_3 + MgO)]$ that measures the extent to which iron has been enriched during differentiation of a magma.
isochemical metamorphism:	Metamorphism that proceeds without changing the bulk composition of the rock, apart from the loss of volatiles.
isochron:	A line on an isotope ratio diagram that identifies a suite of rock or mineral samples which formed at the same time.
isograd:	A line on a map showing the location of a univariant reaction in metamorphic rocks which separates zones of different metamorphic mineral assemblages and which represents a plane of constant metamorphic grade.
isotopes:	Two or more atoms that have the same number of electrons and protons but different numbers of neutrons. Isotopes of a particular element have the same chemical properties but different atomic weights.
isotopic fractionation:	The separation of isotopes of an element during naturally occurring processes as a result of the mass differences of their nuclei.
isotope geochemistry:	The study and application of stable and radiogenic isotopes to geological processes and their timescales.
komatiite:	An ultramafic lava that is common in the Archean Eon but extremely rare in younger rocks.
lapilli:	Pyroclastic material that is between 4 mm and 32 mm in size. Compare with **ash**, **block**, and **bomb**.
large igneous province (LIP):	A very large accumulation of igneous rocks (more than 100 000 km^2) that was emplaced over a very short period of time (a few million years or less).
lava dome:	A mound-shaped body resulting from the slow extrusion of highly viscous felsic lava.
layered mafic intrusion (LMIs):	An intrusive body of ultramafic to mafic rock that contains layers of different composition or texture.
leucosome:	Light-colored component of migmatite, formed by partial melting, See also **migmatite**.
lever rule:	A rule for locating the composition of a phase on a binary or ternary phase diagram.
lineation:	Any linear feature in a metamorphic rock, such as elongated minerals, the axis of a fold, or stretching direction of pebbles.
liquidus:	A line on an igneous phase diagram that shows the composition of the melt at any given temperature. Compare with **solidus**.
mafic:	A term describing relatively dark igneous rock. The term is derived to describe rocks rich in *ma*gnesium and iron (*fe*).
mafic selvage:	A margin to a leucosome that is enriched in ferromagnesian minerals and which is considered to represent residua after the extraction of granitic melt. See also **leucosome**, **migmatite**.
magma:	A term describing a mixture of lava and crystals.
magmatic differentiation:	Any process that causes the chemical composition of a magma to change during crystallization. See **filter pressing**, **flow segregation**.

major elements:	Elements in a rock analysis that are present in the abundance of more than 1.0 percent.
marl:	A carbonate-rich shale.
MASH zone:	A deep crustal zone of a magma system where **m**elting, **a**ssimilation, **s**torage, and **h**omogenization of the magma may occur.
mass-transfer reaction:	A metamorphic reaction, usually a barometer, where the abundances of the various minerals involved change as the reaction proceeds. Compare with **ion-exchange reaction**.
mélange:	A rock unit, usually found in fossil subduction zones, consisting of fragments of various rock types that record a range of different metamorphic facies.
melanosome:	Dark-colored component of migmatite formed by injection or separation of partial melt (leucosome). See **migmatite**.
metabasite:	A term describing a protolith for a metamorphic rock that has the composition of a basalt or gabbro but which shows no primary igneous textures.
metamorphism:	The crystallization or recrystallization of a rock under elevated temperatures below the melting temperature.
metamorphic assemblage diagrams:	Diagrams calculated from thermodynamic databases that show the pressure and temperature fields for metamorphic assemblages in a rock of fixed composition.
metamorphic field gradient:	The range of pressures and temperatures recorded from rocks in a suite of localities. For each locality, the $P–T$ conditions are assumed to have been locked in at the peak of metamorphism.
metaluminous:	An adjective describing rocks that have neither aluminum nor alkalis in excess of the amount needed to make feldpars. See **aluminum-saturation index**.
metasomatism:	Metamorphism that is accompanied by a change in the bulk composition of the rocks.
metastable:	A condition in which a set of unstable minerals persists in a rock because an energy barrier must be overcome before they can react to a stable assemblage.
mid-ocean ridge basalt (MORB):	Tholeiitic basalts with a rather restricted composition range that are erupted along the mid-ocean ridges. See also **ocean island basalt**.
migmatite:	A "mixed" rock consisting of dark layers (**melanosomes**) interlayered with light layers (**leucosomes**). The leucosomes are inferred to be related to melt that has been injected or which formed *in situ*, the melanosomes are thought to represent either the original rock or a residue left after melt has been extracted.
minor elements:	Elements in a rock or mineral analysis that have an abundance of 0.1 to 1.0 weight percent.
mode:	The abundance of minerals in a rock, by volume percent.
modified alkali–lime index (MALI):	A geochemical index calculated from the whole-rock analysis wherein $Na_2O + K_2O – CaO$ is plotted against SiO_2. The SiO_2 content at which $Na_2O + K_2O – CaO = 0.0$ provides the points in the alkali–lime index where the $Na_2O + K_2O$ and CaO curves cross in the alkali–lime index. See **alkali–lime index**.
mole fraction:	The ratio of the number of moles of an element in a given mineral to the total number of moles of all elements in the mineral.
network-forming ions:	Ions in a silicate melt, such as Si^{4+}, Al^{3+}, and P^{5+}, which sit in tetrahedral sites of the melt and which help polymerize the melt.
network-modifying ions:	Ions in a melt, such as Ca^{2+}, Fe^{3+}, Fe^{2+}, Mg^{2+}, and Na^+, which sit in sites other than tetrahedral sites and which help depolymerize the melt.
norm:	The abundance of minerals (in weight percent) that are likely to be present in the rock as calculated from its geochemical composition. The calculations assume that only anhydrous minerals were present, with the result that the norm is not identical to the mode for rocks containing hydrous phases such as micas or amphibole.

normative mineralogy: The mineralogy that would be present in a rock as calculated by the norm. See **norm**.

nucleation: The process by which a solid mineral starts to form from a melt.

nucleus: A cluster of ions in a melt or in a metamorphic rock that is large enough to be stable and to allow growth of the crystallizing mineral to proceed.

ocean island basalt (OIB): Basalt that forms the ocean islands. This encompasses a broader composition range than MORB, ranging from tholeiite to alkali basalt. Compare with **mid-ocean ridge basalt**.

ophiolite: A common association of peridotite (commonly serpentinized), gabbro, pillow basalt, and deep-sea sediments that is assumed to represent ocean floor material that has been incorporated into mountain belts.

orthogneiss: A gneiss that has been derived from an igneous protolith. Compare with **paragneiss**.

pahoehoe: A Hawaiian term for a basaltic lava flow with a smooth, ropy surface. Compare with **aa**.

paragneiss: A gneiss that was derived from a sedimentary protolith. Compare with **orthogneiss**.

partition coefficient: A thermodynamic term that describes the distribution of an element between a magma and a crystallizing mineral. See also **bulk-distribution coefficient.**

pegmatite: A textural term describing a coarse-grained igneous rock with grain size generally larger than 2.5 cm. Most pegmatites are felsic, with compositions similar to granite.

pelite: A term describing a shale or a rock that was derived from a shale. Pelitic schists are characterized by abundant mica and aluminous minerals in addition to garnet. Compare with **psammite**.

penetrative deformation: Deformation in a metamorphic rock that occurs at all scales.

peralkaline: Rocks with higher alkali contents than can be accommodated in feldspars alone. See **alkalinity index**.

peraluminous: An adjective describing rocks with more aluminum than can be accommodated in feldspars alone. See **aluminum-saturation index**.

peritectic: An invariant point in a phase diagram where one or more minerals react with the melt to make one or more new mineral phases.

perthite: A textural term describing the presence of tabular or irregular bands of alkali feldspar in a K-feldspar resulting from subsolidus exsolution.

petrochronology: A field of petrology that uses geochronology to constrain the timing and rates of mineral reactions and petrologic evolution.

petrogenetic grid: A phase diagram showing the locations of reactions governing the stability of assemblages in a rock of a given composition as a function of temperature and pressure or temperature and fluid composition.

phaneritic: A coarse-grained igneous rock where the constituent minerals can be identified by the naked eye.

phase: A chemically homogeneous part of a system that is separated from other phases in the system by a boundary. In petrology, phases include minerals, melt, and fluids. See also **component** and **phase rule**.

phase rule: An equation that tells how many phases must be present to constrain the conditions under which a system formed. The equation is $P + F = C + 2$, where P = number of phases, F = degrees of freedom or variance, and C = number of components. See also **component, phase,** and **variance.**

phenocryst: A mineral in a volcanic or hypabyssal rock that is distinctly larger than minerals in the matrix of that rock.

phreatomagmatic explosion: An explosion that emits steam, volcanic gas, and blocks, which occurs when magma encounters groundwater.

pillow:	A pillow-shaped structure, formed when basalt flows are erupted under water.
platinum-group elements (PGEs):	A group of elements (Pt, Pd, Ir, Os, Ru, Rh) that have similar behavior in the geologic environment, are commonly associated with each other, and which may be locally enriched in layered mafic intrusions.
plinian eruption:	A large pyroclastic eruption characterized by a high column of ash (20–50 km) that spreads over areas up to 10^6 km.
pluton:	A body of igneous rock (less than 100 km^2) that crystallized at depth in the crust.
plutonic:	An adjective describing igneous rocks that crystallized at depth in the crust where cooling was relatively slow and hence the rocks are coarse-grained.
polythermal projection:	A ternary igneous phase diagram where the variation in temperature is shown as contours.
porphyroblast:	A large, generally idioblastic grain in a metamorphic rock that is larger than the minerals in the matrix.
porphyry:	An igneous rock that contains relatively coarse-grained minerals in a finer-grained matrix.
postcumulate:	A term for a mineral that formed late in the crystallization history of a rock and which fills in the interstices around the earlier crystallizing minerals. Compare with **cumulate**.
preferred orientation:	The tendency for minerals in a deformed rock to show a similar planar or linear orientation. See also **lineation** and **schistosity**.
protolith:	A term describing the unmetamorphosed parent to a metamorphic rock.
psammite:	A metamorphosed sandstone that contained abundant clay. During metamorphism this protolith may form a quartz–mica rock with garnet but without other aluminous minerals.
pseudosection:	See **metamorphic assemblage diagram**.
pseudoternary diagram:	A ternary phase diagram that is projected from a system with more than three components.
pumice:	A pyroclastic rock that is composed of compressed glass fragments and commonly is porous enough to float on water.
pyroclastic:	An adjective describing fragmental volcanic rocks produced by explosive volcanic eruptions. Synonymous with volcaniclastic. See also **ash**, **lapilli**, **block**, and **bomb**.
pyroclastic fall deposit:	A pyroclastic deposit that covers the topography uniformly, indicating that it fell from an eruption cloud of great height.
pyroclastic flow deposit:	A pyroclastic deposit that occupies the low areas of topography, indicating that it formed from a flow of debris generated during a volcanic eruption.
rare-earth element (REE):	A group of trace elements with atomic weights ranging from 57 to 71 that have similar chemical behavior. See also **trace element**.
restite:	A rock, commonly rich in aluminum minerals, which is the residua after partial melting.
retrograde metamorphism:	Metamorphism that occurs in response to introduction of water into the crust, which allows minerals indicative of relatively low temperature to form.
reverse graded bedding:	A texture in metamorphosed sedimentary rocks in which clays in the fine-grained top of a graded-bedding sequence have reacted to form a coarser rock than was formed by the quartz-rich, originally coarser bottom portion of the graded-bedding sequence.
rodingite:	A rock rich in calcium silicates that was formed by calcium metasomatism of rocks adjacent to a serpentinite. The protolith is usually basalt or gabbro, but other protoliths are known.

schistosity: A planar feature in a metamorphic rock that is caused by preferred orientation of micas. See also **lineation**.

scoria: A highly vesiculated mafic rock.

seismic anisotropy: The directional dependence of seismic velocities in rocks composed of minerals with a strong preferred orientation or rocks with variations in the spatial distributions of constituent mineral phases.

sheeted dikes: A horizon in an ophiolite composed of dikes that have intruded each other during a spreading event. The dikes commonly are chilled only on one side and do not preserve country rock into which the dikes were intruded. See also **ophiolite**.

sill: A tabular igneous body that was intruded parallel to bedding in a sedimentary rock or, more rarely, parallel to the foliation of a metamorphic rock.

skarn: A calc-silicate rock that was formed by fluids from an igneous intrusion reacting with marble.

solidus: A line on a phase diagram that indicates the composition of the solids as a function of temperature. Compare with **liquidus**.

solvus: A curve on a phase diagram, separating the field of a homogeneous solid solution from a field of several phases that may form by exsolution.

stoping: A magmatic process whereby an intruding magma makes space for itself by breaking blocks off the wall of the intrusion and engulfing them in the ascending magma.

strombolian eruption: A moderate-sized pyroclastic eruption where the plume is relatively low (1–3 km) and the fragments accumulate around the vent.

supercritical fluid: A fluid that occurs at temperatures and pressures above the critical point. See also **critical point**.

suprasubduction-zone ophiolite: An ophiolite that formed in an extensional environment above a subduction zone.

surface energy: The energy that is produced by ions on the surface of a mineral that are not bonded to another element. Surface energy is the amount of work required to form more of a surface and varies with the density and strength of broken atomic bonds on a mineral surface.

system: The part of the universe under consideration. Systems are usually labeled by their chemical constituents. See also **phase rule**.

terminal reaction: A reaction involves only one mineral reacting to two or more other minerals and which represents the thermal maximum for that mineral. Compare with **tie-line flip reaction**.

thermal barrier: A plane represented by the congruent melting of an internal phase on a phase diagram. The plane represents a barrier that the melt cannot cross by any process of igneous differentiation.

thermobarometry: The integration of geothermometry and geobarometry to estimate the pressure and temperature at which a rock equilibrated. See **geothermometry, geobarometry**.

tholeiite: A basalt that contains normative hypersthene. Olivine tholeiite also contains normative olivine; quartz tholeiite contains normative quartz. Compare with **alkali basalt**.

tie-line flip reaction: A metamorphic reaction that contains two or more minerals on either side. Crossing the reaction changes the assemblage that is stable, but all phases remain intrinsically stable. Compare with **terminal reaction**.

topology: A diagram that shows the orientation of univariant curves around an invariant point but which is not calibrated with respect to temperature and pressure.

trace element:	An element that is present within a rock or mineral that makes up less than 0.1 percent of the rock. Abundances are usually given in parts per million, rather than percent.
tuff:	A volcanic ash that has been lithified, usually by compaction and welding. See **welding**.
ultra-high-pressure (UHP) metamorphism:	Metamorphism that has taken place in the coesite stability field, 25 kbar or above.
ultra-high-temperature (UHT) metamorphism:	Metamorphism that has occurred at temperatures above 1000 °C.
ultramafic:	A dark rock with more than 90 percent mafic minerals, commonly olivine and pyroxene.
univariant:	An adjective describing a reaction that contains enough phases that only one variable (usually temperature or pressure) needs be fixed to constrain the system. See **phase rule**.
variance:	The number of variables that must be constrained to determine the pressure and temperature of a reaction. See also **phase rule**, **invariant**, **univariant**, and **divariant**.
Volcanic Explosivity Index (VEI):	An index that estimates the amount of energy that was released by a volcanic eruption based upon the amount of material ejected.
volcanogenic massive sulfide (VMS) deposit:	A metal sulfide ore deposit, mainly copper–zinc, formed by volcanic-associated hydrothermal events in submarine environments.
vesicle:	A spherical cavity in a volcanic flow that was produced by a trapped gas bubble.
viscosity:	A property that indicates how resistant a melt (or less commonly a rock) is to flow.
volcanic neck:	A column of hypabyssal igneous rock, solidified in the conduit of a volcano and that remains after the volcanic cone is eroded away.
volcanic:	An adjective describing igneous rocks that crystallized on the Earth's surface where cooling was relatively fast and hence the rocks are fine-grained.
welding:	Welding is a process by which pumice and ash fragments in a tuff are fused together by solid-state diffusion during compaction.
xenocryst:	A crystal in an igneous rock that is not related to the rock in which it occurs but has instead been incorporated from the rocks through which the magma ascends.
xenolith:	An enclave in an igneous rock that has been extracted from a wall rock that had a different origin than the rock in which it occurs. Compare with **autolith** and **enclave**.

References

Albee, A. L., 1965, A petrogenetic grid for the Fe–Mg silicates of pelitic schists. *American Journal of Science*, 263, 512–36.

Amelin, Y., Li, C., Valeyev, O., and Naldrett, A. J., 2000, Nd–Pb–Sr isotope systematics of crustal assimilation in the Voisey's bay and Mushuau intrusions, Labrador, Canada. *Economic Geology*, 95, 815–30.

Andersen, S. G., Bohse, H., and Steenfelt, A., 1981, A geological section through the southern part of the Ilimaussaq intrusion. Rapport. *Grønlands Geologiske Undersøgeise*, 103, 39–42.

Anderson, I. C., Frost, C. D., and Frost, B. R., 2003, Petrogenesis of the Red Mountain pluton, Laramie anorthosite complex, Wyoming: Implications for the origin of A-type granites. *Precambrian Research*, 124, 243–67.

Anderson, J. G. C., 1937, The Etive granite complex, *Journal of the Geological Society of London*, 93, 487–533.

Anonymous, 1972, Penrose Field Conference on ophiolites. *Geotimes*, 17, 24–5.

Arai, S., 1975, Contact metamorphosed dunite–harzburgite complex in the Chugoku District, western Japan. *Contributions to Mineralogy and Petrology*, 52, 1–16.

Archibald, N. J., Bettenay, L. F., Binns, R. A., Groves, D. I., and Gunthorpe, R. J., 1978, The evolution of Archaean greenstone terrains, Eastern Goldfields Province, Western Australia. *Precambrian Research*, 6, 103–31.

Arnaud, N. O., Vidal, P., Tapponnier, P., Matte, P., and Deng, W. M., 1992, The high K_2O volcanism of northwestern Tibet: Geochemistry and tectonic implications. *Earth and Planetary Science Letters*, 111, 351–67.

Ashwal, L. D., 2010, The temporality of anorthosites. *Canadian Mineralogist*, 48, 711–28.

Atherton, M. P. and Ghani, A. A., 2002, Slab breakoff: A model for Caledonian, Late Granite syn-collisional magmatism in the orthotectonic (metamorphic) zone of Scotland and Donegal, Ireland. *Lithos*, 62, 65–85.

Atkins, F. B., 1969, Pyroxenes of the Bushveld Intrusion, South Africa. *Journal of Petrology*, 10, 222–49.

Bailey, J. C., Gwozdz, R., Rose-Hanses, J., and Sørensen, H., 2001, Geochemical overview of the Ilimaussaq alkaline complex, South Greenland. *Geology of Greenland Survey Bulletin*, 190, 35–53.

Baksi, A. K., 2018, Paraná flood basalt volcanism primarily limited to ~1 Myr beginning at 135 Ma: New $^{40}Ar/^{39}Ar$ ages for rocks from Rio Grande do Sul, and critical evaluation of published radiometric data. *Journal of Volcanology and Geothermal Research*, 355, 66–77.

Baksi, A. K., 2010, Comment to "Distribution and geochronology of the Oregon Plateau (USA) flood basalt volcanism: The Steens basalt revisited" by M. E. Brueseke et al. *Journal of Volcanology and Geothermal Research*, 196, 134–8.

Barberi, F., Ferrara, G., Santacrocs, R., Treuil. M., and Varet, J., 1975, A transitional basalt–pantellerite sequence of fractional crystallization, the Boina centre (Afar Rift, Ethiopia). *Journal of Petrology*, 16, 22–56.

Barker, F., Wones, D. R., Sharp, W. N., and Desborough, G. A., 1975, The Pikes Peak batholith, Colorado Front Range, and a model for the origin of gabbro–anorthosite–syenite–potassic granite suite. *Precambrian Research*, 2, 97–160.

Barrow, G., 1893, On an intrusion of muscovite–biotite gneiss in the southeastern Highlands of Scotland, and its accompanying metamorphism. *Geological Society of London Quarterly Journal*, 49, 330–58.

Barrow, G., 1912, On the geology of Lower Dee-side and the southern Highland border. *Proceedings of the Geologists' Association*, 23, 274–90.

Barry, T. L., Self, S., Kelley, S. P., Reidel, S., Hooper, P., and Widdowson, M., 2010, New $^{40}Ar/^{39}Ar$ dating of the Grande Ronde lavas, Columbia River Basalts, USA: Implications for duration of flood basalt eruption episodes. *Lithos*, 118, 213–22.

Basaltic Volcanism Study Project, 1981, *Basaltic Volcanism on the Terrestrial Planets*. Pergamon Press, Inc., New York, NY.

Batchelor, R. A., 1987, Geochemical and petrological characteristics of the Etive granitoid complex, Argyll. *Scottish Journal of Geology*, 23, 227–49.

Bateman, P. C. and Chappell, B. W., 1979, Crystallization, fractionation, and solidification of the Tuolumne intrusive series, Yosemite National Park, California. *Geological Society of America Bulletin*, 90, 465–82.

Batiza, R., 1982, Abundances, distribution and sizes of volcanoes in the Pacific Ocean and implications for the origin of non-hot spot volcanoes. *Earth and Planetary Science Letters*, 60, 195–206.

Beard, J. S. and Day, H. W., 1988, Petrology and emplacement of reversely zoned gabbro–diorite plutons in the Smartville complex, northern California. *Journal of Petrology*, 29, 965–95.

Bebout, G. E., 2014, Chemical and isotopic cycling in subduction zones. In *Treatise on Geochemistry*, 2nd edn, eds. K. Turekian and H. Holland. Elsevier, Amsterdam, vol. 4, pp. 703–46.

Benioff, H., 1949, Seismic evidence for the fault origin of oceanic deeps. *Geological Society of America Bulletin*, 60, 1837–66.

Berger, A., Schmid, S. M., Engi, M., Bosquet, R., and Wiederker, M., 2011, Mechanisms of mass and heat transport during Barrovian metamorphism: A discussion based on field evidence from the central Alps (Switzerland/northern Italy). *Tectonics*, 30, TC1007, doi 10.1029/2009TC002622.

Berman, R. G., 1988, Internally-consistent thermodynamic data for minerals in the system Na_2O–K_2O–CaO–MgO–FeO–$Fe_2O_3Al_2O_3$–SiO_2–TiO_2–H_2O–CO_2. *Journal of Petrology*, 29, 445–522.

Berman, R. G., 1991, Thermobarometry using multi-equilibrium calculations: A new technique, with petrologcal applications. *Canadian Mineralogist*, 29, 833–55.

Bhaskar Rao, Y. J., Chetty, T. R. K., Janardhan, A. S., and Gopalan, K., 1996, Sm–Nd and Rb–Sr ages and P–T history of the Archean Sittampundi and Bhavani layered meta-anorthosite complexes in Cauvery shear zone, South India: Evidence for Neoproterozoic reworking of Archean crust. *Contributions to Mineralogy and Petrology*, 125, 237–50.

Bianchini, G., Beccaluva, L., and Siena, F., 2008, Post-collisional and intraplage Cenozoic volcanism in the rifted Apennines/Adriatic domain. *Lithos*, 101, 125–40.

Bingen, B., Davis, W. J., and Austrheim, H., 2001, Zircon U–Pb geochronology in the Bergen arc eclogites and their Proterozoic protoliths, and implications for the pre-Scandian evolution of the Caledonides in western Norway. *Geological Society of America Bulletin*, 113, 640–9.

Binns, R. A., Gunthorpe, R. J., and Groves, D. I., 1976, Metamorphic patterns and development of greenstone belts in eastern Yilgarn Block, Western Australia. In *The Early History of the Earth*, ed. B. F. Windley. John Wiley & Sons, New York, NY, pp. 303–13.

Bird, P., 1979, Continental delamination and the Colorado Plateau. *Journal of Geophysical Research*, 84, 7561–71.

Blackman, D. K., Ildefonse, B., John, B. E., Ohara, Y., Miller, D. J., MacLeod, C. J., and the Expedition 304/305 Scientists, 2006, *Proceedings IODP, 304/305*: College Station TX (Integrated Ocean Drilling Program Management International, Inc.). doi:10.2204/ iodp.proc.304305.103.2006.

Blackman, D. K. and 48 others, 2011, Drilling constraints on lithospheric accretion and evolution at Atlantis Massif, Mid-Atlantic Ridge 30° N. *Journal of Geophysical Research*, 116, B07103, doi 10.1029/2010JB007931.

Boria, E. and Conticelli, S., 2007, Mineralogy and petrology of associated Mg-rich ultrapotassic, shoshonitic, and calc-alkaline rocks: The Middle Latin valley, monogenetic volcanos, Roman Magmatic Province, southern Italy. *Canadian Mineralogist*, 45, 1443–69.

Bottinga, Y. and Javoy, M., 1975, Oxygen isotope partitioning among the mineral in igneous and metamorphic rocks. *Reviews of Geophysics and Space Physics*, 13, 401–18.

Boudier, F. and Nicolas, A., 2011, Axial melt lenses at oceanic ridges: A case study in the Oman ophiolite. *Earth and Planetary Science Letters*, 304, 313–25.

Bowen, N. L., 1913, The melting phenomena of the plagioclase feldspars. *American Journal of Science*, 35, 577–99.

Bowen, N. L., 1915, The crystallization of haplobasaltic, haplodioritic, and related magmas. *American Journal of Science*, 40, 161–85.

Bowen, N. L., 1940, Progressive metamorphism of siliceous limestone and dolomite. *Journal of Geology*, 48, 225–74.

Bowen, N. L., 1945, Phase equilibria bearing on the origin and differentiation of alkaline rocks. *American Journal of Science*, 243A, 75–89.

Bowen, N. L. and Andersen, O., 1914, The binary system MgO–SiO_2. *American Journal of Science*, 37, 487–500.

Bowen, N. L. and Schairer, J. F., 1935, The system MgO–FeO–SiO_2. *American Journal of Science*, 29, 151–217.

Bowen, N. L. and Tuttle, O. F., 1950, The system $NaAlSi_3O_8$–$KAlSi_3O_8$–H_2O. *Journal of Geology*, 58, 489–511.

Braun, M. G. and Kelemen, P. B., 2002, Dunite distribution in the Oman Ophiolite: Implications for melt flux through porous dunite conduits. *Geochemistry, Geophysics, Geosystems*, 3, 1–21, doi: 10.1029/2001GC000289.

Brotzu, P., Morbidelli, L., and Piccirillo, E. M., 1983, The basanite to peralkaline phonolite suite of the Pilo-quaternary Nyambeni multicenter volcanic range (East Kenya Plateau). *Neues Jahrbuch fur Mineralogy, Abhanglungen*, 147, 253–80.

Brown, E. H., 1974, Comparison of the mineralogy and phase relations of blueschists from the North Cascades, Washington, and greenschists from Otago, New Zealand. *Geological Society of America Bulletin*, 85, 333–44.

Brown, G. C. and Fyfe, W. S., 1970, The production of granitic melts during ultrametamorphism. *Contributions to Mineralogy and Petrology*, 28, 310–18.

Brown, G. C. and Mussett, A. E., 1981, *The Inaccessible Earth*. Allen and Unwin, London.

Brown, G. M., Holland, J. G., Sigurdsson, H., Tomblin, J. F., and Arculus, R. J., 1977, Geochemistry of the Lesser Antilles volcanic island arc. *Geochimica et Cosmochimica Acta*, 41, 785–801.

Brueseke, M. E., Heizler, M. T., Hart, W. K., and Mertzman, S. A., 2007, Distribution and geochronology of the Oregon Plateau (USA) flood basalt volcanism: The Steens basalt revisited. *Journal of Volcanology and Geothermal Research*, 161, 187–214.

Bryan, S. E., Riley, T. R., Jerram, D. A., Stephens, C. J., and Leat, P. T., 2002, Silicic volcanism: An undervalued component of large igneous provinces and volcanic rifted margins. *Geological Society of America Special Paper*, 362, 97–118.

Bryant, B., McGrew, L. S., and Wobus, R. A., 1981, Geologic map of the Denver 1° by 2° quadrangle, north-central Colorado. United States Geologic Survey Miscellaneous Investigations Map I-1163, scale 1:250,000.

Buchan, K. L., Goutier, J., Hamilton, M. A., Ernst, R. E., and Matthews, W. A., 2007, Paleomagnetism, U–Pb geochronology, and geochemistry of Lac Esprit and other dyke swarms, James Bay area, Quebec and implications for Paleoproterozoic deformation of the Superior province. *Canadian Journal of Earth Sciences*, 44, 643–64.

Bucher, K. and Grapes, R., 2009, The eclogite-facies Allalin Gabbro of the Zermatt–Saas ophiolite, western Alps: A record of subduction zone hydration. *Journal of Petrology*, 50, 1405–42.

Burgess, S. D., Bowring, S. A., Fleming, T. H., and Elliot, D. H., 2015, High-precision geochronology links the Ferrar large igneous province with early-Jurassic ocean anoxia and biotic crisis. *Earth and Planetary Science Letters*, 415, 90–9.

Burnham, C. W., 1979, The importance of volatile constituents. In *The Evolution of Igneous Rocks, 50th Anniversary Perspectives*. Princeton University Press, Princeton, pp. 439–82.

Burnham, C. W., Holloway, J. R., and Davis, N. F., 1969, *Thermodynamic Properties of Water to 1,000 °C and 10,000 bars*. Geological Society of America Special Paper 132, Geological Society of America, Boulder, CO.

Camp, V. E., 1995, Mid-Miocene propagation of the Yellowstone mantle plume head beneath the Columbia River basalt source region. *Geology*, 23, 435–8.

Capitani, G. and Mellini, M., 2004, The modulated crystal structure of antigorite: The m=17 polysome, *American Mineralogist*, 89, 147–58.

Carmichael, D. M., 1978, Metamorphic bathozones and bathograds: A measure of the depth of post-metamorphic uplift and erosion on the regional scale. *American Journal of Science*, 278, 769–97.

Carmichael, I. S. E., 1964, The petrology of Thingmuli, a Tertiary volcano in eastern Iceland. *Journal of Petrology*, 5, 435–60.

Carrington, D. P., 1995, The relative stability of garnet–cordierite and orthopyroxene–sillimanite–quartz assemblages in metapelitic granulites: Experimental data. *European Journal of Mineralogy*, 7, 949–60.

Cashman, K. V., Sparks, R. S. J., and Blundy, J. D., 2017, Vertically extensive and unstable magmatic systems: A unified view of igneous processes. *Science*, 355, eaag3055, doi: 10.1126/science.aag3055.

Cawthorn, R. G., Ellam, R. M., Ashwal, L. D., and Webb, S. J., 2012, A clinopyroxenite intrusion from the Plianesberg Alkaline Province, *South Africa, Precambrian Research*, 198–9, 25–36.

Černý, P., London, D., and Novák, M., 2012, Granitic pegmatites as reflections of their sources. *Elements*, 8, 289–94.

Chao, E. C. T., 1966, Impact metamorphism, In *Researches in Geochemistry*, Volume 2, ed. P. H. Ableson. John Wiley & Sons, New York, NY, pp. 204–33.

Chappell, B. W. and White, A. J. R., 1974, Two contrasting granite types. *Pacific Geology*, 8, 173–4.

Charlier, B., Duchsesne, J. C., Vander Auwera, J., Storme, J.-Y., Marquil, R., and Longhi, J., 2010, Polybaric fractional crystallization of high-alumina basalt parental magmas in the Egersund-Ogna Massif-type anorthosite (Rogalandm, SW Norway) constrained by plagioclase and high-alumina orthopyroxene megacrysts. *Journal of Petrology*, 51, 2515–46.

Chernosky, J. V. Jr., 1974, The stability of clinochlore at low pressures and the free energy of formation of Mg-cordierite. *American Mineralogist*, 59, 496–507.

Chiba, H., Chacko, T., Clayton, R. N., and Goldsmith, J. R., 1989, Oxygen isotope fractionations involving diopside, forsterite, magnetite, and calcite: Application to geothermometry. *Geochimica et Cosmochimica Acta*, 53, 2985–95.

Chidester, A. H., 1962, *Petrology and Geochemistry of Selected Talc-Bearing Ultramafic Rocks and Adjacent Country Rocks in North-Central Vermont*. United States Geological Survey Professional Paper 345, United States Government Printing Office, Washington, DC.

Chopin, C., 1981, Talc–phengite: A widespread assemblage in high-grade pelitic blueschists of the Western Alps. *Journal of Petrology*, 22, 628–50.

Chopin, C., 1984, Coesite and pure pyrope in high-grade blueschists of the Western Alps: A 1st record and some consequences. *Contributions to Mineralogy and Petrology*, 86, 107–18.

Christiansen, E. and McCurry, M., 2007, Contrasting origins of Cenozoic silicic volcanic rocks from the western Cordillera of the United States. *Bulletin of Volcanology*, 70, 251–67.

Christiansen, R. L., Lowenstern, J. B., Smith, R. B., Heasler, H., Morgan, L. A., Nathenson, M., Mastin, L. G., Muffler, P., and Robinson, J. E., 2007, Preliminary assessment of volcanic and hydrothermal hazards in Yellowstone National Park and vicinity. United States Geological Survey Open-file Report 2007-1071.

Christiansen, R. L. and Peterson, D. W., 1981, Chronology of the 1980 eruptive activity. *United States Geological Survey Professional Paper*, 1250, 17–30.

Clague, D. A. and Sherrod, D. R., 2014, Growth and degradation of Hawaiian volcanoes. *United States Geological Survey Professional Paper*, 1801, 97–146.

Clarke, D. B., 1992, *Granitoid Rocks*. Chapman and Hall, London.

Cochran, J. R., 2008, Seamount volcanism along the Gakkel Ridge, Arctic Ocean. *Geophysical Journal International*, 174, 1153–73.

Coffin, M. F. and Eldholm, O., 1992, Volcanism and continental break-up: A global compilation of large igneous provinces. *Geological Society Special Publications*, 68, 17–30.

Colchen, M., LeFort, P., and Pecher, A., 1980, *Carte geologique Annapurna- Manaslu-Ganesh, Himalaya du Nepal 1/200.000*. Centre National de la Recherche Scientifique, Paris.

Coleman, R. G. and Lanphere, M. A., 1971, Distribution and age of high-grade blueschists, associated eclogites, and amphibolites from Oregon and California. *Geological Society of America Bulletin*, 82, 2397–412.

Collerson, K. D. and Fryer, B. J., 1978, The role of fluids in the formation and subsequent development of early continental crust. *Contributions to Mineralogy and Petrology*, 67, 151–67.

Compston, W., Williams, I. S., and Meyer, C., 1984. U–Pb geochronology of zircons from Lunar Breccia 73217 using a Sensitive High Mass-Resolution Ion Microprobe. Proceedings of the Fourteenth Lunar and Planetary Science Conference, Part 2. *Journal of Geophysical Research*, 89, B525–34.

Condie, K. C., 1993, Chemical composition and evolution of the upper continental crust: Contrasting results from surface samples and shales. *Chemical Geology*, 104, 1–37.

Coombs, D. S., 1954, The nature and alteration of some Triassic sediments from Southland, New Zealand. *Transactions of the Royal Society of New Zealand*, 82, 65–109.

Coombs, D. S., 1960, Lawsonite metagraywackes in New Zealand. *American Mineralogist*, 45, 454–55.

Coombs, D. S., Nakamura, Y., and Vuagnat, M., 1976, Pumpellyite–actinolite facies schists of the Taveyanne Formation near Loeche, Valais, Switzerland. *Journal of Petrology*, 17, 440–71.

Corliss, J. B., Dymond, J., Gordon, L. I., Edmond, J. M., Von Herzen, R. P., Ballard, R. D., Green, K., Williams, D., Bainbridge, A., Crane, K., and Van Andel, T. H., 1979, Submarine thermal springs on the Galapagos Rift. *Science*, 203, 1073–83.

Cox, K. G., 1980, A model for flood basalt volcanism. *Journal of Petrology*, 21, 629–50.

Crisp, J. A., 1983, Rates of magma emplacement and volcanic output. *Journal of Volcanology and Geothermal Research*, 20, 177–211.

Cross, W., Iddings, J. P., Pirsson, L. W., and Washington, H. S., 1902, A quantitative chemicomineralogical classification and nomenclature of igneous rocks. *Journal of Geology*, 10, 555–690.

Danckwerth, P. A. and Newton, R. C., 1978, Experimental determination of the spinel peridotites to garnet peridotites reaction in the system $MgO–Al_2O_3–SiO_2$ in the range of 900–1100°C and Al_2O_3 isopleths of enstatite in the spinel field. *Contributions to Mineralogy and Petrology*, 66, 189–201.

Dawson, J. B., 1967, A review of the geology of kimberlite. In *Ultramafic and Related Rocks*, ed. P. J. Wyllie. John Wiley & Sons, New York, NY, pp. 241–51.

Dawson, J. B. and Smith, J. V., 1988, Metasomatised and veined upper-mantle xenoliths from Pello Hill, Tanzania: Evidence for anomalously-light mantle beneath the Tanzanian sector of the East African Rift Valley. *Contributions to Mineralogy and Petrology*, 100, 510–27.

De Paoli, M. C., Clarke, G. L., and Daczko, N. R., 2012, Mineral equilibrium modeling of the granulite–eclogite transition: Effects of whole-rock composition on metamorphic facies type-assemblages. *Journal of Petrology*, 53, 949–70.

Deer, W. A., Howie, R. A., and Zussman, J., 1978, *Rock-Forming Minerals. Volume 2A, Single-Chain Silicates*, 2nd edn. John Wiley & Sons, New York, NY.

Deer, W. A., Howie, R. A., and Zussman, J., 1997, *Rock-Forming Minerals. Volume 2B, Double-Chain Silicates*, 2nd edn. The Geological Society, London.

Di Renzo, V., Di Vito, M. A., Arienzo, I., Carandente, A., Civetta, L., D'Antonio, G. F., Orsi, G., and Tonarini, S., 2007, Magmatic history of Somma-Vesuvius on the basis of new geochemical and isotopic data from a deep borehole (Camalodoli della Torre). *Journal of Petrology*, 48, 753–84.

Dick, H. J. B. and 27 others, 2000, A long *in situ* section of the lower ocean crust: Results of ODP Leg 176 drilling at the Southwest Indian Ridge, *Earth and Planetary Science Letters*, 179, 31–51.

Dick, H. J. B., Natland, J. H., and Ildefonse, B., 2006, Past and future impact of deep drilling in the oceanic crust and mantle. *Oceanography*, 19, 72–80.

Dickinson, W. R., 1970, Relations of andesites, granites, and derivative sandstones to arc-trench tectonics. *Reviews of Geophysics and Space Physics*, 8, 813–60.

Dickinson, W. R., Ojakangas, R. W., and Stewart, R. J., 1969, Burial metamorphism of the Late Mesozoic Great Valley Sequence, Cache Creek, California. *Geological Society of America Bulletin*, 80, 519–26.

Dietz, R. S., 1959, Shatter cones in cryptoexplosion structures (meteorite impact?). *Journal of Geology*, 67, 495–505.

Dobson, D. P., Meredith, P. G., and Boon S. A., 2002, Simulation of subduction zone seismicity by dehydration of serpentine. *Science*, 298, 1407–10.

Duke, J. M., 1988, Magmatic segregation deposits of chromite. In *Ore Deposit Models*, eds. R. G. Roberts and P. A. Sheanhan. Geoscience Canada Reprint Series 3, pp. 133–43.

Dyar, M. D., Gunter, M. E., and Tasa, D., 2008. *Mineralogy and Optical Mineralogy*, Mineralogical Society of America, Chantilly, VA.

Dyment, J., Arkani- Hamed, J., and Ghods, A., 1997, Contribution of serpentinized ultramafics to marine magnetic anomalies at slow and intermediate spreading centres: Insights from the shape of the anomalies. *Geophysics Journal International*, 129, 691–701.

Dzurisin, D., Koyanagi, R. Y., and English, T. T., 1984, Magma supply and storage at Kilauea volcano, Hawaii, 1956–1983. *Journal of Volcanology and Geothermal Research*, 21, 177–206.

Eby, G. N., 1985, Monteregian Hills II. Petrography, major and trace element geochemistry and strontium isotopic chemistry of the eastern intrusions: Mounts Shefford, Brome, and Megantic. *Journal of Petrology*, 26, 418–48.

Economist, 2017, After electric cars, what more will it take for batteries to change the face of energy? Published August 12, 2017, www.economist.com/news/briefing/21726069-no-need-subsidies-higher-volumes-and-better-chemistry-are-causing-costs-plum met-after, accessed March 8, 2018.

Elthon, D., 1989, Pressure of origin of primary mid-ocean ridge basalts. *Geological Society Special Publications*, 42, 125–36.

Engi, M., Lanari, P., and Kohn, M. J., 2017, Significant ages: An introduction to petrochronology. *Reviews in Mineralogy and Geochemistry*, 83, 1–12.

England, P. C. and Thompson, A. B., 1984, Pressure–temperature–time paths of regional metamorphism I. Heat transfer during the evolution of regions of thickened continental crust. *Journal of Petrology*, 25, 894–928.

Eskola, P., 1915, On the relations between the chemical and mineralogical composition in the metamorphic rocks of the Orijärvi region. *Bulletin de la Commission géologique de Finlande*, 44, 1–107; English summary 109–45.

Eskola, P., 1920, The mineral facies of rocks. *Norsk Geologisk Tidsskrift*, Bd 6, 143–94.

Eugster, H. P., 1985, Granites and hydrothermal ore deposits: A geochemical framework. *Mineralogical Magazine*, 49, 7–23.

Evans, B. W., 1977, Metamorphism of alpine peridotite and serpentinite. *Annual Reviews of Earth and Planetary Sciences*, 5, 397–447.

Evans, B. W., 1990, Phase relations of epidote-blueschists. *Lithos*, 25, 3–23.

Evans, B. W. and Brown, E. H., 1987, Comment and reply on blueschists and eclogites: Reply, *Geology*, 15, 773–5.

Faggart, B. E., Jr. and Basu, A. R., 1985, Origin of the Sudbury Complex by meteoritic impact: Neodymium isotopic evidence. *Science*, 230, 436–9.

Fairchild, L. M., Swanson-Hysell, N. L., Ramezani, J., Sprain, C. J., and Bowring, S. A., 2017, The end of Midcontinent Rift magmatism and the paleogeography of Laurentia. *Lithosphere*, 9, 117–33.

Farmer, G. L. and DePaolo, D. K., 1983, Origin of Mesozoic and Tertiary granite in the western United States and implications for pre-Mesozoic crustal structure. I. Nd and Sr isotopic studies in the geocline of the northern Great Basin. *Journal of Geophysical Research*, 88, 3379–402.

Favaro, S., Schuster, R., Handy, M. R., Scharg, A., and Pestal, F., 2015, Transition from orogeny-perpedicular to orogeny-parallel exhumation and cooling during crustal indentation: Key constraints from ^{147}Sm/^{144}Nd and ^{87}Rb/^{87}Sr geochronology (Tauern Window, Alps). *Tectonophysics*, 665, 1–16.

Ferguson, J., 1964, Geology of the Ilimaussaq alkaline intrusion, south Greenland. *Bulletin Grølands Geologiske Undersøgelse*, 39.

Ferry, J. M. and Spear, F. S., 1978, Experimental calibration of the partitioning of Fe and Mg between biotite and garnet. *Contributions to Mineralogy and Petrology*, 66, 113–17.

Fettes, D. and Desmons, J., 2007, *Metamorphic Rocks. A Classification and Glossary of Terms. Recommendations of the International Union of Geological Sciences Subcommission on the Systematics of Metamorphic Rocks.* Cambridge University Press, Cambridge.

Fisher, R. V., 1966, Rocks composed of volcanic fragments. *Earth-Science Reviews*, 1, 287–98.

Fitton, J. G. and Godard, M., 2004, Origin and evolution of magmas on the Ontong Java plateau. *Geological Society of London Special Publications*, 229, 151–78.

Fitton, J. G., Mahoney, J. J., Wallace, P. J., and Saunders, A. D., 2004, Origin and evolution of the Ontong Java plateau: Introduction. *Geological Society of London Special Publications*, 229, 1–8.

Fodor, R. V., Frey, F. A., Bauer, G. R., and Clague, D. A., 1992, Ages, rare-earth element enrichment, and petrogenesis of tholeiitic and alkali basalts from Kahoolawe Island, Hawaii. *Contributions to Mineralogy and Petrology*, 110, 442–62.

Forsyth, D. W., 1977, The evolution of the upper mantle beneath mid-ocean ridges. *Tectonophysics*, 38, 89–118.

Frey, F. A. and Clague, D. A., 1983, Geochemistry of diverse basalt types from Loihi Seamount, Hawaii: Petrographic implications. *Earth and Planetary Science Letters*, 66, 337–55.

Frey, F. A., Walker, N., Stakes, D., Hart, S. R., and Nielsen, R., 1993, Geochemical characteristics of basaltic glasses from the AMAR and FAMOUS axial valleys, Mid-Atlantic Ridge (36°–37° N): Petrogenetic implications. *Earth and Planetary Science Letters*, 115, 117–36.

Frey, M., Hunziker, J. C., Frank, W., Bocquer, J., Dal Piaz, G. V., and Niggli, E., 1974, Alpine metamorphism of the Alps, a review. *Schweizerische Mineralogische und Petrographische Mitteilungen*, 54, 247–90.

Frost, B. R., 1975, Contact metamorphism of serpentinite, chloritic blackwall, and rodingite at Paddy-Go-Easy pass, Central Cascades, Washington. *Journal of Petrology*, 16, 272–313.

Frost, B. R., 1976, Limits to the assemblage forsterite–anorthite as inferred from peridotite hornfelses, Icicle Creek, Washington. *American Mineralogist*, 61, 732–50.

Frost, B. R., 1980, Observations on the boundary between zeolite facies and prehnite–pumpellyite facies. *Contributions to Mineralogy and Petrology*, 73, 365–73.

Frost, B. R., Arculus, R. J., Barnes, C. G., Collins, W. J., Ellis, D. J., and Frost, C. D., 2001, A geochemical classification of granitic rock suites. *Journal of Petrology*, 42, 2033–48.

Frost, B. R. and Beard, J. S., 2007, On silica activity and serpentinization, *Journal of Petrology*, 48, 1351–68.

Frost, B. R. and Chacko, T., 1989, The granulite uncertainty principle: Limitations on thermobarometry in granulites. *Journal of Geology*, 97, 435–50.

Frost, B. R. and Frost, C. D., 1987, CO_2, melts, and granulite metamorphism. *Nature*, 327, 503–6.

Frost, B. R. and Frost, C. D., 2008, A geochemical classification for feldspathic rocks. *Journal of Petrology*, 49, 1955–69.

Frost, B. R. and Lindsley, D. H., 1992, Equilibria among Fe–Ti oxides, pyroxenes, olivine and quartz. 2. Applications. *American Mineralogist*, 77, 1004–20.

Frost, B. R. and Touret, J. L. R., 1989, Magmatic CO_2 and saline melts from the Sybille Monzosyenite, Laramie anorthosite complex, Wyoming. *Contributions to Mineralogy and Petrology*, 103, 178–86.

Frost, C. D. and Frost, B. R., 1997, Reduced rapakivi-type granites: The tholeiite connection. *Geology*, 25, 647–50.

Frost, C. D. and Frost, B. R., 2011, On ferroan (A-type) granites: Their compositional variability and modes of origin. *Journal of Petrology*, 52, 39–53.

Frost, C. D. and Frost, B. R., 2013, Proterozoic ferroan feldspathic magmatism. *Precambrian Research*, 228, 151–63.

Frost, C. D., Frost, B. R., Chamberlain, K. R., and Edwards, B. R., 1999, Petrogenesis of the 1.43 Ga Sherman batholith, SE Wyoming: A reduced, rapakivi-type anorogenic granite. *Journal of Petrology*, 40, 1771–802.

Frost, C. D., Frost, B. R., Chamberlain, K. R., and Hulsebosch, T. P., 1998, The Late Archean history of the Wyoming Province as recorded by granitic magmatism in the Wind River Range, Wyoming. *Precambrian Research*, 89, 145–73.

Frost, C. D., Frost, B. R., Lindsley, D. H., Chamberlain, K. R., Swapp, S. M., and Scoates, J. S., 2010, Geochemical and isotopic evolution of the anorthositic plutons of the Laramie anorthosite complex: Explanations for variations in silica activity and oxygen fugacity of massif anorthosites. *Canadian Mineralogist*, 48, 925–46.

Frost, C. D., McLaughlin, J. F., Frost, B. R., Fanning, C. M., Swapp, S. M., Kruckenberg, S. C., and Gonzalez, J., 2017, Hadean origins of Paleoarchean continental crust in the central Wyoming Province. *Geological Society of America Bulletin*, 129, 259–80.

Frost, C. D. and O'Nions, R. K., 1985, Caledonian magma genesis and crustal recycling. *Journal of Petrology*, 26, 515–44.

Frost, C. D. and Snoke, A. W., 1989, Tobago, West Indies, a fragment of a Mesozoic oceanic island arc: Petrochemical evidence. *Journal of the Geological Society of London*, 146, 953–64.

Gee, D. G., Janák, M., Jaroslaw, M., Robinson, P., and van Roemund, H., 2012, Subduction along and within the Baltoscandian margin during closing of the Iapetus Ocean and Baltica–Laurentia collision. *Lithosphere*, 5, 169–78.

Geological Survey of Italy, 1960, Geological map of Italy, scale 1:250,000.

Giordano, D., Mangiacapra, A., Motuzak, M., Russell, J. K., Romano, C., Dingwell, D. B., and Muro, A. D., 2006, An expanded non-Arrhenian model for silicate melt viscosity: a treatment for metaluminous, peraluminous and peralkaline liquids. *Chemical Geology*, 229, 42–56.

Glover, A. S., Rogers, W. Z., and Barton, J. E., 2012, Granitic pegmatites: Storehouses of industrial minerals. *Elements*, 8, 269–73.

Goff, G., Stimac, J. A., Larocque, A. C. L., Hulen, J. B., McMurtry, G. M., Adams, A. I., Roldan, M. A., Trujillo, P. E., Counce, D., Chipera, S. J. and Mann, D., 1994, Gold degassing and deposition at Galeras Volcano, Colombia, *GSA Today*, 4, 241–7.

Goldsmith, J. R. and Heard, H. C., 1961, Subsolidus phase relations in the system $CaCO_3$–$MgCO_3$. *Journal of Geology*, 69, 45–74.

Gray, J. R. and Yardley, B. W. D., 1979, A Caledonian blueschist from the Irish Dalradian, *Nature*, 278, 736–7.

Green, D. H. and Ringwood, A. E., 1969, The origin of basalt magmas. In *The Earth's Crust and Upper Mantle*, ed. P. J. Hart. Geophysical Monograph 13, American Geophysical Union, Washington, DC, pp. 489–95.

Green, D. H. and Ringwood, A. E., 1972, A comparison of recent experimental data on the gabbro–garnet granulite–eclogite transition. *Journal of Geology*, 80, 277–88.

Green, J. C. and Fitz, T. J. III, 1993, Extensive felsic lavas and rheoignimbrites in the Keweenawan Midcontinent Rift plateau volcanics, Minnesota: Petrographic and field recognition. *Journal of Volcanology and Geothermal Research*, 54, 177–96.

Greenwood, H. J., 1967, Wollastonite: Stability in H_2O–CO_2 mixtures and occurrence in a contact-metamorphic aureole near Salmo, British Columbia, Canada. *American Mineralogist*, 52, 1769–80.

Grove, T. L., Baker, M. B., Price, R. C., Parman, S. W., Elkins-Tanton, L. T., Chatterjee, N., and Müntener, O., 2005, Magnesian andesite and dacite lavas from Mt. Shasta, northern California: Products of fractional crystallization of H_2O-rich mantle melts. *Contributions to Mineralogy and Petrology*, 148, 542–65.

Grove, T. L. and Bryan, W. B., 1983, Fractionation of pyroxene-phyric MORB at low pressure: An experimental study. *Contributions to Mineralogy and Petrology*, 84, 293–309.

Grubenmann, U. and Niggli, P., 1924, *Die Gesteinsmetamorphose*, 3rd and revised edition of *Die Kristallinen Schiefer, I. Allgemeiner Teeil*. Gebrüder Borntraeger, Berlin, pp. 368–413.

Halliday, A. N., Fallick, A. E., Dickin, A. P., Mackenzie, A. B., Stephens, W. E., and Hildreth, W., 1983, The isotopic and chemical evolution of Mount St. Helens. *Earth and Planetary Science Letters*, 63, 241–56.

Hamilton, M. A., Pearson, D. G., Thompson, R. H., Kelley, S. P., and Emeleus, C. H., 1998, Rapid eruption of Skye lavas inferred from precise U–Pb and Ar–Ar dating of the Rum and Cuillin plutonic complexes. *Nature*, 394, 260–3.

Han, B. F., He, G-Q., Wang, X.-C., and Guo, Z-J., 2011, Later Carboniferous collision between the Tarim and Kazakhstan–Yili terranes in the western segment of the South Tian Shan Orogen, central Asia and implications for the Northern Xinjiang, western China. *Earth-Science Reviews*, 109, 74–93.

Harker, A., 1909, *The Natural History of Igneous Rocks*. Macmillan, New York, NY.

Harley, S. L., 1998, On the occurrence and characterization of ultrahigh-temperature crustal metamorphism. *Geological Society London Special Publication* 138, 81–107.

Harpp, K. S., and White, W. M., 2001, Tracing a mantle plume: Isotopic and trace element variations of Galapagos seamounts. *Geochemistry, Geophysics, Geosystems*, 2, paper 2000GC000137.

Harpp, K. S., Wirth, K. R., and Korich, D. J., 2002, Northern Galapagos Province: Hotspot-induced, near-ridge volcanism at Genovesa Island. *Geology*, 30, 399–402.

Hashimoto, M., Igi, S., Seki, Y., Banno, S., and Kojima, G., 1970, *Metamorphic facies map of Japan* (scale 1:2,000,000). Geological Survey of Japan.

Hawkins, J. and Melchior, J., 1983, Petrology of basalts from Loihi Seamount, Hawaii. *Earth and Planetary Science Letters*, 66, 356–68.

Hekinian, R., 1982, *Petrology of the Ocean Floor*. Elsevier, Amsterdam.

Helmold, K. P. and Van de Kamp, P. C., 1984, Diagenetic mineralogy and controls on albitization and laumontite formation in Paleogene arkoses, Santa Ynez Mountains, California. *American Association of Petroleum Geologists Memoir*, 37, 239–76.

Hess, P. C., 1989, *Origins of Igneous Rocks*. Harvard University Press, Cambridge, MA.

Hildreth, W. and Moorbath, S., 1988, Crustal contributions to arc magmatism in the Andes of central Chile. *Contributions to Mineralogy and Petrology*, 98, 455–89.

Hildreth, E. W., Halliday, A. N., and Christiansen, R. L., 1991, Isotopic and chemical evidence concerning the genesis and contamination of basaltic and rhyolitic magma beneath the Yellowstone Plateau volcanic field. *Journal of Petrology*, 32, 63–137.

Himmelberg, G. R. and Ford, A. B., 1976, Pyroxenes of the Dufek intrusion. *Journal of Petrology*, 17, 219–43.

Hirschmann, M. M., Renne, P. R., and McBirney, A. R., 1997, $^{40}Ar/^{39}Ar$ dating of the Skaergaard intrusion. *Earth and Planetary Science Letters*, 146, 645–58.

Hoefs, J., 1987, *Stable Isotope Geochemistry*, 3rd edn. Springer-Verlag, Berlin.

Holdaway, M. J. and Lee, S. M., 1977, Fe–Mg cordierite stability in high-grade pelitic rocks based on experimental, theoretical and natural observations. *Contributions to Mineralogy and Petrology*, 63, 175–98.

Hooper, P. R., 2000, Chemical discrimination of Columbia River basalt flows. *Geochemistry, Geophysics, Geosystems*, 1, doi: 10.1029/2000GC000040.

Hooper, P. R., Camp, V. E., Reidel, S. P., and Ross, M. E., 2007, The origin of the Columbia River flood basalt province: Plume versus non plume models. *Geological Society of America Special Paper*, 470, 635–68.

Hooper, P. R. and Swanson, D. A., 1990, The Columbia River Basalt Group and associated volcanic rocks of the Blue Mountains province. In *Geology of the Blue Mountains Region of Oregon, Idaho, and Washington: Cenozoic Geology of the Blue Mountains region. United States Geological Survey Professional Paper* 1437, 63–99.

Hopson, C. A., 2008, Geologic map of Mount St. Helens, prior to the 1980 eruption. United States Geological Survey Open-File Report 2002–468.

Hutchinson, R. M., 1976, Granite-tectonics of Pikes Peak batholith. *Professional Contributions of the Colorado School of Mines*, 8, 32–43.

Ireland, T. R. and Williams, I. S., 2003, Considerations in zircon geochronology by SIMS. *Reviews in Mineralogy and Geochemistry*, 53, 215–41.

Irvine, T. N., 1974, Petrology of the Duke Island ultramafic complex, southeastern Alaska. *Geological Society of America Memoir*, 138.

Irvine, T. N. and Baragar, W. R. A., 1971, A guide to the classification of the common volcanic rocks. *Canadian Journal of Earth Sciences*, 8, 523–48.

Irwin, W. P. and Coleman, R. G., 1974, Ophiolites and ancient continental margins. In *The Geology of Continental Margins*, eds. C. A. Burk and C. L. Drake. Springer-Verlag, New York, NY, pp. 921–31.

Ishihara, S., 1977, The magnetite-series and ilmenite-series granitic rocks. *Mining Geology*, 27, 293–305.

Ivanov, A. V., Rasskazov, S. V., Feoktistov, G. D., He, H., and Boven, A., 2005, $^{40}Ar/^{39}Ar$ dating of Usol'skii sill in the south-eastern Siberian Traps large igneous province: Evidence for long-lived magmatism. *Terra Nova*, 17, 203–8.

Janak, M., Ravana, E. J. K., and Kullerud, K., 2012, Constraining peak *P–T* conditions in UPH eclogites; calculated phase equilibria in kyanite- and phengite-bearing eclogite of the Tromsø Nappe, Norway, *Journal of Metamorphic Geology*, 30, 377–96.

Jicha, B. R., Singer, B. S., and Sobol, P., 2016, Re-evaluation of the ages of $^{40}Ar/^{39}Ar$ sanidine standards and supereruptions in the western U.S. using a Noblesse multi-collector mass spectrometer. *Chemical Geology*, 431, 54–66.

Johnson, R. W., Mackenzie, D. E., and Smith, I. E. M., 1978, Volcanic rock associations at convergent plate boundaries: Re-appraisal of the concept using case histories from Papua New Guinea. *Geological Society of America Bulletin*, 89, 96–106.

Jolly, W. T., 1978, Metamorphic history of the Archean Abitibi belt. *Geological Survey of Canada Paper*, 78–10, 63–78.

Jourdan, F., Féraud, G., Bertrand, H., Watkeys, M. K., and Renne, P. R., 2007, Distinct brief events in the Karoo large igneous province clarified by new $^{40}Ar/^{39}Ar$ ages on the Lesotho basalts. *Lithos*, 98, 195–209.

Jung, H., 2011, Seismic anisotropy produced by serpentine in mantle wedge. *Earth and Planetary Science Letters*, 307, 535–43.

Kamimura, A., Kasahara, J., Shinohara, M., Hino, R., Shiobara, H., Gujie, G., and Kanazawa, T., 2002, Crustal structure study at the Izu-Bonin subduction zone around 31 °N: Implications of serpentinized materials along the subduction plate boundary. *Physics of the Earth and Planetary Interiors*, 132, 105–29.

Kasbohm, J. and Schoene, B., 2018, Rapid eruption of the Columbia River flood basalt and correlation with the mid-Miocene climate optimum. *Science Advances*, 4, eaat8223.

Kerr, A. C., 2004, Oceanic plateaus. In *Treatise on Geochemistry*, eds. H. D. Holland and K. K. Turekian. Elsevier, Amsterdam, vol. 3, pp. 537–65.

Kerr, A. C., Tarney, J., Marriner, G. F., Nivia, A., and Saunders, A. D., 1997, The Caribbean–Colombian Cretaceous igneous province: The internal anatomy of an oceanic plateau. In *Large Igneous Provinces; Continental, Oceanic and Planetary Flood Volcanism*, eds. J. J. Mahoney and M. Coffin. American Geophysical Union Monograph, pp. 45–93.

King, E. M., Valley, J. W., 2001, The source, magmatic contamination, and alteration of the Idaho batholith. *Contributions to Mineralogy and Petrology*, 142, 72–88.

King, E. M., Valley, J. W., Davis, D. W., and Kowallis, B. J., 2001, Empirical determination of oxygen isotope fractionation factors for titanite with zircon and quartz. *Geochimica et Cosmochimica Acta*, 65, 3165–75.

King, S. D., 1995, The viscosity structure of the mantle. In *Reviews of Geophysics, Supplement, US National Report to International Union of Geodesy and Geophysics 1991–1994*. Wiley, New York, NY, pp. 11–17.

Kistler, R. W. and Peterman, Z. E., 1973, Variations in Sr, Rb, K, Na, and initial $^{87}Sr/^{86}Sr$ in Mesozoic granitic rocks and intruded wall rocks in central California. *Geological Society of America Bulletin*, 84, 3489–512.

Kistler, R. W. and Peterman, Z. E., 1978, *Reconstruction of Crustal Blocks of California on the Basis of Initial Strontium Isotopic Compositions of Mesozoic Granitic Rocks*. United States Geological Survey Professional Paper 1071, United States Government Printing Office, Washington, DC.

Kleeman, G. J. and Twist, D., 1989, The compositionally zoned sheet-like granite pluton of the Bushveld Complex: Evidence bearing on the nature of A-type magmatism. *Journal of Petrology*, 30, 1383–414.

Kohn, M. J., 2016, Metamorphic chronology – a tool for all ages: Past achievements and future prospects. *American Mineralogist*, 101, 25–42.

Kolker, A., Frost, C. D., Hanson, G. N., and Geist, D. J., 1991, Nd, Sr and Pb isotopes in the Maloin Ranch Pluton, Wyoming: Implications for the origin of evolved rocks at anorthosite margins. *Geochimica et Cosmochimica Acta*, 55, 2285–97.

Konnerup-Madsen, J. and Rose-Hansen, J., 1982, Volatiles associated with alkaline igneous rift activity: Fluid inclusions in the Ilimaussaq intrusion and the Gadar granitic complexes (South Greenland). *Chemical Geology*, 37, 79–93.

Koziol, A. M. and Newton, R. C., 1988, Redetermination of the anorthite breakdown reaction and improvement of the plagioclase-garnet-Al_2SiO_5-quartz geobarometer. *American Mineralogist*, 73, 216–23.

Kushiro, I. and Yoder, H. S., Jr., 1966, Anorthite–forsterite and anorthite–enstatite reactions and their bearing on the basalt–eclogite transformation. *Journal of Petrology*, 7, 337–62.

Kurth-Velz, M., Sassen, A., and Galer, J. G., 2004, Geochemical and isotopic heterogeneities along an island arc-spreading ridge intersection: Evidence from the Lewis Hills, Bay of Islands ophiolite, Newfoundland. *Journal of Petrology*, 45, 615–68.

Lackey, J. S., Valley, J. W., Chen, J. H., and Stockli, D. F., 2008, Dynamic magma systems, crustal recycling, and alteration in the central Sierra Nevada Batholith: The oxygen isotope record. *Journal of Petrology*, 49, 1397–426.

Laird, J. and Albee, A. L., 1981, High-pressure metamorphism in mafic schist from northern Vermont. *American Journal of Science*, 281, 97–126.

Land, L.S., Mack, L.E., Milliken, K. L., Lynch, F. L., 1997, Burial diagenesis of argillaceous sediment, south Texas Gulf of Mexico sedimentary basin: A reexamination. *Geological Society of America Bulletin*, 109, 2–15.

Le Fort, P., 1981, Manaslu leucogranite: A collision signature of the Himalaya, a model for its genesis and emplacement. *Journal of Geophysical Research*, 86, 10,545–68.

Le Maitre, R. W., Streckeisen, A., Zanettin, B., LeBas, M.J., Bonin, B., and Bateman, P., 2005, *Igneous Rocks: A Classification and Glossary of Terms*, 2nd edn. Cambridge University Press, Cambridge.

LeBas, M. J., LeMaitre, R. W., Streckeisen, A., and Zanettin, B. A., 1986, Chemical classification of volcanic rocks based on the total alkali-silica diagram. *Journal of Petrology*, 27, 745–50.

LeBas, M. J. and Streckeisen, A. L., 1991, The IUGS systematics of igneous rocks. *Journal of the Geological Society*, 148, 825–33.

Li, Y., 1988, Poly-type model for tungsten deposits and vertical structural zoning model for vein-type tungsten deposits in South China. *Geological Association of Canada Special Paper*, 40, 555–68.

Lindsley, D. H., 1991, Experimental studies of oxide minerals. *Reviews in Mineralogy*, 25, 69–106.

Lindsley, D. H. and Frost, B. R., 1992, Equilibria among Fe-Ti oxides, pyroxenes, olivine, and quartz: Part I. Theory. *American Mineralogist*, 77, 987–1003.

Lindsley, D. H., Frost, B. R., Frost, C. D., and Scoates, J. S., 2010, Petrology, geochemistry, and structure of the Chugwater anorthosite, Laramie Anorthosite complex, S.E. Wyoming, U.S.A. *Canadian Mineralogist*, 48, 887–923.

Linnen, R. L. and Keppler, H., 2002, Melt composition control of Zr/Hf fractionation in magmatic processes. *Geochimica et Cosmochimica Acta*, 66, 3293–301.

Liou, J. G., Maruyama, S., and Cho, M., 1987, Very low-grade metamorphism of volcanic and volcaniclastic rocks: Mineral assemblages and mineral facies. In *Low Temperature Metamorphism*, ed. M. Frey. Blackie, London, pp. 59–113.

Liou, J. G., Tsujimori, T., Zhang, R. Y., Katayama, I., and Maruyama, S., 2004, Global UHP metamorphism and continental subduction/collision: The Himalayan model. *International Geology Review*, 46, 1–27.

Liu, Y., Gu, X., Rolfo, F., and Chen, Z., 2011, Ultrahigh-pressure metamorphism and multistage exhumation of eclogite of the Luotian Dome, North Dabie Complex Zone (central China): Evidence from mineral inclusions and decompression textures. *Journal of Asian Earth Sciences*, 42, 607–17.

Loiselle, M. C. and Wones, D., 1979, Characteristics and origin of anorogenic granites. *Geological Society of America Abstracts with Programs*, 11, 468.

Longhi, J. and Ashwal, L. D., 1985, Two-stage models for lunar and terrestrial anorthosites: Petrogenesis without a magma ocean. Proceedings of the 15th Lunar and Planetary Science Conference, Part 2. *Journal of Geophysical Research*, 90 (supplement): C571–84.

Luth, W. C., Jahns, R. H., and Tuttle, O. F., 1964, The granite system at pressures of 4 to 10 kilobars. *Journal of Geophysical Research*, 69, 759–73.

MacDonald, G. A., 1968, Composition and origin of Hawaiian lavas. *Geological Society of America Memoir*, 116, 477–522.

Mackie, R. A., Scoates, J. S., and Weis, D., 2009, Age and Nd–Hf isotopic constraints on the origin of marginal rocks from the Muskox layered intrusion (Nunavut, Canada) and implications for the evolution of the 1.27 Ga Mackenzie large igneous province. *Precambrian Research*, 172, 46–66.

Manga, M. and 28 others, 2012, Heat flow in the Lesser Antilles island arc and adjacent back arc Grenada basin. *Geochemistry, Geophysics, Geosystems*, 13, Q08007.

Markl, G., 2001, A new type of silicate liquid immiscibility in peralkaline nepheline syenites (lujavrites) of the Ilimaussaq complex, South Greenland. *Contributions to Mineralogy and Petrology*, 141, 458–72.

Markl, G., Marks, M. A. W., and Frost, B. R., 2010, On the controls of oxygen fugacity in the generation and crystallization of peralkaline melts. *Journal of Petrology*, 51, 1831–47.

Marks, M., Vennemann, T., Siebel, W., and Markl, G., 2004, Nd-, O-, and H-isotopic evidence for complex, closed-system fluid evolution of the peralkaline Ilimaussaq intrusion, south Greenland. *Geochimica et Cosmochimica Acta*, 68, 3379–95.

Marsh, B. D. and Carmichael, I. S. E., 1974, Benioff zone magmatism. *Journal of Geophysical Research*, 79, 1196–206.

Mason, B. G., Pyle, D. M., and Oppenheimer, C., 2004, The size and frequency of the largest explosive eruptions on Earth. *Bulletin of Volcanology*, 66, 735–48.

Mathez, E.A. and Kinzler, R. J., 2017, Metasomatic chromitite seams in the Bushveld and Rum Layered Intrusions. *Elements*, 13, 397–402.

Mattinson, J. M., 2005, Zircon U–Pb chemical abrasion ("CA–TIMS") method: Combined annealing and multi-step partial dissolution analysis for improved precision and accuracy of zircon ages. *Chemical Geology*, 220, 47–66.

Mayer, A., Hofmann, A. W., Sinigoi, S., and Morais, E., 2004, Mesoproterozoic Sm–Nd and U–Pb ages for the Kunene anorthosite complex of SW Angola. *Precambrian Research*, 133, 187–206.

McBirney, A. R., 1993, *Igneous Petrology*. Jones and Bartlett Publishers, Boston, MA.

McBirney, A. R. and Williams, H., 1969, Geology and petrology of the Galapagos Islands. *Geological Society of America Memoir*, 118.

McCaig, A., Cliff, R. A., Escartin, J., Fallick, A. E., and MacLeod, C. J., 2007, Oceanic detachment faults focus very large volumes of black smoker fluids. *Geology*, 35, 935–8.

McCurry, M., Hayden, K. P., Morse, L. H., and Mertzman, S., 2008, Genesis of post-hotspot, A-type rhyolite of the Eastern Snake River Plain volcanic field by extreme fractional crystallization of olivine tholeiite. *Bulletin of Volcanology*, 70, 361–83.

McKenzie, D., 1984, The generation and compaction of partially molten rock. *Journal of Petrology*, 25, 713–65.

McKenzie, D. and O'Nions, R. K., 1991, Partial melt distributions from inversion of rare earth element concentrations. *Journal of Petrology*, 32, 1021–91.

McLelland, J. M., Bickford, M. E., Hill, B. M., Clechenko, C. C., Valley, J. W., and Hamilton, M. A., 2004, Direct dating of Adirondack massif anorthosite by U–Pb SHRIMP analysis of igneous zircon: Implications for AMCG complexes. *Geological Society of America Bulletin*, 116, 1299–317.

McMillan, W. J. and Panteleyev, A. 1988, Porphyry copper deposits. In *Ore Deposit Models*, eds. R. G. Roberts and P. A. Sheahan. Geoscience Canada Reprint Series 3, pp. 45–58.

Mehnert, K. R., 1968, *Migmatites and the Origin of Granitic Rocks*. Elsevier, New York, NY.

Memeti, V., Paterson, S. Matzel, J., Mundil, R., and Okaya, D., 2010, Magmatic lobes as "snapshots" of magma chamber growth and evolution in large, composite batholiths: An example from the Tuolumne intrusion, Sierra Nevada, California. *Geological Society of America Bulletin*, 122, 1912–31.

Meshesha, D., Shinjo, R., Matsumura, R., and Chekol, T., 2011, Metasomatized lithospheric mantle beneath Turkana depression in southern Ethiopia (the East Africa rift): Geochemical and Sr–Nd–Pb isotopic characteristics. *Contributions to Mineralogy and Petrology*, 162, 889–907.

Metz, P. and Trommsdorff, V., 1968, On phase equilibria in metamorphosed siliceous dolomites. *Contributions to Mineralogy and Petrology*, 18, 305–9.

Mezger, K., Bohlen, S. R., and Hanson, G. A., 1990, Metamorphic history of the Archean Pikwitonei granulite domain and the Cross Lake subprovince, Superior province, Manitoba, Canada. *Journal of Petrology*, 31, 483–517.

Michibayashi, K., Ina, T. and Kanagawa, K., 2006, The effect of dynamic recrystallization on olivine fabric and seismic anisotropy: Insight from a ductile shear zone, Oman ophiolite. *Earth and Planetary Science Letters*, 244, 695–708.

Miles, A. J., Woodcock, N. H., and Hawkesworth, C. J., 2016, Tectonic controls on post-subduction granite genesis and emplacement: The late Caledonian suite of Britain and Ireland. *Gondwana Research*, 39, 250–60.

Miller, R. B. and Snoke, A. W., 2009, The utility of crustal cross sections in the analysis of orogenic processes in contrasting tectonic settings. *Geological Society of America Special Paper*, 456, 1–38.

Milord, I., Sawyer, E. W., and Brown, M., 2001, Formation of diatexite migmatite and granite magma during anatexis of semi-pelitic metasedimentary rocks: An example from St. Malo, France. *Journal of Petrology*, 42, 487–505.

Miron, G. D., Neuhoff, P. S., and Amthauer, G., 2012, Low-temperature hydrothermal metamorphic mineralization of island-arc volcanics, South Apuseni Mountains, Romania. *Clays and Clay Minerals*, 60, 1–17.

Mitchell, R. H., 1986, *Kimberlites: Mineralogy, Geochemistry, and Petrology*. Plenum Press, New York, NY.

Mitchell, R. H. and Bergman, S. C., 1991, *Petrology of Lamproites*. Plenum Press, New York, NY.

Miyashiro, A., 1961, Evolution of metamorphic belts. *Journal of Petrology*, 2, 277–311.

Miyashiro, A., 1973, The Trodoos complex was probably formed in an island arc. *Earth and Planetary Science Letters*, 19, 218–81.

Miyashiro, A., 1974, Volcanic rocks series in island arcs and active continental margins. *American Journal of Science*, 274, 321–55.

Miyashiro, A., 1975, Classification, characteristics, and origin of ophiolites. *Journal of Geology*, 83, 249–81.

Moecher, D. P., Perkins, D., III, Leier-Englehardt, P. J., and Medaris, L. G. Jr., 1986, Metamorphic conditions of the late Archean high-grade gneisses, Minnesota River valley, U.S.A., *Canadian Journal of Earth Sciences*, 23, 633–45.

Mohriak, W. U., Rosendahl, B. R., Turner, J. P., and Valente, S. C., 2002, Crustal architecture of South Atlantic volcanic margins. *Geological Society of America Special Paper*, 362, 159–202.

Moody, J. B., Meyer, D., and Jenkins, J. E., 1983, Experimental characterization of the greenschist/amphibolite boundary in mafic systems. *American Journal of Science*, 283, 48–92.

Moore, J. N. and Kerrick, D. M., 1976, Equilibria in siliceous dolomites of the Alta aureole, Utah, *American Journal of Science*, 276, 502–24.

Moores, E. M., 1982, Origin and emplacement of ophiolites. *Reviews of Geophysics and Space Physics*, 20, 735–60.

Morse, S. A., 1970, Alkali feldspars with water at 5 kb pressure. *Journal of Petrology*, 11, 221–51.

Morse, S. A., 1980, Kiglapait geochemistry. II. Petrography. *Journal of Petrology*, 20, 394–410.

Muir, I. D. and Tilley, C. E., 1957, Contributions to the petrology of Hawaiian basalts 1. The pircrite-basalts of Kilauea. *American Journal of Science*, 255, 241–53.

Mukherjee, B. K. and Sachan, H. K., 2009, Fluids in coesite-bearing rocks of the Tso Morari Complex, NW Himalaya: Evidence for entrapment during peak metamorphism and subsequent uplift. *Geological Magazine*, 146, 876–89.

Müntener, O., Manatschal, G., Desmurs, L., and Pettke, T., 2010, Plagioclase peridotites in ocean–continent transitions: Refertilized mantle domains generated by melt stagnation in the shallow mantle lithosphere. *Journal of Petrology*, 51, 255–94.

Myers, J. S., 1976, Channel deposits of peridotite, gabbro and chromitite from turbidity currents in the stratiform Fiskenæsset anorthosite complex, southwest Greenland. *Lithos*, 9, 281–91.

Nabelek, P. I., Russ-Nabelek, C., and Denison, J. R., 1992, The generation and crystallization conditions of the Proterozoic Harney Peak leucogranite, Black Hills, South Dakota, USA: Petrologic and geochemical constraints. *Contributions to Mineralogy and Petrology*, 110, 173–91.

Naldrett, A. J., Bray, J. G., Gasparrini, E. L., Podolsky, T., and Rucklidge, J. D., 1970, Cryptic variation and the petrology of the Sudbury nickel irruptive. *Economic Geology*, 65, 122–55.

Nash, W. P., Carmichael, I. S. E., and Johnson, B. W., 1969, The mineralogy and petrology of Mount Suswa, Kenya. *Journal of Petrology*, 10, 409–39.

Neal, C. R., Mahoney, J. J., Kroenke, L. W., Duncan, R. A., and Petterson, M. G., 1997, The Ontong Java plateau. In *Large Igneous Provinces; Continental, Oceanic and Planetary Flood Volcanism*, eds. J. J. Mahoney and M. Coffin. American Geophysical Union Monograph, 100, pp. 183–216.

Nesbitt, B. E. and Kelly, W. C., 1980, Metamorphic zonation of sulfides, oxides, and graphite in and around the orebodies at Ducktown, Tennessee. *Economic Geology*, 75, 1010–21.

Nesse, W. D., 1991, *Introduction to Optical Mineralogy*, 2nd edn. Oxford University Press, Oxford.

Newhall, C. G. Hendley, J. W., and Stauffer, P. H., 1997, The cataclysmic 1991 eruption of Mount Pinatubo, Philippines, United States Geological Survey Fact Sheet 113–97, http://pubs.usgs.gov/fs/1997/fs113-97/, accessed August 28, 2012.

Newhall, C. G., and Self, S., 1982, The volcanic explosivity index (VEI): An estimate of explosive magnitude for historical volcanism. *Journal of Geophysical Research*, 87, 1231–8.

Nikogosian, I. K. and van Bergen, M. J., 2010, Heterogeneous mantle sources of potassium-rich magmas in central–southern Italy: Melt inclusion evidence from Roccamonfina and Ernici (mid Latina Valley). *Journal of Volcanology and Geothermal Research*, 197, 279–302.

Nimis, P. and Trommsdorff, V., 2001, Revised thermobarometry of Alpe Arami and other garnet peridotites from the central Alps. *Journal of Petrology*, 42, 103–15.

Nockolds, S. R. and Allen, R., 1956, The geochemistry of some igneous rock series – III. *Geochimica et Cosmochimica Acta*, 9, 34–77.

Nockolds, S. R., Knox, R. W. O'B, and Chinner, G. A., 1978, *Petrology for Students*. Cambridge University Press, Cambridge.

Noh, J. H. and Boles, J. R., 1993, Origin of zeolite cements in the Miocene sandstones, North Tejon oil fields, California. *Journal of Sedimentary Petrology*, 63, 248–60.

Norman, M. D., Borg, L. E., Nyquist, L. E., and Bogard, D. D., 2003, Chronology, geochemistry, and petrology of a ferroan noritic anorthosite clast from Descartes breccia 67215: Clues to the age, origin, structure, and impact history of the lunar crust. *Meteoritics and Planetary Science*. 38, 645–551.

O'Connor, J. M., Stoffers, P., Wijbrans, J. R., and Worthington, T. J., 2007, Migration of widespread long-lived volcanism across the Galapagos Volcanic Province: Evidence for a broad hotspot melting anomaly? *Earth and Planetary Science Letters*, 263, 339–54.

Oberthür, T., Davis, D. W., Blenkinsop, T. G., and Höhndorf, A., 2002, Precise U–Pb mineral ages, Rb–Sr and Sm–Nd systematics for the Great Dyke, Zimbabwe: Constraints on late Archean events in the Zimbabwe craton and Limpopo belt. *Precambrian Research*, 113, 293–305.

Oliver, G. J. H., Wilde, S. A., and Yushen, W., 2008, Geochronology and geodynamics of Scottish granitoids from the late Neoproterozoic break-up of Rodinia to Palaeozoic collision. *Journal of the Geological Society*, 165, 661–74.

Osborn, E. F. and Tait, D. B., 1952, The system diopside–forsterite–anorthite. *American Journal of Science*, Bowen volume, part 2, 413–33.

Owens, B. E., Dymek, R. F., Tucker, R. D., Brannon, J. C., and Podosek, F. A., 1994, Age and radiogenic isotopic composition of a late- to post-tectonic anorthosite in the Grenville Province: The Labrieville massif, Quebec. *Lithos*, 31, 189–206.

Paces, J. B. and Miller, J. D., 1993, Precise U–Pb ages of Duluth Complex and related mafic intrusions, northeastern Minnesota: Geochronological insights to physical, petrogenetic, paleomagnetic, and tectonomagmatic processes associated with the 1.1 Ga Midcontinent Rift System. *Journal of Geophysical Research Solid Earth*, 98, 13,997–14,013.

Padrón-Navarta, J. A., Lopez Sanchez-Vizcaino, V., Garrido, C. J., and Gomez-Pugnaire, M. T., 2011, Metamorphic record of high-pressure dehydration of antigorite serpentinite to chlorite harzburgite in a subduction setting (Cerro del Almirez, Nevado Filabride Complex, southern Spain). *Journal of Petrology*, 52, 2047–78.

Palme, H. and O'Neill, H. St. C., 2014, Cosmochemical estimates of mantle composition. In *Treatise on Geochemistry*, 2nd edn, eds. K. Turekian and H. Holland. Elsevier, Amsterdam, vol. 3, pp. 1–39.

Parry, W. T., Jasumack, M., and Wilson, P. N., 2002, Clay mineralogy and intermediate argillic alteration at Bingham, Utah. *Economic Geology*, 97, 221–39.

Parsons, I., 1978, Feldspars and fluids in cooling plutons. *Mineralogical Magazine*, 42, 1–17.

Paterson, S., Memeti, V., Mundl, R., and Zak, J., 2016, Repeated, multiscale, magmatic erosion and recycling in an upper-crustal pluton: Implications for magma chamber dynamics and magma volume estimates. *American Mineralogist*, 101, 2176–98.

Patiño Douce, A. E., 1997, Generation of metaluminous A-type granites by low-pressure melting of calc-alkaline granitoids, *Geology*, 25, 743–6.

Patiño Douce, A. E., Humphreys, E. D., and Johnston, A. D., 1990, Anatexis and metamorphism in tectonically thickened continental crust exemplified by the Sevier hinterland, western North America. *Earth and Planetary Science Letters*, 97, 290–315.

Pattison, D. R. M., 1992, Stability of andalusite and sillimanite and the Al_2SiO_5 triple point: Constraints from the Ballachulish aureole, Scotland. *Journal of Geology*, 100, 432–46.

Pattison, D. R. M., 2001, Instability of Al_2SiO_5 "triple-point" assemblages in muscovite + biotite + quartz-bearing metapelites, with implications. *American Mineralogist*, 86, 1414–22.

Pattison, D. R. M., Chacko, T., Farquhar, J., and McFarlane, C. R. M., 2003, Temperatures of granulite-facies metamorphism: Constraints from experimental phase equilibria and thermobarometry corrected for retrograde exchange. *Journal of Petrology*, 44, 867–900.

Peacock, M. A., 1931, Classification of igneous rock series. *Journal of Geology*, 39, 54–67.

Pearce, J. A., Harris, N. B. W., and Tindle, A. G., 1984a, Trace element discrimination diagrams for the tectonic interpretation of granitic rocks. *Journal of Petrology*, 25, 956–83.

Pearce, J. A., Lippard, S. J., and Roberts, S., 1984b, Characteristics and tectonic significance of supra-subduction zone ophiolites. *Geological Society of London Special Publication* 16, 77–94.

Pearce, J. A. and Peate, D. W., 1995, Tectonic implications of the composition of volcanic arc magmas. *Annual Review of Earth and Planetary Sciences*, 23, 251–85.

Pecher, A., 1989, The metamorphism in the central Himalaya. *Journal of Metamorphic Geology*, 7, 31–41.

Percival, J. A., 1989, A regional perspective of the Quetico metasedimentary belt, Superior province, Canada. *Canadian Journal of Earth Sciences*, 26, 677–93.

Percival, J. A., 1991, Granulite-facies metamorphism and crustal magmatism in the Ashuanipi Complex, Quebec–Labrador. *Journal of Petrology*, 32, 1261–97.

Percival, J. A. and McGrath, P. H., 1986, Deep crustal structure and tectonic history of the northern Kapuskasing uplift of Ontario: An integrated petrological–geophysical study. *Tectonics*, 5, 553–72.

Percival, J. A., Mortensen, J. K., Stern, R. A., Card, K. D., and Bégin, N. J., 1992, Giant granulite terranes of northeastern Superior province: The Ashuanipi complex and Minto block. *Canadian Journal of Earth Sciences*, 29, 2287–308.

Perfit, M. R., Fornari, D. J., Smith, M. C., Bender, J. F., Langmuir, C. H., and Haymon, R. M., 1994, Small-scale spatial and temporal variations in mid-ocean ridge crest magmatic processes. *Geology*, 22, 375–9.

Perkins, M. E. and Nash, B. P., 2002, Explosive silicic volcanism of the Yellowstone hotspot: The ash fall tuff record. *Geological Society of America Bulletin*, 114, 367–81.

Peretti, A., 1988, Occurrence and stabilities of opaque minerals in the Malenco serpentinite (Sondrio, northern Italy), PhD dissertation, Swiss Federal Institute of Technology, Zürich.

Peterman, Z. E., Hedge, C. E., and Braddock, W. A., 1968, Age of Precambrian events in the northeastern Front Range, Colorado. *Journal of Geophysical Research*, 73, 2277–96.

Petford, N., 2009, Which effective viscosity? *Mineralogical Magazine*, 73, 167–91.

Petterson, M. G., Neal, C. R., Mahoney, J. J., Kroenke, L. W., Saunders, A. D., Babbs, T. L., Duncan, R. A., Tolia, D., and McGrail B., 1997, Structure and deformation of north and central Malaita, Solomon Islands: Tectonic implications for the Ontong Java plateau–Solomon arc collision, and for the fate of oceanic plateaus. *Tectonophysics*, 283, 1–33.

Pettijohn, F. J., 1975, *Sedimentary Rocks*, 3rd edn. Harper and Row, New York, NY.

Pfaff, K., Krumrei, T., Marks, M., Wenzel, T., Rudolf, T., and Markl, G., 2008, Chemical and physical evolution of the "lower layered sequence" from the nepheline syenitic Ilimaussaq intrusion, South Greenland: Implications for the origin of magmatic layering in peralkaline felsic liquids. *Lithos*, 106, 280–96.

Philpotts, A. R. and Ague, J. J., 2009, *Principles of Igneous and Metamorphic Petrology*, 2nd edn. Cambridge University Press, Cambridge.

Poland, M. P., Miklius, A., Montgomery-Brown, E. K., 2014, Magma supply, storage, and transport at shield-stage Hawaiian volcanoes. *United States Geological Survey Professional Paper*, 1801, 179–234.

Polat, A., Fryer, B. J., Appel, P. W. U., Kalvig, P., Kerrich, R., Dilek, Y., and Yang, Z., 2011, Geochemistry of anorthositic differentiated sills in the Archean (~2970 Ma) Fiskenæsset Complex, SW Greenland: Implications for parental magma compositions, geodynamic setting, and secular heat flow in arcs. *Lithos*, 123, 50–72.

Powell, R., Holland, T., and Worley, B., 1998, Calculating phase diagrams involving solid solutions via non-linear equations, with examples using THERMOCALC. *Journal of Metamorphic Geology*, 16, 577–88.

Premo, W. R. and Loucks, R. R., 2000, Age and Pb–Sr–Nd isotopic systematics of plutonic rocks from the Green Mountain magmatic arc, southeastern Wyoming. *Rocky Mountain Geology*, 35, 51–70.

Rabinowicz, M. and Toplis, M. J., 2009, Melt segregation in the lower part of the partially molten mantle zone beneath an oceanic spreading centre: Numerical modeling of the combined effects of shear segregation and compaction. *Journal of Petrology*, 50, 1071–106.

Rajesh, H. M., Chisonga, B. C., Shindo, K., Beukes, N. J., and Armstrong, R. A., 2013, Petrographic, geochemical, and SHRIMP U–Pb titanite age characterization of the Thabazimbi mafic sills: Extended time frame and a unifying petrogenetic model for the Bushveld Large Igneous Province. *Precambrian Research*, 230, 79–102.

Rassmussen, K. L., Lentz, D. R., Falck, H., and Pattison, D. R. M., 2011, Felsic magmatic phases and the role of late stage aplitic dykes in the formation of world class Cantung tungsten skarn deposit, Northwest Territories, Canada. *Ore Geology Reviews*, 41, 75–111.

Ravana, E. J. K, Andersen, T. B., Jolivet, L., and De Capitani, C., 2010, Cold subduction and the formation of lawsonite eclogite. Constraints from prograde evolution of eclogitized pillow lava from Corsica. *Journal of Metamorphic Geology*, 28, 381–95.

Ren, Z-Y, Hanyu, T., Miyazaka, T., Chang, Q., Kawabata, H., Takahashi, T., Hirahara, Y., Nichols, A. R. L., and Tatsumi, Y., 2009, Geochemical differences of the Hawaiian shield lavas: Implications for melting process in the heterogeneous Hawaiian plume. *Journal of Petrology*, 50, 1553–73.

Renne, P. R., Sprain, C. J., Richards, M. A., Self, S., Vanderkluysen, L., and Pande, K., 2015, State shift in Deccan volcanism at the Cretaceous–Paleogene boundary, possibly induced by impact. *Science*, 350, 76–8.

Rice, J. M., 1977, Progressive metamorphism of impure dolomitic limestone in the Marysville aureole, Montana. *American Journal of Science*, 277, 1–24.

Rivers, T., Ketchum, J., Indares, A., and Hynes, A., 2002, The high-pressure belt in the Grenville Province: Architecture, timing, and exhumation. *Canadian Journal of Earth Sciences*, 39, 867–93.

Rohring, B., 2015, Smartphones, smart chemistry. *ChemMatters*, April/May, 10–12. www.acs.org/chemmatters.

Rosenthal, A., Foley, S. F., Pearson, D. G., Nowell, G. M., and Tappe, S., 2009, Petrogenesis of strongly alkaline primitive volcanic rocks at the propagating tip of the western branch of the East African rift. *Earth and Planetary Science Letters*, 284, 236–48.

Rudnick, R. L. and Gao, S., 2014, Composition of the continental crust. In *Treatise on Geochemistry*, 2nd edn, eds. K. Turekian and H. Holland. Elsevier, Amsterdam, vol. 4, pp. 1–51.

Russell, M. J., Hall, A. J., and Martin, W., 2010, Serpentinization as a source of energy at the origin of life. *Geobiology*, 8, 355–71.

Ryan, M. P., Koyanagi, R. Y, and Fiske, R. S., 1981, Modeling the three-dimensional structure of macroscopic magma transport systems: Application to Kilauea volcano, Hawaii. *Journal of Geophysical Research*, 86, 7111–29.

Saleeby, J. B., 1992, Age and tectonic setting of the Duke Island ultramafic intrusion, southeast Alaska. *Canadian Journal of Earth Sciences*, 29, 506–22.

Saleeby, J., Ducea, M., and Clemens-Knott, D., 2003, Production and loss of high-density batholitic root, southern Sierra Nevada, California. *Tectonics*, 22, doi:10.1029/2002TC001374, 24 p.

Saleeby, J., Farley, K. A., Kistler, R. W., and Fleck, R. J., 2007, Thermal evolution and exhumation of deep-level batholithic exposures, southernmost Sierra Nevada, California. *Geological Society of America Special Paper*, 419, 39–66.

Salisbury, M. H. and Christiansen, N. I., 1985, Olivine fabrics in the Bay of Islands Ophiolite; implications for oceanic mantle structure and anisotropy. *Canadian Journal of Earth Sciences*, 22, 1757–66.

Sandiford, M. and Powell, R., 1986, Deep crustal metamorphism during continental extension: Modern and ancient examples. *Earth and Planetary Science Letters*, 79, 151–8.

Sano, T. and Yamashita, S., 2004, Experimental petrology of basement lavas from Ocean Drilling Program Leg 192: Implications for differentiation processes in Ontong Java plateau magmas. *Geological Society of London Special Publications*, 229, 185–218.

Scarfe, C. M., 1973, Viscosity of basic magmas at varying pressure. *Nature Physical Science*, 241, 101–2.

Schairer, J. F. and Bowen, N. L., 1956, The system $Na_2O–Al_2O_3–SiO_2$. *American Journal of Science*, 254, 129–95.

Schaller, W. T., 1916, Mineralogic notes, series 3. *United States Geological Survey Bulletin*, 610.

Schärer, U., Wilmart, E., and Duchesne, J.-C., 1996, The short duration and anorogenic character of anorthosite magmatism: U–Pb dating of the Rogaland complex, Norway. *Earth and Planetary Science Letters*, 139, 335–50.

Schmidt, M. W. and Poli, S., 2014, Devolatilization during subduction. In *Treatise on Geochemistry*, 2nd edn, eds. K. Turekian and H. Holland. Elsevier, Amsterdam, vol. 4, pp. 669–701.

Schmidt, S. Th., 1993, Regional and local patterns of low-grade metamorphism in the North Shore Volcanic Group, Minnesota, USA. *Journal of Metamorphic Geology*, 11, 401–14.

Schoene, B., Samperton, K. M., Eddy, M. P., Keller, G., Thierry, A., Bowring, S. A., Khadri, S. F. R., and Gertsch, B., 2015, U–Pb geochronology of the Deccan Traps and relation to the end-Cretaceous mass extinction. *Science*, 347, 182–4.

Schroeder, T., John, B., and Frost, B. R., 2002, Geologic implications of seawater circulation through peridotite exposed at slow-spreading mid-ocean ridges. *Geology*, 30, 367–70.

Scoates, J. S. and Friedman, R. M., 2008, Precise age of the platiniferous Merensky reef, Bushveld Complex, South Africa, but the U–Pb zircon chemical abrasion ID–TIMS technique. *Economic Geology*, 103, 465–71.

Scoates, J. S., Frost, C. D., Mitchell, J. N., Lindsley, D. H., and Frost, B. R., 1996, Residual-liquid origin for a monzonite intrusion in a mid-Proterozoic anorthosite complex: The Sybille intrusion, Laramie anorthosite complex, Wyoming. *Geological Society of America Bulletin*, 108, 1357–71.

Scoates, J. S., Lindsley, D. H., and Frost, B. R., 2010, Magmatic and structural evolution of an anorthositic magma chamber: The Poe Mountain intrusion, Laramie anorthosite complex, Wyoming. *Canadian Mineralogist*, 48, 851–5.

Scott, G. R., Taylor, R. B., Epis, R. C., and Wobus, R. A., 1978, Geologic map of the Pueblo 1° by 2° quadrangle, south-central Colorado. United States Geological Survey Miscellaneous Investigations Map 1–1022, scale 1: 250,000.

Shand, S. J., 1943, *The Eruptive Rocks*, 2nd edn. John Wiley & Sons, New York, NY.

Shaw, D. M., 2006, *Trace Elements in Magma: A Theoretical Treatment*. Cambridge University Press, Cambridge.

Sheppard, S. M. F. and Harris, C., 1985, Hydrogen and oxygen isotope geochemistry of Ascension Island lavas and granites: Variation with crystal fractionation and interaction with sea water. *Contributions to Mineralogy and Petrology*, 91, 74–81.

Shumilova, T. G., Isaenko, S. I., Ulyashev, V. V., Kazakov, V. A., and Makeev, B. A., 2018, After-coal diamonds: An enigmatic type of impact diamond. *European Journal of Mineralogy*, 30, 61–76.

Sillitoe, R. H., 2010, Porphyry copper systems. *Economic Geology*, 105, 3–41.

Simmons, W. B., Pezzotta, F., Shigley, J. E, and Beurlen, H., 2012, Granitic pegmatites as sources of colored gemstones. *Elements*, 8, 218–87.

Singer, B. S., Myers, J. D., and Frost, C. D., 1992, Mid-Pleistocene lavas from the Seguam volcanic center, central Aleutian arc: Closed-system fractional crystallization of basalt to rhyodacite. *Contributions to Mineralogy and Petrology*, 110, 87–112.

Sinton, J. M. and Detrick, R. S., 1992, Mid-ocean ridge magma chambers. *Journal of Geophysical Research*, 97(B1), 197–216.

Skippen, G. B., 1971, Experimental data for reactions in siliceous marbles. *Journal of Geology*, 79, 457–81.

Skippen, G. B., 1974, An experimental model for low pressure metamorphism of siliceous dolomitic marble. *American Journal of Science*, 274, 487–509.

Skjerlie, K. P. and Johnston, A. D., 1993, Fluid-absent melting behavior of an F-rich tonalitic gneiss at mid-crustal pressures: Implications for the generation of anorogenic granites. *Journal of Petrology*, 34, 785–815.

Smith, D. R. and Leeman, W. P., 1993, The origin of Mount St. Helens andesites. *Journal of Volcanology and Geothermal Research*, 55, 271–303.

Smith, D. R., Noblett, J., Wobus, R. A., Unruh, D., Douglass, J., Beane, R., Davis, C., Goldman, S., Kay, G., Bustavson, B., Saltoun, B., and Steward, J., 1999, Petrology and geochemistry of late-stage intrusions of the A-type mid-Proterozoic Pikes Peak batholith (Central Colorado, USA): Implications for petrogenetic models. *Precambrian Research*, 98, 271–305.

Smithies, R. H., Champion, D. C., and Cassidy, K. F., 2003, Formation of the Earth's early Archean continental crust. *Precambrian Research*, 127, 89–101.

Snoke, A. W., Rowe, D. W., Yule, D. J., and Wadge, G., 2001, *Petrologic and Structural History of Tobago, West Indies: A Fragment of the Accreted Mesozoic Oceanic Arc of the Southern Caribbean.* Geological Society of America Special Paper, 354, Geological Society of America, Boulder, CO.

Sørensen, H., 1974, Introduction. In *The Alkaline Rocks*, ed. H. Sørensen. John Wiley & Sons, New York, NY, pp. 1–11.

Spear, F. S., 1981, An experimental study of hornblende stability and compositional variability in amphibolite. *American Journal of Science*, 281, 697–734.

Spear, F. S. and Pattison, D. R. M., 2017, The implications of overstepping for metamorphic assemblage diagrams (MADs). *Chemical Geology*, 457, 38–46.

Spera, F. J. and Bergman, S. C., 1980, Carbon dioxide in igneous petrogenesis: I. *Contributions to Mineralogy and Petrology*, 74, 55–66.

Spray, J. G. and Boonsue, S., 2018, Quartz-coesite-stishovite relations in shocked metaquartzites from the Vredefort impact structure, South Africa. *Meteoritics and Planetary Science*, 53, 93–109.

Stakes, D. S., Shervais, J. W., and Hopson, C. A., 1984, The volcanic–tectonic cycle of the FAMOUS and AMAR valleys, Mid-Atlantic Ridge (36° 47' N): Evidence from basalt glass and phenocryst compositional variations for a steady state magma chamber beneath the valley midsections, AMAR 3. *Journal of Geophysical Research*, 89, 6995–7028.

Starkey, R. J. and Frost, B. R., 1990, Low-grade metamorphism of the Karmutsen Volcanics, Vancouver Island, British Columbia. *Journal of Petrology*, 31, 167–95.

Stern, C. R. and de Wit, M. J., 2003, Rocas Verdes ophiolites, southernmost South America: Remnants of progressive stages of development of oceanic-type crust in a continental margin back-arc basin. In *Ophiolites in Earth History*, eds. Y. Dilek and P. T. Robinson. Geological Society of London Special Publication, 218, London.

Storey, C. D., Jeffries, T. E., and Smith, M., 2004. Common lead-corrected laser ablation ICP-MS U–Pb systematics and geochronology of titanite. *Chemical Geology*, 227, 37–52.

Streckeisen, A., 1976, To each plutonic rock its proper name. *Earth-Science Reviews*, 12, 1–33.

Strong, D. F. and Hanmer, S. K., 1981, The leucogranites of southern Brittany: Origin by faulting, frictional shearing, fluid flux, and fractional melting. *Canadian Mineralogist*, 19, 163–76.

Swanson, D. A., Casadevall, T. J., Dzurisin, D., Holcomb, R. T., Newhall, C. G., Malone, S. D., and Weaver, C. S., 1985, Forecasts and predictions of eruptive activity at Mount St. Helens, USA. *Journal of Geodynamics*, 3, 397–423.

Swanson, D. A., Wright, T. L., Hooper, P. R., and Bentley, R. D., 1979, Revisions in stratigraphic nomenclature of the Columbia River basalt group. *United States Geological Survey Bulletin*, 1457–G.

Tate, M. C., Norman, M. D., Johnson, S. E., Fanning, C. M., and Anderson, J. L., 1999, Generation of tonalite and trondjhemite by subvolcanic fractionation and partial melting in the Zarza intrusive complex, western Peninsular Ranges Batholith, northwestern Mexico. *Journal of Petrology*, 40, 983–1010.

Taylor, G. J., 2009, Ancient lunar crust: Origin, compositions, and implications. *Elements*, 5, 17–22.

Taylor, H.P., Jr., 1968, The oxygen isotope geochemistry of igneous rocks. *Contributions to Mineralogy and Petrology*, 19, 1–71.

Taylor, H.P., Jr., 1974, The application of oxygen and hydrogen isotope studies to problems of hydrothermal alteration and ore deposition. *Economic Geology*, 69, 843–83.

Taylor, H. P., Jr. and Sheppard, S. M. F., 1986, Igneous rocks: I. Processes of isotopic fractionation and isotope systematics. *Reviews in Mineralogy*, 16, 227–71.

Taylor, S. R., 1975, *Lunar Science: A Post-Apollo View*. Pergamon Press, New York, NY.

Teagle, D. and others, 2011, Integrated Ocean Drilling Program Expedition 335 preliminary report: Superfast Spreading Rate Crust 4: Drilling gabbro in intact ocean crust formed at a superfast spreading rate: April 13–June 3. *Preliminary Report of the Integrated Ocean Drilling Program*, 335.

Teng, F.-Z., Dauphas, N., and Watkins, J. M., 2017. Non-traditional stable isotopes: Retrospective and prospective. *Reviews in Mineralogy and Geochemistry*, 82, 1–26.

Theye, T., Seidel, E., and Vidal, O., 1992, Carpholite, sudoite, and chloritoid in low-grade high-pressure metapelites from Crete and the Peloponnese, Greece. *European Journal of Mineralogy*, 4, 487–507.

Thirlwall, M. F., Gee, M. A. M., Taylor, R. N., and Murton, B. J., 2004, Mantle components in Iceland and adacent ridges investigated using double-spike Pb isotope ratios. *Geochimica et Cosmochimica Acta*, 68, 361–86.

Thompson, A. B. and England, P. C., 1984, Pressure–temperature–time paths of regional metamorphism II: Their inference and interpretation using mineral assemblages in metamorphic rocks. *Journal of Petrology*, 25, 929–55.

Thompson, J. B., 1957, The graphical analysis of mineral assemblages in pelitic schists. *American Mineralogist*, 42, 842–58.

Thurston. P. C., 2002, Autochthonous development of the Superior province greenstone belts. *Precambrian Research*, 115, 11–36.

Thurston, P. C. and Breaks, F. W., 1978, Metamorphic and tectonic evolution of the Uchi-English River subprovince. *Canadian Geological Survey Paper 78-10*, 49–62.

Tilley, C. E., 1925, Metamorphic zones in the southern Highlands of Scotland. *Quarterly Journal of the Geological Society of London*, 81, 100–12.

Tilley, C. E., 1948, Earlier stages in the metamorphism of siliceous dolomites. *Mineralogical Magazine*, 28, 272–6.

Tilling, R. I., 1989, Volcanic hazards and their mitigation: Progress and problems. *Reviews of Geophysics*, 27, 237–69.

Tödheide, K. and Frank, E. U., 1963, Das Zweiphasengebiet und die Kritsche Kurve im System Kohlendioxide–Wasser bis zu Drucken von 3500 bar. *Zeitschrift für Physikalische Chemie Neue Folge*, 37, 387–401.

Trail, D., Bindeman, I. N., Watson, E. B., and Schmitt, A. K., 2009, Experimental calibration of oxygen isotope fractionation between quartz and zircon. *Geochimica et Cosmochimica Acta*, 73, 7110–26.

Trommsdorff, V. and Evans, B. W., 1972, Progressive metamorphism of antigorite schist in the Bergell tonalite aureole (Italy). *American Journal of Science*, 272, 423–37.

Trommsdorff, V. and Evans, B. W., 1977, Antigorite-ophicarbonates; phase relations in a portion of the system $CaO–MgO–SiO_2–H_2O–CO_2$. *Contributions to Mineralogy and Petrology*, 60, 39–56.

Turner, F. J., 1948, Mineralogical and structural evolution of the metamorphic rocks. *Geological Society of America Memoir 30*.

Turner, F. J., 1968, *Metamorphic Petrology, Mineralogical and Field Aspects*. McGraw Hill, New York, NY.

USGS, 2016, A world of minerals in your mobile device. *USGS General Information Product 167*.

Valley, J. W., Kinny, P. D., Schulze, D. J., and Spicuzza, M. J., 1998, Zircon megacrysts from kimberlite: Oxygen isotope variability among mantle melts. *Contributions to Mineralogy and Petrology*, 133, 1–11.

Van Kranendonk, M. J., Hickman, A. H., Smithies, R. H., Nelson, D. R., and Pike, G., 2002, Geology and tectonic evolution of the Archean North Pilbara terrain, Pilbara craton, Western Australia. *Economic Geology*, 97, 695–732.

Vantongeren, J. A., Mathez, E. A., and Kelemen, P. B., 2010, A felsic end to Bushveld differentiation. *Journal of Petrology*, 51, 1891–912.

Varne, R., Brown, A. V., and Faloon, T., 2000, Macquarie Island: Its geology, structural history, and the timing and tectonic setting of its N-MORB to E-MORB magmatism. *Geological Society of America Special Paper*, 349, 53–64.

Vauchez, A. and Garrido, C., 2001, Seismic properties of an asthenospherized lithospheric mantle: Constraints from lattice preferred orientations in peridotites from the Ronda Massif. *Earth and Planetary Science Letters*, 192, 235–49.

Vermaak, C. F. and Von Gruenewaldt, G., 1981, *Guide to the Bushveld Excursion (28th of June to the 4th of July, 1981)*. Geological Society of South Africa, Marshalltown.

Vernon, R. H., 2004, *A Practical Guide to Rock Microstructures*. Cambridge University Press, Cambridge.

Vikre, P. G., 1989, Fluid–mineral relations in the Comstock Lode. *Economic Geology*, 84, 1574–613.

Voordouw, R., Gutzmer, J., and Beukes, N. J., 2009, Intrusive origin for Upper Group (UG1, UG2) stratiform chromitite seams in the Dwars River area, Bushveld Complex, South Africa. *Mineralogy and Petrology*, 97, 75–94.

Wager, L. R. and Brown, G. M., 1967, *Layered Igneous Rocks*, W.H. Freeman and Company, San Francisco, CA.

Wall, C. J., Scoates, J. S., Weis, D., Friedman, R. M., Amini, M., and Meurer, W. P., 2018, The Stillwater Complex: Integrating zircon geochronological and geochemical constraints on the age, emplacement history and crystallization of a large, open-system layered intrusion. *Journal of Petrology*, 59, 153–190.

Watson, E. B. and Harrison, T. M., 1983, Zircon saturation revisited; temperature and composition effects in a variety of crustal magma types. *Earth and Planetary Science Letters*, 64, 295–304.

Watts, A. B. and Talwani, M., 1974, Gravity anomalies seaward of deep-sea trenches and their tectonic implications. *Geophysical Journal of the Royal Astronomical Society*, 36, 57–90.

West, H. B., Garcia, M. O., Gerlach, D. C., and Romano, J., 1992, Geochemistry of tholeiites from Lanai, Hawaii. *Contributions to Mineralogy and Petrology*, 112, 520–42.

Westphal, M., Schumacher, J. C., and Boschert, S., 2003, High-temperature metamorphism and the role of magmatic heat sources at the Rogaland Anorthosite Complex in southwestern Norway. *Journal of Petrology*, 44, 1145–62.

Wickham, S. M. and Oxburgh, E. R., 1985, Continental rifts as a setting for regional metamorphism. *Nature*, 318, 330–4.

Wignall, P., The link between large igneous provinces and mass extinctions. *Elements*, 1, 293–7.

Wilkinson, J. F. G. and Hensel, H. D., 1988, The petrology of some picrites from Mauna Loa and Kilauea volcanoes. *Hawaii, Contributions to Mineralogy and Petrology*, 98, 326–45.

Willemse, J., 1969, The vanadiferous magnetic iron ore of the Bushveld igneous complex. *Economic Geology Monograph*, 4, 187–208.

Williams, I. S., 1998, U-Th-Pb geochronology by ion microprobe. *Reviews in Economic Geology*, 7, 1–35.

Wilson, M., 1989, *Igneous Petrogenesis: A Global Tectonic Approach*. Unwin Hyman, London.

Winkler, H. G. F., 1976, *Petrogenesis of Metamorphic Rocks* (English editor, E. Froese). Springer-Verlag, New York, NY.

Woodhead, J. D., Horstwood, M. S. A., and Cottle, J. M., 2016, Advances in isotope ratio determination by LA–ICP–MS. *Elements*, 12, 317–22.

Wright, T. L., 1984, Origin of Hawaiian tholeiite: A mesomatic model. *Journal of Geophysical Research*, 89, 3233–52.

Yoder, H. S. and Tilley, C. E., 1962, Origin of basalt magmas: An experimental study of natural and synthetic rock systems. *Journal of Petrology*, 3, 342–529.

Yu, Y. and Morse, S. A., 1992, Age and cooling history of the Kiglapait intrusion from an ^{40}Ar/^{39}Ar study. *Geochimica et Cosmochimica Acta*, 56, 2471–85.

Zeh, A., Gerdes, A., Barton, J., and Klemd, R., 2010, U-Th–Pb and Lu–Hf systematics of zircon from TTGs, leucosomes, meta-anrothosites, and quartzites of the Limpopo Belt (South Africa): Constraints for the formation, recycling and metamorphism of Paleoarchean crust. *Precambrian Research*, 179, 50–68.

Zen, E.-A., 1986, Aluminum enrichment in silicate melts by fractional crystallization: Some mineralogic and petrographic constraints. *Journal of Petrology*, 27, 1095–117.

Zen, E.-A., 1988, Phase relations of peraluminous granitic rocks and their petrogenetic implications. *Annual Review of Earth and Planetary Sciences*, 16, 21–52.

Zhang, Y., 2010, Diffusion in minerals and melts: Theoretical background. *Reviews in Mineralogy and Geochemistry*, 72, 5–59.

Zhang, Y., Ni, H., and Chen, Y., 2010, Diffusion data in silicate melts. *Reviews in Mineralogy and Geochemistry*, 72, 311–408.

Index

Bold = key term; italics = figure or map; * = mineral described in the appendix

Printed in the United States
by Baker & Taylor Publisher Services